Physics of Space Plasma Activity

Space plasma is so hot that the atoms break up into charged particles which then become trapped and stored in magnetic fields. When critical conditions are reached the magnetic field breaks up, releasing a large amount of energy and causing dramatic phenomena. A prominent example is the magnetospheric substorm occurring in the Earth's magnetosphere. It involves plasma and magnetic field structures extending from 100 km to tens of Earth radii, and can be seen as strong intensifications of the northern and southern lights. The largest space plasma activity events observed in the Solar System occur on the Sun, when coronal mass ejections expel several billion tons of plasma mass into space.

Physics of Space Plasma Activity provides a coherent and detailed treatment of the physical background of large plasma eruptions in space. It provides the background necessary for dealing with space plasma activity, and allows the reader to reach a deeper understanding of this fascinating natural event. The book employs both fluid and kinetic models, and discusses the applications to magnetospheric and solar activity.

This book will form an interesting reference for graduate students and academic researchers in the fields of astrophysics, space science and plasma physics.

KARL SCHINDLER is an Emeritus Professor with the Faculty of Physics and Astronomy, Ruhr University Bochum, Germany, and a distinguished theorist in the field of space plasma physics. In 2001 he was granted the Orson Anderson Scholarship at Los Alamos National Laboratory by the Institute of Geophysics and Planetary Physics (IGPP). Dr Schindler is also a Fellow of the American Geophysical Union.

Physics of Space Plasma Activity

Karl Schindler

Ruhr-Universität Bochum, Germany

CAMBRIDGE UNIVERSITY PRESS
Cambridge, New York, Melbourne, Madrid, Cape Town, Singapore,
São Paulo, Delhi, Dubai, Tokyo

Cambridge University Press
The Edinburgh Building, Cambridge CB2 8RU, UK

Published in the United States of America by Cambridge University Press, New York

www.cambridge.org
Information on this title: www.cambridge.org/9780521142366

First published 2007
This digitally printed version 2010

A catalogue record for this publication is available from the British Library

ISBN 978-0-521-85897-7 Hardback
ISBN 978-0-521-14236-6 Paperback

To Erika

Contents

vii

Preface

A major motivation for writing this book is the strong fascination that visible signatures of plasma activity are able to generate. This goes along with considerable professional research interest in this area. Also, those who have admired spectacular pictures or video presentations on the internet displaying spacecraft observations of auroral activity or of solar eruptions, are often motivated to learn more about their physical background.

In the early days of spacecraft observations, the understanding of dynamical phenomena such as geomagnetic storms and solar flares was considered as poor and high up on the list of particularly challenging problems. Remarkably, this is still true today. The observational database has increased dramatically and important new phenomena were discovered, such as coronal mass ejections and manifestations of the global nature of magnetospheric substorms involving large regions of the magnetosphere. There are many more aspects than envisaged originally, and today we have good reasons to use the comprehensive notions of *solar* and *magnetospheric activity*, which in this book are combined under the working term *space plasma activity*. The desire to understand these complex phenomena has mobilized considerable research efforts, but due to the overwhelming complexity that one encounters, our present understanding is still far from being satisfactory.

One might ask, whether in this situation it is appropriate to write a book that concentrates on space plasma activity. Would it not be more reasonable to wait until our understanding of the underlying physical processes has settled down more solidly?

The main reason for writing this book at this time is the fact that during past decades a substantial wealth of theoretical tools has been developed, which can be expected to remain useful, even if many of the final answers are still to be found. In fact, good knowledge and further development of those tools could well help to accelerate progress in this field. The situation appears to be similar to that of the Earth's lower atmosphere, where several phenomena associated with atmospheric disturbances are still not well understood; on the other hand, there is little doubt that gas-dynamics methods play an important role in present and future investigations in that area. Regarding that the electrodynamic interactions make plasma dynamics considerably more complicated than gas dynamics, it is obvious that there is a strong need for reviewing and, where possible, for improving the existing tools and for developing new ways of approach. Therefore, after the phenomenological survey given in Part I, the relevant methods and plasma properties are addressed in Parts II and III, starting from basic plasma models. In this way an updated (although necessarily incomplete) *toolbox* for the study of space plasma activity arises. I hope that this will be found useful for research in space plasma activity.

A further aim is to meet the needs of scientists or graduate students trying to enter the field of large-scale space plasma phenomena. They often ask for coherent descriptions of the theoretical methods that would allow them to discriminate between conclusions that can safely be drawn and concepts that are more of a speculative nature.

There are also indications suggesting that the occupation with topics related to space plasma activity have enjoyed increasing attractiveness since it became clear that such phenomena have aspects falling under notions of modern theoretical physics, such as nonlinear dynamics, spontaneous processes or catastrophe theory.

The separation between the theoretical tools (Parts II and III) and the applications (Part IV) was chosen for several reasons. First, this separation allows a systematic and coherent presentation of the theory. Further, it makes it possible to present the applications of Part IV in such a way that, to some extent, they can be understood without detailed knowledge of Parts II and III. Lastly, the separation suggests itself in view of the speculative elements that necessarily play a more important role in Part IV than in Parts II and III.

The reader should have knowledge of physics and basic mathematical techniques, as is commonly available after, say, four years' study of physics or astronomy, mathematics, or engineering. Some knowledge of plasma physics, as drawn from textbooks (e.g., Sturrock, 1994; Boyd and Sanderson, 2003; Cravens, 2004), would help the reader to understand the basic plasma models and to follow the formal developments of Parts II and III. Selected background material, tailored to the requirements of this book, is available on the internet; the addresses are inserted in the text where appropriate. Part IV is more descriptive and needs less background in mathematics and plasma physics.

It is a pleasure to acknowledge the invaluable help and support that I received from many sides before and during the preparation of this book. A substantial part of the material originates from the research carried out before the mid 1990s by our group *Theoretische Physik IV* at the Ruhr-University of Bochum. It still fills me with joy when I remember its lively and creative atmosphere. Particularly, I feel indebted to Joachim Birn and Michael Hesse, with whom fruitful and pleasant collaboration has continued until today. I also profited greatly from discussions with many colleagues of the plasma, space and solar physics communities, too numerous to name all of them. But I wish to mention those with whom I had particularly valuable contacts during my extended struggle with space science, namely Ian Axford, Dieter Biskamp, Jörg Büchner, Peter Gary, Akira Hasegawa, Ed Hones, Jim McKenzie, Eric Priest, Philip Rosenau, Roald Sagdeev, Reinhard Schlickeiser, George Siscoe, Bengt Sonnerup, Ted Speiser and Vytenis Vasyliunas. I am grateful to the space science group of the Los Alamos National Laboratory for their warm hospitality during numerous visits.

For their thoughtful comments that led to many important improvements of the manuscript I am deeply thankful to Joachim Birn, Terry Forbes, Michael Hesse, Gunnar Hornig, Michael Kiessling, Thomas Neukirch, Antonius Otto and Heinz Wiechen. I thank Angelika Schmitz, Isabelle Tissier and Heike Neukirch for their competent assistance in the early days of the project.

I am particularly grateful to my wife Erika for her continuous understanding support and valuable help.

Bochum, August, 2006   *Karl Schindler*
                                         ks@tp4.ruhr-uni-bochum.de

1

Introduction

Space plasma phenomena have attracted particular interest since the beginning of the exploration of space about half a century ago. Already a first set of pioneering observations (e.g., Ness, 1969) discovered that matter and electromagnetic fields in space have a complex structure, which was largely unpredicted. Terrestrial and, particularly, spacecraft observations of solar plasmas and fields point in the same direction. In fact, our present picture of the plasma and the electromagnetic fields in space throughout the solar system (and beyond) is that of an extremely complex medium with spatial and temporal variations on large ranges of scales. The wealth of dynamical phenomena observed in space plasmas has steadily increased as more and more refined observational techniques have become available, and it can be expected that important processes still await their detection.

An outstanding class of space plasma phenomena is addressed here under the notion of *space plasma activity*. Quite generally, in the area of space and astrophysical plasmas the term *activity* is used for a set of particular magnetospheric, stellar or galactic phenomena, which, although vastly different regarding their space and time scales and their dominant physical processes, have an important characteristic property in common. In all cases they show sudden transitions from relatively quiet states with less pronounced time-dependence to dynamic states in a strongly time-dependent evolution. (Note that this property by no means is restricted to plasma phenomena, volcanic activity being a prominent example from another discipline.)

The term *activity* is commonly used in two different ways. In a narrow sense *activity* refers to the strongly time-dependent dynamic phase alone. In a wider sense, it means the entire phenomenon including the relevant quiescent intervals. The latter meaning is adopted for the title of this book and, to a large extent, also for the text. It will be clear from the context, when, occasionally, we will use the narrower meaning.

Strictly speaking, as is the Earth's atmosphere, the plasma in space is always in a time-dependent evolution. Therefore, in a strict sense it is impossible to identify intervals where the space plasma (in some region) is *quiet*. However, as in the atmosphere, it often does make sense to speak of quiet and dynamic plasma conditions in an approximate way. There is a qualitative difference between a situation where during an atmospheric storm a strong gust blows across a countryside and the comparatively quiet state of the air before the gust arrives. It is in this sense that we will speak of *quiescent* and of *dynamic* space plasma states. Also, as we will see, the notion of *quiescence* can be an important theoretical tool even if the real system considered has a level of superimposed time-dependent phenomena.

Generally, for systems with multiple time scales *quiescence* and *dynamics* are relative terms; what counts is that the processes that one compares occur on different, well-separated time scales. A more precise definition of activity in the present context does not seem to be available, nor is it necessary for our purposes. From a phenomenological point of view we simply refer to the processes described in Chapter 2.

For magnetospheric activity, the most direct visual evidence is provided by auroral light emission. Here, a corresponding black and white reproduction (Fig. 1.1) should suffice to indicate strong temporal variations of the

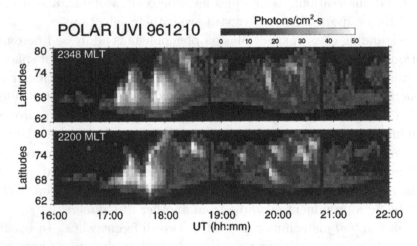

Fig. 1.1 Auroral luminosity enhancement at two magnetic meridians during a magnetospheric substorm approximately lasting from 17:10 to about 19:00 Universal Time (reproduced from Sergeev *et al.* (2001) by permission of the American Geophysical Union).

aurora occurring in connection with magnetospheric activity. A quiescent state ends near 17:10 UT (*Universal Time*), when the dynamic phase starts. Such strong enhancement of auroral emissions are important signatures of *magnetospheric substorms* (Akasofu, 1968), to be addressed in detail later. There are many manifestations of such changes in the signatures of characteristic plasma quantities, monitored by spacecraft in the Earth's magnetosphere (see Section 2.1).

A most spectacular class of activity processes in the solar system involves large plasma outbursts from the solar corona into interplanetary space. Such events are called *coronal mass ejections*. An example is shown in Fig. 2.8; a brief survey is given in Section 2.2.

Part I sets the scene with regard to the phenomenological background and to the basic plasma models. Concerning the latter, the kinetic description follows from basic principles, specifically Newton's mechanics, Maxwell's theory of electromagnetism and statistical mechanics, while the fluid pictures involve additional simplifications. The models are presented without derivations, but their physical meaning is outlined. For more on the foundations the reader should consult introductions to plasma physics.

Parts II and III are devoted to theoretical tools, specific to space plasma activity. Their splitting into two parts reflects the fact that it makes sense to distinguish the quiescent from the dynamic phases not only in their phenomenological appearance but also in the choice of appropriate theoretical modelling.

The present tasks are complicated by the fact that the plasmas that we want to study are spatially inhomogeneous. This excludes a considerable fraction of the available methods in space plasma physics, such as the theory of waves, instabilities and wave–particle interaction on a homogeneous background. Such processes will be considered only if a special motive for doing so arises. Our present purpose leads us to consider space plasma processes with background gradients playing an important role.

We mostly deal with situations where the gradients are supported by large-scale magnetic forces. In addition, the present scope does include external gravity in a few instances, but self-gravitation is excluded. Thus, although small scale galactic magnetic field structures may be covered in principle, active galactic nuclei are outside the present scope.

Our approach takes into account that it has proven appropriate to deal with inhomogeneous space plasmas by considering systems with both two and three spatial dimensions, profiting from their characteristic advantages. 3D systems are more realistic, but general analytical results are scarce, so that in many cases numerical simulations are required. For models with two spatial dimensions a considerable wealth of analytical techniques is available, but the results are less realistic. Still, they are often indispensable for providing a qualitative understanding of complex phenomena. Also, 2D results can provide valuable guidance for the interpretation of numerical simulations.

In Part IV it is attempted to discuss particular aspects of magnetospheric and solar activity in the light of Parts II and III. It will become apparent that in some areas the theoretical results are able to provide a deeper physical understanding. In other domains the discussion reveals a strong need for further theoretical investigations.

The provided references should be regarded as typical examples, complete referencing would have exceeded the scope of this book.

Part I

Setting the scene

2

Sites of activity

Here we will give a qualitative overview on major activity processes in the solar system. Since our main aim is to concentrate on basic aspects and on theoretical results, a full account of the observational background is outside our present scope. However, in the following sections we will summarize the main observational facts that are relevant for our later discussion. For details the reader is referred to the literature. Note that in the present chapter we will largely refrain from giving physical interpretations. They will be discussed in Part IV using the tools provided in Parts II and III.

2.1 Geospace

Magnetospheric activity comprises the major global dynamical phenomena of the Earth's magnetosphere including ionospheric processes. It results from the interaction of the solar wind with the Earth's magnetosphere (Fig. 2.1).

The solar wind is characterized by a fast (supersonic) plasma flow from the Sun into interplanetary space. The magnetosphere is the region above the ionosphere that is dominated by the geomagnetic field. The solar wind compresses the Earth's magnetic field on the day-side and stretches it out to a long tail (*magnetotail*) on the night-side of the Earth (Fig. 2.1). A bow shock wave stands in front of the magnetosphere, which has a rather thin boundary, the *magnetopause*. Its thickness, which varies considerably, can become as small as a few hundred km.

The magnetotail consists of *tail lobes*, where the magnetic field energy density dominates, and the *plasma sheet*, which in its central part (*central or inner plasma sheet*) is dominated by the energy density of the plasma. Since the plasma sheet is particularly important for magnetospheric activity, typical values of plasma sheet parameters are listed in Table 2.1. Because of substantial spatial and temporal variability these numbers can provide only

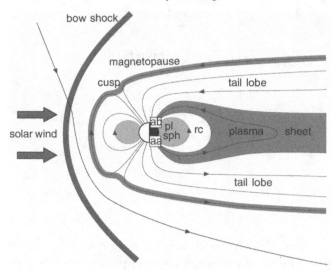

Fig. 2.1 A qualitative sketch of major features of the Earth's magnetosphere. Here 'pl sph', 'rc', 'aa' and 'ab' stand for *plasma sphere, ring current, Aurora Australis (southern lights)* and *Aurora Borealis (northern lights)*, respectively. An interplanetary magnetic field line is shown for a case with a southward field component.

Table 2.1 *Characteristic properties of the plasma sheet.*

plasma sheet	thickness	20 000 km
	length	500 000 km
central plasma sheet	number density	$2 \times 10^5 \, \mathrm{m}^{-3}$
	ion temperature	5×10^7 K
	electron temperature	10^7 K
	magnetic field strength	2 nT
magnetic lobes	magnetic field strength	20 nT

a general orientation. Temperatures are not thermodynamic temperatures, they simply measure kinetic energy of random motion. Fig. 2.2 shows a cross-section of the magnetotail.

The solar wind transfers energy into the magnetosphere. Correlation studies (e.g., Bargatze *et al.*, 1985) have established that the energy flux is particularly strong when the interplanetary magnetic field component perpendicular to the ecliptic plane points southward (Fig. 2.1). The magnetosphere responds to the energy input through a set of complex dynamical phenomena.

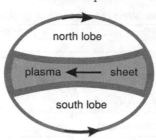

Fig. 2.2 A qualitative sketch of a cross section of the magnetotail. The view is from the tail to the Earth. Black arrows indicate the perpendicular current pattern of the tail. The plasma sheet is surrounded by boundary layers (broad line). The outer boundary is the magnetopause.

Fig. 2.3 gives an example (Sergeev *et al.*, 2001) showing detailed measurements obtained from spacecraft and ground instruments on 10 December 1996. The time interval is the same as that of Fig. 1.1. The panels a and b show the solar wind dynamic pressure P_D and a parameter (*Eps3*), which strongly emphasizes the occurrence of a southward component of the interplanetary magnetic field, and which can be regarded as a measure of the energy flux directed into the magnetosphere. The panels c and d show the ground observation indices *D*st and *AE*. Here, *D*st measures the disturbance of the mid-latitude magnetic field component parallel to the dipole axis, so that it monitors the intensity of the magnetospheric ring current. The *AE* index (in panel c shown as a stackplot of magnetograms from several stations) measures magnetic signatures caused by the auroral electrojet (an east-west electric current), which intensifies and changes its location during dynamic periods. The solid curves in panels e, f, g show measurements made aboard the GEOTAIL spacecraft, located at about 25 R_E (Earth radii) geocentric distance on the night-side of the Earth. Panel e shows a pressure parameter P_T, which is the sum of (scalar) kinetic and magnetic pressure and can serve as a rough measure of the energy density of the magnetotail. The parameters *IPS* and *LOBE/BLPS* indicate whether the satellite was in the inner plasma sheet or in the lobe or plasma sheet boundary layer regions (see Fig. 2.2), respectively. Panel f gives the magnetic field component B_z perpendicular to the mid-plane of the plasma sheet (positive in the northward direction) and panel g shows the plasma velocity component in the earthward direction, v_x. Vertical dashed lines show approximate onsets of enhanced auroral activity.

Before 17:10 UT the magnetosphere was in a comparatively quiescent state with a substantial energy flux entering it. A significant fraction of the energy

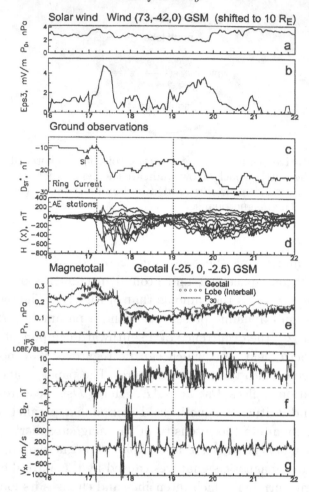

Fig. 2.3 Ground-based and satellite measurements during a dynamic period of the magnetosphere, details are explained in the text (reproduced from Sergeev *et al.* (2001) by permission of the American Geophysical Union).

accumulates in the magnetotail, indicated by a corresponding increase of P_T. Near 17:10 UT the magnetosphere suddenly turns into a different state, which is much more dynamic and involves the entire system consisting of the magnetosphere and the ionosphere. This is a *magnetospheric substorm* (Akasofu, 1968).

Under suitable conditions sequences of substorms can be accompanied by a gradual build-up of the ring current (Reeves and Henderson, 2001). A significant increase of $|Dst|$ indicates a *magnetic storm* (Chapman and Bartels, 1940). This means that understanding magnetic storms requires insight into the substorm process. Therefore, the magnetospheric substorm is widely regarded as the basic element of magnetospheric activity.

Substorms, however, cannot account for all large scale activity processes in the magnetosphere. This is clear from the plots in Fig. 2.3 also. In fact, shortly after 19:00 UT a second dynamic period begins. This period does not show the typical substorm signatures. Detailed studies indicate that such periods show features consistent with quasi-steady phases of considerable plasma flow. The plasma flow velocity includes a significant component perpendicular to the magnetic field, the corresponding plasma transport being referred to as *convection* (Axford and Hines, 1961). This has led to the term *convection bay*. (The term *bay* refers to the shape of the magnetograms.)

In the following we discuss a few further properties of magnetospheric substorms. Note that in view of our overall topic we deliberately concentrate on substorms, leaving aside many magnetospheric phenomena that would be of interest from other viewpoints.

One distinguishes three phases of a magnetospheric substorm (McPherron *et al.*, 1973; Russell and McPherron, 1973):

(i) the *growth phase*, which coincides with the quiescent phase before onset, where energy is accumulated in the magnetotail (before 17:10 UT in Fig. 2.3),

(ii) the *expansion phase*, which is the dynamical phase following substorm onset (near 17:10 UT in Fig. 2.3),

(iii) the *recovery phase*, which – at least in a fraction of the cases – can be identified as the phase during which the magnetosphere returns to a more quiescent state. (In the example of Fig. 2.3 the recovery phase does not fully develop as it goes over into the convection bay.)

These and many similar findings have led to the interpretation that during the growth phase (predominantly magnetic) energy is loaded into the magnetotail and is released, i.e., turned into heat, kinetic energy of directed flow and energetic particles, during the expansion phase (Baker *et al.*, 1985). The overall energy transfer that occurs during a substorm has been estimated as amounting to 10^{14}–10^{15} J. The duration of a growth phase is of the order of an hour but shows large variability. The dynamic processes observed after onset have a broad spectrum of time scales, the largest typically being of the order of 10 min.

In the late stages of a growth phase often a new feature occurs in the near-Earth tail, described as the *formation of thin current sheets* (McPherron *et al.*, 1987; Kaufmann, 1987; Mitchell *et al.*, 1990; Sergeev *et al.*, 1990;

Pulkkinen *et al.*, 1994). Here, the electric current in the plasma sheet develops a considerable concentration in a rather small region which may well develop widths of the order of a few 100 km or even smaller, which is small compared with typical plasma sheet widths (Table 2.1).

During the substorm expansion phase one typically observes significant earthward flow in the near-Earth tail region with flow velocities of a few hundred km/s. Also, the magnetic field, which becomes considerably stretched during the growth phase, relaxes into a more dipolar shape, a process called *dipolarization* (Slavin *et al.*, 1997; Nakai and Kamide, 2000). This process is connected with the development of a wedge-shaped structure in the pattern of the electric current (McPherron *et al.*, 1973), the *current wedge*, which forms as a sizable fraction of the near-Earth tail current, is deviated through the ionosphere. A particular current wedge event that was investigated in detail by Angelopoulos *et al.* (1996) occurred late in the expansion phase. The substorm wedge is well correlated with the intensification of the *auroral electrojet* (e.g., Baumjohann, 1982). The change in the magnetospheric current pattern has also been described as *current disruption* (Lopez *et al.*, 1993).

In the more distant parts of the magnetotail, say beyond 30 R_E, large *tailward* plasma flows are characteristic for expansion phases. Large scale structures are embedded in these flows, which are described as *plasmoids* (Hones, 1977; Moldwin and Hughes, 1992; Ieda *et al.*, 1998, 2001). The internal magnetic field of plasmoids is often found to have a helical or *flux rope* structure (Slavin *et al.*, 1989).

The early observations of plasma flow and magnetic fields have led to the phenomenological concept of the formation of a *near-Earth magnetic neutral line* (Hones, 1973; Nishida and Nagayama, 1973; Nishida and Hones, 1974; Hones, 1977), as a magnetic island, the magnetic signature of the plasmoid, begins to form in the magnetotail. For details we have to refer to Sections 11.2.10, 11.4.2 and 13.3.2.

It is also typical of magnetospheric substorms that their onset is closely associated with sudden increases of energetic particle fluxes in the inner magnetosphere (e.g., Reeves *et al.*, 1990). They are typically observed by geostationary satellites (circling the Earth at a geocentric distance of 6.6 R_E). Such *injections* are largely dispersionless (within a few hours of local midnight), which excludes that the particles are accelerated elsewhere and then move to the observation location with speeds corresponding to their energy.

The magnetospheric current system plays an important role in the coupling of the magnetospheric dynamics with ionospheric dynamics (*magnetosphere/ionosphere coupling*). One of the direct manifestations of that coupling is the enhancement of auroral activity during substorms (Fig. 1.1). Thus, the understanding of the auroral emissions necessarily requires insight into the magnetosphere/ionosphere coupling. In view of this complicated coupling, it is not surprising that in spite of considerable progress made during the last decades many questions about this fascinating phenomenon are still open. Historically, the Aurora Borealis has provided the first evidence of magnetospheric activity.

From the point of view of thermodynamics or statistical mechanics, the magnetosphere is a nonequilibrium system. This is manifested by the presence of considerable pressure gradients and electric currents (see Table 2.1, Fig. 2.1 and Fig. 2.2). It is apparent from Fig. 2.1 that the plasma is largely organized by the magnetic field. For instance, the boundary of the (quiescent) plasma sheet is a magnetic flux surface, at least approximately. The spatial inhomogeneity of the plasma is sustained by the magnetic forces.

However, the thermodynamic nonequilibrium nature of space plasmas does not exclude phases of approximate equilibrium in the sense of force balance between mechanical and electromagnetic forces. This is the notion of equilibrium that will play an important role in our discussions.

Observations of plasmoids and of the dipolarization indicate that magnetospheric activity involves large scale reconfigurations of the magnetic field, with indications of changes in the magnetic topology. The central plasma physics process that leads to changes of magnetic topology is *magnetic reconnection* (Dungey, 1961; Vasyliunas, 1975; Biskamp, 2000). Naturally, considerable attention will be paid to that process.

We have already mentioned a convection bay as an example for a plasma sheet process that is not directly related to a genuine substorm. Other examples are pseudo-breakups (Koskinen *et al.*, 1993; Partamies *et al.*, 2003), which seem to initiate like a substorm, without however reaching the large scale signatures of a substorm.

With the exception of the thin current sheets, the phenomena described so far have spatial scales of the order of an Earth radius or larger and time scales of the order of minutes or larger. This, however, by far is not the full space plasma picture. In fact, magnetospheric dynamics involves a broad spectrum of wave phenomena with a rich set of wave-particle interaction processes. This is a wide field with many detailed observational results (e.g., Gurnett and Frank, 1977; Tsurutani *et al.*, 1998). We will return to particular aspects of waves and turbulence at several points in this book. Here, it seems sufficient to point out that, for example, waves have been observed in the entire frequency range (from 5 Hz to 311 kHz) by the *Polar* satellite with a correspondingly wide range of wavelengths (e.g., Tsurutani *et al.*, 1998). Most pronounced wave activity is often seen in the magnetospheric boundary layers such as the plasma sheet boundary layer and the magnetopause (see Fig. 2.2).

The existence of the broad wave spectrum is closely related to intrinsic length and time scales of a magnetized plasma. By intrinsic scales we mean characteristic times and lengths that depend on particle mass or charge (see the discussion farther below). Table 2.2 gives a list of typical intrinsic plasma scales for central plasma sheet conditions. For details and for the exact definitions of the quantities listed in the table, we refer to the appropriate sections. (The relevant equations are quoted in the table.) Here, the following qualitative explanations should suffice to bring out the main points. The electron plasma frequency is a frequency characteristic for electric plasma oscillations and the gyroradius r_g (for particles moving with *thermal velocity* corresponding to temperature) and the gyrofrequency ω_g are quantities characteristic for particle orbits in magnetic fields. In a sufficiently strong magnetic field the particles perform helical orbits around the magnetic field lines (they *gyrate*) with radius r_g and frequency ω_g. The electron inertial length plays the role of an inertia-based skin depth for electromagnetic disturbances. The Debye length λ_D is a characteristic shielding length for electric fields in a plasma. The table includes the plasma parameter Λ, which is the number of electrons in a sphere with radius λ_D.

Table 2.2 *Intrinsic plasma scales and the plasma parameter for central plasma sheet conditions. The numbers in brackets are generated by replacing the magnetic field strength of the plasma sheet by the corresponding lobe value. Where appropriate, the last column gives the number of the defining equation.*

ion (proton)	plasma frequency	$\omega_{pi} = 600/s$	(3.24)
	gyroradius	$r_{gi} = 6000$ km (600 km)	(3.4)
	gyrofrequency	$\omega_{gi} = 0.2/s$ (2/s)	(3.5)
	inertial length	$c/\omega_{pi} = 500$ km	
electron	plasma frequency	$\omega_{pe} = 3 \times 10^4/s$	(3.24)
	gyroradius	$r_{ge} = 60$ km (6 km)	(3.4)
	gyrofrequency	$\omega_{ge} = 400/s$ (4000/s)	(3.5)
	inertial length	$c/\omega_{pe} = 10$ km	
Debye length		$\lambda_D = 500$ m	(3.23)
plasma parameter		$\Lambda = 10^{14}$	(3.25)

A large value of Λ means that the fluctuation level caused by the presence of discrete particles is small and indicates that the plasma is approximately collisionless.

At this point we draw only two general conclusions from the table. First we observe that length and time scales cover wide ranges over several orders of magnitude. This is qualitatively consistent with a wide observed fluctuation spectrum, because the length and time scales of the table enter the dispersion relations as characteristic wavelengths and frequencies of waves and instabilities (Gary, 1993). The second conclusion is that the plasma is collisionless, indicated by the large value of Λ.

Here, an additional discussion of the meaning of the term *quiescence* is necessary. The background of a first remark is the difference in the scales contained in Tables 2.1 and 2.2. The first are macroscopic in nature, they exclusively are determined by the interaction with the surrounding medium, namely the solar wind and the Earth with its magnetic dipole and its atmosphere. Clearly, the shape of the magnetosphere will be determined by solar wind conditions, by the strength of the Earth's magnetic dipole and by ionospheric conditions. In contrast to the macroscopic scales of Table 2.1 the scales of Table 2.2 depend on particle charge or mass in addition to local plasma properties. Since they tend to be considerably smaller than the macroscopic scales they are also called *microscopic scales*. We can now define the notion of *quiescence* more precisely, by associating it with a low level of *macroscopic variations* only. Microscopic turbulence can still be present. This definition of quiescence is motivated by the fact that observations convey the general impression that during macroscopically quiet times the coupling between micro- and macro-dynamics is rather weak. This is in contrast to dynamic periods, when that coupling seems to be strong or at least non-negligible. We return to this important aspect in Part IV.

The second remark concerns the relevance of quiescent states. Even under the restricted definition, introduced just above, one might argue that the central plasma sheet often is not in a quiescent state even when large scale dynamics is absent. In particular, *bursty bulk flows* (Angelopoulos *et al.*, 1992, 1993) are not uncommon. Given these time-dependent flows, one might question the relevance of the concept of *quiescence* altogether. For an answer let us consider the two cases, concerning the role of bursty bulk flows in large scale dynamics. If they play a relevant role, it is essential to study their origin. One way of doing so would be to start out from a hypothetical quiescent state and perform a stability analysis. Possibly one finds an instability that grows into a bursty bulk flow. Alternatively, if the bursty bulk flows are irrelevant for large scale dynamics, one might decide to ignore them in that context and treat the plasma sheet before onset of the dynamic phase as quiescent. In both cases, the quantitative theoretical approach would require introducing a quiescent state. Thus, the notion of quiescence can still be important as a theoretical tool even if the real system has a time-varying feature. Note that the same situation applies to the theory of gas-dynamic turbulence, where an important approach is to consider the stability of quiescent states such as a stratified atmosphere in a gravity field or a steady state with sheared bulk flow.

This brief outline of major characteristics of magnetospheric activity is meant as a morphological introduction. At this stage we have refrained from providing detailed physical interpretations. Although it is one of the main purposes of this book to provide such interpretations, we have to develop appropriate methods before we can do so. Therefore, we postpone a more profound discussion of magnetospheric activity; it will be given in Part IV.

2.2 Solar atmosphere

Solar activity is the most spectacular manifestation of plasma activity in the solar system. Compared with magnetospheric activity, solar activity has the advantage that spacecraft such as SOHO can take pictures of the entire phenomenon. (In magnetospheric research first promising steps have been taken in the same direction, using emission of neutral atoms generated from plasma ions by charge exchange. However, because of the large differences in intensity, the results cannot compete with solar observations.) Traditionally, it has been regarded as a clear disadvantage of solar research that *in situ* observations are not available. However, as we will see, today's spacecraft instruments can well carry out *in situ* observations of structures resulting from large scale solar eruptions, as they propagate outward from the solar atmosphere into the interplanetary medium.

Fascinating documentation of solar activity can be found in the large collection of corresponding pictures and movies available on the internet. Here we can discuss only a few fundamental aspects in connection with the main theme of this book. Be aware that the present grey-scale pictures are poor compared with original colour graphs.

As in the magnetosphere, the plasma of the solar atmosphere is organized by the magnetic field. This aspect is evident in all figures of this section. For example, Fig. 2.4 illustrates the association between sunspots and active regions. Obviously, the two phenomena are closely related.

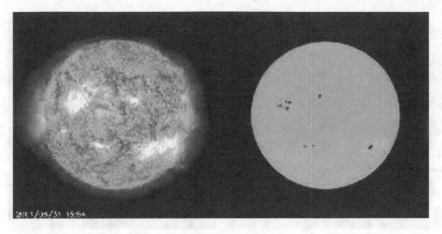

Fig. 2.4 Two pictures of the Sun taken at about the same time in early September 2001. The right image shows visible light emission from the photosphere. Sunspots, regions with stronger magnetic field and lower plasma temperature compared with the surroundings, appear as dark spots. On the left, the Sun is shown in ultraviolet emission corresponding to a plasma temperature of about 60 000 K. The brighter parts are traces of magnetic activity. Source: SOHO (ESA&NASA).

The classical manifestation of solar activity is the *solar flare*. Its characteristic properties are well documented (e.g., Tandberg-Hanssen and Emslie, 1988; Forbes, 2000b). The name stems from the sudden appearance of photospheric light emission, observed in the H_α-line (see the left image of Fig. 2.5). Today, flares are observed in a broad range of the electromagnetic spectrum. Within minutes the emission sharply increases and the plasma is heated up to 10^7 K.

Several types of flares have been distinguished. In the *two-ribbon flare* two bright ribbons appear in the photospheric light emission; the flare shown in the left image of Fig. 2.5 is of that type. Typically, the ribbons are located on opposite sides and roughly oriented along a *polarity reversal line*, where the magnetic field component normal to the surface of the Sun reverses.

Magnetic flux tubes connecting to the flare site typically stand out as *X-ray loops* in their X-ray emission. Fig. 2.6 shows the X-ray loops as part of a particular flare model. Such loops can exist for some time after the optical flare, then called *post-flare loops*; the right image of Fig. 2.5 shows an example. The *compact flare* is characterized by a spot-like emission coming from *bright points* in the solar corona. For detailed observations of a large flare observed by the Yohkoh spacecraft see Phillips (2004).

Another type of structure relevant for solar eruptions are filaments. Filaments are localized clouds of high density and low temperature plasma held

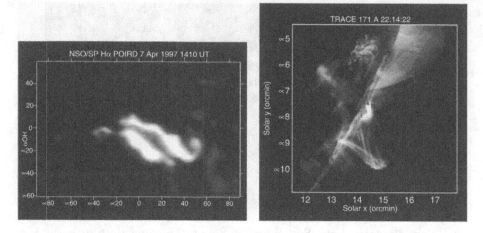

Fig. 2.5 Solar flares. The image on the left shows a two-ribbon flare in H_α emission observed on April 7, 1997; the right image shows post-flare loops, observed on April 3, 2001 at 701 Å. Source: National Solar Observatory/AURA/NSF (left image) and TRACE consortium (right image); TRACE is a mission of the Stanford-Lockheed Institute for Space Research, and part of the NASA Small Explorer program.

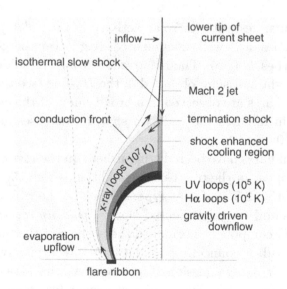

Fig. 2.6 Flare model exhibiting various physical properties (courtesy of T. G. Forbes).

together and kept above the solar surface by magnetic forces (Tandberg-Hanssen, 1994). Filaments are often associated with flares. A filament may stably exist over a time interval of the order of a solar rotation period, and in this form constitutes a remarkable manifestation of quiescence. Nevertheless, a quiescent filament may suddenly erupt, in other words, it suddenly disappears as a filament, a process called *disparition brusque.*

At the solar limb, filaments can appear as large bright structures against the sky, which are called *prominences.* The overall diameter of a prominence can be of the order of 100 000 km. Sometimes erupting solar prominences show strong emissions from neutral atoms, indicating that, in addition to the plasma, chromospheric gas is launched into space. Fig. 2.7, apart from three bright active regions, shows a *sweeping prominence.*

The most spectacular manifestation of solar activity is the *coronal mass ejection* (CME), in the course of which huge amounts of mass and energy, typically in the ranges of $2 \cdot 10^{11}$–$4 \cdot 10^{13}$ kg and 10^{22}–10^{25} J, respectively, are ejected into interplanetary space (Howard *et al.*, 1982; Gosling, 1990; Hundhausen *et al.*, 1994; Hundhausen, 1999). CMEs assume velocities in

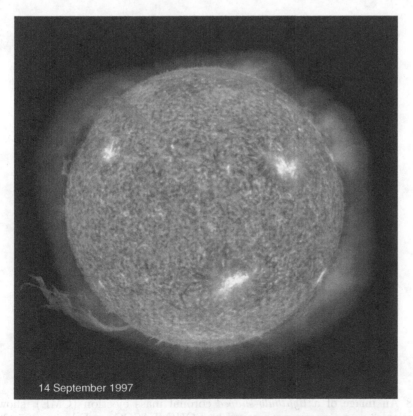

14 September 1997

Fig. 2.7 A *sweeping prominence* appearing in the lower-left part of the image of the Sun's atmosphere, observed by SOHO/EIT at 304 Å. Source: SOHO (ESA&NASA).

the range of several tens to 1000 km/s (Hundhausen *et al.*, 1994). When the CME disturbance reaches the Earth the magnetosphere becomes strongly active (Brueckner *et al.*, 1998). Space missions, such as SOHO and Yohkoh, have been providing us with most valuable material on this phenomenon. For example, Fig. 2.8 shows a large CME propagating in interplanetary space. It is a beautiful example of a *lightbulb CME*.

It seems that *coronal streamers* play an important role in the formation of coronal mass ejections. Streamers reach far out into the solar wind (for instance indicated by the bright structures seen in the lower-right of Fig. 2.8). Frequently, coronal mass ejections involve a multiple streamer configuration.

The different manifestations of solar activity that we have introduced here in this brief overview, are not independent of each other. However, their connection and causal relationship is not yet clear. This is an important issue, because the answers would set constraints to theoretical models. Although in many cases CMEs can be associated with filament eruptions,

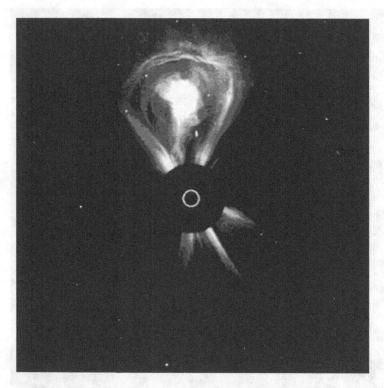

Fig. 2.8 An image of a *lightbulb-shaped* coronal mass ejection (CME), showing the classical 3-part structure, observed by SOHO/LASCO on February 27, 2000. The Sun is blocked out by an occulting disk, the white circle indicates the size of the Sun. Source: SOHO (ESA&NASA).

there does not seem to be a 1:1 relationship (Choudhary and Moore, 2003). The same applies to the relationships between filament eruptions and flares and to CMEs and flares. Gosling (1993) pointed out that observations suggest that large flares are caused by CMEs, contradicting earlier suggestions. An analysis of more than 300 flares revealed that 40% of (M-class) flares did not have a CME association (Andrews, 2003). It seems that the structure of the magnetic field in the relevant region plays an important role in determining what type of activity or what combination of processes results (Švestka, 2001).

A property of solar activity that is widely accepted is the release of magnetic energy, previously accumulated in the solar chromosphere and corona. Magnetic energy is largely turned into heat and kinetic energy of plasma flow; a fraction also goes into electromagnetic and particle radiation. Regarding the overall changes in magnetic structures, the net effect seems to be a reduction in complexity. Often, solar flares are described as releasing energy that was associated with magnetic shear or twist in configurations with current sheets (Priest, 1981). For a two-ribbon flare Asai *et al.* (2004) estimated the released energy as $6 \cdot 10^{23}$ J, using a model based on reconnection in a current sheet. These processes will be the topic of Chapter 14.

Solar activity exhibits an 11-year cycle in its intensity, quantitatively monitored by sunspot counts. Although a final explanation is not yet available, the solar cycle is generally believed to be a signature of the solar magnetic dynamo, which is driven by the differential rotation of the Sun.

Typical parameters of the plasma and magnetic field in the solar corona, where most of the activity takes place, are highly variable. Therefore the values of Table 2.3 can only serve as a rough guideline for general orientation.

Table 2.3 *Characteristic parameters of the solar photosphere and corona.*
For n and T see Guenther et al. (1992), *the photospheric magnetic field refers to a sunspot, the coronal magnetic field to an active region (Lin, 2003).*

	Photosphere	Corona
Number density n [m^{-3}]	8×10^{22}	1×10^{15}
Temperature T [K]	6×10^3	2×10^6
Magnetic field strength B [T]	2×10^{-1}	1×10^{-2}
Length L [m]	1×10^6	3×10^7
Plasma parameter	2	1×10^8

(For instance, the magnetic field strength varies by more than two orders of magnitude.) Note that the table indicates a large difference in the values of the plasma parameter. While the coronal plasma is essentially collisionless the photospheric plasma is substantially influenced by collisions.

We will resume the topic of solar activity in Part IV, where we attempt to address the underlying physical processes.

2.3 Other sites

The phenomenon of activity is by no means confined to the Solar System.

Of the many planets that one speculates to be present in the Universe – an increasing number is being detected in our galaxy – a significant fraction will be in conditions favourable for magnetospheric activity. Necessary conditions are a stellar wind and a sufficiently strong magnetic dipole moment. In the solar system this applies to Mercury, Earth, Jupiter, Saturn, Uranus and Neptune. The size of the magnetosphere, measured by the distance between the subsolar point of the magnetopause and the centre of the planet, is determined by force balance between the solar wind and the planetary magnetic field. Measured in units of the radius of the planet, the size varies from 1.4–1.6 for Mercury to 50–100 for Jupiter (S. Curtis and P. Clark, lep694.gsfc.nasa.gov).

Substorm-like phenomena have been observed on several planets, particularly in the Jovian magnetosphere (Nishida, 1983; Russell *et al.*, 2000). As Kronberg *et al.* (2005) concluded from measurements of the spacecraft Galileo, there is a slow loading phase, during which the Jovian plasma sheet is in a stable configuration. The onset of the dynamic phase takes place in a thinned plasma sheet and leads to inward and outward plasma flows and plasmoid ejection. The authors envisage that the main difference, compared with the Earth, is that the dominant energy source is plasma loading of fast rotating flux tubes rather than supply from the solar wind.

Stellar flares (e.g., Bopp and Moffett, 1973), analogous to solar flares, occur on many stars (Shibata and Yokoyama, 2002). Stellar flares are observed typically on stars with a sufficiently hot corona (e.g., Feldman *et al.*, 1995; Shibata and Yokoyama, 2002). Also, starspots seem to be a common feature of sun-like stars (Solanki and Unruh, 2004).

Flares also have been suggested to originate from coronae of accretion disks (Galeev *et al.*, 1979) and from stellar objects interacting with their accretion disks (Hayashi *et al.*, 1996). Again, the invoked processes show close similarities with corresponding solar phenomena.

The term *activity* is also applied to galactic nuclei. There is strong evidence indicating that large plasma jets are emitted from active galactic nuclei.

Finally, let us have a brief look at the brightest astrophysical flare events that are observed on Earth. The sources are galactic X-ray stars, belonging to the variety of soft-γ-ray repeaters (SGRs). They are believed to be magnetars, which are neutron stars with exceptionally strong magnetic fields, with typical field strengths of the order of 10^{11} T (Duncan and Thompson, 1992; Kouveliotou *et al.*, 1998). SGRs emit *giant flares* in a sporadic way. The largest event (ever seen so far) was observed on 27 December 2004 from SRG 1806-20 (Hurley *et al.*, 2005).

In that event a short initial pulse, lasting 0.2 sec, contained the main part of the total energy of the order of 10^{39} J, corresponding to a photon energy pulse of about 10^{-3} J/m^2. That pulse makes the event the most intense of all solar or cosmic transients observed on Earth before (Hurley *et al.*, 2005). The total energy that the magnetar emitted during 0.2 sec equals the energy the Sun radiates in about 300 000 years.

It is assumed (e.g., Thompson *et al.*, 2002) that magnetars accumulate energy in a twisted magnetic field configuration. The eruption is believed to involve the breaking of the crust of the neutron star, so that mechanical stresses play an important role. It is interesting to note that these extreme phenomena are likely to have important aspects in common with the eruptions characteristic of space plasma activity in the Solar System.

3
Plasma models

By *plasma model* we denote a set of equations governing the temporal evolution of a plasma under a given set of boundary and initial conditions. Ideally, plasma models should be based on first principles. Unfortunately, these are not yet available, at least not from a strict point of view. In any event, simplifications are necessary to keep applications feasible.

For describing space plasma dynamics, it is largely appropriate to ignore quantum effects. The condition for this assumption to be satisfied is that typical values of the action, such as *momentum* × *length* or *energy* × *time* are much larger than Planck's constant $h = 6.63 \cdot 10^{-34}$ Js. This condition is well satisfied for typical space dynamical processes. Therefore, we will discuss models based on classical elementary or statistical mechanics and electromagnetism. Radiation reaction is ignored.

After the foundations were laid by Isaac Newton, James Clerk Maxwell, Ludwig Boltzmann, Albert Einstein and others, these theories have been extremely successful within their ranges of applicability and form the classical basis of modern technology. Thus, the difficulties that we will be facing, to a large extent, do not lie in uncertainties about the foundations but rather in the complexity of the interactions, which require further simplifying assumptions. Therefore, a number of different plasma models exist, each of which has its characteristic range of applicability. It is the aim of this chapter to summarize the plasma models that are relevant for our purposes.

Regarding particle dynamics, the plasma models are formulated in their non-relativistic limit, because the velocities involved mostly are considerably smaller than the speed of light. Occasional use of relativistic generalizations will be discussed where needed.

For completeness, we begin with single particle motion in prescribed electromagnetic fields. This model uses the mechanical equation of motion alone and it is useful for understanding particle motion if a realistic model of the electromagnetic field is available, e.g., from observations, from separate theoretical analysis or from computer simulations.

However, in many cases such a field model is not available. Then the electromagnetic field vectors have to be included in the set of unknowns. In other words, by including Maxwell's equations the field vectors are determined in a self-consistent way. Typically, self-consistent plasma models are formulated as *fluid models* or *kinetic models*. Mixtures of both types result in *hybrid models*. (Thus, the fields are macroscopic approximations of the microscopic fields.)

In the present chapter the plasma models are simply listed with a few comments on properties and physical significance added. For derivations the reader should consult the basic plasma physics literature quoted in each section. Readers with some plasma physics background may find this part still useful as a formulary.

The meaning of the symbols is explained in the List of Symbols. In the text definitions are given only for the more important variables.

3.1 Single particle motion

A characteristic property of a plasma is the presence of charged particles moving in electromagnetic fields. If the electromagnetic field vectors $\boldsymbol{E}(\boldsymbol{r}, t)$ and $\boldsymbol{B}(\boldsymbol{r}, t)$ are prescribed, the motion of a particle with mass m, charge q and velocity $\boldsymbol{w}(t)$ in the non-relativistic limit is determined by the equation of motion together with the definition of velocity,

$$m\frac{\mathrm{d}\boldsymbol{w}}{\mathrm{d}t} = q\big(\boldsymbol{E}(\boldsymbol{r}, t) + \boldsymbol{w} \times \boldsymbol{B}(\boldsymbol{r}, t)\big) - m\nabla\psi(\boldsymbol{r}) \qquad (3.1)$$

$$\frac{\mathrm{d}\boldsymbol{r}}{\mathrm{d}t} = \boldsymbol{w}, \qquad (3.2)$$

where we have added an external gravity force represented by the potential ψ. Equations (3.1) and (3.2) are two, typically nonlinear, ordinary differential equations for the location of the particle $\boldsymbol{r}(t)$ and its velocity $\boldsymbol{w}(t)$ at time t. We summarize a number of properties.

By scalar multiplication with $\boldsymbol{w}$ one finds from the equation of motion (3.1) an energy equation of the form

$$\frac{\mathrm{d}}{\mathrm{d}t}\left(\frac{1}{2}mw^2\right) = q\boldsymbol{w} \cdot \boldsymbol{E} - m\boldsymbol{w} \cdot \nabla\psi. \qquad (3.3)$$

The magnetic part of the Lorentz force in (3.1) is perpendicular to w and therefore does not change kinetic energy.

The most pronounced property of the particle motion in the presence of a magnetic field is the *gyration*. For its illustration let us set $E = 0$, $\nabla\psi = 0$ and let B be a constant magnetic field. Then, the particle moves along a helix oriented along the magnetic field direction with constant velocity. The *gyroradius* r_g (radius of the helix) and the *gyrofrequency* ω_g (angular frequency of the gyration) are given by

$$r_g = \frac{m\, w_\perp}{|q|B} \tag{3.4}$$

$$\omega_g = \frac{|q|B}{m}, \tag{3.5}$$

where $w_\perp$ is the velocity component perpendicular to the magnetic field. A particle with positive (negative) charge q performs a left (right) sense rotation with respect to the positive field direction.

If a constant electric field E directed perpendicular to B is present, the helix moves with a drift velocity

$$u_E = \frac{E \times B}{B^2}. \tag{3.6}$$

In other words, this motion corresponds to gyration in a reference frame moving with velocity u_E, which is easily verified explicitly by a corresponding Galilean transformation applied to (3.1).

The gyration property qualitatively persists for electromagnetic fields that vary slowly in space and time compared with the spatial and temporal gyro-scales defined by (3.4) and (3.5). This regime is referred to as *adiabatic particle motion*.

Time variations and gradients of E and B lead to drifts in addition to (3.6). The total drift of the centre of the instantaneous gyro-motion (*gyrocentre*) is obtained to first order in an expansion with respect to typical values of $1/(\omega_g t_f)$ and r_g/r_f, where t_f and r_f are characteristic time and length scales of the electromagnetic field. Treating both quantities as of the order of a smallness parameter, say ϵ, and assuming that the components of the electric field and the gravity force parallel to the magnetic field are of order ϵ also, one finds for the *gyrocentre velocity* to zeroth and first order,

$$u_d = u_\parallel b + u_E + \frac{\mu}{qB^2}B \times \nabla B + \frac{m}{qB^2}B \times \left(u_\parallel \frac{db}{dt} + \frac{du_E}{dt} + \nabla\psi \right), \tag{3.7}$$

where b is the unit vector associated with the magnetic field direction and the parallel (with respect to B) drift velocity component $u_\parallel$ satisfies the differential equation

$$m\frac{\mathrm{d}u_\parallel}{\mathrm{d}t} = qE_\parallel - \mu b \cdot \nabla B + m u_E \cdot \frac{\mathrm{d}b}{\mathrm{d}t} - m b \cdot \nabla\psi \tag{3.8}$$

with

$$\frac{\mathrm{d}}{\mathrm{d}t} = \frac{\partial}{\partial t} + u_\parallel b \cdot \nabla + u_E \cdot \nabla. \tag{3.9}$$

Here, μ is the magnetic moment associated with the gyration (area × current),

$$\mu = \frac{\frac{1}{2}m w_\perp^2}{B}. \tag{3.10}$$

An important property is that μ is an *adiabatic invariant*. This means that, within the present approximation, μ stays constant over time periods of order $1/\epsilon$, during which the electromagnetic field develops changes of order 1. (For an outline of adiabatic invariance see www.tp4.rub.de/~ks/tb.pdf.)

Additional adiabatic invariants exist for special magnetic field geometries. For instance, if particles are trapped between regions of enhanced magnetic field strength, the integral

$$I_2 = \int u_\parallel \,\mathrm{d}s \tag{3.11}$$

is an adiabatic invariant, where the integration is carried out along the arc length of the field line, on which the gyrocentre is located instantaneously, and is limited by the mirror points. In a slowly time-varying field with rotational invariance, the magnetic flux encircled by the gyrocentre trajectory is another adiabatic invariant.

There is a simple intuitive interpretation of the expressions (3.7) and (3.8). Consistent with (3.6) a force F acting on the particle causes a drift velocity $(F/q) \times B/B^2$. The magnetic force, considered in the limit of small gyroradii, is given by $-\mu\nabla B$, the inertia and gravity forces give corresponding contributions. This explains (3.7). To understand (3.8), one applies the same argument to the parallel (to B) component of the equation of motion.

For details of adiabatic particle motion and derivations see the literature (e.g., Northrop, 1963; Balescu, 1988; Boyd and Sanderson, 2003).

For later reference we add the *Hamiltonian* formulation of single particle motion. This is an equivalent representation of the problem (3.1) with (3.2) using electromagnetic potentials $\phi(r,t)$ (electric potential) and $A(r,t)$

(magnetic vector potential) instead of the field vectors (Jackson, 1998). In terms of the potentials the field vectors are obtained as

$$\boldsymbol{E} = -\nabla\phi - \frac{\partial \boldsymbol{A}}{\partial t} \tag{3.12}$$

$$\boldsymbol{B} = \nabla \times \boldsymbol{A}. \tag{3.13}$$

This representation is valid for any arbitrary electromagnetic field satisfying the homogeneous subset of Maxwell's equations (see (3.28) and (3.29)).

In Cartesian coordinates (r_1, r_2, r_3) the Hamiltonian is given by

$$H(\boldsymbol{P}, \boldsymbol{r}) = \frac{1}{2m} \left(\boldsymbol{P} - q\boldsymbol{A}(\boldsymbol{r}, t)\right)^2 + q\phi(\boldsymbol{r}, t) + m\psi(\boldsymbol{r}) \tag{3.14}$$

where $\boldsymbol{P} = m\boldsymbol{w} + q\boldsymbol{A}$ is the *canonical momentum*. The equations of motion then are obtained from *Hamilton's equations*

$$\frac{\mathrm{d}r_i}{\mathrm{d}t} = \frac{\partial H}{\partial P_i} \qquad i = 1 \ldots 3 \tag{3.15}$$

$$\frac{\mathrm{d}P_i}{\mathrm{d}t} = -\frac{\partial H}{\partial r_i} \qquad i = 1 \ldots 3\,. \tag{3.16}$$

These are ordinary differential equations for $\boldsymbol{r}(t)$ and $\boldsymbol{P}(t)$. For details see standard texts (e.g., Landau and Lifshitz, 1963; Goldstein, 2002).

3.2 Kinetic models

Consider a plasma consisting of several particle species, such as electrons and a number of different ions. To introduce kinetic models we first look at the simple case where the particles interact only through an electromagnetic field of the form $\boldsymbol{E}(\boldsymbol{r}, t), \boldsymbol{B}(\boldsymbol{r}, t)$. In other words, the particles do not interact directly with each other so that the trajectory of each particle is governed by the equations of single particle motion. Then, Hamilton's equations of motion (3.15) and (3.16) imply a *Liouville equation* of the form

$$\frac{\partial F_s}{\partial t} + \frac{\partial F_s}{\partial \boldsymbol{r}} \cdot \frac{\partial H_s}{\partial \boldsymbol{P}} - \frac{\partial F_s}{\partial \boldsymbol{P}} \cdot \frac{\partial H_s}{\partial \boldsymbol{r}} = 0\,. \tag{3.17}$$

Under the present assumptions, $F_s(\boldsymbol{r}, \boldsymbol{P}, t)$ may be understood as the *distribution function* of particles of species s in 6-dimensional phase space spanned by $\boldsymbol{r}$ and $\boldsymbol{P}$. Equation (3.17) simply expresses Liouville's theorem (Balescu, 1975; Goldstein, 2002) in single particle phase space, stating that the distribution function is constant on particle trajectories.

If particle interactions, such as collisions, have to be taken into account, Liouville's theorem no longer applies to single particle phase space but to

$6N$-dimensional phase space, where N is the number of particles. For sufficiently weak collisional coupling, one can still find a description in single particle space with an additional term, the *collision term*. The equation is then called *kinetic equation* or *transport equation*. For plasmas, the particle phase space density, traditionally, is expressed in terms of velocity rather than of canonical momentum, i.e., it has the form $f(\boldsymbol{r}, \boldsymbol{w}, t)$, usually called *single particle distribution function*. The kinetic equation then assumes the form

$$\frac{\partial f_s}{\partial t} + \boldsymbol{w} \cdot \frac{\partial f_s}{\partial \boldsymbol{r}} + \left(\frac{q_s}{m_s}(\boldsymbol{E} + \boldsymbol{w} \times \boldsymbol{B}) - m_s \nabla \psi \right) \cdot \frac{\partial f_s}{\partial \boldsymbol{w}} = \left. \frac{\partial f_s}{\partial t} \right|_{\mathrm{c}}, \qquad (3.18)$$

where the term on the right side is the collision term, and $\boldsymbol{E}$ and $\boldsymbol{B}$ are the macroscopic fields. Such an equation exists for each particle species s. Often, equation (3.18) is named after the specific form of the collision term, examples being *Boltzmann, Fokker–Planck or Lenard–Balescu equations* (Balescu, 1975).

The distribution function is normalized such that the integration over velocity space gives particle number density n_s,

$$n_s = \int f_s \, \mathrm{d}^3 w \,. \qquad (3.19)$$

Then, bulk velocity $\boldsymbol{v}_s$ is

$$\boldsymbol{v}_s = \frac{1}{n_s} \int \boldsymbol{w} f_s \, \mathrm{d}^3 w \,. \qquad (3.20)$$

The equations (3.18) are coupled to Maxwell's equations, where charge density σ and current density $\boldsymbol{j}$ are given by

$$\sigma = \sum_s \sigma_s, \quad \sigma_s = \sum_s q_s n_s \qquad (3.21)$$

$$\boldsymbol{j} = \sum_s \boldsymbol{j}_s, \quad \boldsymbol{j}_s = \sum_s q_s n_s \boldsymbol{v}_s \,. \qquad (3.22)$$

Equations (3.18) and Maxwell's equations form a closed set of equations only if the collision term, at least approximately, can be expressed by $f_s, s = 1, 2, \ldots$ alone. If the plasma has this property, it is said to be in a *kinetic regime*.

In general, the collision term involves the distribution functions in two-particle phase space. Therefore, equations for the two-particle distribution functions are needed, too. However, these equations couple to even higher order distribution functions. The result is an infinite hierarchy of distribution functions and corresponding equations. If the interaction can approximately be represented by an electrostatic force, which suffices for most purposes, this hierarchy can be derived from Liouville's equation in the full $6N$-dimensional phase space by standard techniques of statistical mechanics (Ichimaru, 1973; Balescu, 1975).

In fully ionized plasmas, collision terms arise from Coulomb collisions between the charged particles. They are based on electric fluctuations in the Debye sphere, a sphere with the Debye length

$$\lambda_D = \sqrt{\epsilon_0 k_B T_e / e^2 n_e} = \frac{1}{\omega_p} \sqrt{\frac{k_B T_e}{m_e}} \qquad (3.23)$$

as its radius. Here, n_e is electron density and, in local thermodynamic equilibrium, T_e electron temperature. For plasmas in nonequilibrium states $k_B T_e$ measures the energy of random electron motion. Further, ω_p is the (electron) plasma frequency (the ion plasma frequency is defined analogously)

$$\omega_p = \sqrt{\frac{e^2 n_e}{\epsilon_0 m_e}}. \qquad (3.24)$$

Typical Coulomb collision terms scale as $\log(\Lambda_p)/\Lambda_p$, where

$$\Lambda_p = \frac{4\pi}{3} \lambda_D{}^3 n_e \qquad (3.25)$$

is the plasma parameter, which equals the number of electrons in the Debye sphere. Clearly, more particles cause less collective fluctuations, at least in a local thermodynamic equilibrium picture.

As the plasma parameter scales as $T_e^{3/2}/n_e^{1/2}$, Coulomb collisions are negligible for sufficiently high temperatures and low densities, the regime of *collisionless plasmas*. This regime applies to most plasmas in space, which show extremely large plasma parameters (see Tables 2.2, 2.3 and 9.1). The regime where the collision term in (3.18) is neglected is usually referred to as the *Vlasov regime*. With the collision term all consequences of the presence of discrete particles have disappeared. Electric charge and mass can be regarded as smeared out continuously in phase space.

For reference reasons we list the equations of the Vlasov theory:

$$\frac{\partial f_s}{\partial t} + \boldsymbol{w} \cdot \frac{\partial f_s}{\partial \boldsymbol{r}} + \left(\frac{q_s}{m_s} (\boldsymbol{E} + \boldsymbol{w} \times \boldsymbol{B}) - m_s \nabla \psi \right) \cdot \frac{\partial f_s}{\partial \boldsymbol{w}} = 0 \qquad (3.26)$$

$$\nabla \cdot \boldsymbol{E} = \frac{1}{\epsilon_0} \sum_s q_s \int f_s \, \mathrm{d}^3 w \qquad (3.27)$$

$$\nabla \times \boldsymbol{E} = -\frac{\partial \boldsymbol{B}}{\partial t} \tag{3.28}$$

$$\nabla \cdot \boldsymbol{B} = 0 \tag{3.29}$$

$$\nabla \times \boldsymbol{B} = \mu_0 \sum_s q_s \int \boldsymbol{w} f_s \, \mathrm{d}^3 w + \frac{1}{c^2} \frac{\partial \boldsymbol{E}}{\partial t} . \tag{3.30}$$

This is a set of nonlinear integro-differential equations for the unknowns $f_s(\boldsymbol{r}, \boldsymbol{w}, t)$, $\boldsymbol{E}(r, t)$, $\boldsymbol{B}(r, t)$. It consists of a kinetic equation (3.26) for each species and Maxwell's equations (3.27)–(3.30).

For prescribed electromagnetic fields, the problem of solving (3.26) for the distribution function is equivalent to solving the problem of single particle motion. This is based on the fact that (3.1) and (3.2) are the characteristic equations of the partial differential equation (3.26).

The Vlasov theory is the theory best suited for describing space phenomena occurring in collisionless, fully ionized plasmas.

An important aspect of kinetic theories of plasmas is that the separation of the electromagnetic field in averaged fields and fluctuation fields is not unambiguous, which has consequences for the meaning of the collision term in (3.18). Here we have chosen the classical point of view, where the collision term contains only the fluctuations resulting from particle discreteness in the sense of two-particle correlations. All other effects are described by $\boldsymbol{E}$ and $\boldsymbol{B}$.

In the presence of turbulence there is another possibility, which is applicable if the time scales of the turbulence and of the averaged quantities are well separated. Then, after suitable averaging, the turbulence gives rise to an effective collision term even in Vlasov theory. (Separate equations are needed for describing local turbulence properties.) To distinguish the resulting transport effects from the transport due to discrete particle collisions, one speaks of *turbulent transport*. More details of this aspect are discussed in connection with collective interactions in Section 9.3.2.

We mention some important properties and consequences of Vlasov theory. Integrating (3.26) over velocity space gives the conservation of particles,

$$\frac{\partial n_s}{\partial t} + \nabla \cdot (n_s \boldsymbol{v}_s) = 0 \tag{3.31}$$

which implies conservation of mass with mass density given by $\rho_s = m_s n_s$

$$\frac{\partial \rho_s}{\partial t} + \nabla \cdot (\rho_s \boldsymbol{v}_s) = 0 \tag{3.32}$$

and conservation of electric charge,

$$\frac{\partial \sigma_s}{\partial t} + \nabla \cdot \boldsymbol{j}_s = 0. \tag{3.33}$$

By multiplying (3.26) by $m_s \boldsymbol{w}_s$ and integrating over velocity space and then making use of (3.31) one finds the momentum equation for species s,

$$\rho_s \frac{\partial \boldsymbol{v}_s}{\partial t} + \rho_s \boldsymbol{v}_s \cdot \nabla \boldsymbol{v}_s = -\nabla \cdot \mathcal{P}_s + \boldsymbol{j}_s \times \boldsymbol{B} + \sigma_s \boldsymbol{E} - \rho_s \nabla \psi. \tag{3.34}$$

Here, $\mathcal{P}_s$ is the pressure tensor of species s with components

$$\mathcal{P}_{s,ik} = \int m_s (w_i - v_{s,i})(w_k - v_{s,k}) f_s(\boldsymbol{r}, \boldsymbol{w}, t)\, \mathrm{d}^3 w \tag{3.35}$$

or, using the dyadic tensor formulation,

$$\mathcal{P}_s(\boldsymbol{r}, t) = m_s \int (\boldsymbol{w} - \boldsymbol{v}_s)(\boldsymbol{w} - \boldsymbol{v}_s) f_s(\boldsymbol{r}, \boldsymbol{w}, t)\, \mathrm{d}^3 w. \tag{3.36}$$

Equation (3.34) is an equation of motion analogous to (3.1), however, it applies to a spatial plasma element rather than to a single particle and therefore contains particles with different velocities, which generates the pressure term.

The energy equation is obtained by applying the factor $m w^2 / 2$ to (3.26) before integrating over velocity space,

$$\frac{\partial e_{\mathrm{kin},s}}{\partial t} + \nabla \cdot \boldsymbol{Q}_s = \boldsymbol{E} \cdot \boldsymbol{j}_s - \rho_s \boldsymbol{v}_s \cdot \nabla \psi \tag{3.37}$$

where

$$e_{\mathrm{kin},s} = \int \frac{m_s w^2}{2} f_s\, \mathrm{d}^3 w \tag{3.38}$$

$$\boldsymbol{Q}_s = \int \frac{m_s w^2}{2} \boldsymbol{w} f_s\, \mathrm{d}^3 w \tag{3.39}$$

denote kinetic energy density and the energy flux density of species s. By taking the sum of (3.37) over plasma species one finds the energy conservation law

$$\frac{\partial}{\partial t}(e_{\mathrm{kin}} + e_{\mathrm{mag}} + e_{\mathrm{el}} + \rho\psi) + \nabla \cdot (\boldsymbol{Q} + \boldsymbol{S} + \rho \boldsymbol{v}\psi) = 0 \tag{3.40}$$

where

$$e_{\mathrm{kin}} = \sum_s e_{\mathrm{kin},s}, \quad \boldsymbol{Q} = \sum_s \boldsymbol{Q}_s, \quad e_{\mathrm{mag}} = \frac{B^2}{2\mu_0}, \quad e_{\mathrm{el}} = \frac{\epsilon_0 E^2}{2} \tag{3.41}$$

with e_{mag} and e_{el} being magnetic and electric energy densities and

$$S = \frac{1}{\mu_0} E \times B \qquad (3.42)$$

the electromagnetic energy flux density (Poynting vector). In deriving (3.40) the Poynting theorem

$$\frac{\partial}{\partial t}(e_{\mathrm{mag}} + e_{\mathrm{el}}) + \nabla \cdot S = -E \cdot j \qquad (3.43)$$

has been used, which is a consequence of Maxwell's equations (Jackson, 1998).

An equation for the pressure tensor is obtained by multiplying (3.26) by the tensor $m(w - v_s)(w - v_s)$ and integrating over velocity space. With the help of (3.32) and (3.34) one finds

$$\frac{\partial \mathcal{P}_s}{\partial t} + v_s \cdot \nabla \mathcal{P}_s + \mathcal{P}_s \nabla \cdot v_s + \mathcal{P}_s \cdot \nabla v_s + [\mathcal{P}_s \cdot \nabla v_s]^{\mathrm{T}}$$
$$+ \frac{q_s}{m_s} B \times \mathcal{P}_s + \frac{q_s}{m_s}[B \times \mathcal{P}_s]^{\mathrm{T}} + \nabla \cdot \mathcal{Q}_s = 0, \qquad (3.44)$$

where

$$\mathcal{Q}_s = m_s \int (w - v_s)(w - v_s)(w - v_s) f_s(r, w, t)\, \mathrm{d}^3 w \qquad (3.45)$$

is the (third-order) heat flow tensor. Note that $e_{\mathrm{kin},s} = \mathrm{Tr}(\mathcal{P}_s)/2 + \rho_s v_s^2/2$. Also, Q is related to the heat conduction vector $q_s = \int (m_s/2)(w - v_s)^2(w - v_s) f_s\, \mathrm{d}^3 w$ by

$$Q_s = e_{\mathrm{kin},s} v_s + v_s \cdot \mathcal{P}_s + q_s . \qquad (3.46)$$

A typical simplification consists in the assumption of quasi-neutrality, which allows one to approximate the Poisson equation (3.27) by the condition of charge neutrality,

$$\sigma = 0. \qquad (3.47)$$

This approximation is best understood as an asymptotic expression in the limit of small ratio of $\lambda_{\mathrm{D}}/r_{\mathrm{f}}$. In a suitable dimensionless form, the square of that ratio appears as a factor in front of $\nabla \cdot E$ in (3.27), such that one obtains (3.47) in lowest order of a singular perturbation scheme. Note that, in this asymptotic sense, (3.47) does not imply that $\nabla \cdot E$ vanishes.

A useful physical picture here is that electric fields in plasmas are coupled to the random particle motion. An energy balance shows that strong charge separation, driven by that motion, can occur only over distances of the order of the Debye length. For distances large compared with the Debye length charge separation becomes negligible. Keeping $\sigma = 0$ in the presence of particle species with different masses or charges, typically requires an electric field. Thus, (3.47) can still be understood as imposing a constraint on the electric field, although $\boldsymbol{E}$ does not appear explicitly. Similarly, externally applied electric fields, in typical situations, are shielded from the plasma (assumed to be stable) by a layer that has a thickness of a few Debye lengths. If, within the plasma volume, structures on the Debye scale develop, quasi-neutrality breaks down and the exact form of Poisson's equation is needed.

3.3 Fluid models

Fluid models result from attempts to formulate a self-consistent plasma description exclusively by observables that are defined as functions of coordinates and time (*fluid variables*), such as $\rho(\boldsymbol{r}, t)$ or $\boldsymbol{v}(\boldsymbol{r}, t)$. (Here, the term *fluid* is meant to include compressible media, a terminology largely used in plasma physics, although in other fields it implies incompressibility.) Primarily, fluid variables arise as integrals of the form $M_n = \int g_n(\boldsymbol{w}) f_s(\boldsymbol{r}, \boldsymbol{w}, t) \, \mathrm{d}^3 w$ where $g_n(\boldsymbol{w})$ is a suitable (dyadic) polynomial of particle velocity $\boldsymbol{w}$, n being the order of the polynomial. The integral is called *velocity moment of the distribution function of order n*. Equations for the temporal evolution of the moments are obtained by multiplying the transport equation (3.18) (or (3.26) for collisionless plasmas) by $g_n(\boldsymbol{w})$ and integrating over velocity space. The resulting equations are called *moment equations*. Examples are (3.31), (3.34) and (3.44).

Typically, moment equations are coupled and do not form a finite closed set. Any given moment equation for a moment of order n involves a moment of order $n + 1$. (Note that the equation (3.31) for n_s involves $\boldsymbol{v}_s$, the equation (3.34) for $\boldsymbol{v}_s$ involves $\mathcal{P}_s$ and the equation (3.44) for $\mathcal{P}_s$ involves $\mathcal{Q}_s$.) Thus, the moment equations form an infinite hierarchy. A finite set of fluid equations arises by truncation of that hierarchy such that a closed set of equations is obtained.

A serious drawback of standard truncation procedures is that often the manipulations that lead to closure are difficult to justify rigorously for given plasma conditions. Therefore, fluid equations play the role of model equations on their own. It seems that a safe way to assess their validity is to gain experience with applications and to compare the results with those of other approaches.

A set of fluid equations for each particle species typically is obtained by assuming quasi-neutrality, ignoring heat flux ($Q_s = 0$) and assuming an isotropic pressure tensor

$$\mathcal{P}_s = p_s \mathcal{I}, \tag{3.48}$$

where $\mathcal{I}$ is the unit tensor and p_s the scalar pressure of species s. The assumption of isotropy requires an isotropization process, and indeed observations show that in particular space plasma regions such processes must be active. Also, let us assume that collisions between the particles of a given species are mass, momentum, and energy conserving, which excludes chemical reactions. Under these conditions fluid models take the form

$$\frac{\partial \rho_s}{\partial t} + \nabla \cdot (\rho_s v_s) = 0 \tag{3.49}$$

$$\rho_s \frac{\partial v_s}{\partial t} + \rho_s v_s \cdot \nabla v_s = -\nabla p_s + j_s \times B + \sigma_s E - \rho_s \nabla \psi + M_{cs} \tag{3.50}$$

$$\frac{\partial p_s}{\partial t} + v_s \cdot \nabla p_s + \frac{5}{3} p_s \nabla \cdot v_s + N_{cs} = 0 \tag{3.51}$$

$$\sum_s \sigma_s = 0 \tag{3.52}$$

$$\nabla \times E = -\frac{\partial B}{\partial t} \tag{3.53}$$

$$\nabla \cdot B = 0 \tag{3.54}$$

$$\nabla \times B = \mu_0 \sum_s j_s . \tag{3.55}$$

M_{cs} and N_{cs} result from the collision term in (3.18). The equation (3.49) is a consequence of (3.31); (3.50) and (3.51) can be obtained from (3.34) and from the trace of (3.44) using the truncation assumptions. In (3.55) the displacement current is neglected, which is consistent with quasi-neutrality, for which charge conservation simply is given by $\nabla \cdot j = 0$.

The following specializations and modifications are of particular interest.

3.3.1 Momentum equations of two-fluid models

Let us consider a plasma consisting of electrons (subscript e) and a single ion species (subscript i) with $q_i = e$. In that case it is convenient to rearrange the model equations and to use variables representing the fluid as a whole instead of the ion variables. The condition of quasi-neutrality gives $n_e = n_i = n$ and the conservation properties of the collisions $M_{ce} + M_{ci} = 0$, $N_{ce} + N_{ci} = 0$. Density is defined as $\rho = \rho_e + \rho_i$, bulk velocity as $v = (\rho_e v_e + \rho_i v_i)/\rho$ and pressure as $p = p_e + p_i$. Taking the sum of the momentum equations (3.34) for electrons and ions, one finds

$$\rho \frac{\partial v}{\partial t} + \rho v \cdot \nabla v = -\nabla p + j \times B - \rho \nabla \psi . \qquad (3.56)$$

As the second momentum equation one chooses the momentum equation of the electrons, where in the $v_e \times B$-term one uses $v_e = v_i - j/en_e$ and expresses v_i by v. For $m_e \ll m_i$ the result can be written as

$$E + v \times B = -\frac{1}{en_e} \nabla p_e + \frac{1}{en_e} j \times B + \eta j. \qquad (3.57)$$

The second term on the right side is called the *Hall term*. It is relevant for phenomena occurring on intrinsic ion scales (see Section 3.6). The term ηj, where η denotes resistivity, stems from M_{ce}, which for moderate resistivities can be assumed to be proportional to the relative velocity between electrons and ions, i.e., to the current density j.

The equation (3.57) is written in the form of *Ohm's law*. As pointed out above, for typical space plasmas, the collisional resistivity is negligible. However, an analogous term can take into account effects resulting from micro-turbulence. In that case one refers to η as the *turbulent resistivity* η_{turb}, see Section 9.3.2.

3.3.2 Ideal magnetohydrodynamics

The equations of *ideal magnetohydrodynamics* may be obtained from a two-fluid picture with Ohm's law of the form (3.57) by assuming that scale lengths are small as compared to the ion gyro- and inertia scales and that resistive effects are ignorable. One then finds from (3.57) Ohm's law in its ideal form $E + v \times B = 0$. The resulting model equations are

$$\frac{\partial \rho}{\partial t} + \nabla \cdot (\rho v) = 0 \qquad (3.58)$$

$$\rho \frac{\partial v}{\partial t} + \rho v \cdot \nabla v = -\nabla p + j \times B - \rho \nabla \psi \qquad (3.59)$$

$$E + v \times B = 0 \qquad (3.60)$$

$$\left(\frac{\partial}{\partial t} + v \cdot \nabla \right) \left(\frac{p}{\rho^\gamma} \right) = 0 \qquad (3.61)$$

$$\nabla \times E = -\frac{\partial B}{\partial t} \qquad (3.62)$$

$$\nabla \cdot B = 0 \qquad (3.63)$$

$$\nabla \times B = \mu_0 j. \qquad (3.64)$$

Regarding (3.61), an additional simplification has been introduced. For $N_{cs} = 0$ the equations (3.49) and (3.51) would yield an *adiabatic law* of the type (3.61) for each species separately. However, as ideal MHD does not refer to individual species, one postulates a corresponding equation for the fluid as a whole, in analogy to the standard entropy conservation law of neutral gases. In (3.61) the *polytropic index* γ is left open as a parameter, for most purposes one chooses 5/3, corresponding to local thermodynamic equilibrium of a gas of particles with 3 degrees of freedom. With internal energy $u = p/(\gamma - 1)$, one may rewrite (3.61) as

$$\frac{\partial u}{\partial t} + \nabla \cdot (u\boldsymbol{v}) = -p\nabla \cdot \boldsymbol{v}, \qquad (3.65)$$

which indicates that internal energy changes by compression.

The basic equations (3.58)–(3.64) do not involve the plasma temperature explicitly. A temperature, which is not defined in the thermodynamic sense but as measuring the random part of kinetic energy per particle, is conveniently introduced by the ideal gas law

$$p = nk_{\mathrm{B}}T, \qquad (3.66)$$

where number density n is related to density ρ by $\rho = nm$, m being an effective ion mass.

3.3.3 Resistive magnetohydrodynamics

If resistive effects cannot be ignored, the equations of ideal MHD are modified in two ways. Ohm's law now reads

$$\boldsymbol{E} + \boldsymbol{v} \times \boldsymbol{B} = \eta\boldsymbol{j}, \qquad (3.67)$$

and in (3.65) resistive dissipation has to be taken into account by adding the term ηj^2 on the right side. This corresponds to the following modification of (3.61)

$$\left(\frac{\partial}{\partial t} + \boldsymbol{v} \cdot \nabla\right)\left(\frac{p}{\rho^{\gamma}}\right) = \frac{\gamma - 1}{\rho^{\gamma}}\eta j^2, \qquad (3.68)$$

which keeps total energy conserved. For sufficiently small values of $|j|$, the quadratic dissipation term in (3.68) may be ignored, such that the only modification of the ideal MHD equations is the $\eta\boldsymbol{j}$ term in Ohm's law.

3.4 Chew–Goldberger–Low momentum equation

In cases where an isotropic pressure tensor is not available, there are still a number of useful simplifications. In particular, this applies to the limit of strong magnetic fields, where the form of the pressure tensor $\mathcal{P}_s$ simplifies considerably. Its form is found from the condition that the sum of the terms involving $\boldsymbol{B}$ in (3.44) must vanish, assuming that all velocity moments are bounded for large B. One finds that $\mathcal{P}_s$ must have the form

$$\mathcal{P}_s = p_{\perp s}\mathcal{I} + (p_{\|s} - p_{\perp s})\boldsymbol{bb} \tag{3.69}$$

where $p_{\|s}$ and $p_{\perp s}$ denote pressures associated with directions parallel and perpendicular to the magnetic field direction, and $\mathcal{I}$ is the unit tensor. Chew *et al.* (1956) derived equation (3.69), which is called the CGL momentum equation, by a corresponding large-B expansion of the equations of the Vlasov theory, i.e., (3.26)–(3.30), and averaging over the gyration. They also showed that in the absence of heat flux $p_{\|s}$ and $p_{\perp s}$ satisfy the equations

$$\left(\frac{\partial}{\partial t} + \boldsymbol{v}_s \cdot \nabla\right)\left(\frac{p_{\perp s}}{\rho_s B}\right) = 0 \tag{3.70}$$

$$\left(\frac{\partial}{\partial t} + \boldsymbol{v}_s \cdot \nabla\right)\left(\frac{p_{\|s}B^2}{\rho_s^{\,3}}\right) = 0\,. \tag{3.71}$$

These equations are readily obtained from the trace and the $\boldsymbol{bb}$-component of (3.44), using the relationship

$$\frac{1}{B}\frac{\mathrm{d}B}{\mathrm{d}t} - \frac{1}{\rho}\frac{\mathrm{d}\rho}{\mathrm{d}t} = \boldsymbol{b} \cdot (\nabla \boldsymbol{v}_s) \cdot \boldsymbol{b}. \tag{3.72}$$

The latter relationship is based on the fact that in lowest order guiding centre theory one finds (consistent with (3.6))

$$\boldsymbol{E} + \boldsymbol{v}_s \times \boldsymbol{B} = 0 \tag{3.73}$$

which by (3.28) implies

$$\frac{\partial \boldsymbol{B}}{\partial t} - \nabla \times (\boldsymbol{v}_s \times \boldsymbol{B}) = 0. \tag{3.74}$$

(3.72) is then obtained from the $\boldsymbol{b}$-component of (3.74), after elimination of $\nabla \cdot \boldsymbol{v}_s$ with the help of mass conservation.

To obtain an anisotropic generalization of the ideal MHD model one postulates the form of (3.69) with (3.70) and (3.71) for the one-fluid pressure tensor $\mathcal{P}$. Then, the quantities $p_\perp/\rho B$ and $p_\|B^2/\rho^3$ are constant on streamlines defined by the plasma velocity $\boldsymbol{v}$ and play the role of adiabatic invariants, the former invariant being consistent with the invariance of the magnetic moment (3.10). The CGL model has proven useful for describing space plasmas with anisotropic pressures (e.g., Shuo and Wolf, 2003).

The isotropic limit of (3.70) and (3.71) is not straightforward, because of the additional effects that must be at work to keep the pressure tensor isotropic on long time scales. Nevertheless, by eliminating B from the two invariants, one finds the invariant $p_\parallel p_\perp^2/\rho^5$, which for $p_\parallel = p_\perp = p$ is consistent with (3.61) for $\gamma = 5/3$. Then, the isotropy condition replaces the adiabatic invariance of the magnetic moment, which requires that the isotropizing fluctuations are in an appropriate frequency range.

3.5 On the validity of MHD models for space plasmas

Fluid models face the difficulty that, by the various simplifications, their applicability for a given space plasma phenomenon often is not obvious. Sometimes it is only after a sequence of studies of a characteristic sample process using different techniques that one develops a feeling for the range in which fluid models are useful. Clearly, the answer also depends on the rigour that one wishes to achieve. Even in domains where fluid models are considered as useful, their main strength is to provide information on the qualitative behaviour of a process by comparatively simple means. Quantitatively exact results generally require a kinetic treatment. It is in that sense that fluid models have proven useful even for collisionless plasmas under a variety of circumstances.

There are a few general guidelines indicating the range where fluid models can be expected to give useful results.

By assuming an isotropic pressure one typically eliminates resonant interactions between waves and particles. This is intuitively clear considering that the resonance strongly modifies the distribution function in the region of resonance in phase space, resulting in a complex, typically non-isotropic, pressure tensor. Resonances associated with frequencies near intrinsic frequencies of the plasma become unimportant for plasmas with length and time scales much larger than the intrinsic ion scales (see the discussion in Section 2.1). This *MHD limit* is a domain where fluid models are generally expected to be useful.

Many non-resonant features of waves are qualitatively correctly described in the MHD limit, compared with a kinetic treatment of waves on a comparable background. A quantitative comparison typically shows that the truncation method, such as the omission of heat flux, can lead to incorrect numerical factors in dispersion relations.

The question of validity of MHD is still more complicated than it might seem so far. The reason is that there exist other resonances, which are not associated with the intrinsic scales and which do not necessarily disappear in the MHD limit. A particle moving along a magnetic field line with a velocity close to the phase velocity of a wave propagating in the same direction is in resonance with that wave even in the MHD limit. This type of resonance typically causes *Landau damping* (Gary, 1993); for example, it eliminates ion acoustic waves in two-component plasmas with $T_i > T_e$.

Nevertheless, MHD models have proven quite useful for fostering our understanding of space plasma dynamics on large scales. But one has to keep their limitations in mind. Uncritical application of fluid models to collisionless plasmas can lead to drastic errors!

Occasionally, fluid models can be useful tools even for processes occurring on small scales, as the example of the following section demonstrates.

3.6 Electron magnetohydrodynamics, Hall current

Consider a non-resonant process in a plasma consisting of electrons and a single ion species, with length and time scales small compared with the corresponding ion scales. Under such conditions one can expect a fluid theory to apply, in which the ions are treated as unaffected by the process. A particularly simple model arises if the scales are still larger than the electron scales, such that electron inertial effects are unimportant, and if the electrons are sufficiently cold for electron pressure effects to be negligible. Then, in a local frame where the ions are at rest, one finds that in the non-resistive case Ohm's law (3.57) reduces to

$$E = \frac{1}{en_e} j \times B \qquad (3.75)$$

where by quasi-neutrality $n_e = q_i n_i / e$, with n_i being the unperturbed ion density. Here, the current is entirely carried by electrons,

$$j = -en_e v_e \qquad (3.76)$$

and Ohm's law can also be written as $E + v_e \times B = 0$; the perpendicular (with respect to B) component of j becomes

$$j_\perp = -en_e \frac{E \times B}{B^2}, \qquad (3.77)$$

which is called the *Hall current density*. This regime is a simple version of *electron magnetohydrodynamics (EMHD)*.

If n_e is set to a constant, the only non-trivial two-fluid equations that survive in addition to (3.75) are Maxwell's equations.

This regime is often called the *whistler regime*, named after the wave mode of EMHD (e.g., Gary, 1993).

3.7 Initial and boundary conditions

All models discussed above require the specification of initial and boundary conditions. Clearly, the initial conditions of the observables must satisfy the equations that do not contain time derivatives. Examples are (3.29), (3.52) and (3.64). Also, any physical constraint, such as positiveness of a particle distribution function or of a plasma mass density or pressure, must be taken into account.

Rigorous results on the sets of initial and boundary conditions that lead to unique solutions are extremely difficult to obtain and not generally available. Even when available, such a classical dynamics point of view often is of little use in practice. For instance, in a bifurcation study of a spontaneous dynamical process the non-uniqueness of solutions can be more relevant than uniqueness. Also, most numerical simulations give best results with simple boundary conditions, chosen ad hoc; resulting boundary layers do not seem to pose a severe problem. In practice, the boundary values are dealt with on a case by case basis.

3.8 Conservation laws

An important key to the understanding of space plasma activity is the fact that ideal MHD implies a rich set of conservation laws. It will turn out that often ideal MHD conservation laws typically suppress the onset of an activity process. In those cases the occurrence of activity requires the breaking of at least one of the ideal MHD conservation laws.

3.8.1 Conservation of mass and energy in ideal MHD

Here we consider ideal MHD and begin with conservation of mass, which is expressed by the continuity equation (3.58). The conservation property is more directly seen in integral form, obtained by integrating (3.58) over a spatial domain D with a non-moving boundary S,

$$\frac{\mathrm{d}}{\mathrm{d}t} \int_{\mathrm{D}} \rho \, \mathrm{d}^3 r = - \oint_{S} \rho \boldsymbol{v} \cdot \boldsymbol{n} \, \mathrm{d}S \tag{3.78}$$

where $\boldsymbol{n}$ is the unit normal vector of the boundary. This means that the total mass inside the domain changes only by inflow of mass through the boundary. There is no volume source of matter.

Let us now turn to the energy balance. Multiplying (3.59) by $\boldsymbol{v}$ and making use of (3.62), (3.64) and (3.65) one finds

$$\frac{\partial}{\partial t}\left(\frac{\rho v^2}{2} + u + \frac{B^2}{2\mu_0}\right)$$

$$+\nabla \cdot \left(\frac{\rho v^2}{2}\mathbf{v} + (u+p)\mathbf{v} + \frac{1}{\mu_0}\mathbf{E} \times \mathbf{B}\right) = 0 \, , \tag{3.79}$$

which expresses conservation of energy. Equation (3.79) is the ideal MHD version of (3.40). Note that the kinetic energy density has been split into internal and bulk flow energy densities and that (by assumption) there is no heat flow and (by quasi-neutrality) no electric energy density.

3.8.2 Magnetic flux, line and topology conservation

Here three conservation laws are formulated without specifying a particular plasma model. In each case the result is applied to ideal MHD.

We begin with conservation of magnetic flux. Consider a singly connected surface S with normal $\boldsymbol{n}$ and the boundary l, which moves with a smooth (differentiable) velocity $\boldsymbol{V}$ (Fig. 3.1). The smoothness of $\boldsymbol{V}$ keeps the surface singly connected during its motion. The rate at which the magnetic flux connected with S changes is

$$\frac{\mathrm{d}}{\mathrm{d}t}\int_S \boldsymbol{n} \cdot \boldsymbol{B}\,\mathrm{d}S = \int_S \boldsymbol{n} \cdot \frac{\partial \boldsymbol{B}}{\partial t}\,\mathrm{d}S - \oint_l \boldsymbol{V} \times \boldsymbol{B} \cdot \mathrm{d}\boldsymbol{l}$$

$$= -\int_S \nabla \times (\boldsymbol{E} + \boldsymbol{V} \times \boldsymbol{B}) \cdot \boldsymbol{n}\,\mathrm{d}S \, . \tag{3.80}$$

Setting this change to zero for arbitrary choices of S one finds that the velocity field $\boldsymbol{V}$ conserves magnetic flux if and only if

$$\nabla \times (\boldsymbol{E} + \boldsymbol{V} \times \boldsymbol{B}) = 0 \, . \tag{3.81}$$

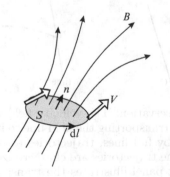

Fig. 3.1 Magnetic flux through an arbitrary surface S, the boundary moving with velocity $\boldsymbol{V}$.

With (3.60) it follows immediately that in ideal MHD the plasma velocity v conserves magnetic flux. In other words, the magnetic flux connected with an arbitrary surface that moves with the plasma velocity is conserved.

Let us now turn to line and topology conservation, which are closely related. It is important to understand under what conditions the evolution of a magnetic field can be described in terms of moving field lines (Newcomb, 1958; Stern, 1966; Vasyliunas, 1972). In this concept any given field line, identified at an initial time, may undergo displacement and deformation, but it can be traced through the evolution and thereby it preserves its identity. A velocity field w that transports magnetic field lines in that way is said to have the property of *magnetic line conservation*, or shorter w is *line-conserving*. If for a given smooth magnetic field $B(r, t)$ there exists a smooth line-conserving velocity, B is said to conserve its *magnetic topology*. (Throughout this book *smooth* means differentiable if not stated otherwise.)

Naturally, for a velocity to be line-conserving it must satisfy a certain requirement. To obtain a corresponding criterion we begin by describing a field line at time t by the vector $\boldsymbol{F}_B(\boldsymbol{r}_0, \sigma, t)$ tracing the field line, where $\boldsymbol{r}_0$ is a fixed point on the field line (identifying the line) and σ a parameter determining the running position on that line (Fig. 3.2), with $\sigma = 0$

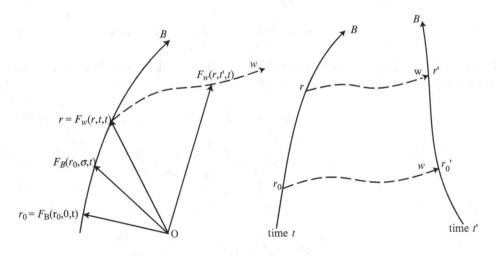

Fig. 3.2 Magnetic line conservation: The velocity field w generates the temporal magnetic field evolution by transporting the field line to its new position. Magnetic field lines are represented by full lines, trajectories associated with w by broken lines. The field lines and the trajectories are characterized by the vectors $\boldsymbol{F}_B$ and $\boldsymbol{F}_w$ (left panel). The right panel illustrates the transport of a field line from its position at time t to its position at time t'. For line conservation the characteristic requirement is that the points $\boldsymbol{r}_0'$ and $\boldsymbol{r}'$ lie on the same field line.

corresponding to r_0. Obviously, $\partial F_B/\partial\sigma$ is tangential to the field line, so that we can choose σ such that

$$B = \frac{\partial F_B}{\partial \sigma} . \tag{3.82}$$

The same procedure is applied to the trajectories associated with the transport of the field line. The trajectory beginning at the point r at time t is characterized by the vector $F_w(r,t',t)$, with time t' being the parameter that fixes the points on the trajectory and $r = F_w(r,t,t)$. Then, the velocity is represented as

$$w = \frac{\partial F_w}{\partial t'} . \tag{3.83}$$

(The field line velocity w should not be confused with the single particle velocity denoted by the same symbol in other sections.) The characteristic property of magnetic line conservation is that the points r_0' and r' lie on the same field line for arbitrary σ, t, t'. This means that r' is connected with r_0 by field line and trajectory sections via r and via r_0'. Accordingly, we can express r' in the following two ways:

$$r' = F_w(r,t',t) = F_w(F_B(r_0,\sigma,t),t',t), \tag{3.84}$$
$$r' = F_B(r_0',\sigma',t') = F_B(F_w(r_0,t',t),\sigma',t') , \tag{3.85}$$

where the parameters σ and σ' belong to r and r', respectively; σ' depends on (r_0,t,σ,t').

Equating (3.84) and (3.85), differentiating the resulting equation with respect to t' and σ and then setting $t' = t$ and $\sigma = 0$, gives

$$B \cdot \nabla w = w \cdot \nabla B + \frac{\partial B}{\partial t} + \lambda_1 B , \tag{3.86}$$

where $\lambda_1 = \partial^2\sigma'/\partial\sigma\partial t'$. Finally, one rewrites (3.86) as

$$\nabla \times (E + w \times B) = \lambda B , \tag{3.87}$$

where $\lambda = \lambda_1 - \nabla \cdot w$ and (3.62) was used together with the identity

$$\nabla \times (w \times B) = B \cdot \nabla w - w \cdot \nabla B - B\nabla \cdot w . \tag{3.88}$$

As λ depends on w (although rather implicitly), for a given $B(r,t)$ the equation (3.87) is a condition on w. Therefore, one defines a velocity field $w(r,t)$ as line-conserving if and only if (3.87) is satisfied.

It is tempting to eliminate λ by taking the cross-product of (3.87) with B, which gives

$$B \times (\nabla \times (E + w \times B)) = 0 . \tag{3.89}$$

This equation was chosen by several authors to define line conservation (Newcomb, 1958; Stern, 1966; Vasyliunas, 1972). For many purposes (3.87) and (3.89) are indeed equivalent. However, there is a subtle difference (Hornig and Schindler, 1996a). At a magnetic neutral point ($B = 0$) the condition (3.87) requires that $\nabla \times (E + w \times B) = 0$, while (3.89) does not necessarily impose that constraint. That difference, however, vanishes if the neutral point is connected with its environment by a field line. (In two dimensions an x-type neutral point is connected with the environment by each of the two separatrices, while for the o-point there is no such connection (Fig. 3.3). In the presence of a separatrix one can argue as follows. At any point with $B \neq 0$, (3.89) is equivalent to

$$\nabla \times (E + w \times B) = \Lambda B , \qquad (3.90)$$

where Λ is a scalar field, which must be smooth at any point with $B \neq 0$. What can we say about Λ at the neutral point? Away from the neutral point the divergence of (3.90) gives $B \cdot \nabla \Lambda = 0$, so that Λ is constant on field lines. Let $\Lambda = \Lambda_s$ on the separatrix, where Λ_s at most is a (smooth) function of time. We conclude that the right side of (3.90) becomes arbitrarily small as, at any fixed time, the neutral point is approached along the separatrix. Then, by continuity one finds that $\nabla \times (E + w \times B) = 0$ is satisfied at the neutral point.

To include the remaining cases (with no separatrix), we reformulate the necessary and sufficient condition for w to be line conserving as

$$\begin{aligned} B \times (\nabla \times (E + w \times B)) &= 0 \\ \nabla \times (E + w \times B) &= 0 \quad \text{where } B = 0 . \end{aligned} \qquad (3.91)$$

The second condition in (3.91) adds the conservation of neutral points. Note that at a neutral point $\nabla \times (E + w \times B) = 0$ is equivalent to $\partial B/\partial t + w \cdot \nabla B = 0$, which describes the transport of the neutral point.

The necessary and sufficient criterion for conservation of magnetic topology then becomes the existence of a smooth velocity field w satisfying (3.91).

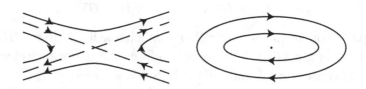

Fig. 3.3 Neutral points of two-dimensional magnetic fields. The x-type neutral point (left) is the crossing point of two field lines (separatrices, broken lines), the o-type (right) has no separatrices.

Equivalently, using (3.87), the criterion requires the existence of a smooth vector field $\boldsymbol{w}_\perp$ and a smooth scalar field λ. Here $\boldsymbol{w}_\perp$ is perpendicular to $\boldsymbol{B}$, the parallel component enters through λ.

In ideal MHD, (3.91) is solved by $\boldsymbol{w} = \boldsymbol{v}$, so that magnetic topology is conserved.

Depending on the application, it can be appropriate to consider line conservation for a restricted class of transport velocities. The latter will play an important role for magnetic reconnection (Chapter 11).

Comparing the requirements for flux, line and topology conservation one finds the following hierarchy

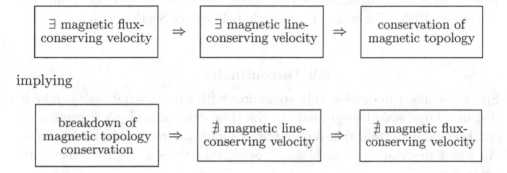

implying

Note that we defined flux and line conservation for a selected $\boldsymbol{w}$, while topology conservation is defined on the space of all admissible velocities.

For ideal MHD we can summarize that all three properties hold.

3.8.3 Conservation of magnetic helicity

Magnetic helicity K is a measure of linkage or twist of magnetic flux tubes. For a given domain D and suitable boundary conditions (discussed below) it is defined as

$$K = \int_D \boldsymbol{A} \cdot \boldsymbol{B}\, \mathrm{d}^3 r \tag{3.92}$$

where $\boldsymbol{A}$ is the magnetic vector potential, which generates the magnetic field by $\boldsymbol{B} = \nabla \times \boldsymbol{A}$. Helicity K plays a central role in describing the magnetic field topology (Berger and Field, 1984).

For a non-moving boundary S, Maxwell's equations imply

$$\frac{\mathrm{d}K}{\mathrm{d}t} + \oint_S \boldsymbol{n} \cdot (\phi \boldsymbol{B} - \boldsymbol{A} \times \boldsymbol{E})\, \mathrm{d}S = -2 \int_D \boldsymbol{E} \cdot \boldsymbol{B}\, \mathrm{d}^3 r \,. \tag{3.93}$$

This equation is obtained by applying the time-differentiation to the integrand, using (3.12) and (3.62) to express $\partial \boldsymbol{A} / \partial t$ and $\partial \boldsymbol{B} / \partial t$.

In view of (3.60) the term $\boldsymbol{E} \cdot \boldsymbol{B}$ vanishes for ideal MHD systems and (3.93) assumes the form of a continuity equation in integral representation. Then, helicity is conserved if the surface integral in (3.93) vanishes. For instance, this is the case for sets of isolated closed flux tubes. One finds that K vanishes for a single closed flux tube without twist, while K assumes a finite value for two linked flux tubes or a single flux tube with a twisted field structure (Moffatt, 1978; Berger and Field, 1984).

It should be noted that K as given by (3.92) is not necessarily independent of the electromagnetic gauge, unless special conditions prevail. One such condition is that $\boldsymbol{n} \cdot \boldsymbol{B} = 0$ on the boundary. Modified definitions of helicity have been introduced with different invariance properties (Finn and Antonsen, 1985). For more on magnetic helicity see Section 11.5.3.

3.9 Discontinuities

Space plasmas can develop thin structures with small spatial scales embedded in a large-scale background. In the thin structures physical properties can become important that do not play a significant role in the background. An exact description of such double-scale structures is rather difficult to achieve.

Luckily, in some cases the large-scale behaviour can be approximated by discontinuous solutions, where the thin structures are reduced to discontinuities. An appropriate set of equations provides *jump conditions* relating the quantities on both sides of the discontinuity to each other. To derive the jump conditions, some knowledge of the properties of the thin structure is still required. For instance, one has to decide what quantities are allowed to become singular (e.g., current density) or remain bounded (e.g., magnetic field or pressure). Also, a careful choice of the equations is necessary. For instance, in ideal MHD it is equivalent to use entropy conservation (3.61) or energy conservation (3.79). As one finds that entropy can change across a discontinuity, e.g., by viscous dissipation, while energy can be expected to be conserved, it is essential to use the energy and not the entropy equation.

Consider an equation of the (conservation) form

$$\frac{\partial f}{\partial t} + \nabla \cdot \boldsymbol{g} = 0 \tag{3.94}$$

and assume that the scalar f and the vector $\boldsymbol{g}$ remain bounded. Then, for a discontinuity at rest the 'pillbox' argument leads to the jump condition

$$[\boldsymbol{n} \cdot \boldsymbol{g}] = 0 \,, \tag{3.95}$$

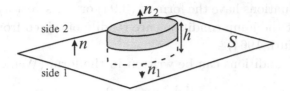

Fig. 3.4 Box-shaped region ('pillbox') as the integration domain for deriving the jump condition from an equation of the type (3.94) at a discontinuity surface (S). The normal of the discontinuity is identified with the normal n_2 after applying Gauss's theorem and taking the limit $h \to 0$.

where $[a] = a_2 - a_1$ with a_1 and a_2 denoting the values of a on the two sides of the continuity, and $\mathbf{n}$ is the unit normal of the discontinuity, pointing from side 1 to side 2 (Fig. 3.4).

To derive (3.95) one considers a cylindrical box-type volume extending across the discontinuity as illustrated in Fig. 3.4. One integrates the equation (3.94) over the box, applies Gauss's theorem, takes the limit $h \to 0$ and finally chooses the radius of top and bottom surfaces sufficiently small so that $\mathbf{g}$ can be considered as being constant over those surfaces (Landau and Lifshitz, 1963). The contribution from the time derivative vanishes in the limit $h \to 0$. The jump condition (3.95) remains valid when f is replaced by a vector and $\mathbf{g}$ by a tensor.

For obtaining the jump conditions of ideal MHD it is appropriate to start from the MHD equations written in the form

$$\frac{\partial \rho}{\partial t} + \nabla \cdot (\rho \mathbf{v}) = 0 \tag{3.96}$$

$$\frac{\partial \rho \mathbf{v}}{\partial t} + \nabla \cdot \left(\rho \mathbf{v}\mathbf{v} + (p + \frac{B^2}{2\mu_0})\mathcal{I} - \frac{\mathbf{B}\mathbf{B}}{\mu_0} \right) = 0 \tag{3.97}$$

$$\frac{\partial}{\partial t} \left(\frac{\rho v^2}{2} + \frac{p}{\gamma - 1} + \frac{B^2}{2\mu_0} \right)$$
$$+ \nabla \cdot \left(\frac{\rho v^2}{2}\mathbf{v} + \frac{\gamma p}{\gamma - 1}\mathbf{v} - \frac{1}{\mu_0}(\mathbf{v} \times \mathbf{B}) \times \mathbf{B} \right) = 0 \tag{3.98}$$

$$\frac{\partial \mathbf{B}}{\partial t} + \nabla \cdot (\mathbf{v}\mathbf{B} - \mathbf{B}\mathbf{v}) = 0 \tag{3.99}$$

$$\nabla \cdot \mathbf{B} = 0 . \tag{3.100}$$

All of these equations have the form of (3.94) or of its vector/tensor generalization, so that the jump conditions are readily obtained from (3.95) (e.g.,
Jeffrey and Taniuti, 1964).

The resulting conditions can be written in the form (Weitzner, 1983):

$$m[\tau] - [v_n] = 0 \tag{3.101}$$

$$m[\boldsymbol{v}] + \left[p + \frac{B^2}{2\mu_0}\right]\boldsymbol{n} - \frac{B_n}{\mu_0}[\boldsymbol{B}] = 0 \tag{3.102}$$

$$m\left(\left[\frac{\tau p}{\gamma - 1}\right] + [\tau]\langle p \rangle + [\tau]\frac{[\boldsymbol{B}]^2}{4\mu_0}\right) = 0 \tag{3.103}$$

$$m\langle\tau\rangle[\boldsymbol{B}] + m[\tau]\langle\boldsymbol{B}\rangle - B_n[\boldsymbol{v}] = 0 . \tag{3.104}$$

Here $v_n = \boldsymbol{n} \cdot \boldsymbol{v}$, $B_n = \boldsymbol{n} \cdot \boldsymbol{B}$, $m = \rho_1 v_{n1} = \rho_2 v_{n2}$, which is the mass flow
through the discontinuity. We choose $v_n \geq 0$ so that for $v_n \neq 0$ the plasma
flows from side 1 (upstream region) to side 2 (downstream region). Further,
$\langle q \rangle = (q_1 + q_2)/2$ for any quantity q. The specific volume $\tau = 1/\rho$ is used
instead of the density ρ. The continuity of B_n resulting from (3.100) has
already been incorporated. Note that $[qr] = [q]\langle r \rangle + [r]\langle q \rangle$ for any q and r.

One distinguishes linear and nonlinear discontinuities. A discontinuity is
called linear or nonlinear if $[v_n] = 0$ or $[v_n] \neq 0$, respectively. Further classification uses properties of the tangential magnetic field $\boldsymbol{B}_t$. By considering
the tangential contributions of (3.102) and (3.104) and eliminating $\boldsymbol{v}_t$ one
finds

$$(m^2\langle\tau\rangle - B_n^2/\mu_0)[\boldsymbol{B}_t] = -m^2[\tau]\langle\boldsymbol{B}_t\rangle . \tag{3.105}$$

By (3.101) linear solutions correspond to $m[\tau] = 0$. Three cases of that
type can be distinguished, which are listed under (a), (b) and (c) below.
Several distinct classes of nonlinear solutions are listed under (d).

(a) *Contact discontinuities* have $m = 0$, $[\tau \neq 0]$, $B_n \neq 0$. The jump conditions give $[p] = 0$, $[\boldsymbol{v}_t] = 0$, $[\boldsymbol{B}_t] = 0$. So density is the only
   non-continuous quantity.

(b) *Tangential discontinuities* have $m = 0$, $[\tau \neq 0]$, $B_n = 0$. The only
   nontrivial condition comes from (3.102),

$$\left[p + \frac{B^2}{2\mu_0}\right] = 0, \tag{3.106}$$

   the pressure balance across the discontinuity. No restriction is imposed on the tangential velocity $\boldsymbol{v}_t$.

(c) *Rotational discontinuities* are characterized by $m \neq 0$ and $[\tau] = 0$, $B_n \neq 0$. The jump conditions give

$$v_n = \frac{|B_n|}{\sqrt{\mu_0 \rho}} \quad [\boldsymbol{v}] = \pm \frac{[\boldsymbol{B}]}{\sqrt{\mu_0 \rho}} \quad [B^2] = 0 \quad [p] = 0 , \qquad (3.107)$$

where $\pm$ represents the sign of B_n. Through the discontinuity the magnetic field rotates from $\boldsymbol{B}_1$ to $\boldsymbol{B}_2$, so that for fixed $\boldsymbol{B}_1$ the tip of $\boldsymbol{B}_2$ can lie on any point of the circle defined by the constancy of B_n and B. The plasma crosses the discontinuity in the normal direction with the Alfvén velocity with respect to B_n.

(d) *Shock waves* have $m \neq 0$ and $[\tau] \neq 0$. Thermodynamics requires that fluid elements experience an increase of entropy when crossing the shock. This condition implies that shocks must be compressive, i.e., $[\tau] < 0$.

For *slow shocks* and *fast shocks* $\boldsymbol{B}_t$ does not vanish and points into the same direction on both sides. From (3.105) one concludes that $m^2 \langle \tau \rangle - B_n^2/\mu_0 \neq 0$. One defines

$$\textit{fast shock} \text{ for } (m^2 \langle \tau \rangle - B_n^2/\mu_0) > 0 \qquad (3.108)$$
$$\textit{slow shock} \text{ for } (m^2 \langle \tau \rangle - B_n^2/\mu_0) < 0 . \qquad (3.109)$$

In both cases (3.105) confirms that $\boldsymbol{B}_{t1}$ and $\boldsymbol{B}_{t2}$ are either parallel or antiparallel. This implies that $\boldsymbol{n}, \boldsymbol{B}_1, \boldsymbol{B}_2$ lie in a plane (*coplanarity theorem*), so that $\boldsymbol{B}_t$ can be expressed by a single component (B_t). Note that tangential and rotational discontinuities are not subject to the coplanarity constraint.

From (3.108) and (3.109) together with the properties of B_t one finds that for fast (slow) shocks $v_n^2 > a_n^2$ ($v_n^2 < a_n^2$) holds on both sides, where

$$a_n^2 = \frac{B_n^2}{\rho \mu_0}, \qquad (3.110)$$

a_n being the Alfvén velocity with respect to the normal component.

To obtain a further difference between fast and slow shocks one uses scalar multiplication of (3.105) with $\langle \boldsymbol{B}_t \rangle$,

$$\frac{1}{2}[B_t^2] = -\frac{m^2 [\tau] \langle \boldsymbol{B}_t \rangle^2}{m^2 \langle \tau \rangle - B_n^2/\mu_0} . \qquad (3.111)$$

With $[\tau] < 0$ this implies that for fast shocks $[B_t^2] > 0$ such that, in view of the continuity of B_n, one finds that $[B] > 0$, i.e., the magnetic field strength increases across the shock. For slow shocks (3.111) gives the opposite result, the magnetic field strength decreases. So,

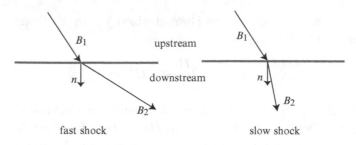

Fig. 3.5 Magnetic fields at fast and slow MHD shock waves.

fast shocks deflect the magnetic field away from the normal direction, while the slow shocks deflect it toward the normal (Fig. 3.5).

A limiting case of fast shocks with $B_{t1} = 0$ is the *switch-on shock* requiring $v_{n2} = |B_n|/\sqrt{(\mu_0 \rho_2)}$. The corresponding slow limit (*switch-off shock*) has $B_{t2} = 0$ and $v_{n1} = |B_n|/\sqrt{(\mu_0 \rho_1)}$. Although switch-on and switch-off shocks can be understood as limiting cases of fast and slow shocks, strictly speaking, they do not fall under the definition of fast or slow shocks, because several properties, such as the condition $\boldsymbol{B}_t \neq 0$, are violated. For $\boldsymbol{B}_t$ vanishing on both sides the magnetic field drops out of the problem, so that the result is an ordinary gas dynamic shock.

The above classification leaves the case $(m^2 \langle \tau \rangle - B_n^2/\mu_0) = 0$ as the final (nontrivial) possibility. Such solutions are called *intermediate shocks*. There is an ongoing debate about the physical significance of intermediate shocks. On the one hand, typical intermediate shocks can be shown to be *non-evolutionary* (Jeffrey and Taniuti, 1964), which implies that they are structurally unstable with respect to small changes in the initial conditions. On the other hand, intermediate shocks have been observed in two-dimensional simulations, e.g., by Brio and Wu (1988). Results obtained by Falle and Komissarov (2004) suggest that the two-dimensional case has singular mathematical properties, which cause the intermediate wave to be evolutionary while the three-dimensional case is non-evolutionary based on the larger class of initial conditions of the perturbations. This and further arguments led Falle and Komissarov (2004) to dismiss the intermediate shocks as unphysical, except for very short time intervals during which they might exist before they decay.

For the solutions with $B_n \neq 0$ there is a frame of reference, called the *de Hoffmann–Teller frame* (de Hoffmann and Teller, 1950) in which the velocity (v') becomes aligned with the magnetic field. With $v' = v - u$, where u is the transformation velocity with $u_n = 0$, the alignment condition gives

$$u = v_t - \frac{v_n}{B_n} B_t \ . \tag{3.112}$$

Note that, in view of (3.104), $[u] = 0$. The transformed electric field vanishes. For example, rotational discontinuities have

$$v' = \pm \frac{B}{\sqrt{\mu_0 \rho}} \tag{3.113}$$

and $u = v_t \mp B_t / \sqrt{\mu_0 \rho}$.

Treating a plasma as incompressible does not necessarily imply the absence of shock waves. As discussed above, there may be good reasons for describing the physical conditions inside and outside a thin structure with different model equations. So, a thin structure may well involve compressible dynamics internally while for the smooth regions incompressibility may be a good approximation.

If the plasma pressure tensor is not isotropic, the jump conditions and the resulting discontinuity properties have to be generalized accordingly (Hudson, 1970; Walthour *et al.*, 1994).

Situations that are more complex than discussed here arise when two discontinuities merge to form a compound structure, which has become known as double discontinuity. Refined plasma models beyond ideal MHD seem to be necessary to describe such structures. Whang (2004) showed that a double discontinuity consisting of a slow shock and a rotational discontinuity can exist in a plasma described by Hall-MHD.

Part II
Quiescence

4
Introduction

As outlined in Chapter 1, it is widely understood that an essential aspect
of space plasma activity is that quiescent plasma configurations suddenly
turn into a state of fast dynamic evolution. Examples for activity-relevant
quiescent structures are the magnetotail during the growth phase of a magnetospheric substorm and a preflare configuration in the solar atmosphere.

Fundamental questions arise about the conditions under which such transitions take place. Since they start out from a quiescent state it is important
to obtain detailed knowledge about those states, before one can tackle the
transitions themselves. The present part is devoted to that task.

Even during quiescence a system often undergoes significant changes due
to external driving forces. Planetary magnetospheres are driven by the solar
wind, solar chromospheric or coronal structures by subphotospheric convective motions. But any *snapshot* that one would take during a sufficiently
slow evolution would approximately satisfy the equations of a steady state.
Thus, quiescence is an asymptotic concept, applying to situations where the
time scale of external driving is large compared with any relevant dynamical
time scale of the system considered.

We illustrate these aspects more precisely for fluid models. Let the characteristic time of external forces be t_c and the largest dynamical time scale
be t_d. Here, t_d typically is the time it takes for the slowest wave mode to
travel across the system. The present assumption is that $\epsilon = t_d/t_c$ is much
smaller than 1, such that the two time scales are well separated. In that
case one is led to use a formulation which exhibits the different time scales
explicitly. In the simplest version of such a (multiple time scale) procedure
it is assumed that any fluid quantity q may be written as

$$q(\boldsymbol{r}, t) = Q(\boldsymbol{r}, t_0, t_1, \epsilon) \qquad (4.1)$$

57

where $t_0 = t$ and $t_1 = \epsilon t$, such that the presence of two time scales becomes explicit. In general, a system will vary on both time scales. However, here we are interested in the limiting behaviour, where – at least approximately – the dependence with respect to t_0 is absent, such that q depends on t only through t_1. (We will return to the t_0-dependence below.) This implies that time derivatives are small, because

$$\frac{\partial q}{\partial t} = O(\epsilon) \tag{4.2}$$

such that, to zeroth order in ϵ, the system is approximately found in a steady state at any time t_1. Such a steady state is what we mean by a snapshot. Nevertheless, over a sufficiently long period T of time, say, $T = 1/\epsilon$, small changes might secularly accumulate to a change of order 1.

Under these conditions, the secular evolution is appropriately described as a sequence of steady states, where a careful handling of conserved quantities is required. This is a standard procedure in thermodynamics and statistical mechanics. However, space plasmas are far from any thermodynamic equilibrium state so that the construction of sequences of steady states is a more complicated problem.

Note that we have restricted the multiple scale argument to time-dependence. For spatial variations a corresponding separation of scales is available only under particular circumstances. An example is the plasma sheet in the Earth's magnetotail, which will be addressed later.

The spatial structure of space plasmas requires inhomogeneous steady state models. This is a rather involved topic, most difficulties arising from the nonlinearity of the underlying equations. Here, it is a pleasing fact that, in the past, appropriate solutions of the nonlinear equations have been obtained, which can be used to model quiescent states of space plasmas and associated sequences in considerable complexity.

A point of potential weakness of describing space plasmas by steady state models results from the presence of fluctuations, causing t_0-dependent observables q. (For the relevance of non-thermal fluctuations see the corresponding discussions in Chapters 2 and 3.) Frequently, however, the fluctuation level is small enough for not entering the steady state properties explicitly. In other cases the effect of the fluctuations can be taken into account by imposing a suitable constraint on the steady state. An important example is the case where observations indicate that the fluctuations keep the pressure tensor close to isotropic. Then it may turn out to be a reasonable approach to choose an isotropic pressure tensor and to ignore any other effects of the fluctuations.

Even in the presence of a non-negligible fluctuation level, steady state models without fluctuations are needed to study the origin of the fluctuations if they are excited by a microinstability (Gary, 1993). A typical stability analysis requires a steady state background configuration. The same applies to most computer simulation studies. In fluid models, if necessary, small scale fluctuations may be taken into account through anomalous transport terms such as anomalous resistivity (see Section 9.3.2). Thus, there is ample motivation for using steady state models for describing quiescent configurations.

Sets of basic equations, describing steady states, are readily obtained from the time-dependent models of Chapter 3 by setting all partial time derivatives to zero. For later reference we here list two cases explicitly:

(a) Steady state magnetohydrodynamics for an ideal plasma in electromagnetic and (external) gravity fields, derived from (3.58)–(3.64),

$$\nabla \cdot (\rho \boldsymbol{v}) = 0 \tag{4.3}$$

$$\rho \boldsymbol{v} \cdot \nabla \boldsymbol{v} = -\nabla p + \boldsymbol{j} \times \boldsymbol{B} - \rho \nabla \psi \tag{4.4}$$

$$\boldsymbol{E} + \boldsymbol{v} \times \boldsymbol{B} = 0 \tag{4.5}$$

$$\boldsymbol{v} \cdot \nabla (p/\rho^\gamma) = 0 \tag{4.6}$$

$$\nabla \times \boldsymbol{E} = 0 \tag{4.7}$$

$$\nabla \cdot \boldsymbol{B} = 0 \tag{4.8}$$

$$\nabla \times \boldsymbol{B} = \mu_0 \boldsymbol{j} \,, \tag{4.9}$$

(b) steady state Vlasov theory, i.e., the steady state version of (3.26)–(3.30),

$$\boldsymbol{w} \cdot \frac{\partial f_s}{\partial \boldsymbol{r}} + \left(\frac{q_s}{m_s} (\boldsymbol{E} + \boldsymbol{w} \times \boldsymbol{B}) - m_s \nabla \psi \right) \cdot \frac{\partial f_s}{\partial \boldsymbol{w}} = 0 \tag{4.10}$$

$$\nabla \cdot \boldsymbol{E} = \frac{1}{\epsilon_0} \sum_s q_s \int f_s \, \mathrm{d}^3 w \tag{4.11}$$

$$\nabla \times \boldsymbol{E} = 0 \tag{4.12}$$

$$\nabla \cdot \boldsymbol{B} = 0 \tag{4.13}$$

$$\nabla \times \boldsymbol{B} = \mu_0 \sum_s q_s \int \boldsymbol{w} f_s \, \mathrm{d}^3 w. \tag{4.14}$$

Because of (4.7) or (4.12) the fields $\boldsymbol{E}$ can be derived from the electric potential alone, $\boldsymbol{E} = -\nabla\phi$.

The physical significance of the individual equations of both models are covered by the discussions of Chapter 3. The simplifications discussed there, such as the quasi-neutrality assumption, and the generalizations, such as the CGL pressure tensor, in principle, can also be applied to the present steady state equations. But in view of the small Debye lengths of the envisaged plasmas it seems justified to employ the quasi-neutrality approximation throughout this part.

Even if large sets of snapshot solutions are available, it is not a straightforward procedure to combine them to a secular evolution. The snapshots contain parameters or free functions that are functions of t_1. Their choice must be consistent with all physical constraints, such as conservation of particle number or of entropy.

Chapters 5–7 deal with steady state snapshots; the problem of secular evolution is addressed in Chapter 8.

Some of the properties that are discussed in this part also hold for time-dependent systems.

5
Magnetohydrostatic states

In quiescent states that are candidates for becoming active, the plasma flow velocity is often small compared to typical magnetohydrodynamic wave velocities. Then, at any given time during the slow quiescent evolution the plasma and the fields can be approximately described as a static state. A corresponding ideal MHD model is provided by the magnetohydrodynamic equations (4.3)–(4.9) with v set to zero. The resulting equations define *magnetohydrostatics*. Typically, for space plasmas that are trapped in regions of closed magnetic fields in stellar and planetary environments magnetohydrostatics turns out to be a useful model on spatial scales large compared with the intrinsic plasma scales.

5.1 General properties

Although the model of magnetohydrostatics is already considerably simplified as compared to general steady state plasma models, it is still complicated enough, and its application spectrum is wide enough, to call for a discussion of its major properties.

5.1.1 Basic equations

If bulk velocity v is negligible, the equations of magnetohydrodynamics reduce to the equations of magnetohydrostatics (*MHS*),

$$-\nabla p + j \times B - \rho \nabla \psi = 0 \tag{5.1}$$

$$\nabla \times B = \mu_0 j \tag{5.2}$$

$$\nabla \cdot B = 0 . \tag{5.3}$$

Frequently, a further simplification is possible in that the gravity term in (5.1) can be neglected. The condition for this approximation to be applicable follows from a comparison between the gravity force and the pressure force, where the gradient is expressed by a characteristic scale length L and $-\nabla\psi$ is identified as the gravity acceleration $\boldsymbol{g}$,

$$\frac{|\rho\nabla\psi|}{|\nabla p|} = \frac{\rho g}{p/L} = \frac{L}{h} \ll 1 \tag{5.4}$$

where

$$h = \frac{p}{\rho g} \tag{5.5}$$

is the gravity scale height. Neglecting gravity is largely justified for magnetospheric plasmas, for solar plasmas (5.4) may or may not apply, depending on circumstances.

For reasons of reference we rewrite the MHS equations without the gravity term,

$$-\nabla p + \boldsymbol{j} \times \boldsymbol{B} = 0 \tag{5.6}$$

$$\nabla \times \boldsymbol{B} = \mu_0 \boldsymbol{j} \tag{5.7}$$

$$\nabla \cdot \boldsymbol{B} = 0 . \tag{5.8}$$

In spite of the seemingly simple structure of (5.6)–(5.8), the nonlinearity in (5.6) causes considerable difficulties. In fact, a general mathematical existence theory is not available. There are, however, a number of interesting properties that are readily accessible.

A first property that follows from (5.6) is that $\boldsymbol{B} \cdot \nabla p = 0$ implying that the pressure is constant on magnetic field lines. Next, consider the Lorentz force density $\boldsymbol{f}_{\mathrm{L}} = \boldsymbol{j} \times \boldsymbol{B}$. Eliminating the current density by (5.2), and expanding the double vector product we find

$$\begin{aligned}
\boldsymbol{f}_{\mathrm{L}} &= \frac{1}{\mu_0}(\nabla \times \boldsymbol{B}) \times \boldsymbol{B} \\
&= -\nabla\frac{B^2}{2\mu_0} + \frac{1}{\mu_0}\boldsymbol{B} \cdot \nabla\boldsymbol{B} .
\end{aligned} \tag{5.9}$$

The first term has the same effect as a scalar pressure, therefore $B^2/2\mu_0$ is called *magnetic pressure*. The second part is anisotropic, involving the direction of the magnetic field explicitly. Physical insight is gained by asking under what conditions the anisotropic part vanishes. Writing $\boldsymbol{B} = B\boldsymbol{b}$, where $\boldsymbol{b}$ is the unit vector in the direction of $\boldsymbol{B}$, we find

$$\boldsymbol{B} \cdot \nabla\boldsymbol{B} = B^2\,\boldsymbol{b} \cdot \nabla\boldsymbol{b} + B\,\boldsymbol{b}\boldsymbol{b} \cdot \nabla B . \tag{5.10}$$

Noting that the two terms on the right of (5.10) are perpendicular to each other, both terms must vanish for the anisotropic part of f_L to be zero (unless B vanishes identically). Since $|b \cdot \nabla b| = 1/R$, where R is the local radius of curvature of the field line, we find that magnetic fields must be straight lines and that the field magnitude B must be constant on each field line. A simple example of a magnetic field satisfying these conditions is a vector field B which (in Cartesian coordinates) has no x-component and the two other components depend on x only.

The identity (5.10) implies that strongly curved magnetic field lines will exert a large magnetic force on the plasma. The Lorentz force acting on a finite part of the plasma covering a domain D (cut out arbitrarily) with surface S and outward pointing normal vector n is obtained by noting that the Lorentz force density can also be written as

$$f_L = \nabla \cdot \mathcal{M}, \tag{5.11}$$

where $\mathcal{M}$ is (the magnetic part of) Maxwell's stress tensor

$$\mathcal{M} = \frac{B^2}{2\mu_0}\mathcal{I} - \frac{BB}{\mu_0} \tag{5.12}$$

with $\mathcal{I}$ denoting the unit tensor. Thus, we obtain for the total Lorentz force

$$\begin{aligned} F_L &= \int_D \nabla \cdot \mathcal{M} \, d^3r \\ &= \oint_S n \cdot \mathcal{M} \, d^2S \\ &= \frac{1}{\mu_0} \oint_S (\frac{B^2}{2}n - n \cdot BB) \, d^2S. \end{aligned} \tag{5.13}$$

This indicates that the magnetic stress exerted locally on the surface consists of a part directed along the normal vector of the surface and a part that locally is directed along the field line. In the absence of other forces, the latter force causes a tendency for the magnetic field lines to shorten. (In that sense magnetic fields behave like rubber bands, however, that analogy is rather limited otherwise.)

In (5.6) the pressure and the current density may be eliminated by taking the curl of that equation, and using (5.7). Then the MHS equations assume the form

$$\nabla \times (B \cdot \nabla B) = 0 \tag{5.14}$$

$$\nabla \cdot B = 0. \tag{5.15}$$

Furthermore, (5.15) may, of course, be eliminated by deriving $\boldsymbol{B}$ from a vector potential $\boldsymbol{B} = \nabla \times \boldsymbol{A}$, such that (5.14) becomes

$$\nabla \times [(\nabla \times \boldsymbol{A}) \cdot \nabla (\nabla \times \boldsymbol{A})] = 0 . \tag{5.16}$$

A further representation of magnetic fields is considered in the following section.

5.1.2 Euler potentials

Under suitable conditions (defined below) a magnetic field $\boldsymbol{B}$ may be represented in the form

$$\boldsymbol{B}(\boldsymbol{r}) = \nabla \alpha(\boldsymbol{r}) \times \nabla \beta(\boldsymbol{r}) \tag{5.17}$$

where α and β are the *Euler potentials* (also called *Clebsch potentials*), which are functions of position $\boldsymbol{r}$ (Stern, 1970).

As in the case of a vector potential, the presentation of a magnetic field by Euler potentials ensures that $\nabla \cdot \boldsymbol{B} = 0$. In fact, Euler potentials generate a vector potential $\boldsymbol{A}$ in the form

$$\boldsymbol{A} = \alpha \nabla \beta. \tag{5.18}$$

The Euler representation (5.17) implies that each field line is the intersection line of two surfaces, on which α and β are constant, respectively (Fig. 5.1).

Although Euler potentials can be quite useful for representing magnetohydrostatic states, it must be kept in mind that not all magnetic fields can be represented by Euler potentials in the entire domain of interest. Euler potentials, however, do exist under the following circumstances (Stern, 1970).

Let $\boldsymbol{B} \neq 0$ in the entire domain of interest and consider a surface S which is intersected by magnetic field lines unidirectionally and assume that every field line, which intersects that surface, intersects it only once (Fig. 5.2). Then, every field line can be uniquely labelled by a pair of coordinates

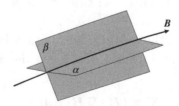

Fig. 5.1 In the Euler representation a magnetic field line is the intersection line of two surfaces with constant potentials α and β.

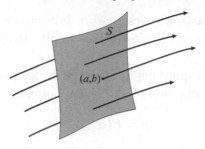

Fig. 5.2 A magnetic field can be represented by Euler potentials if each field line passes through a surface S once and only once. The point at which a given field line intersects S is specified by a pair of surface coordinates (a, b).

a, b marking the intersection point on S. For sufficiently smooth magnetic fields $\boldsymbol{B}(\boldsymbol{r})$, the set of all points connected with S by a field line define a domain D_S. Every point $\boldsymbol{r} \in D_S$ is uniquely determined by specifying the intersection coordinates a, b and the arc length s measured along the field line (increasing in the direction of $\boldsymbol{B}$) with an arbitrary choice of s on S. Under the present conditions, the vector function $\boldsymbol{r} = \boldsymbol{r}(a, b, s)$ can be inverted, yielding $a(\boldsymbol{r}), b(\boldsymbol{r}), s(\boldsymbol{r})$. Here, the values of a and b are simply transported from the surface S along field lines into the domain D_S.

Since $\boldsymbol{B}$ is perpendicular to both ∇a and ∇b, $\boldsymbol{B}$ can be represented as

$$\boldsymbol{B} = g(a, b)\nabla a \times \nabla b . \tag{5.19}$$

Here, the condition that $\nabla \cdot \boldsymbol{B} = 0$ forces the multiplying scalar function g to depend on a and b only. A suitable nonsingular transformation $a(\alpha, \beta), b(\alpha, \beta)$ leads to the desired form (5.17) of $\boldsymbol{B}$.

Particular care must be taken for the case of toroidal magnetic fields. Obviously, for domains containing a magnetic flux torus, a surface cutting the torus in general cannot be identified with the surface S in the sense discussed above, because, generically, a field line returns to any surface that it intersects with multiple intersection points. In fact, there may be infinitely many intersection points, such that the notion of magnetic field lines becomes questionable. Regions may exist where field lines become chaotic in the sense of Hamiltonian mechanics. In such cases, one either must restrict the domain of interest suitably or abandon (a single set of) Euler potentials altogether. It seems, however, that many space plasma structures are non-toroidal and do admit the use of Euler potentials. For instance, in models of the magnetic field in the Earth's magnetotail or of simple magnetic loops above the solar photosphere recurrent fields do not play an important role. Therefore, using Euler potentials is much more common in space and astrophysics than in laboratory plasma physics. We will employ them in several places in this book. However, the limitations must be kept in mind.

To express the MHS equations in terms of Euler potentials α and β, we write the pressure as a function of the Euler potentials

$$p = P(\alpha, \beta), \tag{5.20}$$

which is possible because the pressure is constant on field lines. Inserting (5.17) and (5.20) into (5.6) and collecting the coefficients of $\nabla\alpha$ and $\nabla\beta$ gives

$$\boldsymbol{j} \cdot \nabla\beta \;=\; \frac{\partial P}{\partial \alpha} \tag{5.21}$$

$$\boldsymbol{j} \cdot \nabla\alpha \;=\; -\frac{\partial P}{\partial \beta}\,. \tag{5.22}$$

With the help of (5.7) the equations (5.21) and (5.22) assume the form

$$\nabla\beta \cdot \nabla \times (\nabla\alpha \times \nabla\beta) = \mu_0 \frac{\partial P(\alpha, \beta)}{\partial \alpha} \tag{5.23}$$

$$\nabla\alpha \cdot \nabla \times (\nabla\alpha \times \nabla\beta) = -\mu_0 \frac{\partial P(\alpha, \beta)}{\partial \beta}\,. \tag{5.24}$$

Euler potentials can be interpreted as *magnetic flux coordinates* in the following sense. The magnetic flux F_Σ through a surface Σ with unit normal vector $\boldsymbol{n}$ is

$$F_\Sigma = \int_\Sigma \boldsymbol{n} \cdot \boldsymbol{B}\, dS = \int_{\hat{\Sigma}} \pm\, \mathrm{d}\alpha\, \mathrm{d}\beta \tag{5.25}$$

where the first integral extends over Σ and the second over the corresponding domain $\hat{\Sigma}$ in α, β-space, the $+(-)$ sign corresponding to $\boldsymbol{n} \cdot \boldsymbol{B} > 0\,(< 0)$. Thus, magnetic flux is simply the (directed) area in α, β-space. Magnetic flux coordinates have proven quite useful for modelling magnetic fields (e.g., Zweibel and Boozer, 1985).

Note that $P(\alpha, \beta)$ can be chosen freely, as long as P is positive. This freedom is related to the free choice of initial conditions in a time-dependent case, where the initial pressure can be specified arbitrarily.

Finally we remark that Euler potentials, representing a given magnetic field, are not unique. Like the vector potential they are subject to a gauge transformation. From (5.18) it is clear that

$$\boldsymbol{A}' = \alpha\nabla\beta + \nabla\varphi \tag{5.26}$$

is a possible vector potential, too, for an arbitrary choice of the gauge functions φ. If, however, $\boldsymbol{A}'$ is to have the form $\alpha'\nabla\beta'$, φ is subject to a restriction. Defining

$$\chi = \varphi + \alpha\beta \qquad (5.27)$$

we find from (5.26) and (5.27) after eliminating φ

$$\alpha'\nabla\beta' = -\beta\nabla\alpha + \nabla\chi, \qquad (5.28)$$

which implies that it must be possible to write χ as a function of α and β' alone such that $\nabla\chi = \partial\chi/\partial\alpha\nabla\alpha + \partial\chi/\partial\beta'\nabla\beta'$. Inserting this expression into (5.28) one concludes that

$$\beta = \frac{\partial\chi(\alpha, \beta')}{\partial\alpha}, \qquad \alpha' = \frac{\partial\chi(\alpha, \beta')}{\partial\beta'}. \qquad (5.29)$$

This means that the gauge transformation for α and β is formally identical with a canonical transformation with a generating function $\chi(\alpha, \beta')$ as known from Hamiltonian mechanics (Goldstein, 2002). In other words, any arbitrary $\chi(\alpha, \beta')$ generates a new set of Euler potentials α', β' which can be computed by solving (5.29) for α' and β'. In α, β-space the class of gauge transformations $(\alpha, \beta) \rightarrow (\alpha', \beta')$ represents the class of all incompressible maps. This condition is equivalent to $|J| = 1$, where

$$J = \frac{\partial\alpha'}{\partial\alpha}\frac{\partial\beta'}{\partial\beta} - \frac{\partial\alpha'}{\partial\beta}\frac{\partial\beta'}{\partial\alpha} \qquad (5.30)$$

is the Jacobian of the map.

We add that by (5.27) a transformation generated by χ leads to an electric potential

$$\phi' = \phi - \frac{\partial\varphi}{\partial t} = \phi + \frac{\partial}{\partial t}(\alpha\beta - \chi). \qquad (5.31)$$

(Note that the assumption of a steady-state condition for the electromagnetic field does not necessarily exclude certain forms of time-varying potentials.)

5.1.3 Boundary conditions and a variational principle

The magnetohydrostatic equations (5.6)–(5.8) require boundary conditions for p and $\boldsymbol{B}$. These conditions are rather difficult to formulate. The main problem is that the boundary condition for the pressure must be consistent with the property that pressure is constant on field lines, which, however, are not known beforehand. Furthermore, because of the nonlinearity of the equations, it is not clear a priori whether for a given set of boundary conditions a solution does exist at all. (We return to this difficulty in the context of bifurcation properties in Part III.) A more formal discussion of the boundary conditions requires an analysis of the characteristic manifolds (e.g., Parker, 1979). Here we pursue a different line, based on Euler potentials. In that case, with $P(\alpha, \beta)$ specified, an obvious choice of magnetic boundary conditions are Dirichlet conditions, i.e.,

$$\alpha(\boldsymbol{r}) \text{ and } \beta(\boldsymbol{r}) \text{ prescribed for } \boldsymbol{r} \text{ on the boundary.} \qquad (5.32)$$

Physically, this boundary condition prescribes the location where a field line with labels α, β cuts through the boundary. Thus, this boundary condition fixes the footpoints of flux tubes, as, for instance, it is appropriate for modelling quiescent states in the solar atmosphere.

On some boundaries (or sections of boundaries) homogeneous Neumann conditions

$$\boldsymbol{n} \cdot \nabla \alpha = 0, \qquad \boldsymbol{n} \cdot \nabla \beta = 0 \qquad (5.33)$$

may also be appropriate; (5.33) means that the tangential magnetic field component vanishes (e.g., as a consequence of an imposed symmetry). This is easily seen by expressing the tangential part of the magnetic field $\boldsymbol{B}_{\mathrm{t}} = \boldsymbol{n} \times (\boldsymbol{B} \times \boldsymbol{n})$ by Euler potentials,

$$\boldsymbol{B}_{\mathrm{t}} = \boldsymbol{n} \times \nabla \beta \, \boldsymbol{n} \cdot \nabla \alpha - \boldsymbol{n} \times \nabla \alpha \, \boldsymbol{n} \cdot \nabla \beta \, . \qquad (5.34)$$

One might wonder whether boundary value problems are useful for quiescent space plasmas at all. Although in some cases a boundary or a section of a boundary may actually be identified with a physical boundary (e.g., the magnetopause), other model boundaries are artificial, because the actual plasma extends smoothly beyond the domain chosen in the model. Clearly, any extended field assumes certain values on the surface representing the boundary in a model description. If such realistic values were used to formulate boundary conditions, the field inside the domain could be expected to provide a useful model for the actual field quantities. The obvious problem is that, although realistic boundary conditions exist, there is usually no way to get hold of them exactly. This means that it takes a good deal of physical intuition to construct appropriate models of space plasmas. (This problem is less stringent for laboratory plasmas where physical boundaries exist. Note, however, that even there, e.g., in studies of plasma-wall interaction processes, the idealization of a wall as a smooth surface might not be good enough.)

As it will become clear later, it is of considerable interest that the equilibrium equations (5.23) and (5.24) may be derived from a variational principle.

This is straightforward for Dirichlet or Neumann boundary conditions for α and β.

Consider the functional $F[\alpha, \beta]$ defined as

$$F[\alpha, \beta] = \int_{\mathrm{D}} \left[\frac{1}{2\mu_0} (\nabla\alpha \times \nabla\beta)^2 - P(\alpha, \beta) \right] \mathrm{d}^3 r. \tag{5.35}$$

One obtains (5.23) and (5.24) from the condition that the variation of $F[\alpha, \beta]$ vanishes, i.e., $\delta F = 0$ with boundary conditions either of the form (5.32) or (5.33). In fact, for an arbitrary domain D with boundary $\partial\mathrm{D}$ the condition $\delta F = 0$ can be written as

$$\oint_{\partial\mathrm{D}} \left[\delta\alpha(\boldsymbol{n} \cdot \nabla\alpha(\nabla\beta)^2 - \boldsymbol{n} \cdot \nabla\beta\nabla\alpha \cdot \nabla\beta) \right.$$

$$+ \delta\beta(\boldsymbol{n} \cdot \nabla\beta(\nabla\alpha)^2 - \boldsymbol{n} \cdot \nabla\alpha\nabla\alpha \cdot \nabla\beta) \right] \mathrm{d}S$$

$$+ \int_{\mathrm{D}} \left[\delta\alpha(\nabla\beta \cdot \nabla \times (\nabla\alpha \times \nabla\beta) - \mu_0\frac{\partial P}{\partial\alpha}) \right.$$

$$\left. - \delta\beta(\nabla\alpha \cdot \nabla \times (\nabla\alpha \times \nabla\beta) + \mu_0\frac{\partial P}{\partial\beta}) \right] \mathrm{d}^3 r = 0 . \tag{5.36}$$

The surface integral vanishes due to the boundary conditions, and the vanishing of the volume integral for arbitrary $\delta\alpha$ and $\delta\beta$ (satisfying the boundary conditions) implies (5.23) and (5.24). As in many other cases, this variational principle is not only a handy tool for deriving the equilibrium equations, but it also provides a stability criterion. We return to this point in Part III.

5.1.4 Field-aligned currents and current closure

It is instructive to observe that in general the current density $\boldsymbol{j}$ has components $\boldsymbol{j}_\parallel$ and $\boldsymbol{j}_\perp$ parallel and perpendicular to the magnetic field $\boldsymbol{B}$. Useful expressions for $\boldsymbol{j}_\perp$ and $\boldsymbol{j}_\parallel$ are obtained in the following way.

Taking the vector product of equation (5.6) with $\boldsymbol{B}$, we find

$$\boldsymbol{j}_\perp = \frac{1}{B^2} \boldsymbol{B} \times \nabla p . \tag{5.37}$$

The parallel component of $\boldsymbol{j}$ follows from the equation $\nabla \cdot \boldsymbol{j} = 0$ which is a consequence of (5.7), giving

$$\nabla \cdot \boldsymbol{j}_\parallel = -\nabla \cdot \boldsymbol{j}_\perp . \tag{5.38}$$

By using $\nabla \cdot \boldsymbol{B} = 0$, by writing $j_\parallel = j_\parallel \boldsymbol{B}/B$ and noting that $(\boldsymbol{B} \cdot \nabla)/B = \partial/\partial s$, where s is the arc length on a given field line, we find

$$\frac{j_\parallel}{B} = \left[\frac{j_\parallel}{B} \right]_{s_0} - \int_{s_0}^{s} \frac{\nabla \cdot \boldsymbol{j}_\perp}{B} \, \mathrm{d}s' \tag{5.39}$$

with s_0 being the origin of s on each field line.

The expression (5.39) assumes a particularly interesting form when Euler potentials are used. Noting that the divergence of an arbitrary vector field $\boldsymbol{u}$ may be written as

$$\nabla \cdot \boldsymbol{u} = B \left[\frac{\partial}{\partial \alpha} \left(\frac{\boldsymbol{u} \cdot \nabla \alpha}{B} \right) + \frac{\partial}{\partial \beta} \left(\frac{\boldsymbol{u} \cdot \nabla \beta}{B} \right) + \frac{\partial}{\partial s} \left(\frac{\boldsymbol{u} \cdot \nabla s}{B} \right) \right] \tag{5.40}$$

we find

$$\frac{j_\parallel}{B} = \left[\frac{j_\parallel}{B} + \frac{\boldsymbol{j}_\perp \cdot \nabla s}{B} \right]_{s_0} - \frac{\boldsymbol{j}_\perp \cdot \nabla s}{B}$$

$$- \frac{\partial}{\partial \alpha} \int_{s_0}^{s} \frac{\boldsymbol{j}_\perp \cdot \nabla \alpha}{B} \, \mathrm{d}s' - \frac{\partial}{\partial \beta} \int_{s_0}^{s} \frac{\boldsymbol{j}_\perp \cdot \nabla \beta}{B} \, \mathrm{d}s' . \tag{5.41}$$

Using (5.21) and (5.22) this expression assumes the form (Vasyliunas, 1970)

$$\frac{j_\parallel}{B} = \left[\frac{j_\parallel}{B} + \frac{\boldsymbol{j}_\perp \cdot \nabla s}{B} \right]_{s_0} - \frac{\boldsymbol{j}_\perp \cdot \nabla s}{B} + \frac{\partial P}{\partial \beta} \frac{\partial V}{\partial \alpha} - \frac{\partial P}{\partial \alpha} \frac{\partial V}{\partial \beta} \tag{5.42}$$

where $V = V(\alpha, \beta, s)$ is the differential flux tube volume

$$V(\alpha, \beta, s) = \int_{s_0}^{s} \frac{\mathrm{d}s'}{B(\alpha, \beta, s')}. \tag{5.43}$$

The expression (5.42) is particularly useful, when the magnetic field is close to a potential field, so that pressure gradients and currents can be treated as small perturbations. Then, one can use the potential field in (5.42) to evaluate the parallel current density in the leading order. The result may be improved by further iterations (Vasyliunas, 1970).

For field lines passing through the boundary, in (5.42) the term which is to be evaluated at $s = s_0$ will require a boundary condition. Here, we encounter an example of the general difficulty of specifying boundary conditions that was discussed above. How do we know whether the outside regime can cope with an assumed parallel electrical current? This problem, for instance, arises for magnetospheric models where parallel currents exist, such as Birkeland currents, manifesting the coupling between magnetospheric and ionospheric processes. It is generally necessary to deal with the closure of the electric currents (Alfvén, 1972). Fully satisfactory models must cover sufficiently large domains, such that the currents close inside. Note that there is no problem of current continuity inside the model domain itself, because $\nabla \cdot \boldsymbol{j} = 0$ is satisfied everywhere in an MHD system.

5.1.5 Force-free fields

Here we consider the limiting situation where the pressure gradient is uniformly small compared to $|\boldsymbol{j}||\boldsymbol{B}|$. Neglecting pressure gradients, (5.6) implies that $\boldsymbol{j} \times \boldsymbol{B} = 0$ or $\boldsymbol{j} \parallel \boldsymbol{B}$ which may be expressed as

$$\nabla \times \boldsymbol{B} = \kappa \boldsymbol{B} \tag{5.44}$$

where κ is a scalar function of $\boldsymbol{r}$. Taking the divergence of (5.44) and using (5.8) one finds

$$\boldsymbol{B} \cdot \nabla \kappa = 0 \,, \tag{5.45}$$

implying that κ is constant on magnetic field lines.

Solutions $\boldsymbol{B}$ of equation (5.44) with (5.45) are called *force-free fields*. Note that neither j nor p is required to be small. The quantities that must be small are $j_\perp$ and $|\nabla p|$. The force-free field approximation is thought to be applicable to large regions of the solar corona.

The magnitude of the current density is tied to the magnitude of the magnetic field because

$$\frac{\partial}{\partial s} \left(\frac{j_\parallel}{B} \right) = 0 \tag{5.46}$$

which follows directly from $\nabla \cdot \boldsymbol{j} = 0$.

A simple example, often used for analytical representations of force-free fields, is obtained by choosing κ to be a constant. Then, (5.45) is satisfied automatically and (5.44) is a linear differential equation for $\boldsymbol{B}(\boldsymbol{r})$ which for a number of simple domains can be solved analytically. This case, however, does not only serve as a mathematical simplification. As we shall see, constant-κ force-free fields are believed to be the result of a particular relaxation process, which conserves magnetic helicity. This is based on the following property.

Force-free fields with constant κ minimize the magnetic energy under the constraint of helicity conservation and with the normal component kept fixed on the boundary $\partial\Omega$ of a finite domain Ω (Woltjer, 1958). The corresponding variational problem is

$$\delta \int_\Omega \left(\frac{(\nabla \times \boldsymbol{A})^2}{2\mu_0} - \frac{\kappa_0}{2\mu_0} \boldsymbol{A} \cdot \nabla \times \boldsymbol{A} \right) \mathrm{d}^3 r = 0 \,, \tag{5.47}$$

where (3.92) was used for the helicity and $\kappa_0/(2\mu_0)$ is the Lagrangian multiplier. Carrying out the variation and using Gauss's law, with the surface terms vanishing due to the boundary condition, one finds

$$\nabla \times (\nabla \times \boldsymbol{A}) = \kappa_0 \nabla \times \boldsymbol{A} \tag{5.48}$$

so that indeed the minimizing field is force-free with constant κ.

5.1.6 Tangential discontinuities

It is of considerable interest from both the theoretical and the practical point of view to explore whether the MHS equations (5.6)–(5.8) admit discontinuous solutions.

So let us specialize the general discontinuities of ideal MHD, as found in Section 3.9 for MHS conditions. Setting the plasma velocity to zero one finds that the only remaining discontinuity is the tangential discontinuity so that

$$\left[p + \frac{B^2}{2\mu_0} \right] = 0, \tag{5.49}$$

so that the sum of kinetic and magnetic pressure is continuous. There is no magnetic field component normal to the discontinuity.

By (5.7) a tangential discontinuity carries a surface current with density

$$\boldsymbol{K} = \frac{1}{\mu_0} \boldsymbol{n} \times (\boldsymbol{B}_2 - \boldsymbol{B}_1) \tag{5.50}$$

where it is important to understand $\boldsymbol{n}$ as pointing from side 1 to side 2 of the discontinuity. Equation (5.50) is derived in a way similar to the derivation of the jump conditions of Section 3.9; instead of the pillbox-integration a line integral is taken along a closed contour intersecting the test surface.

5.1.7 Virial theorem and generalizations

Here we address conditions that are necessary for an isolated magnetic flux tube to exist in MHS equilibrium. This is an application of the *virial theorem* (Chandrasekhar, 1961; Priest, 1982; Low, 1999). Although conditions derived from the virial theorem are not sufficient for equilibrium, the virial approach has proven quite useful; its particular strength lies in its potential to classify certain test configurations as not being in MHS equilibrium.

First, let us consider a test equilibrium in which the magnetic flux is isolated in the sense that $\boldsymbol{B}$ vanishes outside a bounded domain d embedded in a larger domain D where the pressure p assumes a constant value p_0. The surface ∂d of the domain d may be a tangential discontinuity.

A set of useful relations is derived in the following way. We start from the MHS momentum balance (5.6), which with the help of (5.12) is written as

$$\nabla \cdot \mathcal{T} = 0, \tag{5.51}$$

where

$$\mathcal{T} = -(p + \frac{B^2}{2\mu_0})\mathcal{I} + \frac{BB}{\mu_0} . \tag{5.52}$$

Let us multiply (5.51) by an arbitrary differentiable vector field $V(r)$ and integrate the result over the domain d,

$$\int_d V \cdot (\nabla \cdot \mathcal{T}) \, d^3 r = 0 . \tag{5.53}$$

Integration by parts (using Gauss's theorem) gives

$$\oint_{\partial d} V \cdot \mathcal{T} \cdot n_d \, dS - \int_d \mathcal{T} : (\nabla V) \, d^3 r = 0 \tag{5.54}$$

where n_d is the outward-pointing normal of ∂d and the colon denotes complete contraction of the two tensors. Using (5.52) we find

$$V \cdot \mathcal{T} \cdot n = -\left(p + \frac{B^2}{2\mu_0}\right) V \cdot n + \frac{1}{\mu_0} V \cdot BB \cdot n \tag{5.55}$$

$$\mathcal{T} : (\nabla V) = -\left(p + \frac{B^2}{2\mu_0}\right) \nabla \cdot V + \frac{1}{\mu_0} B \cdot (B \cdot \nabla V). \tag{5.56}$$

The surface integral over ∂d is evaluated as follows,

$$\begin{aligned}
\int_{\partial d} V \cdot \mathcal{T} \cdot n_d \, dS &= \int_{\partial d} \left(-p - \frac{B^2}{2\mu_0}\right) V \cdot n_d \, dS \\
&= -p_0 \int_{\partial d} V \cdot n_d \, dS \\
&= -p_0 \int_d \nabla \cdot V \, d^3 r
\end{aligned} \tag{5.57}$$

where, for the case of ∂d being a tangential discontinuity, (5.49) has to be invoked. Thus, we obtain from (5.54)

$$\int_d \left[\left(p - p_0 + \frac{B^2}{2\mu_0}\right) \nabla \cdot V - \frac{1}{\mu_0} B \cdot (B \cdot \nabla V)\right] d^3 r = 0 . \tag{5.58}$$

The equation (5.58) also remains valid if the equilibrium contains additional tangential discontinuities inside d. In that case integration by parts would give rise to additional surface integrals, which cancel each other.

For a given MHS equilibrium of the present type, equation (5.58) has to be satisfied for arbitrary choices of $\boldsymbol{V}$. Here we explore cases where $\boldsymbol{V}$ depends linearly on the coordinates,

$$\boldsymbol{V} = \mathcal{A} \cdot \boldsymbol{r} \tag{5.59}$$

where $\mathcal{A}$ is a constant tensor. From (5.58) we obtain

$$\int_{\mathrm{d}} \left[\left(p - p_0 + \frac{B^2}{2\mu_0} \right) \mathrm{Tr}\, (\mathcal{A}) - \frac{1}{\mu_0} \boldsymbol{B} \cdot \mathcal{A} \cdot \boldsymbol{B} \right] \mathrm{d}^3 r = 0 \,. \tag{5.60}$$

Particularly interesting are the special choices

$$\mathcal{A} \;=\; \mathcal{I} \qquad\qquad \Rightarrow \left\langle \frac{B^2}{2\mu_0} \right\rangle_{\mathrm{d}} = 3 \langle p_0 - p \rangle_{\mathrm{d}} \tag{5.61}$$

$$\mathcal{A} \;=\; \boldsymbol{e}_x \boldsymbol{e}_x + \boldsymbol{e}_y \boldsymbol{e}_y \;\Rightarrow\; \left\langle \frac{B_z^2}{2\mu_0} \right\rangle_{\mathrm{d}} = \langle p_0 - p \rangle_{\mathrm{d}} \tag{5.62}$$

$$\mathcal{A} \;=\; \boldsymbol{e}_x \boldsymbol{e}_y \qquad\quad \Rightarrow \langle B_x B_y \rangle_{\mathrm{d}} = 0 \tag{5.63}$$

where $\langle \ldots \rangle_{\mathrm{d}}$ means $\int_{\mathrm{d}} \ldots \mathrm{d}^3 x / \int_{\mathrm{d}} \mathrm{d}^3 x$.

Since the z-axis is not particularly preferred, equation (5.62) also holds for B_z replaced by B_x or B_y. Similarly, (5.63) remains valid for $B_x B_y$ replaced by $B_y B_z$ or $B_z B_x$. Equations (5.61)–(5.63) imply that there is equipartition in the large of energy in different Cartesian magnetic field components, and that there is no spatial correlation between those components.

A major conclusion to be drawn from (5.61) is that the magnetic field and the pressure must vanish for $p_0 = 0$. This implies nonexistence of magneto-hydrostatic self-confinement, i.e., confinement of a static plasma (described by MHS) by its own magnetic field. Although for $p_0 \neq 0$ nontrivial solutions are no longer excluded by the virial expressions that were considered here, there is no guarantee for their existence. In fact, the existence of three-dimensional MHS solutions defined in entire space is by no means trivial (see Section 5.4.4).

Now let us turn to configurations which are less likely to face existence problems. Consider the case where the domain d is cut by a surface Q. An example is a plane surface cutting a bent flux tube with intersections q_1 and q_2 (Fig. 5.3).

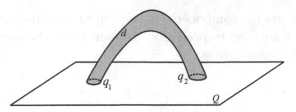

Fig. 5.3 A magnetic flux tube cut by a plane surface Q.

Let Q be parallel to the x, y-plane of a Cartesian coordinate system. Only the upper half space needs to be in MHS equilibrium conditions, a different physical situation may prevail in the lower half. Then, instead of (5.61) one finds

$$\int_d \left[3\,(p - p_0) + \frac{B^2}{2\mu_0} \right] \mathrm{d}^3 r - \int_{q_1,q_2} \frac{1}{\mu_0} B_z (x B_x + y B_y)\, \mathrm{d}S = 0 \, . \qquad (5.64)$$

Here one uses

$$\int_{q_1,q_2} \left[-(p - p_0 + \frac{B^2}{2\mu_0})\boldsymbol{n}_\mathrm{d} + \frac{1}{\mu_0}\boldsymbol{n}_\mathrm{d} \cdot \boldsymbol{B}\,\boldsymbol{B} \right] \mathrm{d}S = 0 \qquad (5.65)$$

which is obtained by integrating (5.51) over d. Equation (5.65) also makes sure that (5.64) does not depend on the choice of the origin of the coordinate system. Note that for a force-free field ($p = p_0$) the magnetic energy contained in the flux tube is determined uniquely by the boundary conditions on the surface Q.

The same method can be applied to find virial expressions for a domain d covering full 3D space, however, a finite region being excluded. From (5.54) with $\boldsymbol{V} = \boldsymbol{r}$ one finds (assuming the existence of the integrals)

$$\int_d \left[3p + \frac{B^2}{2\mu_0} \right] \mathrm{d}^3 r - \int_{\partial d} \left[\left(p + \frac{B^2}{2\mu_0} \right) \boldsymbol{r} \cdot \boldsymbol{n} - \frac{1}{\mu_0} \boldsymbol{r} \cdot \boldsymbol{B}\boldsymbol{B} \cdot \boldsymbol{r} \right] \mathrm{d}S = 0 \, . \quad (5.66)$$

Note that the normal $\boldsymbol{n}$ points into the excluded domain. Specializing for a sphere of radius R gives

$$\int_{r>R} \left[3p + \frac{B^2}{2\mu_0} \right] \mathrm{d}^3 r = R \int_{r=R} \left[\frac{B_n^2}{2\mu_0} - \frac{B_t^2}{2\mu_0} - p \right] \mathrm{d}S \, , \qquad (5.67)$$

where B_n and B_t are the components normal (positive outward) and tangential to the sphere surface, respectively. As seen later, the expression (5.67) is relevant for the solar corona.

5.2 Symmetric MHS states

Quite generally, symmetries provide valuable tools for solving problems which otherwise are hopelessly complicated. This is most obvious for translational invariance with respect to a Cartesian coordinate because here the invariance simply means that one of the coordinates can be ignored.

Fortunately, there exist configurations of space plasmas for which a two-dimensional model can be regarded at least as a crude approximation. Examples are magnetic arcades in the solar corona (Fig. 14.1) or the quiescent magnetotail configuration in the Earth's magnetosphere. Fig. 5.4 illustrates how a model with (internal) translational invariance represents the plasma sheet of the Earth's magnetotail (Fig. 2.2). Two-dimensional tail models where the cross-tail coordinate is ignorable inside the magnetosphere have proven successful in various respects.

Another example is rotational invariance, which is outlined only briefly.

The question, whether invariance can also be found for other coordinates, is answered in Section 5.2.4.

For simplicity, the following discussion is confined to static systems without external gravity force.

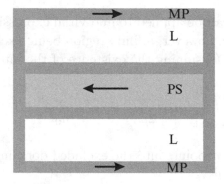

Fig. 5.4 Sketch of the cross-section of a two-dimensional version of a magnetotail model; shown are the magnetopause (MP), the lobes (L), the plasma sheet (PS) with its boundary layers, and the tail current system (arrows). Internal quantities are invariant with respect to translation in the horizontal direction.

5.2.1 MHS systems with translational invariance

The standard example of a symmetry in MHD is translational invariance with respect to a Cartesian coordinate, say z. In magnetohydrostatics, the invariance is then defined by z-independence of the observable quantities p, j, and B. (Note that electromagnetic and Euler potentials are not directly observable and particular forms of linear z-dependence of the potentials may be admitted without destroying the invariance of the electromagnetic fields E and B.)

It is appropriate to decompose vector fields with translational invariance in a *poloidal* component comprising the x- and y-components, and the remaining z-component, denoted as the *toroidal* part. (This terminology (Moffatt, 1978) is borrowed from rotational symmetry.) For the magnetic field this decomposition reads

$$B = B_p(x, y) + B_t(x, y) \tag{5.68}$$

where

$$B_p = B_x e_x + B_y e_y \tag{5.69}$$
$$B_t = B_z e_z . \tag{5.70}$$

Note that the divergence of B_p and of B_t vanishes separately. Therefore, we can introduce a vector potential for B_p alone. By a suitable gauge transformation, that vector potential can be put in the form $A(x, y)e_z$ such that B may be written as

$$B = \nabla A \times e_z + B_z e_z . \tag{5.71}$$

Note that here we did not include B_z in the representation by a vector potential. Instead, the field B is described by the two scalar quantities A and B_z, which often is a convenient representation. Since $B \cdot \nabla A = B_p \cdot \nabla A = 0$, as follows from (5.71), the function A is constant both on the field lines of B and on the field lines of B_p.

The current density is obtained by inserting (5.71) into (5.7),

$$j = -\frac{1}{\mu_0} \Delta A\, e_z + \frac{1}{\mu_0} \nabla B_z \times e_z . \tag{5.72}$$

With (5.71) and (5.72) the momentum balance (5.6) assumes the form

$$-\nabla p - \frac{1}{\mu_0} \Delta A \nabla A - \frac{1}{\mu_0} (\nabla B_z \times e_z) \cdot \nabla A\, e_z - \frac{1}{\mu_0} B_z \nabla B_z = 0 . \tag{5.73}$$

The z-component of (5.73) reads

$$(\nabla B_z \times \nabla A) \cdot e_z = 0 , \tag{5.74}$$

which implies that B_z depends on x, y only via $A(x,y)$,

$$B_z = B_z(A), \qquad \nabla B_z = \frac{\mathrm{d} B_z}{\mathrm{d} A}\nabla A . \tag{5.75}$$

With these properties in mind we take the vector product of (5.73) with ∇A which gives

$$\nabla p \times \nabla A = 0 \tag{5.76}$$

implying

$$p = p(A) . \tag{5.77}$$

Using (5.75) and (5.77) we finally take the dot product of (5.73) with ∇A which gives (e.g., Lüst and Schlüter, 1957; Grad and Rubin, 1958; Shafranov, 1958)

$$-\frac{1}{\mu_0}\Delta A = \frac{\mathrm{d}}{\mathrm{d} A}\left(p(A) + \frac{B_z(A)^2}{2\mu_0}\right) . \tag{5.78}$$

This equation is usually called the 'Grad–Shafranov equation'.

In a strict sense, defining the functions $B_z(A)$ and $p(A)$ it is assumed that a given value of A corresponds to a single-branch field line. In situations where two or more spatially separated branches with the same value of A exist, a label would have to be attached to the functions $B_z(A)$ and $p(A)$ that identifies the branch, so that different functions can be assigned to different branches. Here we allow for that possibility, but the label is suppressed in the notation.

In (5.78) the toroidal magnetic field component B_z simply generates a magnetic pressure $B_z^2/2\mu_0$ which adds to the plasma pressure p. Because of curvature effects, the poloidal field, represented by the function A, enters in a different way.

With (5.75) and (5.78) the electric current density (5.72) can be written as

$$\boldsymbol{j} = -\frac{1}{\mu_0}\Delta A\,\boldsymbol{e}_z + \frac{1}{\mu_0}\frac{\mathrm{d} B_z}{\mathrm{d} A}\boldsymbol{B}_{\mathrm{p}} \tag{5.79}$$

$$= \frac{\mathrm{d} p}{\mathrm{d} A}\boldsymbol{e}_z + \frac{1}{\mu_0}\frac{\mathrm{d} B_z}{\mathrm{d} A}\boldsymbol{B} \tag{5.80}$$

where (5.79) separates $\boldsymbol{j}$ in toroidal and poloidal components, and (5.80) best illustrates the force-free limit.

In the present formulation it is straightforward to switch to Euler potentials. Obviously, the function A itself can be chosen as one of the

Euler potentials, because A is constant on magnetic field lines. Thus, we identify

$$\alpha = A . \tag{5.81}$$

The second Euler potential β is easily found by writing

$$\beta = z + \tilde{\beta}(x, y) \tag{5.82}$$

and setting $\boldsymbol{B} = \nabla\alpha \times \nabla\beta$ in (5.71) we find

$$
\begin{aligned}
B_z &= \nabla A \times \nabla\tilde{\beta} \cdot \boldsymbol{e}_z \\
&= -\boldsymbol{B}_{\mathrm{p}} \cdot \nabla\tilde{\beta} \\
&= -B_{\mathrm{p}} \frac{\partial\tilde{\beta}}{\partial\sigma} \tag{5.83}
\end{aligned}
$$

where σ is the arc length on poloidal field lines. Equation (5.83) is easily inverted,

$$\tilde{\beta} = -B_z \int^\sigma \frac{\mathrm{d}\sigma}{B_{\mathrm{p}}} \tag{5.84}$$

$$= -B_z \int^s \frac{\mathrm{d}s}{B} \tag{5.85}$$

where we have used that on a given field line $\mathrm{d}\sigma/\mathrm{d}s$ equals B_{p}/B and B_z is constant. The integrations in (5.84) and (5.85) are carried out along poloidal and full field lines, respectively.

Since A and β are Euler potentials, they are magnetic flux coordinates in the sense of (5.25). Similarly, A and z are flux coordinates of the poloidal field, which implies that $A(x_2, y_2) - A(x_1, y_1)$ is the poloidal magnetic flux (per unit length with respect to the z-coordinate) passing through a line connecting the points (x_1, y_1) and (x_2, y_2). Therefore, A is also called the (poloidal) *flux function*.

The force-free limit of the Grad–Shafranov equation (5.78) is obtained by setting $\mathrm{d}p/\mathrm{d}A$ to zero. Note that, by (5.80), indeed $\boldsymbol{j} \times \boldsymbol{B}$ vanishes.

Finally, we remark that every choice of $p(A) + B_z(A)^2/2\mu_0$ defines an equilibrium problem in terms of a Grad–Shafranov equation (5.78). An important particular choice is discussed in the following section.

5.2.2 Liouville's solutions

Here we consider a special form of the Grad-Shafranov equation (5.78).
Suppose

$$p(A) + \frac{B_z^2(A)}{2\mu_0} = \frac{1}{2}p_c\, e^{-2\frac{A}{A_c}} \tag{5.86}$$

$$B_z(A) = 0, \tag{5.87}$$

where p_c and A_c are constants. Normalizing x- and y-coordinates by
$A_c/\sqrt{\mu_0 p_c}$ and A by A_c, we find from (5.86)

$$\Delta A = e^{-2A} \tag{5.88}$$

where, for simplicity, we keep the symbols for the normalized quantities
unchanged.

The equation (5.88) has important applications also in areas other than
space plasma physics; Joseph Liouville took a strong interest in that equation
in the context of fluid dynamics (Liouville, 1853). He found that solutions of
(5.88) can be generated by arbitrary analytical (complex) functions, called
generating functions. To express this property appropriately, let us replace
the real coordinates x and y by the complex variables $\zeta = x + iy$ and its
complex conjugate $\bar{\zeta}$, such that A may be understood as a function of ζ
and $\bar{\zeta}$. With this substitution the Grad–Shafranov equation (5.88) takes the
form

$$4\frac{\partial^2 A}{\partial\zeta\partial\bar{\zeta}} = e^{-2A}. \tag{5.89}$$

In this form it is straightforward to verify that any conformal mapping

$$\zeta' = \zeta'(\zeta) \tag{5.90}$$

with $\zeta' = x' + iy'$ generates a new solution $A'(x', y')$ of (5.89) from a known
solution $A(x, y)$ with

$$A' = A + \ln\left|\frac{d\zeta'}{d\zeta}\right|. \tag{5.91}$$

Obviously, if any solution exists at all, infinitely many solutions can be
readily obtained by conformal mapping.

These solutions can be parameterized by the analytical functions. If
$f(\zeta) = a(x, y) + ib(x, y)$ is an analytical function, then

$$A(x, y) = \ln\frac{1 + |f(\zeta)|^2}{2\left|\frac{df}{d\zeta}\right|} \tag{5.92}$$

is a solution of the Grad–Shafranov equation (5.88), which easily can be verified explicitly.

We have introduced (5.88) in an ad hoc way, such that one may ask, whether that particular choice has any physical significance at all. Fortunately, the choice (5.88) indeed is of considerable importance from the physical point of view, too. For constant B_z it is associated with *local thermodynamic equilibrium*. This aspect will be dealt with in Section 6.2.2, where we will look at static states from the kinetic point of view.

For explicit examples of Liouville solutions see Section 5.3.4.

5.2.3 Rotational invariance

A symmetry which also deserves attention within our context is rotational invariance. This symmetry is best described in cylindrical or spherical coordinates. Here we choose cylindrical coordinates r, φ, z. We assume that observable quantities do not depend on φ. Then, we may proceed as in the case of translational invariance; the only difference is the appearance of metric coefficients, which are powers of r. In analogy with (5.71) we write

$$\boldsymbol{B} = \nabla \times (A_\varphi \boldsymbol{e}_\varphi) + B_\varphi \boldsymbol{e}_\varphi \qquad (5.93)$$

$$= \nabla \psi \times \nabla \varphi + H \nabla \varphi \qquad (5.94)$$

where $\psi = r A_\varphi$ and $H = r B_\varphi$. This is the decomposition of $\boldsymbol{B}$ in poloidal and toroidal components, similar to (5.71). The toroidal component of $\boldsymbol{j} \times \boldsymbol{B}$ must vanish, which gives

$$H = H(\psi) \qquad (5.95)$$

in analogy with (5.75).

The poloidal component of the force balance (5.6) implies that the pressure must be of the form $p = p(\psi)$ and also gives the final form of the Grad–Shafranov equation in cylindrical coordinates,

$$-\frac{1}{\mu_0} \left\{ \frac{1}{r} \frac{\partial}{\partial r} \left(\frac{1}{r} \frac{\partial \psi}{\partial r} \right) + \frac{1}{r^2} \frac{\partial^2 \psi}{\partial z^2} \right\} = \frac{dp}{d\psi} + \frac{H}{\mu_0 r^2} \frac{dH}{d\psi} . \qquad (5.96)$$

Obviously, this equation has a structure similar to that of (5.78), except for the appearance of the metric coefficients.

The Grad–Shafranov equation in the form of (5.96) has been applied successfully to many astrophysical objects, as well as to laboratory plasmas.

From the structure of (5.94) it is obvious that ψ and φ can be regarded as Euler potentials of the poloidal field component. Euler potentials for the total field can be constructed in a way similar to the Cartesian case,

$$\boldsymbol{B} = \nabla\psi \times \nabla\beta \tag{5.97}$$

$$\beta = \varphi + \tilde{\beta} \tag{5.98}$$

where

$$\tilde{\beta} = -H \int^{\sigma} \frac{\mathrm{d}\sigma}{r^2 B_{\mathrm{p}}} . \tag{5.99}$$

Here, as in (5.84), σ denotes the arc length on the poloidal field lines.

5.2.4 General invariance

As seen in Sections 5.2.1–5.2.3 a suitable invariance property may lead to a considerable simplification, consisting of a reduction from three to two independent variables. This raises the question whether there exist further symmetries of that nature. We first assume that such further symmetry exists and write down the generalized equations describing the corresponding equilibria. In a second step we will then describe the general class of equilibria with symmetry (Solov'ev, 1975; Edenstrasser, 1980b,a).

This section requires more formal mathematics than is needed for most other sections. On the other hand symmetry is a fundamental physical aspect which, in the present context, should be covered at least by a brief overview. (A more detailed presentation is given in www.tp4.rub.de/~ks/tb.pdf.)

Since non-orthogonal coordinate systems are included, covariant and contravariant vectors and tensors are used which, in standard notation, are denoted by upper and lower indices, respectively. Also, the summation convention is applied, such that the simultaneous appearance of an index in lower and in upper position in a product implies summation over that index from 1 to 3.

Let ξ^1, ξ^2, ξ^3 denote general coordinates, which may be defined by a coordinate transformation $x^i(\xi^1, \xi^2, \xi^3)$ where x^1, x^2, x^3 are Cartesian coordinates. The central quantity characterizing general coordinates is the metric tensor which is defined in terms of the arc length element $\mathrm{d}s$ expressed in the general coordinates,

$$\mathrm{d}s^2 = g_{ik}\,\mathrm{d}\xi^i\mathrm{d}\xi^k, \tag{5.100}$$

where g_{ik} denote the covariant components of the metric tensor, explicitly given by

$$g_{ik} = \delta_{rs} \frac{\partial x^r}{\partial \xi^i} \frac{\partial x^s}{\partial \xi^k} , \tag{5.101}$$

where δ_{rs} denotes the components of the unit tensor (Kronecker's symbol).

For the formulation of the MHS equations in coordinates that exhibit a symmetry, we follow the analysis of Edenstrasser (1980a) and Edenstrasser (1980b). Let us assume invariance with respect to ξ^3 in the sense that observables do not depend on ξ^3. To achieve that property we have to make sure that appropriate components of g_{ij} do not depend on ξ^3 as well, because they enter the expressions for the divergence and for the curl of the magnetic field. In fact, all g_{ij} must be independent of ξ^3. This is necessary to guarantee that all covariant vector components obey the symmetry if the contravariant components have that property and vice versa.

Thus we define the present invariance property by the condition

$$\frac{\partial g_{ij}}{\partial \xi^3} = 0. \tag{5.102}$$

If this condition is satisfied, all physical quantities can be written in invariant form. For instance, the pressure and the magnetic field take the form

$$p = p(\xi^1, \xi^2) \tag{5.103}$$
$$B^i = B^i(\xi^1, \xi^2) . \tag{5.104}$$

The current density j and the pressure gradient ∇p then are also symmetric. The general form of the Grad–Shafranov equation is found by a procedure analogous to that applied to translational invariance above. One finds

$$\frac{1}{\sqrt{g}} \frac{\partial}{\partial \xi^1} \left(\frac{g^{11} \sqrt{g}}{g_{33}} \frac{\partial A}{\partial \xi^1} + \frac{g^{12} \sqrt{g}}{g_{33}} \frac{\partial A}{\partial \xi^2} \right)$$

$$+ \frac{1}{\sqrt{g}} \frac{\partial}{\partial \xi^2} \left(\frac{g^{21} \sqrt{g}}{g_{33}} \frac{\partial A}{\partial \xi^1} + \frac{g^{22} \sqrt{g}}{g_{33}} \frac{\partial A}{\partial \xi^2} \right)$$

$$+ \frac{B_3}{\sqrt{g}} \left[\frac{\partial}{\partial \xi^2} \left(\frac{g_{13}}{g_{33}} \right) - \frac{\partial}{\partial \xi^1} \left(\frac{g_{23}}{g_{33}} \right) \right]$$

$$+ \frac{1}{2g_{33}} \frac{dB_3(A)^2}{dA} + \mu_0 \frac{dp(A)}{dA} = 0 , \tag{5.105}$$

where g is the determinant of the metric tensor. The equation (5.105) was derived by Edenstrasser (1980b) based on work by Solov'ev (1975).

The second step is concerned with the class of coordinates that satisfy the symmetry condition (5.102). This is a classical problem of differential geometry, often referred to as the search for *Lie symmetry*. Here, we briefly outline the main steps. For details the reader may consult texts on differential geometry or on general relativity. The problem is considerably simplified by the fact that in the present non-relativistic context the geometry is Euclidian.

The form of the condition (5.102) is not useful as it stands, because it is expressed in terms of an unknown metric. For a more appropriate formulation one introduces a vector $\boldsymbol{u}(\xi^1, \xi^2, \xi^3)$ tangent to the lines of constant ξ^1, ξ^2 (*Killing vector*),

$$\boldsymbol{u} = \frac{\partial \boldsymbol{r}}{\partial \xi^3} \tag{5.106}$$

which defines the direction of invariance. In the present Euclidian geometry, Killing vectors satisfy the following differential equation (*vanishing Lie derivative*)

$$\nabla \boldsymbol{u} + (\nabla \boldsymbol{u})^{\mathrm{T}} = 0 \tag{5.107}$$

where the superscript T stands for transpose.

The general solution of (5.107) is readily obtained as

$$\boldsymbol{u} = \omega \times \boldsymbol{r} + \boldsymbol{w} \tag{5.108}$$

where ω and $\boldsymbol{w}$ are constant vectors. Equation (5.108) states that the general isometric displacement of Euclidian geometry is simply a combination of rigid rotation and translation.

Inserting (5.108) into (5.106) we find the differential equation for $\boldsymbol{r}(\xi^1, \xi^2, \xi^3)$

$$\frac{\partial \boldsymbol{r}}{\partial \xi^3} = \omega \times \boldsymbol{r} + \boldsymbol{w} \tag{5.109}$$

where ξ^1 and ξ^2 enter as integration constants.

Solving that equation we find the most general form of the metric tensor that allows for a symmetry. The solution is conveniently expressed in a polar coordinate system r, φ, z where the z-axis is oriented parallel to ω and the location of the origin is adjusted such that $\boldsymbol{w}$ points in the z-direction also, allowing us to write $\boldsymbol{w} = \nu \boldsymbol{e}_z$. Then, a sufficiently general solution

$r(\xi^1, \xi^2, \xi^3)$ of (5.109) gives after inversion

$$\xi^1 = r \tag{5.110}$$
$$\xi^2 = \eta \tag{5.111}$$
$$\xi^3 = \frac{z}{\nu} + m(r, \eta) \tag{5.112}$$

where $\eta = \nu\phi - \omega z$. The arbitrary function $m(r, \eta)$ is left unspecified for convenience.

Inserting the corresponding metric into the general form of the Grad–Shafranov equation (5.105), we obtain

$$\frac{1}{r}\frac{\partial}{\partial r}\left(\frac{r}{\nu^2 + \omega^2 r^2}\frac{\partial A}{\partial r}\right) + \frac{1}{r^2}\frac{\partial^2 A}{\partial \eta^2} - \frac{2\nu\omega B_3(A)}{(\nu^2 + \omega^2 r^2)^2}$$
$$+ \frac{1}{2(\nu^2 + \omega^2 r^2)}\frac{\mathrm{d}B_3(A)^2}{\mathrm{d}A} + \mu_0\frac{\mathrm{d}p(A)}{\mathrm{d}A} = 0 . \tag{5.113}$$

This equation essentially represents all symmetric magnetohydrostatic equilibria. We remark that the function $m(r, \eta)$ has dropped out.

The geometric properties of the general case can be visualized in terms of the coordinate surfaces. While the surfaces of constant ξ^1 are simply cylinders, the surfaces of constant ξ^2 have a helical structure.

The limits of translational and rotational invariance as discussed in the previous sections are easily recovered. In both cases the ξ^2-surfaces reduce to planes. Translational invariance is found for

$$\omega = 0, \quad \nu = 1, \quad m(\eta, r) = 0 \tag{5.114}$$

and rotational invariance is obtained in the limit

$$\omega = 1, \quad m(\eta, r) = \frac{\eta}{\nu}, \quad \nu \to 0 . \tag{5.115}$$

In most cases where symmetric equilibria have been applied to describe quiescent states of space plasmas these limiting cases were considered. In a few cases, however, helical equilibria were already addressed (Ali and Sneyd, 2002, and S. Titov, private communication, 2003).

5.3 Examples of exact solutions

In this section we provide a collection of explicit examples of magnetohydrostatic configurations, which are relevant for the modelling of quiescent states of space plasmas under a variety of different conditions. We start out with extremely simple, however widely used, one-dimensional magnetohydrostatic structures. These cases have only local significance, for instance as a model for the local magnetopause structure. Structures on larger scales require variation in at least two spatial directions, while fully satisfactory models need three-dimensional descriptions. Relevant three-dimensional examples are rare and typically require approximations. The present examples, with one exception, are without a gravity force.

5.3.1 One-dimensional systems with magnetic fields of the form $(B_x(z), B_y(z), 0)$

Let all observables depend on the Cartesian z-coordinate only and let B_z vanish:

$$\boldsymbol{B} = (B_x(z), B_y(z), 0), \tag{5.116}$$

then $\boldsymbol{B} \cdot \nabla \boldsymbol{B} = 0$ such that the MHS momentum equation (5.6) with (5.9) reduces to

$$\frac{\mathrm{d}}{\mathrm{d}z}(p + \frac{B^2}{2\mu_0}) = 0. \tag{5.117}$$

Since $\nabla \cdot \boldsymbol{B} = 0$ is satisfied trivially, there is no further condition. Thus, any arbitrary choice of $p(z) > 0$, $B_x(z)$ and $B_y(z)$ with constant total pressure $p(z) + (B_x(z)^2 + B_y(z)^2)/2\mu_0$ is a magnetohydrostatic solution.

If the pressure p is constant separately, the magnetic field is force-free, (5.44) is satisfied with

$$\kappa = \frac{1}{\mu_0 B_y}\frac{\mathrm{d}B_x}{\mathrm{d}z} = -\frac{1}{\mu_0 B_x}\frac{\mathrm{d}B_y}{\mathrm{d}z}. \tag{5.118}$$

We briefly discuss two special cases.

Rotating field

A simple example for a field of type (5.116) is given by the choice

$$\begin{aligned} B_x &= B_0 \sin(hz) \\ B_y &= B_0 \cos(hz), \end{aligned} \tag{5.119}$$

B_0, h and the pressure p are constants. (5.119) is a force-free field with $\kappa = h/\mu_0$.

The field direction rotates as z varies, resulting in *magnetic shear*.

5.3.2 Harris sheet

Here we choose

$$p = \frac{B_0{}^2}{2\mu_0 \cosh^2(hz)}$$

$$B_x = -B_0 \tanh(hz)$$

$$B_y = B_z = 0,$$

(5.120)

such that the pressure balance (5.117) is satisfied. At $z = 0$ this field has a *neutral sheet*, where B vanishes. The neutral sheet separates regions of oppositely directed fields (see Fig. 5.5). Originally, this solution was obtained by Harris (1962) from a particle point of view, see Section 6.2.2.

The Harris sheet can also be regarded as a one-dimensional solution of the Liouville problem (5.88). If the flux function varies only in the z-direction, (5.88), written in dimensionless form, reduces to

$$\frac{\mathrm{d}^2 A(z)}{\mathrm{d}z^2} = \mathrm{e}^{-2A(z)}.$$

(5.121)

This equation is solved by the magnetic flux function

$$A = \ln \cosh(z)$$

(5.122)

generating the magnetic field

$$B_x = -\tanh(z)$$

(5.123)

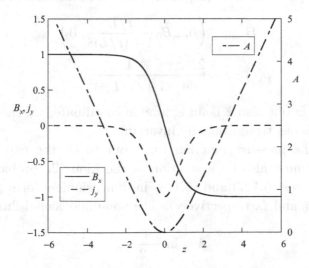

Fig. 5.5 Harris sheet: Flux function $A(z)$, magnetic field $B_x(z)$ and electric current density $j_y(z)$ in non-dimensional form.

and the current density

$$j_y = -\frac{1}{\cosh^2(z)} ,\qquad (5.124)$$

consistent with (5.120).

It is instructive to find out what function $f(\zeta)$, introduced in Section 5.2.2, generates the solution (5.122). The answer is

$$f(\zeta) = e^{i\zeta} \qquad (5.125)$$

as is readily verified by inserting (5.125) into (5.92), identifying ζ with $x+iz$.

The Harris sheet is a standard model of a one-dimensional structure with a neutral sheet. It describes one of the simplest situations where a plasma is confined (here in one direction only) by a magnetic field.

5.3.3 Cylindrical configurations, Bennett pinch

Let us choose cylindrical coordinates (r, φ, z) and let $\boldsymbol{B} = (0, B_\varphi(r), B_z(r))$. Unlike the Cartesian case, the $\boldsymbol{B} \cdot \nabla \boldsymbol{B}$ term does not vanish. Instead of (5.117) the momentum balance gives

$$\frac{\mathrm{d}}{\mathrm{d}r}\left(p + \frac{B_\varphi^2}{2\mu_0} + \frac{B_z^2}{2\mu_0}\right) + \frac{B_\varphi^2}{\mu_0 r} = 0. \qquad (5.126)$$

This condition describes *pinch*-configurations.

The Bennett pinch solution (Bennett, 1934) is characterized by the choice

$$\boldsymbol{B} = \left(0, 2B_0\frac{r/L}{1 + (r/L)^2}, 0\right) \qquad (5.127)$$

$$p(r) = \frac{2B_0^2}{\mu_0}\frac{1}{(1 + (r/L)^2)^2} . \qquad (5.128)$$

Since not only φ but also z is an ignorable coordinate, we can understand this solution also as translationally invariant.

As, again, the pressure function turns out to be the exponential, the Bennett pinch must also be a solution of the Liouville problem. In fact, $f(\zeta) = \zeta$ generates (5.127) and (5.128) in dimensionless form (normalizing B_ϕ and r by B_0 and L, respectively). The non-dimensional flux function is

$$A = \ln\frac{1 + r^2}{2}. \qquad (5.129)$$

Bennett (1934) seems to be the first to emphasize the capacity of magnetic fields to confine plasmas.

5.3.4 Examples of Liouville's solutions

Here we discuss two further examples of the general method described in Section 5.2.2.

Kelvin's cat's eyes

A two-dimensional generalization of the Harris solution is generated by the function

$$f(\zeta) = \frac{1}{\sqrt{1-p^2}} \left(e^{i\zeta} + p \right), \tag{5.130}$$

where p is a real parameter with $0 \le p < 1$. The choice $p = 0$ corresponds to the Harris sheet. For $0 < p < 1$ one finds a structure, which is periodic in x with period 2π. The flux function is

$$A = \ln \left(\frac{\cosh(y) + p \cos(x)}{\sqrt{1-p^2}} \right) \tag{5.131}$$

where $\zeta = x + iy$ such that z is the ignorable coordinate. Fig. 5.6 shows the magnetic field lines (contours of constant A) for $p = 0.3$ extending over two periods. Such structures were originally discussed by Kelvin in a corresponding fluid-dynamics context and have been termed *Kelvin's cat's eye solution*.

The case (5.130) is one of the few known choices of $f(\zeta)$ that does not suffer from singularities. However, even a solution that contains singularities may still be useful in nonsingular regions. After all, the regions in space where

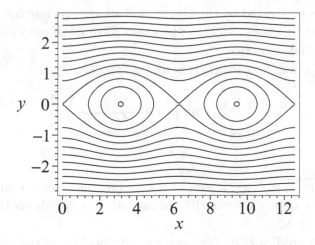

Fig. 5.6 Magnetic field lines of the cat's eye solution for $p = 0.3$.

MHS is a reasonable approximation are finite also. The following example illustrates that point.

Two-dimensional magnetosphere model

Here we discuss a simple two-dimensional magnetohydrostatic model of the Earth's magnetosphere (Section 2.1). Besides morphological description, various other purposes require a quantitative background configuration, for instance, the study of wave propagation, of stability properties and of single particle motion.

We will make use of non-dimensional variables. Coordinates are normalized by a characteristic scale length L, magnetic field components by a characteristic field strength B_0, flux functions by $B_0 L$ and pressures by B_0^2/μ_0, however, without changing the notation. The net effect is simply that μ_0 is set to 1. As customary for 2D magnetosphere models, y is chosen as the ignorable coordinate. The B_y-component is set to zero. Thus, we are looking for a solution $A(x, z)$ of the Grad–Shafranov equation (5.78). The pressure function is chosen as an exponential such that, in the chosen normalization, the Grad–Shafranov equation reduces to the Liouville equation (5.88).

The problem is to find a suitable generating function $f(\zeta)$. An interesting choice, which for a number of purposes provides an acceptable description of the nightside $(x > 0)$ magnetosphere, is based on the generating function

$$f(\zeta) = e^{i(\zeta + \sqrt{\zeta/\epsilon})}, \quad \zeta = x + iz \ . \tag{5.132}$$

This choice was motivated by the aim to find an appropriate modification of the Harris sheet function (5.125). The exact form of the modification then is a matter of trial and error.

The corresponding flux function is found from (5.92),

$$A(x, z) = \ln \frac{\cosh\left[\left(1 + \frac{1}{\sqrt{2(r_1 + x_1)}}\right) z\right]}{\sqrt{1 + \frac{\sqrt{r_1 + x_1}}{\sqrt{2}\, r_1} + \frac{1}{4 r_1}}} \tag{5.133}$$

where $r_1 = \sqrt{x_1^2 + z_1^2}$, $x_1 = \epsilon x$, $z_1 = \epsilon z$. The parameter ϵ measures the strength of the x-dependence in the distant tail. Fig. 5.7 shows the magnetic field lines for $\epsilon = 0.2$.

The pressure profiles (Fig. 5.8) are strongly peaked in the plasma sheet. Beyond, say, $x = 9$ the profiles show only moderate x-dependence.

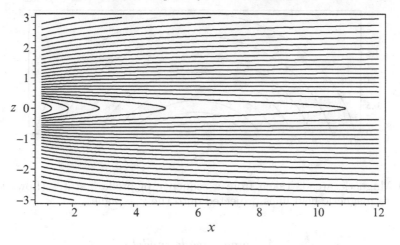

Fig. 5.7 Magnetic field lines from the magnetic flux function (5.133) with $\epsilon = 0.2$.

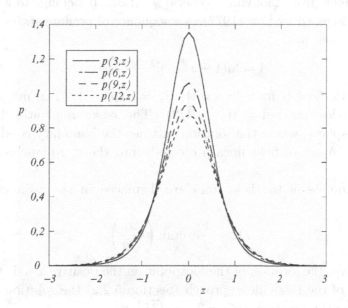

Fig. 5.8 Plasma pressure from the flux function (5.133) with $\epsilon = 0.2$. The figure shows pressure p vs. z for four values of x.

The singularity at the origin corresponds to a line current rather than to a line dipole. Nevertheless, the qualitative increase of the magnetic field strength with decreasing radial distance is well represented.

A similar solution, however, showing a more rapid tailward decay, was presented by Kan (1973).

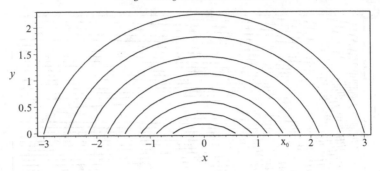

Fig. 5.9 Magnetic arcade taken from a sequence of preflare states (Low, 1977). Plotted are magnetic field lines projected onto the x, y-plane.

A simple arcade model

A simple model for an arcade-like magnetic structure in the solar atmosphere is also available from Liouville's general solution. It belongs to a set of 2D equilibria discussed by Low (1977) as a sequence of preflare states. The flux function is given by

$$A = \ln(1 + x^2 + y^2 - \sqrt{3}y) \,. \tag{5.134}$$

The magnetic field is force-free with $B_z = \exp(-A)$. The magnetic field lines are circles centered at $(0, -\sqrt{3}/2)$. The x-axis is identified with the solar photosphere, where the solution satisfies the boundary condition $A = \ln(1 + x^2)$. A set of field lines projected onto the x, y-plane is shown in Fig. 5.9.

The footpoints of the field lines are displaced in the z-direction by a distance

$$Dz = \frac{1}{2} \arctan\left(\frac{2x_0}{\sqrt{3}}\right) \tag{5.135}$$

where x_0 gives the location of the footpoint on the positive x-axis (Fig. 5.9).

In terms of the Liouville approach (Section 5.2.2) the solution (5.134) is generated by the function $f(\zeta) = 2\zeta - \sqrt{3}\,\mathrm{i}$.

5.3.5 A linear two-dimensional solution

Although in typical cases the Grad–Shafranov equation is nonlinear, there is an interesting linear exception. Let us choose $p(A) + B_z(A)^2/2\mu_0 = cA^2/2$ in (5.78) with c being a constant. This gives the linear differential equation

$$\frac{1}{\mu_0}\Delta A + cA = 0 \,. \tag{5.136}$$

Analytical solutions of that equation have been used to model the Earth's magnetotail (Voigt, 1986). We will return to a solution of (5.136) in the context of stability (see (10.103)).

5.3.6 A sheet supported by gravity

Here is a simple example with a gravity force taken into account. Suppose we consider a vertical magnetohydrostatic sheet with a gravity force $(-ge_z)$ corresponding to $\psi = gz$, which has been suggested as a model for the inner region of a solar prominence (Kippenhahn and Schlüter, 1957) in a potential field environment. Let the sheet (in Cartesian coordinates) be one-dimensional with non-ignorable coordinate x, such that the relevant observables to be determined are $p(x)$, $\rho(x)$ and $\boldsymbol{B} = (B_x, B_y, B_z(x))$ where $B_x \neq 0$ and B_y are constants. A further observable is temperature T, which may be obtained from the perfect gas law (3.66).

Then, after eliminating current density j, the equations (5.1)–(5.3) reduce to

$$\frac{\mathrm{d}}{\mathrm{d}x}\left(p(x) + \frac{B_z(x)^2}{2\mu_0}\right) = 0 \tag{5.137}$$

$$\frac{B_x}{\mu_0}\frac{\mathrm{d}B_z}{\mathrm{d}x} - \rho(x)g = 0. \tag{5.138}$$

Here (5.137) is simply the pressure balance of a one-dimensional sheet without gravity and (5.138) states that the gravity force is balanced by the z-component of the $\boldsymbol{j} \times \boldsymbol{B}$ force. These equations are solved by the following choice

$$B_z = B_0 \tanh\left(\frac{x}{d}\right) \tag{5.139}$$

$$p = \frac{B_0^2}{2\mu_0}\frac{1}{\cosh\left(\frac{x}{d}\right)^2} \tag{5.140}$$

$$\rho = \frac{B_x B_0}{\mu_0 g d}\frac{1}{\cosh\left(\frac{x}{d}\right)^2} \tag{5.141}$$

with $k_B T = mghB_0/2B_x$. The magnetic field lines projected in the x, z-plane are shown in Fig. 5.10.

5.4 Asymptotic expansions

The class of exact solutions of the MHS equations (5.6)–(5.8), although useful for selected purposes, is insufficient for modelling all relevant quiescent

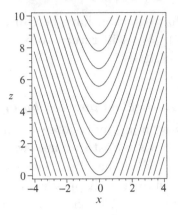

Fig. 5.10 Inner part of the Kippenhahn–Schlüter sheet.

states of space plasmas. Therefore, there is a strong demand for approximate solutions.

A situation that is particularly suited for an asymptotic approach is an equilibrium with two distinct length scales, prominent examples being tails of planetary magnetospheres or solar streamer configurations.

Particular care has to be taken to ensure validity of the approximation over the relevant spatial range. Consider a function f that depends on the coordinates x and z with length scales L_x and L_z, respectively, and suppose that $L_z/L_x = \epsilon \ll 1$. Then f may be expressed in the form

$$f = f(\epsilon x, z, \epsilon) \, . \tag{5.142}$$

If both coordinates are normalized with L_z the x-scale L_x becomes $1/\epsilon$ and any approximation must be valid on that scale. This is ensured by employing the standard multiple-scale expansion widely used in boundary layer theory (Eckhaus, 1973). We briefly summarize the basic idea adjusted to the present (particularly simple) case.

It is tempting to try to expand f in a power series with respect to ϵ. However, the straightforward power expansion of $f(\epsilon x, z, \epsilon)$ fails to hold on the x-scale $1/\epsilon$, because the expansion of the ϵx-dependence is applicable only in the range $\epsilon x \ll 1$. This difficulty is avoided by excluding the ϵx-dependence from the expansion, which is achieved by introducing the new variable $x_1 = \epsilon x$ and expanding the function $f(x_1, z, \epsilon)$ in a power series with respect to ϵ,

$$f = \sum_n f_n(x_1, z)\epsilon^n \, . \tag{5.143}$$

We will apply this method in the following two sections to two- and three-dimensional configurations.

5.4.1 Systems with translational invariance

Here we employ the method outlined in the previous section to obtain an asymptotic expansion for an equilibrium with translational invariance with respect to the y-coordinate. It is assumed that the equilibrium is characterized by different length scales in z- and x-directions with ratio ϵ (Schindler, 1972; Birn et al., 1975). Thus, our aim is to find a corresponding approximate solution of the Grad–Shafranov equation (5.78) which we write in the form

$$\Delta A + J(A) = 0, \tag{5.144}$$

where

$$J = \mu_0 j_y = \mu_0 \frac{\mathrm{d}}{\mathrm{d}A}\left(\frac{B_y(A)^2}{2\mu_0} + p(A)\right). \tag{5.145}$$

Choosing weak spatial variation with respect to the x-direction,

$$A = A(x_1, z, \epsilon), \quad x_1 = \epsilon x \tag{5.146}$$

the Grad–Shafranov equation becomes

$$\epsilon^2 \frac{\partial^2 A}{\partial x_1^2} + \frac{\partial^2 A}{\partial z^2} + J(A) = 0. \tag{5.147}$$

Inserting the formal power expansion of type (5.143)

$$A = \sum_n A_n(x_1, z)\epsilon^n \tag{5.148}$$

into (5.147) and expanding $J(A)$ also in powers of ϵ, we obtain an ordinary differential equation for every power of ϵ. Writing out the first five of these equations explicitly one finds

$$\frac{\partial^2 A_0}{\partial z^2} + J(A_0) = 0 \tag{5.149}$$

$$\frac{\partial^2 A_1}{\partial z^2} + J'(A_0)A_1 = 0 \tag{5.150}$$

$$\frac{\partial^2 A_2}{\partial z^2} + J'(A_0)A_2 = -\frac{\partial^2 A_0}{\partial x_1^2} - \frac{1}{2}J''(A_0)A_1^2 \tag{5.151}$$

$$\frac{\partial^2 A_3}{\partial z^2} + J'(A_0)A_3 = -\frac{\partial^2 A_1}{\partial x_1^2} - \frac{1}{6}J'''(A_0)A_1^3 - J''(A_0)A_1 A_2$$

$$\tag{5.152}$$

$$\frac{\partial^2 A_4}{\partial z^2} + J'(A_0)A_4 = -\frac{\partial^2 A_2}{\partial x_1^2} - \frac{1}{24}J''''(A_0)A_1^4 - J''(A_0)A_1 A_3$$

$$-\frac{1}{2}J''(A_0)A_2^2 - \frac{1}{2}J'''(A_0)A_1^2 A_2 \tag{5.153}$$

$$\cdots$$

Here, the prime symbol denotes the derivative with respect to A_0. Note that in each one of these equations the right side may be regarded as known, because it is determined by the previous equations.

Since there is no partial derivative with respect to x_1, the asymptotic solution A depends on x_1 in a parametric way, determined by suitable (x-dependent) boundary conditions. Let us choose a domain as shown in Fig. 5.11, which is typical for applications of the present approach. In that case boundary conditions are to be prescribed at the upper and lower boundaries b and d. It is a remarkable property of the system (5.149)–(5.153) that it is not possible to specify boundary conditions on the lateral boundaries (a and c in Fig. 5.11). The reason is that the expansion is not appropriate for arbitrary boundary conditions. In fact, a more general expansion would be of the form

$$A = \sum_n A_n(x, x_1, z)\epsilon^n \tag{5.154}$$

allowing for strong variation in the x-direction also. Solutions that show a weak x-dependence are characterized by particular values of A on the lateral boundaries. Thus, by insisting in solutions with weak dependence on

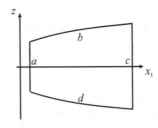

Fig. 5.11 Domain for two-scale expansion.

x, we have lost the freedom to specify boundary conditions on the lateral boundaries. Formally, this behaviour is a consequence of the fact that we are dealing with a problem of singular perturbation theory (Eckhaus, 1973).

The system (5.149)–(5.153) is solved as follows. First, (5.149) together with suitable boundary conditions yields $A_0(x_1, z)$. Conveniently, the exact boundary conditions on the boundaries b and d are applied to A_0, such that all A_n with $n > 0$ will have to vanish on those boundaries. The next equation, (5.150), is of a different nature. It is appropriately discussed in terms of the eigenvalue problem

$$\frac{\partial^2 a}{\partial z^2} + J'(A_0)a = \lambda a \qquad (5.155)$$

where a is to vanish on the upper and lower boundaries. For any x_1 this is a one-dimensional Schrödinger equation. Since the domain is finite, we expect a discrete spectrum of eigenvalues λ_n, which depend parametrically on x_1. Clearly, the existence of an eigenvalue vanishing for all x_1 is highly exceptional, because it would require specially chosen boundary conditions on the upper and lower boundaries. Here we will assume the generic case, where the λ_n vanish at most for isolated values of x_1. As a consequence, (5.150) implies that A_1 must vanish. Then, A_3 vanishes because of (5.152). In fact, this argument continues and all A_n of odd order n can be shown to vanish.

Thus, the system (5.149)–(5.153) reduces to

$$\frac{\partial^2 A_0}{\partial z^2} + J(A_0) \quad = \quad 0 \qquad (5.156)$$

$$\frac{\partial^2 A_2}{\partial z^2} + J'(A_0)A_2 \quad = \quad -\frac{\partial^2 A_0}{\partial x_1^2} \qquad (5.157)$$

$$\frac{\partial^2 A_4}{\partial z^2} + J'(A_0)A_4 \quad = \quad -\frac{\partial^2 A_2}{\partial x_1^2} - \frac{1}{2}J''(A_0)A_2^2 \qquad (5.158)$$

$$\cdots$$

So far, our procedure is only formal and does not guarantee convergence, nor can we be sure that the expansion at least is an asymptotic representation of the actual solution. However, by a detailed analysis it has been established (Kopp and Schindler, 1991) that, for sufficiently small values of ϵ, the series (5.148) does converge to an exact solution of the Grad–Shafranov equation.

Note that the vanishing of A_1 implies that A_0 represents an exact solution within an error of order ϵ^2, which is sufficient for most purposes.

It is a convenient property that (5.156) can be integrated in general form, without specifying $J(A)$. We illustrate this procedure for the case where $B_y = 0$, such that $J = \mu_0\, dp(A)/dA$. Using that property after multiplying (5.156) by $\partial A/\partial z$ one finds

$$\frac{\partial}{\partial z}\left[\frac{1}{2}\left(\frac{\partial A}{\partial z}\right)^2\right] + \mu_0\frac{dp(A)}{dA}\frac{\partial A}{\partial z} = 0 \ . \tag{5.159}$$

Integration of (5.159) with respect to z gives

$$\frac{1}{2\mu_0}\left(\frac{\partial A}{\partial z}\right)^2 + p(A) = \hat{p}(x_1) \ , \tag{5.160}$$

where $\hat{p}(x)$ is an arbitrary function of x. This equation describes the local pressure balance perpendicular to the central plane (Siscoe, 1972; Schindler, 1972), which indicates that the lowest approximation consists of considering each cross-section (fixed x) as a one-dimensional structure (Section 5.3).

After separating variables, further integration gives the solution in the form

$$z = z_0(x_1) + \int_{\hat{A}(x_1)}^{A}\frac{dA'}{\sqrt{2\mu_0(\hat{p}(x_1) - p(A'))}} \ . \tag{5.161}$$

Here $z_0(x_1)$ describes a surface (line in the x_1, z-plane) on which A assumes the boundary values $\hat{A}(x_1)$. For systems with a mirror symmetry with respect to the plane $z = 0$ the function $z_0(x_1)$ vanishes and $\hat{A}$ is determined by the condition $p(\hat{A}(x_1)) = \hat{p}(x_1)$. This symmetric form of (5.161) has been widely used for making simple two-dimensional asymptotic models of the tail of the Earth's magnetosphere during quiescent intervals.

We have introduced the present problem as a boundary value problem with a boundary condition prescribed on the boundaries b and d (Fig. 5.11). In this interpretation the functions $p(A)$ and $\hat{p}(x_1)$ have to be chosen consistent with that boundary condition. However, the form of (5.161) suggests the alternative procedure of prescribing $p(A)$ and $\hat{p}(x_1)$ instead of a boundary condition. This approach has proven useful for a number of applications, as in the following example. (We return to the boundary value problem in the next section.)

An example

Consider (5.161) and, for example, let us choose $p(A) = p_c \exp(-2A/A_c)/2$, the same pressure function that was used in Section 5.2.2. With the same normalization that was used there and choosing $z_0 = 0$ we find from (5.161) (Birn *et al.*, 1975)

$$z = \int_{\hat{A}(x_1)}^{A} \frac{\mathrm{d}A'}{\sqrt{2(\hat{p}(x_1) - p(A'))}} \tag{5.162}$$

with $p(A) = \exp(-2A)/2$ and $\hat{A}(x_1) = -\ln(2\hat{p}(x_1))/2$. Carrying out the integration and solving for A gives

$$A(x_1, z) = \ln \left[\frac{\cosh\left(\sqrt{2\hat{p}(x_1)}\, z\right)}{\sqrt{2\hat{p}(x_1)}} \right]. \tag{5.163}$$

This expression has become a familiar tool for the modelling of tail-like configurations, with $\hat{p}(x_1)$ adapted from observations. Therefore, it seems worthwhile to test its applicability by comparing the exact solution (5.133) with its asymptotic representation for small ϵ.

As expected one finds that to lowest order (5.133) reduces to the asymptotic form (5.163) with

$$\hat{p}(x_1) = \frac{1}{2}\left(1 + \frac{1}{2\sqrt{x_1}}\right)^2. \tag{5.164}$$

Fig. 5.12 illustrates the usefulness of the asymptotic expression for $\epsilon = 0.2$. In a large region (bounded by the thick lines) the relative difference between the exact flux function and its asymptotic representation is smaller than 0.01.

A final remark concerns the use of the variable $x_1 = \epsilon x$. It was introduced to develop the ϵ-expansion and asymptotic expressions. Predominantly, we dealt with cases with a single x-scale, and relevant quantities Q of the form $Q(x_1, z)$. In those cases one can absorb ϵ in the function Q and use quantities of the form $Q(x, z)$ in the approximate expressions, where, however, the dependence on x must be kept much weaker than the z-dependence. We will use the latter form in some of the applications.

5.4.2 A boundary value problem of two-dimensional MHS

Here we return to the boundary value problem formulated in the previous section and give an outline of the theory of Birn (1991).

Let us assume a configuration that is symmetric with respect to the plane $z = 0$ and let us choose a boundary condition which prescribes the shape of a bounding field line in the form $z = a(z)$. We also prescribe $\hat{p}(x_1)$ as

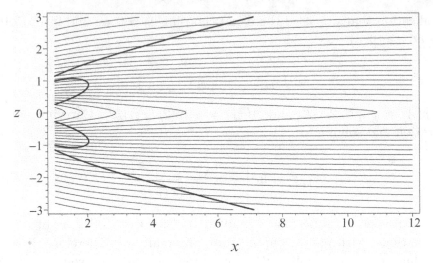

$$x$$

Fig. 5.12 Thin lines: field lines of exact magnetosphere solution for $\epsilon = 0.2$ (same as Fig. 5.7); thick lines: curves on which the relative difference between the asymptotic solution (5.163) and the exact solution is 0.01; the relative difference is smaller than 0.01 between the upper and lower thick lines except for the regions on the concave side of the loop-shaped thick lines on the left.

a monotonic function and try to find a monotonic function $p(A)$ matching these assumptions. Let the pressure on the bounding field line be p_b.

To find $p(A)$, we write (5.161) in the form

$$z(x_1, A) = \int_{p(A)}^{\hat{p}(x_1)} \left(-\frac{dA'(p')}{dp'} \right) \frac{dp'}{\sqrt{2\mu_0(\hat{p}(x_1) - p')}} \,, \qquad (5.165)$$

where the integration variable A' has been replaced by p', obtained by inverting the equation $p' = p(A')$. The boundary condition then takes the form

$$a(x_1) = \int_{p_b}^{\hat{p}(x_1)} \left(-\frac{dA'(p')}{dp'} \right) \frac{dp'}{\sqrt{2\mu_0(\hat{p}(x_1) - p')}} \,. \qquad (5.166)$$

Since $\hat{p}(x_1)$ is monotonic, one may express a in (5.166) as a function of $\hat{p}$. Then (5.166) assumes the form of Abel's integral equation (Whittaker and Watson, 1973) which can be solved for the unknown function dA/dp

$$\frac{dA}{dp} = -\frac{\sqrt{2\mu_0}}{\pi} \frac{d}{dp} \int_{p_b}^{p} a(\hat{p}) \frac{d\hat{p}}{\sqrt{p - \hat{p}}} \,. \qquad (5.167)$$

By integration we find the inverse of $p(A)$

$$A = -\frac{\sqrt{2\mu_0}}{\pi} \int_{p_b}^{p} a(\hat{p}) \frac{d\hat{p}}{\sqrt{p - \hat{p}}} , \quad (5.168)$$

an arbitrary integration constant is chosen such that $A(p_b) = 0$. As long as $A(p)$ turns out to be a monotonic function, inversion of (5.168) gives $p(A)$. For non-monotonic functions a more refined treatment is necessary.

With the present procedure the freedom of choosing $p(A)$ and $\hat{p}(x_1)$ is replaced by the freedom of choosing $a(x_1)$ and $\hat{p}(x_1)$.

After determining $p(A)$ from (5.168) one can obtain $A(x_1, z)$ from (5.165).

The procedure described here requires that all field lines are closed in the sense that they cross the plane $z = 0$. However, the approach can be generalized to include an open field line region (Birn, 1991).

5.4.3 Birn's asymptotic solutions of three-dimensional MHS

Although the 2D theory described in the previous two sections has proven useful for a variety of purposes, its limitation to two spatial dimensions is a severe restriction which, in some cases, may not be acceptable. For instance, the inclusion of realistic field-aligned currents in MHS models of magnetospheric structures requires three-dimensional models. It is the purpose of this section to describe a two-scale asymptotic approach to MHS, which generalizes the theory of Section 5.4.1 to fully three-dimensional configurations (Birn, 1991). The price that one has to pay for achieving this generalization is that the solution is not completely explicit, even not to lowest order. However, the problem is reduced to solving two coupled ordinary differential equations. Given the complexity of the MHS equations, this represents a rather drastic and useful simplification.

We start out from the MHS equations (5.6)–(5.8) without gravity, which with the help of (5.9) we write in the form

$$\boldsymbol{B} \cdot \nabla \boldsymbol{B} = \mu_0 \nabla \left(p + \frac{B^2}{2\mu_0} \right) \quad (5.169)$$

$$\nabla \cdot \boldsymbol{B} = 0 . \quad (5.170)$$

We generalize the scaling of Section 5.4.1 by allowing for a weak y-dependence (in addition to the weak x-dependence),

$$B_z, \ \partial/\partial x, \ \partial/\partial y = O(\epsilon)$$
$$p, \ B_x, \ B_y, \ \partial/\partial z = O(1) . \quad (5.171)$$

For L_x, L_y, L_z denoting characteristic lengths in the x, y, z directions, respectively, this ordering means that L_z/L_x and L_z/L_y are of order ϵ. This approximation applies to highly stretched three-dimensional configurations and provides realistic descriptions of magnetospheric tails or stretched coronal structures.

Here we aim for an approximation to lowest order in ϵ, corresponding to the zeroth order solution A_0 of Section 5.4.1. (For dealing with lowest order only, there is no need to introduce new variables corresponding to x_1 in the 2D case.)

Since $\mathbf{B} \cdot \nabla B_z$ is $O(\epsilon^2)$, i.e., vanishing in the present approximation, we find from the z-component of (5.169) with (5.171) that to lowest order

$$p + \frac{B_x{}^2 + B_y{}^2}{2\mu_0} = \hat{p}(x, y) \qquad (5.172)$$

where $\hat{p}(x, y)$ is arbitrary. Birn's approach (Birn, 1991) is to represent $\mathbf{B}$ by Euler potentials α and β

$$\mathbf{B} = \nabla\alpha \times \nabla\beta \qquad (5.173)$$

(see Section 5.1.2). A field line may then be represented by the position vector

$$\mathbf{r} = \mathbf{r}(x, \alpha, \beta) \qquad (5.174)$$

where x is used as the parameter marking points along the field line.

Since in MHS theory the pressure p is constant on magnetic field lines (see Section 5.1.1), p can be expressed as $p(\alpha, \beta)$.

Let

$$q = \frac{B_y}{B_x} \qquad (5.175)$$

be a new variable defining the slope of any given magnetic field line projected into the x, y-plane. Then one finds from (5.174)

$$\frac{\partial y(x, \alpha, \beta)}{\partial x} = q(x, \alpha, \beta) . \qquad (5.176)$$

We use these properties to rewrite the y-component of (5.169)

$$\begin{aligned}
\mu_0 \frac{\partial \hat{p}}{\partial y} &= \mathbf{B} \cdot \nabla B_y \\
&= \mathbf{B} \cdot \nabla(q B_x) \\
&= \mu_0 q \frac{\partial \hat{p}}{\partial x} + B_x \mathbf{B} \cdot \nabla q \\
&= \mu_0 q \frac{\partial \hat{p}}{\partial x} + B_x{}^2 \frac{\partial q(x, \alpha, \beta)}{\partial x} \qquad (5.177)
\end{aligned}$$

where we have used the x-component of (5.169) also.

Eliminating B_y from (5.172) and (5.175) we find an expression for B_x which is inserted into (5.177). Solving for $\partial q / \partial x$ gives

$$\frac{\partial q(x, \alpha, \beta)}{\partial x} = \frac{1 + q^2}{2(\hat{p} - p)} \left(\frac{\partial \hat{p}}{\partial y} - q \frac{\partial \hat{p}}{\partial x} \right). \tag{5.178}$$

Considering the functions $\hat{p}(x, y)$ and $p(\alpha, \beta)$ as given, the equations (5.176) and (5.178) represent two first order differential equations for y and q as a function of x for any fixed pair of α, β values. Since both equations involve derivatives with respect to x only, they are ordinary differential equations with independent variable x and parameters α and β.

Solving (5.176) and (5.178) we find $y(x, \alpha, \beta)$ and $q(x, \alpha, \beta)$ including arbitrary functions of α and β, the integration constants with respect to the x-integration.

By solving $y(x, \alpha, \beta)$ for β we can express β as a function of x, y, and α. It remains to determine α. For this purpose we note that (5.173) with $\beta = \beta(x, y, \alpha)$ gives

$$B_x = -\frac{\partial \alpha(x, y, z)}{\partial z} \frac{\partial \beta(x, y, \alpha)}{\partial y} \tag{5.179}$$

$$B_y = \frac{\partial \alpha(x, y, z)}{\partial z} \frac{\partial \beta(x, y, \alpha)}{\partial x}. \tag{5.180}$$

Inserting these expressions into pressure balance (5.172) we get

$$\left(\frac{\partial \alpha}{\partial z} \right)^2 |\nabla_2 \beta|^2 = 2\mu_0(\hat{p} - p) \tag{5.181}$$

where $|\nabla_2 \beta|^2 = (\partial \beta(x, y, \alpha)/\partial x)^2 + (\partial \beta(x, y, \alpha)/\partial y)^2$. Equation (5.181) can be integrated (for fixed x and y),

$$z = z_0(x, y) + \int_{\alpha_0(x,y)}^{\alpha} \pm |\nabla_2 \beta|_{\alpha'} \frac{d\alpha'}{\sqrt{2\mu_0[\hat{p}(x, y) - \hat{p}(\alpha', \beta(x, y, \alpha'))]}} \tag{5.182}$$

where $z_0(x, y)$ is an arbitrary function and $\alpha_0(x, y) = \alpha(x, y, z_0(x, y))$. Solving (5.182) for α gives the solution in the form $\alpha(x, y, z)$, which allows us to write β as a function of x, y and z also.

Summarizing, Birn's solution method reduces the equilibrium problem to solving the two ordinary differential equations (5.176) and (5.178) and to inverting $y(x, \alpha, \beta)$ to find $\beta(x, y, \alpha)$ and evaluating the integral in (5.182). This is possible by straightforward procedures for finite domains.

In the two-dimensional case of Section 5.4.1, where all observables are independent of y and B_y vanishes, it is appropriate to set

$$\alpha = A(x, z) \tag{5.183}$$

$$\beta = y \tag{5.184}$$

$$q = 0. \tag{5.185}$$

Then, the differential equations for q and β are satisfied trivially and (5.182) reduces to (5.161), where z_0 is chosen independent of y and p independent of β and where α_0 is identified with $\hat{A}$.

5.4.4 On nonequilibrium theorems and their limitations

In view of the rareness of exact three-dimensional solutions of the MHS equations it seems natural to try to start from an exact two-dimensional solution and then add a three-dimensional perturbation in the form of a power expansion with respect to a small parameter measuring the amplitude of the perturbation. Surprisingly, such expansions do not exist in general.

Parker (1979) has investigated the analytical neighbourhood of a number of one- and two-dimensional equilibria under a variety of conditions and proved several nonexistence theorems. A typical result can be summarized as follows (Rosner and Knobloch, 1982). Consider a magnetic field $\boldsymbol{B}_0$ and a pressure p_0 with one ignorable Cartesian coordinate, say, z, satisfying the MHS equations (5.6)–(5.8) in the domain $-L/2 < z < L/2$, where L is allowed to go to infinity. Stresses exerted on the boundaries are ignored. The perturbations

$$\delta\boldsymbol{B}(x, y, z) = \sum_{n=1}^{\infty} \epsilon^n \boldsymbol{b}_n(x, y, z)$$

$$\delta p(x, y, z) = \sum_{n=1}^{\infty} \epsilon^n p_n(x, y, z) \tag{5.186}$$

with $\epsilon \ll 1$ define the perturbed quantities

$$\boldsymbol{B}(x, y, z) = \boldsymbol{B}_0(x, y) + \delta\boldsymbol{B}(x, y, z)$$
$$p(x, y, z) = p_0(x, y) + \delta p(x, y, z) . \tag{5.187}$$

If $\boldsymbol{B}(x, y, z)$ and $p(x, y, z)$ satisfy the MHS equations (5.6)–(5.8) and remain bounded as $L \to \infty$ then the coefficients $\boldsymbol{b}_n$ can be shown to satisfy the additional constraint

$$\frac{\partial \boldsymbol{b}_n}{\partial z} = 0. \tag{5.188}$$

This means that under the imposed conditions the perturbed magnetic field necessarily has the same symmetry as the original state and an analytic non-symmetric neighbourhood is not available. This result was generalized to include plasma flow by Tsinganos (1982).

These exciting findings have led to the (more general) question, whether three-dimensional magnetohydrostatic states can exist at all. Nonexistence would imply that a symmetric system that somehow is forced into a non-symmetric state would lose equilibrium and would become dynamic.

This question has been answered positively. Non-symmetric solutions of the MHS equations have been shown to exist (Rosner and Knobloch, 1982). Also, note that Birn's 3D solutions discussed in Section 5.4.3 are not constrained to symmetry.

How can these different results be reconciled? A partial answer is based on the property that in finite domains, and for unrestricted boundary conditions, non-symmetric states of the form (5.187) with (5.186) can be found and explicit examples have been constructed (Arendt and Schindler, 1988). This possibility is based on formal differences between the cases of infinite and finite domains. For instance, consider a term of the type $a(x, y)z$. For an infinite domain boundedness requires that a vanishes, while this is not the case if the domain remains bounded. Similarly, $\Delta g(x, y, z) = 0$ implies $g = 0$ for infinite domains and bounded g, while g does not vanish identically for almost all boundary conditions if the domain remains finite. Using these properties one finds that the constraint (5.188) can be avoided for finite domains. Notably, in the applications to be discussed (Part IV) the regions where magnetohydrostatics is applicable, at least approximately, are bounded.

Van Ballegooijen (1985) found non-symmetric equilibria without restricting the domain, however, applying an expansion scheme different from Parker's. The existence of 3D MHS solutions was also confirmed by Neukirch (1997).

6

Particle picture of steady states

As discussed in Part I, the limitations of MHD are difficult to assess in general and seem to vary considerably from application to application. In particular, phenomena occurring on spatial and temporal scales near the intrinsic plasma scales (see Chapter 2) require a kinetic description based on the particle picture. Therefore, in this chapter steady states are explored from that point of view. We will restrict the discussion to plasmas for which binary particle collisions are negligible, so that the equations of motion (3.1) with (3.2) apply and a self-consistent description of steady states is provided by the Vlasov theory in the form (4.10)–(4.14).

6.1 General properties

Even if time-dependence is absent, steady state Vlasov theory still is a formidable nonlinear system of integro-differential equations, and a general solution method is not available. Luckily, there exist a number of special cases, for which important steps of the solution procedure can be carried out by analytical techniques. Typically, one exploits constants of motion of particle dynamics.

We briefly illustrate the significance of constants of motion for Vlasov theory. Although the method is applicable to the general (time-dependent) case, in view of the present context we specialize the argument for steady states.

The point is simply that (4.10) means that the distribution function $f_s(\boldsymbol{r}, \boldsymbol{w})$ of species s is constant on particle trajectories in phase space, so that

$$\frac{\mathrm{d}f_s}{\mathrm{d}t} = 0 \,, \qquad (6.1)$$

where the time derivative is taken along the particle orbit. This is easily verified by noting that

$$\frac{\mathrm{d}f_s}{\mathrm{d}t} = \frac{\partial f_s}{\partial \boldsymbol{r}} \cdot \frac{\mathrm{d}\boldsymbol{r}}{\mathrm{d}t} + \frac{\partial f_s}{\partial \boldsymbol{w}} \cdot \frac{\mathrm{d}\boldsymbol{w}}{\mathrm{d}t} \,. \tag{6.2}$$

Inserting $\mathrm{d}\boldsymbol{r}/\mathrm{d}t$ and $\mathrm{d}\boldsymbol{w}/\mathrm{d}t$ from (3.2) and (3.1) into (6.2) one immediately finds that (6.1) and (4.10) are equivalent.

Suppose now that $K_s(\boldsymbol{r}, \boldsymbol{w})$ is a constant of motion for particle species s,

$$\frac{\mathrm{d}K_s}{\mathrm{d}t} = 0 \,. \tag{6.3}$$

Obviously, this implies that any function $f_s(\boldsymbol{r}, \boldsymbol{w})$ that can be written in the form

$$f_s = F(K_s) \tag{6.4}$$

is a solution of (4.10), where F is an arbitrary (differentiable) function of K_s.

For the particle motion in steady state force fields a Hamiltonian (typically identical with energy) can be found, which does not depend on time explicitly. Such a Hamiltonian is a constant of motion. This is because the Hamiltonian equations of motion (3.15) and (3.16) imply that $\mathrm{d}H_s/\mathrm{d}t = \partial H_s/\partial t$. Thus, for $\partial H_s/\partial t = 0$ any choice $f_s = F(H_s)$ is a solution of (4.10). However, solutions of that form, which are successfully used in statistical mechanics (for the entire many-body system) are of little use for space plasmas. The main reason is that the electric current density, which plays an important role in space plasmas, would vanish.

An obvious way out of this difficulty is to look for situations where at least one more constant of motion is available (e.g., Schindler *et al.*, 1973). This is the case for translational or rotational invariance (or for the more general invariance discussed in Section 5.2.4).

6.2 Systems with translational invariance

Consider a system for which the Hamiltonian (3.14) is independent of the Cartesian z-coordinate. This implies the existence of an additional constant of motion, because from (3.16) one immediately concludes that $P_{z,s} = m_s w_z + q_s A_z(\boldsymbol{r})$ is a constant of motion.

It is important to note that translational invariance of the electromagnetic field vectors $\boldsymbol{E}$ and $\boldsymbol{B}$ does not necessarily imply translational invariance of the Hamiltonian. For steady state electric fields $\boldsymbol{E} = -\nabla\phi$ the condition $\partial \boldsymbol{E}/\partial z = 0$ leads to $\nabla(\partial\phi/\partial z) = 0$, implying $\phi = -E_z z + \phi_0(x, y)$, where E_z is the (constant) z-component of $\boldsymbol{E}$ and $\phi_0(x, y)$ is an arbitrary function.

Clearly, for translational invariance of the Hamiltonian, requiring translational invariance of ϕ, it is necessary that $E_z = 0$. At first sight, a similar argument seems to apply to the magnetic field. However, one can always find a gauge transformation that leads to a vector potential of the form $\boldsymbol{A}(x, y)$ for any $\boldsymbol{B}(x, y)$. We conclude that for achieving translational invariance of the Hamiltonian with respect to z we impose translational invariance on $\boldsymbol{E}$ and $\boldsymbol{B}$ with the additional restriction

$$E_z = 0 \,. \tag{6.5}$$

(Note that in a steady state with $\partial H_s/\partial t = 0$ it is not possible to avoid the z-dependence of ϕ by deriving E_z from a time-dependent vector potential.) For reasons of simplicity (rather than necessity) we ignore the gravity force. Then, the Hamiltonian (3.14) takes the form

$$H_s(x, y, \boldsymbol{P}) = \frac{1}{2m_s}(\boldsymbol{P} - q_s\boldsymbol{A}(x, y))^2 + q_s\phi(x, y). \tag{6.6}$$

This means that

$$f_s(x, y, \boldsymbol{P}) = F_s(H_s(x, y, \boldsymbol{P}), P_z) \tag{6.7}$$

is a solution of (4.10) for any choice of F_s.

Returning to $x, y, \boldsymbol{w}$ as the independent variables, we find

$$f(x, y, \boldsymbol{w}) = F_s\left(\frac{m_s w^2}{2} + q_s\phi(x, y), \; m_s w_z + q_s A_z(x, y)\right). \tag{6.8}$$

Without loss of generality we can choose the Coulomb gauge

$$\nabla \cdot \boldsymbol{A} = 0 \,. \tag{6.9}$$

Having solved the kinetic equation (4.10), it remains to consider steady state Maxwell's equations (4.11)–(4.14), which determine the potentials A_z and ϕ. In the following, Maxwell's equations first are discussed in their exact form and then in quasi-neutral formulation.

6.2.1 The equations for A and ϕ

First we note that for distribution functions of the form (6.8) the current density components j_x and j_y vanish. By (4.14) and (6.9) this means that $\Delta A_x = \Delta A_y = 0$. Again using (6.9), this implies that B_z is a constant. As the symbols A_x, A_y, j_x, j_y are no longer needed, we set $A_z = A$ and

$j_z = j$. Under those conditions steady state Maxwell's equations (4.11)–(4.14) assume the (dimensional) form

$$-\Delta\phi = \frac{1}{\epsilon_0}\sigma(A,\phi) \tag{6.10}$$

$$-\Delta A = \mu_0 j(A,\phi), \tag{6.11}$$

where

$$\sigma(A,\phi) = \sum_s q_s \int F_s\left(\frac{m_s w^2}{2} + q_s\phi, m_s w_z + q_s A\right) d^3 w \tag{6.12}$$

$$j(A,\phi) = \sum_s q_s \int w_z F_s\left(\frac{m_s w^2}{2} + q_s\phi, m_s w_z + q_s A\right) d^3 w \tag{6.13}$$

are the electrical charge density and the z-component of the electrical current density, respectively, j_x and j_y vanish. For prescribed distribution functions F_s, the equations (6.10) and (6.11) are two nonlinear partial differential equations for $A(x,y)$ and $\phi(x,y)$.

At first sight one might have the impression that the freedom to choose the functions $F_s(H_s, P_{z,s})$ arbitrarily corresponds to an arbitrary choice of $\sigma(A,\phi)$ and $j(A,\phi)$. However, this is not the case, because σ and j are subject to an important restriction, namely

$$\frac{\partial\sigma(A,\phi)}{\partial A} + \frac{\partial j(A,\phi)}{\partial\phi} = 0. \tag{6.14}$$

This condition is readily obtained in the following way

$$\frac{\partial\sigma}{\partial A} + \frac{\partial j}{\partial\phi}$$

$$= \sum_s q_s\left(\frac{\partial}{\partial A}\int F_s\, d^3 w + \frac{\partial}{\partial\phi}\int w_z F_s\, d^3 w\right)$$

$$= \sum_s q_s^2 \int\left(\frac{\partial F_s}{\partial P_z} + w_z\frac{\partial F_s}{\partial H_s}\right) d^3 w$$

$$= \sum_s \frac{q_s^2}{m_s}\int\frac{\partial f_s(x,y,\boldsymbol{w})}{\partial w_z}\, d^3 w$$

$$= 0. \tag{6.15}$$

The final step uses that f_s has to vanish at large $|\boldsymbol{w}|$ for σ and j to exist.

The condition (6.14) implies that σ and j can be derived from a potential $p(A,\phi)$, which is found as

$$p(A,\phi) = \sum_s \int\frac{m_s}{2}(w_x^2 + w_y^2)f_s\, d^3 w, \tag{6.16}$$

such that

$$\sigma(A, \phi) = -\frac{\partial p(A, \phi)}{\partial \phi} \tag{6.17}$$

$$j(A, \phi) = \frac{\partial p(A, \phi)}{\partial A}, \tag{6.18}$$

where we have used integration by parts and $d^3w = (2\pi/m_s)d(w_\perp^2/2)\,dP$.

The function $p(A, \phi)$ has an obvious physical interpretation. Noting that the bulk velocity

$$v_s = \frac{\int \boldsymbol{w} f_s \, d^3 w}{\int f_s \, d^3 w} \tag{6.19}$$

has vanishing x- and y-components and that f_s depends on w_x and w_y only through $w_x^2 + w_y^2$, we can identify $p(A, \phi)$ with the xx-component P_{xx} of the pressure tensor $\mathcal{P} = \sum_s \mathcal{P}_s$, given by

$$
\begin{aligned}
P_{xx} &= \sum_s \int m_s w_x^2 f_s \, d^3 w \\
&= \sum_s \int \frac{m_s}{2}(w_x^2 + w_y^2) f_s \, d^3 w \\
&= p(A, \phi) \, ;
\end{aligned} \tag{6.20}
$$

for the definition of the pressure tensor $\mathcal{P}_s$ see (3.35). Analogously, one finds $P_{yy} = p(A, \phi)$; the non-diagonal components vanish. Thus, the pressure tensor is isotropic in the x, y-plane.

It is instructive to realize that (6.17) and (6.18) guarantee the validity of the momentum balance, specialized from (3.34) for steady states and for vanishing poloidal bulk velocity (projection onto the x, y-plane) and gravity, summed over particle species,

$$-\nabla \cdot \mathcal{P} + \boldsymbol{j} \times \boldsymbol{B} + \sigma \boldsymbol{E} = 0. \tag{6.21}$$

Indeed, using pressure isotropy in the x, y-plane, translational invariance together with $\boldsymbol{B} = \nabla A \times \boldsymbol{e}_z$ and $j_x = j_y = 0$, we get

$$
\begin{aligned}
-\nabla \cdot \mathcal{P} &+ \boldsymbol{j} \times \boldsymbol{B} + \sigma \boldsymbol{E} \\
&= -\nabla p + j \nabla A - \sigma \nabla \phi \\
&= \left(j - \frac{\partial p}{\partial A}\right) \nabla A - \left(\sigma + \frac{\partial p}{\partial \phi}\right) \nabla \phi \\
&= 0.
\end{aligned}
$$

Appropriate boundary conditions for solving the system (6.10) and (6.11) are Dirichlet conditions, Neumann conditions or suitable mixtures.

We also note that (6.10) and (6.11) can be derived from a variational principle where the variation functional is given by

$$V(A, \phi) = \int \left(\frac{(\nabla A)^2}{2\mu_0} - \frac{\epsilon_0 (\nabla \phi)^2}{2} - p(A, \phi) \right) \mathrm{d}x \, \mathrm{d}y \,. \tag{6.22}$$

It is easy to verify that (6.10) and (6.11) are the corresponding Euler–Lagrange equations for the above-mentioned boundary conditions.

The approach presented in this section is exact for arbitrary z-independent systems with $E_z = 0$ and constant B_z. No approximation is made and the analysis is based on exact particle orbits.

One may wonder how such a rather simple formulation is possible, given the complications of determining the particle orbits. Obviously, equilibrium theory requires only partial knowledge of the orbits and, fortunately, that knowledge is available in analytical form.

6.2.2 Quasi-neutral systems and local Maxwellians

As in fluid theory we may use the quasi-neutrality approximation (see Section 3.2) if all length scales are much larger than the Debye length. Correspondingly, (6.10) is replaced by

$$\sigma(A, \phi) = 0 \,, \tag{6.23}$$

which in many interesting cases can be inverted to give $\phi(A)$.

The equation (6.11) then reduces to

$$-\Delta A = \mu_0 \frac{\mathrm{d}p(A)}{\mathrm{d}A} \,, \tag{6.24}$$

where we have written $p(A, \phi(A))$ as $p(A)$. We also replaced $\partial p / \partial A$ by $\mathrm{d}p/\mathrm{d}A$ because of the following property,

$$\begin{aligned}
\frac{\mathrm{d}p(A)}{\mathrm{d}A} &= \frac{\partial p}{\partial A} + \frac{\partial p}{\partial \phi} \frac{\mathrm{d}\phi}{\mathrm{d}A} \\
&= \frac{\partial p}{\partial A} - \sigma \frac{\mathrm{d}\phi}{\mathrm{d}A} \tag{6.25} \\
&= \frac{\partial p}{\partial A} \,. \tag{6.26}
\end{aligned}$$

Obviously, (6.24) agrees with the Grad–Shafranov equation (5.78) for constant B_z. Thus, the quasi-neutral limit of present Vlasov theory for systems with translational invariance and constant B_z reduces to a corresponding magnetohydrostatic description. This fact is easily explained by the pressure isotropy in the poloidal plane and by the fact that setting σ to zero eliminates the $\sigma \boldsymbol{E}$ term in the momentum equation (6.21) such that this equation reduces to the corresponding MHD version.

We illustrate the present procedure for a commonly used choice of the distribution functions F_s (Harris, 1962), given by

$$F_s(H_s, P_{z,s}) = C_s \exp(-\alpha_s P_{z,s} - \beta_s H_s), \tag{6.27}$$

where C_s, α_s and $\beta_s = 1/k_B T_s$ are constants. If expressed in terms of velocity $\boldsymbol{w}$, (6.27) reads

$$f_s(x, y, \boldsymbol{w}) = C_s \exp \left(\frac{1}{2} \beta_s m_s u_s^2 - \beta_s q_s \phi(x,y) - \alpha_s q_s A(x,y) \right.$$
$$\left. - \frac{1}{2} \beta_s m_s (w_x{}^2 + w_y{}^2 + (w_z - u_s)^2) \right). \tag{6.28}$$

Obviously, this is a local Maxwellian centred at velocity $u_s = -\alpha_s/\beta_s$ in the w_z-direction.

The pressure p is obtained from (6.16)

$$p = \sum_s \hat{n}_s k_B T_s \exp \left(-\frac{q_s}{k_B T_s} (\phi - u_s A) \right), \tag{6.29}$$

where $\hat{n}_s = C_s (2\pi k_B T_s/m_s)^{3/2} \exp(m_s \beta_s u_s{}^2/2)$. From (6.17) and (6.18) we find σ and j as

$$\sigma = \sum_s q_s \hat{n}_s \exp \left(-\frac{q_s}{k_B T_s} (\phi - u_s A) \right) \tag{6.30}$$

$$j = \sum_s q_s \hat{n}_s u_s \exp \left(-\frac{q_s}{k_B T_s} (\phi - u_s A) \right). \tag{6.31}$$

The potentials A and ϕ are found by solving (6.10) and (6.11) with (6.30) and (6.31).

For a plasma consisting of electrons (species e) and a single ion species (i) with charge Ze, the quasi-neutrality approximation yields

$$\hat{n}_e = Z \hat{n}_i \tag{6.32}$$

$$\phi = \frac{T_i u_e + Z T_e u_i}{T_i + Z T_e} A. \tag{6.33}$$

Inserting (6.32) and (6.33) into (6.31) we find

$$j = e \hat{n}_e (u_i - u_e) \exp \left(\frac{Ze(u_i - u_e)}{k_B T_i + Z k_B T_e} A \right). \tag{6.34}$$

Thus, from (6.11) we obtain a Grad–Shafranov equation in the form considered earlier (see Section 5.2.2)

$$\Delta a = e^{-2a} \tag{6.35}$$

with

$$a = -\frac{Ze(u_i - u_e)}{2(k_B T_i + Z k_B T_e)} A \tag{6.36}$$

and where the coordinates are normalized by the length

$$L = \left(\frac{2(k_B T_i + Z k_B T_e)}{\mu_0 e^2 Z \hat{n}_e (u_i - u_e)^2} \right)^{1/2} . \tag{6.37}$$

We conclude that the exponential choice for $p(A)$, which is frequently used for applications and which we also encountered in connection with Liouville's approach, corresponds to the choice of local Maxwellians in a quasi-neutral plasma.

A particularly simple case results from choosing the parameters, such that

$$T_i u_e + Z T_e u_i = 0. \tag{6.38}$$

Then ϕ vanishes identically and we find (6.35) with

$$a = \frac{e u_e}{2 k_B T_e} A, \quad L = \left(\frac{2 Z k_B^2 T_e^2}{\mu_0 e^2 \hat{n}_e u_e^2 (k_B T_i + Z k_B T_e)} \right)^{1/2} , \tag{6.39}$$

which is an exact description of a charge neutral equilibrium. The condition (6.38) can be achieved by a Galilean transformation in the z-direction.

Clearly, all solutions of (6.35) that we discussed in Sections 5.2.2 and 5.3 from the MHD point of view also apply to the present kinetic picture, as long as the magnetic field component in the invariant direction is kept constant.

A remark seems necessary about the surprising property that there is a parameter choice for which the electric field vanishes. This property is closely tied to the choice (6.27) of the distribution function. Non-Maxwellian choices, which also are of interest for collisionless plasmas, generally lead to solutions with non-vanishing electric fields. Choosing a Maxwellian with (6.38), one must be aware that one does not obtain a representative electric field signature. This aspect plays an important role for kinetic models of thin current sheets (see Section 8.5).

6.3 Adiabatic particle motion

As discussed in the previous sections, the use of constants of motion in addition to the Hamiltonian gives us a powerful method for constructing current-carrying kinetic equilibria. On the other hand, such constants, if based on a symmetry, seem to exist only under rather special circumstances. It is clear from the discussions in Section 5.2.4 that we can find symmetry-based constants of motion only for helical invariance with the limiting cases of translational and rotational invariance.

Fortunately, in an approximate sense, there exist constants of motion that do not require an exact symmetry of the fields. These are the adiabatic invariants (see www.tp4.rub.de/~ks/tb.pdf).

The most widely used example of an adiabatic invariant is the magnetic moment (3.10) of a particle gyrating in a field with spatial and temporal scales large compared with the gyroscales. Other adiabatic invariants are associated with near-periodic motion of the gyrocentre. If a particle oscillates along magnetic field lines between regions of enhanced field magnitude (*magnetic mirrors*) there is a second invariant given by (3.11).

For adiabatic particle motion in weakly varying fields, the zeroth order gyrocentre velocity is simply $\boldsymbol{E} \times \boldsymbol{B}/B^2$ (see (3.6)), which implies magnetic line conservation. For a magnetic field possessing Euler potentials this means that potentials α, β can be found which, approximately, are constants of motion.

Thus, for adiabatic particle motion we have several additional quantities, that can serve to construct models of steady states. As an example, let us consider a set of adiabatic particles where the only near-periodic motion is the gyration and where there is no bulk plasma flow. Let us also assume that the magnetic field possesses Euler potentials α, β. Then, H, μ, α, β is an appropriate set of (exact or approximate) constants of motion and

$$f = F(H, \mu, \alpha, \beta) \tag{6.40}$$

is a steady state distribution function for arbitrary functions F to lowest order in the smallness parameter.

For particles that are not reflected in their motion along field lines, the distribution function F can be specified differently for particles moving in different directions along the field. In other words, the sign of $w_{||}$, denoted by $\mathrm{sgn}(w_{||})$, is an additional constant of motion. In that case we can even allow for a bulk flow component parallel to the magnetic field by choosing

$$f = F(H, \mu, \alpha, \beta, \mathrm{sgn}(w_{||})) \tag{6.41}$$

as the distribution function.

The pressure tensor that one obtains from (6.40) or (6.41) has the form of the CGL tensor (3.69).

6.4 Regular and chaotic particle motion

As we have discussed in Section 3.1, the motion of particles in electromagnetic fields may be formulated as a problem of Hamiltonian mechanics. Here we discuss the case of negligible gravity force and steady state electromagnetic fields. For convenience, we repeat the corresponding Hamiltonian in Cartesian coordinates

$$H(\boldsymbol{r},\boldsymbol{p}) = \frac{1}{2m}(\boldsymbol{P} - q\boldsymbol{A}(\boldsymbol{r}))^2 + q\phi(\boldsymbol{r}) \qquad (6.42)$$

and the equations of motion

$$\begin{aligned}
\frac{\mathrm{d}\boldsymbol{r}}{\mathrm{d}t} &= \frac{\partial H}{\partial \boldsymbol{P}} \\[2mm]
\frac{\mathrm{d}\boldsymbol{P}}{\mathrm{d}t} &= -\frac{\partial H}{\partial \boldsymbol{r}} \ .
\end{aligned} \qquad (6.43)$$

It is well known that under suitable conditions Hamiltonian systems are subject to the phenomenon of *deterministic chaos*. For details we refer to the literature on this subject (e.g., Arnold, 1979; Lichtenberg and Liebermann, 1983). In the present context we cannot do more than make a few comments and illustrate the main aspects in terms of a simple example that is relevant for later applications.

First we have to distinguish between *integrable* and *nonintegrable* systems, where the term integrable essentially means that solving the dynamical problem can be reduced to evaluating a number of integrals. For a system of n degrees of freedom to be integrable it is necessary that n constants of motion can be found. (Sufficiency requires that these constants of motion are *isolating* (Lichtenberg and Liebermann, 1983) which, in many cases, does not apply to constants involving initial conditions. However, this will not cause a problem for the cases that we are going to discuss here.)

Integrable systems have well-ordered (regular) trajectories, while nonintegrable systems (i.e., systems with an insufficient number of constants of motion) have regions of phase space where the trajectories are irregular in the sense that small variations of the initial conditions lead to much larger differences between the trajectories at later times than for integrable systems. It is mainly that strong dependence on initial conditions that characterizes chaos.

Even chaotic systems can have large regular domains of phase space. A quantitative formulation of this property is provided by the famous KAM theorem. (For a historical note and original references see Lichtenberg and Liebermann, 1983.)

Chaotic orbits play an important role in space plasma physics. The best studied examples belong to the systems with translational invariance as discussed in Section 5.2. Since in most cases the invariance is chosen with respect to the y-coordinate we will adopt this choice here. Then, setting $B_y = 0$, the vector potential may be chosen as

$$\mathbf{A} = A(x, z)\mathbf{e}_y . \tag{6.44}$$

We start out from systems where A depends on z only. As we will show, these systems are integrable. This choice (together with the assumed vanishing of B_y) corresponds to a one-dimensional configuration with $\mathbf{B}$ having an x-component only. Of particular interest are cases with a *neutral sheet*, which means that B vanishes for some value of z. Such fields show a particularly rich variety of different types of particle orbits (Sonnerup, 1971).

A simple example is the choice

$$A = -B_0 \frac{z^2}{2L} , \tag{6.45}$$

where B_x varies linearly with z changing its sign at $z = 0$. Another example is the Harris sheet (see Section 5.3) with

$$A(z) = -B_0 L \ln \left(\cosh \left(\frac{z}{L} \right) \right) . \tag{6.46}$$

Here, B_x changes its sign at the plane $z = 0$ also, but the field becomes asymptotically homogeneous for large $|z|$.

In these and other cases with $A = A(z)$ the Hamiltonian (6.42) assumes the form

$$H = \frac{1}{2m} \left(P_x{}^2 + P_z{}^2 + (P_y - qA(z))^2 \right) , \tag{6.47}$$

where the electric potential was set to zero, which for the Harris sheet is justified by the choice (6.38).

Since H does not depend on the variables t, x, y, we see from the property $dH/dt = \partial H/\partial t$ and from (6.43) that our system possesses three constants of motion, namely H, P_x, P_y. This property makes our system integrable. In fact, x and y are ignorable coordinates so that, formally, our system has one degree of freedom with a Hamiltonian of the form $H(P_z, z)$. The equations of motion can be integrated simply by noting that $P_z = m \, dz/dt$, which can be inserted into the Hamiltonian. Straightforward integration gives

$$t = \int_{z_0}^{z} \pm \frac{dz'}{\sqrt{\frac{2}{m} \left(H - \frac{P_x{}^2}{2m} - \frac{1}{2m} (P_y - qA(z'))^2 \right)}} , \tag{6.48}$$

where z_0 is the value of z at $t = 0$ and the sign in the integrand is fixed by
the initial condition and continuity. Here we see explicitly that our system
is integrable. (In fact it is a general result that every autonomous system
with one degree of freedom is integrable.)

Integrability does by no means imply that the orbits have a simple structure. Depending on the initial conditions, the orbits show different qualitative properties (e.g., Sonnerup, 1971). Orbits of particles with sufficiently
small energies and initial position sufficiently far away from the neutral sheet
are simple gyration orbits with a ∇B-drift superimposed (see (3.7)). Particles that cross the neutral sheet have rather exotic orbits; Fig. 6.1 shows an
example for the magnetic flux function given by (6.45).

A transition from integrable to nonintegrable systems takes place through
an additional x-dependence of the flux function, such that $A = A(x, z)$.
Then, P_x is no longer a constant of motion, the system becomes nonintegrable, and regions with chaotic particle motion can be expected to exist in
phase space.

Physically, this transition means the addition of a B_z-component. In
typical cases this eliminates exact neutral sheets. However, for sufficiently
small $|B_z|$ the field configuration is geometrically close to the neutral sheet
case and often the term *neutral sheet* is also used for that case. Since one of
the constants of motion is missing, the particle orbits are no longer tied to a
line in the P_z, z-plane (as it is the case for $B_z = 0$) and the intersection points

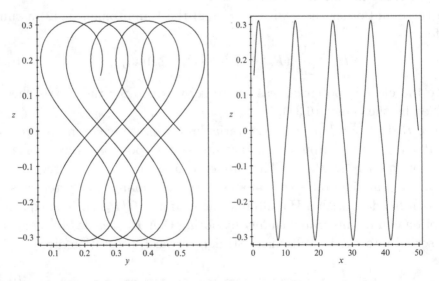

Fig. 6.1 An orbit of a particle crossing a one-dimensional neutral sheet projected
in the y, z-plane (left) and x, z-plane (right).

with an appropriately chosen P_z, z-plane (*surface of section*) may scatter and cover certain areas of that plane more densely, indicating the chaotic regions. One of the most striking consequences is the strong dependence on initial conditions.

In a simple model case B_z is a constant, say B_n. Adding a corresponding term to (6.45) the magnetic flux function reads

$$A = -B_0 \frac{z^2}{2L} + B_n x , \qquad (6.49)$$

the magnetic field lines (level curves of A) are parabolas.

Particle orbits in two-dimensional neutral sheet configurations were studied by Speiser (1965), Chen and Palmadesso (1986), Büchner and Zelenyi (1989), and others. Here we emphasize the strong dependence on initial conditions. For an illustration we use (6.49), so that the Hamiltonian (6.42) becomes

$$H = \frac{P_x^2}{2m} + \frac{P_z^2}{2m} + \frac{1}{2m} \left(P_y + qB_0 \frac{z^2}{2L} - qB_n x \right)^2 . \qquad (6.50)$$

Obviously, the constant P_y can be eliminated by transforming to the new x-coordinate $x - P_y/qB_n$. Normalizing H by mw^2, where w is the magnitude of the particle velocity, P_x, P_z by mw and x, z by $\sqrt{Lr_{g0}}$ with $r_{g0} = mw/|q|B_0$ being the gyroradius of a particle moving with velocity w in the field B_0, we obtain (leaving, however, the notation of the variables unchanged)

$$H = \frac{P_x^2}{2} + \frac{P_z^2}{2} + \frac{1}{2} \left(\frac{z^2}{2} - \lambda x \right)^2 , \qquad (6.51)$$

where

$$\lambda = \frac{B_n}{B_0} \sqrt{\frac{L}{r_{g0}}} . \qquad (6.52)$$

An orbit with energy E is confined between the (bounding) field lines

$$\frac{z^2}{2} = \lambda x \pm \sqrt{2E} \qquad (6.53)$$

and performs an oscillatory motion around the centre field line $z^2/2 = \lambda x$, on which w_y vanishes. This field line carries the guiding centre in the limit of small gyroradius. The orbits depend on a single parameter, λ. This parameter has a direct physical meaning: λ^2 is the ratio of the radius of curvature of the field lines and the gyroradius, both taken at $z = 0$ (Büchner and Zelenyi, 1989).

For small values of λ one expects relatively small chaotic effects, the chaotic regions in phase space are strongly localized. Over large sections the orbits are similar to those of the case where B_n vanishes. However, between these sections the orbit may switch between the different types, e.g., from a gyration orbit to the orbit shown in Fig. 6.1.

Orbits with large values of λ satisfy the condition that the gyroradius is small compared to the smallest scale length L of the field, such that the magnetic moment μ associated with that particle is an adiabatic invariant (Section 6.3). Thus, in the large λ limit we have an additional (approximate) constant of motion and the system becomes integrable.

Pronounced chaotic motion requires values of λ not too far from 1. Fig. 6.2 shows two particle orbits for $\lambda = 0.2$. The initial conditions are chosen such that the bounding and centre field lines coincide. In both cases the particle starts at the same point (with $z = 3$) on the centre field line. The remaining freedom is used to impose a 1% difference of initial velocity components for the two orbits. This small initial difference has a very large effect on the orbits: after the same run-time the upper orbit has returned into a gyrating mode (with a magnetic moment different from the initial one), while the lower orbit was reflected back into the chaotic domain near $z = 0$. The end

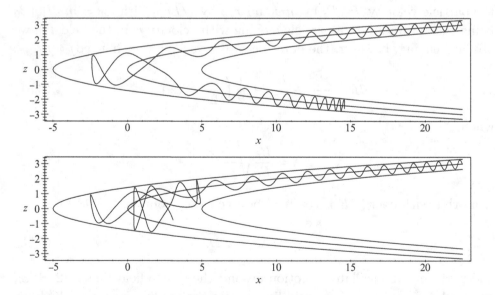

Fig. 6.2 Two particle orbits in a parabolic magnetic field with Hamiltonian (6.51) for $\lambda = 0.2$, illustrating strong dependence on initial conditions. The orbits are projected into the x, z-plane. Also shown are the field lines that bound the accessible domain and the centre field line, which are the same in both cases. The initial conditions differ only by a one per cent difference in the velocity.

points lie in entirely different regions. This is the signature of chaos that we emphasize in the present context.

Surfaces of section for orbits with the Hamiltonian (6.51) were studied by Büchner and Zelenyi (1989) and by Chen and Palmadesso (1986). See also Chapman and Watkins (1996). The role of chaotic motion in slowly time-dependent sheets is discussed in Section 8.5.

Due to the small field magnitude near $z = 0$ the particle is displaced by rather large distances along the y-axis. (The particle spends rather long periods of time on a given side of the centre field line, where w_y has a given sign.) If there is an additional y-component of the electric field, strong energetic coupling between chaotic particles and the electromagnetic field can arise. In a steady state field configuration particles that are energized during their meandering motion are ejected along field lines with high velocities. This effect was intensively studied by Speiser (1965), and the corresponding orbits are called *Speiser orbits*. This kind of coupling is also relevant for collisionless instabilities (see Section 10.4).

7

A unified theory of steady states

Earlier in this part we have encountered several different forms of equations describing steady states. The simplest case is magnetohydrostatics, physically richer versions include an anisotropic pressure tensor, directed flow or gravity.

In this chapter we consider a generalized formulation for a class of steady states that includes such generalizations. The method reduces the original fluid equations to two field equations, which can be understood as generalizations of the MHS equations (5.21) and (5.22).

In the first part the effect of an external gravity force is ignored; however, it will be incorporated in the second part. In all cases we assume that the magnetic field possesses Euler potentials. Again, it should be kept in mind that in space physics environments Euler potentials have considerable applicability (Section 5.1.2).

Remarkably, it turns out that in each of the cases that we consider the steady state problem can be reduced to solving the field equations for the Euler potentials. The form of these equations is uniquely determined by a single scalar function, which also serves as a Lagrangian generating the field equations.

However, it should be kept in mind that this is not a general theory of steady states. Besides the existence of Euler potentials this procedure excludes bulk flow perpendicular to the magnetic field. Is is only for a particular symmetric configuration that we show how a perpendicular flow component can be included (Section 7.3.2).

Here we outline the main properties of our unified formulation. More details and derivations are given in Appendix 1.

7.1 Steady states without gravity

Ignoring the gravity force we consider the following form of the momentum equation,

$$\nabla \cdot \mathcal{M} = 0 \tag{7.1}$$

with the stress tensor $\mathcal{M}$ having the structure

$$\mathcal{M} = R\mathcal{I} + S\boldsymbol{B}\boldsymbol{B} \tag{7.2}$$

where R and S are scalar expressions and $\mathcal{I}$ is the unit tensor.

The stress tensor (7.2) is oriented with respect to the direction of the magnetic field and keeps local isotropy around that direction. The main physical restrictions are absence of nonideal transport effects, such as viscous stresses, of radiation losses and of perpendicular flow. Pressure anisotropy and plasma flow parallel to the magnetic field are included.

The main aim is to derive the field equations, i.e., an appropriate set of differential equations for the Euler potentials α and β. For our procedure it is appropriate to understand R and S as functions of α, β and the magnitude of the magnetic field B, which formally replace the coordinates x, y, z. We assume that the Jacobian $|\boldsymbol{B} \cdot \nabla B|$ of that transformation will not vanish identically. (Cases where the Jacobian vanishes everywhere have a rather simple field structure and may be discussed by more explicit means.) Thus, R and S are assumed to be given in the form $R(\alpha, \beta, B)$ and $S(\alpha, \beta, B)$.

It turns out (see Appendix 1) that (7.1) implies that the functions R and S are subject to the constraint

$$R_B + B^2 S_B + SB = 0 \tag{7.3}$$

where here and throughout this chapter differentiation is indicated by subscripts. Physically, (7.3) states momentum balance in the direction of ∇B. Using that constraint one finds the remarkable fact that the problem (7.1) with (7.2) can be characterized by the single function

$$T = \boldsymbol{b} \cdot \mathcal{M} \cdot \boldsymbol{b} = R + B^2 S, \tag{7.4}$$

where $\boldsymbol{b} = \boldsymbol{B}/B$ is the unit vector in the magnetic field direction. We refer to T as the *steady state potential*. From the potential T, the functions R and S can be computed by partial differentiation,

$$S = \frac{T_B}{B} \tag{7.5}$$

$$R = T - BT_B, \tag{7.6}$$

which immediately follows from (7.3) and (7.4). The final form of $T(\alpha, \beta, B)$ has to satisfy appropriate constraints where the set of constraints consists of (7.5) or (7.6) (the other one being satisfied automatically) and additional constraints such as conservation laws.

Then the field equations for the Euler potentials assume the following form (see Appendix 1),

$$\nabla \beta \cdot \left[\nabla \times \left(\frac{T_B}{B} \nabla \alpha \times \nabla \beta \right) \right] + T_\alpha = 0$$

$$-\nabla \alpha \cdot \left[\nabla \times \left(\frac{T_B}{B} \nabla \alpha \times \nabla \beta \right) \right] + T_\beta = 0 \,.$$

(7.7)

Here, B, which so far played the formal role of a variable, is reinstalled in its original meaning $|\nabla \alpha \times \nabla \beta|$, such that the field equations become two (nonlinear) partial differential equations for α and β.

The field equations (7.7) can be derived from the variational principle

$$\delta \int T(\alpha, \beta, B) \, \mathrm{d}^3 r = 0$$

(7.8)

where the independent functions to be varied are α and β. The field equations (7.7) are the Euler–Lagrange equations of the variation problem (7.8). Thus, T also plays the role of a Lagrangian of the field equations.

The existence of a variational principle for steady state conditions is of considerable interest in connection with questions regarding the existence and the stability of solutions. For stability properties see Part III.

The present formulation is also well suited to incorporate additional constraints that are part of the original plasma model. For instance, this applies to mass conservation and entropy conservation in the presence of plasma bulk flow along the magnetic field (see the corresponding example discussed below).

7.1.1 Examples

Here we give the final form of the steady state potential $T(\alpha, \beta, B)$ for a number of different plasma models. With T known, the field equations are available from their general form (7.7). Details are given in Appendix 1.

Magnetohydrostatics with isotropic pressure

For MHS defined by (5.6)–(5.8) one finds

$$T(\alpha, \beta, B) = \frac{B^2}{2\mu_0} - p(\alpha, \beta).$$

(7.9)

The constraint (7.5) reduces to $p_B = 0$ which is already taken into account in (7.9).

Magnetohydrostatics with CGL pressure tensor

For a MHS model with a CGL pressure tensor $\mathcal{P} = \sum_s \mathcal{P}_s$, where $\mathcal{P}_s$ has the form of (3.69), one obtains

$$T(\alpha, \beta, B) = \frac{B^2}{2\mu_0} - p_{||}(\alpha, \beta, B). \tag{7.10}$$

Here (7.5) yields a relationship between $p_\perp$ and $p_{||}$ (Cowley, 1978),

$$p_{||} - p_\perp = B \frac{\partial p_{||}}{\partial B}. \tag{7.11}$$

If no further constraints are imposed, the function $p_{||}(\alpha, \beta, B)$ can be chosen arbitrarily, $p_\perp$ is then determined by (7.11).

Steady state with isotropic pressure and parallel flow

Here we consider the steady state version of ideal MHD, defined by (4.3)–(4.9) with $\boldsymbol{v}$ parallel to $\boldsymbol{B}$ and negligible gravity force. Parallel velocity implies $\boldsymbol{E} = 0$.

The continuity equation (4.3) is integrated and gives

$$\frac{\rho v_{||}}{B} = m(\alpha, \beta) \tag{7.12}$$

where $m(\alpha, \beta)$ is arbitrary. Similarly, the adiabatic law (4.6) gives $p/\rho^\gamma = k(\alpha, \beta)$ with arbitrary $k(\alpha, \beta)$. We generalize this property and assume that the pressure is an arbitrary function of the form

$$p = P(\rho, \alpha, \beta) . \tag{7.13}$$

Then one finds

$$T(\alpha, \beta, B) = \frac{B^2}{2\mu_0} - P(\rho, \alpha, \beta) - \frac{m(\alpha, \beta)^2 B^2}{\rho} . \tag{7.14}$$

Here, ρ is understood as a function of α, β, B, which is determined by the equation

$$\frac{B^2 m(\alpha, \beta)^2}{2\rho^2} + \int^\rho \frac{P_\rho(\rho, \alpha, \beta)}{\rho} \, d\rho = C(\alpha, \beta), \tag{7.15}$$

where $C(\alpha, \beta)$ is arbitrary. Equation (7.15) is Bernoulli's equation for the flow in each magnetic flux tube.

This model has several aspects that are standard in fluid dynamics, such as the appearance of Bernoulli's equation. Here, the emphasis is placed on embedding these fluid dynamics techniques into a self-consistent electromagnetic picture in a unified way.

If the adiabatic law (4.6) is replaced by the incompressibility law

$$\nabla \cdot \boldsymbol{v} = 0 \tag{7.16}$$

one finds from (7.16) and (4.3) that ρ is an arbitrary function of α and β and $v_{\parallel} = n(\alpha, \beta)B$ with $n(\alpha, \beta)$ arbitrary. Then the steady state potential becomes

$$T(\alpha, \beta\, B) = \frac{B^2}{2\mu_0} - \frac{1}{2}\rho(\alpha, \beta)n(\alpha, \beta)^2 B^2 - D(\alpha, \beta) \tag{7.17}$$

where the pressure was eliminated by Bernoulli's law, which here takes the form

$$p(\alpha, \beta) + \frac{1}{2}\rho(\alpha, \beta)n(\alpha, \beta)^2 B^2 = D(\alpha, \beta), \tag{7.18}$$

where $D(\alpha, \beta)$ is arbitrary.

7.2 Inclusion of the gravity force

To include a gravity force we replace (7.1) by

$$\nabla \cdot \mathcal{M} - \rho\nabla\psi = 0. \tag{7.19}$$

We will introduce functions $\hat{R}$, $\hat{S}$ and $\hat{T}$ instead of R, S and T, where the hat-label on a variable indicates that the variable formally is understood as a function of α, β, B, ψ. Accordingly, we define

$$\mathcal{M} = \hat{R}\mathcal{I} + \hat{S}\boldsymbol{B}\boldsymbol{B} \tag{7.20}$$
$$\hat{T} = \hat{R} + B^2\hat{S}. \tag{7.21}$$

The ψ-dependence of $\hat{T}$ is fixed by the condition

$$\hat{T}_{\psi} = \rho. \tag{7.22}$$

The conditions (7.5) and (7.6) are replaced by

$$\hat{S} = \frac{\hat{T}_B}{B} \tag{7.23}$$
$$\hat{R} = \hat{T} - B\hat{T}_B. \tag{7.24}$$

Again, only one of the two latter equations is needed, the other one is satisfied automatically.

If the function $\hat{T}(\alpha, \beta, B, \psi)$ satisfying all constraints has been determined, as shown in Appendix 1, the field equations then assume the form

$$\nabla\beta \cdot \left[\nabla \times \left(\frac{\hat{T}_B}{B} \nabla\alpha \times \nabla\beta \right) \right] + \hat{T}_\alpha = 0$$

$$-\nabla\alpha \cdot \left[\nabla \times \left(\frac{\hat{T}_B}{B} \nabla\alpha \times \nabla\beta \right) \right] + \hat{T}_\beta = 0.$$

(7.25)

Here, for B and ψ one has to insert $|\nabla\alpha \times \nabla\beta|$ and the known gravity potential $\psi(\,r)$, respectively, so that again one obtains two (nonlinear) partial differential equations for α and β.

The variational principle, from which (7.25) may be derived, is

$$\delta \int \hat{T} \, \mathrm{d}^3 r = 0 \,,$$

(7.26)

where again the two Euler potentials are the functions to be varied independently.

As an illustration of this formalism we revisit two earlier examples and include a gravity force.

MHS with gravity

From the model equations (5.1)–(5.3) one readily finds

$$\hat{T}(\alpha, \beta, B, \psi) = \frac{B^2}{2\mu_0} - p(\alpha, \beta, \psi)$$

(7.27)

with

$$\rho = -\frac{\partial p}{\partial \psi}.$$

(7.28)

If one uses (7.28) to determine ρ, the only condition on the function $p(\alpha, \beta, \psi)$ is that the derivative with respect to ψ is negative.

MHD with parallel flow and gravity

As shown in Appendix 1 we find in this case

$$\hat{T}(\alpha, \beta, B, \psi) = \frac{B^2}{2\mu_0} - P(\rho, \alpha, \beta) - \frac{m(\alpha, \beta)^2 B^2}{\rho}.$$

(7.29)

Here, ρ is a function of α, β, B, ψ determined by Bernoulli's equation, which now reads

$$\psi + \frac{B^2 m^2}{2\rho^2} + \int^\rho \frac{P_\rho}{\rho} \, \mathrm{d}\rho = C(\alpha, \beta),$$

(7.30)

where $C(\alpha, \beta)$ is arbitrary.

7.3 Symmetric states

In Section 5.2 we have described steady state theory for symmetric magnetohydrostatic states. Here we generalize that description from the present unified point of view. Details and derivations are given in Appendix 1. Here we only give the unified form of the Grad–Shafranov equation for translational and rotational invariance. The case of helical invariance is given in Appendix 1. The gravity force is ignored, it can be included by the general method outlined above.

As in Section 5.2 we assume translational invariance with respect to the Cartesian z-component and write the magnetic field as

$$\boldsymbol{B} = \nabla A(x,y) \times \boldsymbol{e}_z + B_z(x,y)\boldsymbol{e}_z, \tag{7.31}$$

where $A(x,y)$ is the magnetic flux function, and observe that $\boldsymbol{B}$ may also be expressed by the Euler potentials

$$\alpha = A(x,y) \tag{7.32}$$

$$\beta = z + \tilde{\beta}(x,y), \tag{7.33}$$

where

$$B_z = \boldsymbol{e}_z \cdot \nabla A \times \nabla \tilde{\beta}. \tag{7.34}$$

Symmetry requires that the steady state potential T is independent of z which implies $T = T(A, B)$. From the field equations we find (unless ∇A vanishes identically)

$$B_z = \frac{BG(A)}{T_B} \tag{7.35}$$

where $G(A)$ is arbitrary and the unified Grad–Shafranov equation

$$-\nabla \cdot \left(\frac{T_B}{B}\nabla A\right) - \frac{B}{T_B}G(A)G'(A) + T_A = 0. \tag{7.36}$$

For magnetohydrostatics with isotropic pressure this equation reduces to equation (5.78).

For rotational invariance we use cylindrical coordinates r, φ, z and as in Section 5.2.3 assume that the observables do not depend on φ. In this case the Euler potentials are chosen as

$$\alpha = U(r,z) \tag{7.37}$$

$$\beta = \varphi + \tilde{\beta}(r,z), \tag{7.38}$$

implying that

$$B_\varphi = e_\varphi \cdot \nabla U \times \nabla \tilde{\beta} . \qquad (7.39)$$

The steady state potential has the form $T(U, B)$. We obtain

$$B_\varphi = \frac{K(U)}{rS} \qquad (7.40)$$

where $K(U)$ is arbitrary, and the unified form of the Grad–Shafranov equation becomes

$$-\frac{1}{r}\frac{\partial}{\partial r}\left(\frac{T_B}{rB}\frac{\partial U}{\partial r}\right) - \frac{1}{r^2}\frac{\partial}{\partial z}\left(\frac{T_B}{B}\frac{\partial U}{\partial z}\right) - \frac{BK(U)K'(U)}{r^2 T_B} + T_U = 0 . \qquad (7.41)$$

Helical invariance is discussed in Appendix 1. Inclusion of a gravity force leads to the same field equations as found without gravity except for replacing $T(\alpha, \beta, B)$ by $\hat{T}(\alpha, \beta, B, \psi)$ and determining the ψ-dependence of $\hat{T}$ by the condition $\hat{T}_\psi = \rho$.

7.3.1 Gravity supported sheets revisited

For an example we revisit the problem of magnetohydrostatic structures supported by gravity. A simple explicit model was given in Section 5.3.6.

Consider a configuration with translational invariance with respect to the Cartesian z-coordinate. For this case (7.27) and (7.28) reduce to

$$\hat{T}(A, B, \psi) = \frac{B^2}{2\mu_0} - p(A, \psi) \qquad (7.42)$$

with

$$\rho = -\frac{\partial p(A, \psi)}{\partial \psi} . \qquad (7.43)$$

The field equation for A is found from (7.36) as

$$\frac{1}{\mu_0}\Delta A + \frac{d}{dA}\frac{B_z(A)^2}{2\mu_0} + p_A = 0 . \qquad (7.44)$$

After inserting $|\nabla A|$ for B and the actual gravity potential $\psi(x, y)$ for ψ, the differential equation (7.36) can be solved for $A(x, y)$.

An interesting simplification consists in setting $\psi = -gy$ and choosing constant values for B_x and B_z. Then $A(x, y) = \tilde{A}(x) + B_x y$. Further let us choose $p(A, \psi) = F(A + B_x\psi/g)$. Then, p reduces to $p = F(\tilde{A}(x))$. The y-dependence has dropped out and the field equation (7.44) becomes an equation for $\tilde{A}(x)$. For any monotonically decreasing function F a sheet structure arises in which the y-component of the Lorentz force balances the gravity force. The special choice

$$F(\tilde{A}) = \frac{B_0^2}{2\mu_0} \, e^{-2\frac{\tilde{A}}{B_0 d}} \qquad (7.45)$$

leads to the particular sheet model discussed in Section 5.3.6.

7.3.2 A case with a perpendicular flow component

The unified formulation, as discussed so far, restricts the plasma bulk velocity to be directed parallel to the magnetic field. Here we present a case, where it is possible to include a perpendicular velocity component. The geometry is restricted to translational invariance with respect to the Cartesian z-coordinate. The magnetic field and the plasma bulk velocity are decomposed in their poloidal (x, y) and their toroidal (z)-components,

$$\boldsymbol{B} = \boldsymbol{B}_{\mathrm{p}} + B_z \boldsymbol{e}_z, \qquad \boldsymbol{v} = \boldsymbol{v}_{\mathrm{p}} + v_z \boldsymbol{e}_z, \qquad (7.46)$$

with the familiar representation $\boldsymbol{B}_{\mathrm{p}} = \nabla A \times \boldsymbol{e}_z$. For concreteness let us assume that the perpendicular flow stems from ideal Ohm's law (4.5). Present symmetry would imply an electric potential of the form $\phi = \Phi(x, y) - E_z z$. Choosing $E_z = 0$ gives

$$\boldsymbol{v} \times \boldsymbol{B} = \nabla \Phi(x, y). \qquad (7.47)$$

Models with these properties typically generate stress tensors of the type

$$\mathcal{M} = \hat{R}\mathcal{I} + \hat{S}\boldsymbol{B}_{\mathrm{p}}\boldsymbol{B}_{\mathrm{p}} + \hat{V}(\boldsymbol{B}_{\mathrm{p}}\boldsymbol{e}_z + \boldsymbol{e}_z \boldsymbol{B}_{\mathrm{p}}) + \hat{U}\boldsymbol{e}_z\boldsymbol{e}_z . \qquad (7.48)$$

We follow the same procedure as in the case of parallel flow with a gravity force included, except that we use variables $A, \boldsymbol{B}_{\mathrm{p}}, \psi$ instead of A, B, ψ (see also Appendix 1). Correspondingly, we define

$$\hat{T}(A, B_{\mathrm{p}}, \psi) = \hat{R}(A, B_{\mathrm{p}}, \psi) + B_{\mathrm{p}}^2 \hat{S}(A, B_{\mathrm{p}}, \psi), \qquad (7.49)$$

fixing the ψ-dependence by

$$\hat{T}_\psi = \rho. \qquad (7.50)$$

Then one finds

$$-\nabla \cdot \left(\frac{\hat{T}_{B_{\mathrm{p}}}}{B_{\mathrm{p}}} \nabla A \right) + \hat{T}_A = 0 \qquad (7.51)$$

$$\hat{V} = \hat{V}(A). \qquad (7.52)$$

The coefficient $\hat{U}$ has dropped out of the problem because $\nabla \cdot (\hat{U} e_z e_z)$ vanishes. Besides (7.50), the function $\hat{T}$ has to satisfy the condition

$$\frac{\hat{T}_{B_\mathrm{p}}}{B_\mathrm{p}} = \hat{S}. \tag{7.53}$$

For symmetric static fields (7.51) is equivalent to (7.36); the different structure results from using the variables A, B_p in (7.51) and A, B in (7.36). Additional constraints such as conservation laws have to be incorporated also.

To illustrate this procedure let us deal with a steady state ideal MHD model. (Ideal Ohm's law has already been anticipated.) Let $\rho = \rho(A)$ and $\psi = \psi(x, y)$. The continuity equation is integrated to represent v_p by a scalar function D,

$$v_\mathrm{p} = \frac{1}{\rho} \nabla D \times e_z. \tag{7.54}$$

Ohm's law (7.47) implies that $D = D(A)$ and $\Phi = \Phi(A)$

$$v_z - \frac{1}{\rho} D_A B_z = \Phi_A. \tag{7.55}$$

From (7.48) specialized for the present example, one then finds

$$\frac{B_z}{\mu_0} - D_A v_z = \hat{V}. \tag{7.56}$$

From (7.55) and (7.56) we conclude that $B_z = B_z(A)$ and $v_z = v_z(A)$. The condition (7.53) gives Bernoulli's equation in the form

$$p + \frac{B_z^2}{2\mu_0} + \frac{\rho v_\mathrm{p}^2}{2} + \rho\psi = C(A) \tag{7.57}$$

where (7.50) was used also. The final form of $\hat{T}$ with all constraints taken into account is obtained as

$$\hat{T}(A, B_\mathrm{p}, \psi) = \frac{B_\mathrm{p}^2}{2\mu_0} - \frac{D_A(A)^2 B_\mathrm{p}^2}{2\rho(A)} + \rho(A)\psi - C(A) \tag{7.58}$$

and the field equation (7.51) takes the form

$$\left(1 - \frac{\mu_0 D_A^2}{\rho}\right) \Delta A - \frac{(\nabla A)^2}{2} \frac{\mathrm{d}}{\mathrm{d}A}\left(\frac{\mu_0 D_A^2}{\rho}\right) - \mu_0 \psi(x, y)\frac{\mathrm{d}\rho}{\mathrm{d}A} + \mu_0 \frac{\mathrm{d}C}{\mathrm{d}A} = 0. \tag{7.59}$$

This equation was derived directly from the MHD equations by Tsinganos (1981). (In a comparison one should be aware that the functions $C(A), \hat{V}(A), D(A)$ are defined differently in Tsinganos' work.) Goedbloed (2004) obtained a corresponding equation from a variational approach devised for axisymmetric systems.

8

Quasi-static evolution and the formation of thin current sheets

8.1 Introduction

In many ways the simplest description of a quiescent plasma state is provided by a steady state model. We dealt with such models in detail in previous chapters, especially emphasizing static states. However, as discussed in Chapter 4, we cannot expect that exact steady states exist in nature, a more realistic picture of quiescence being that of slow temporal evolution. A convenient description of such a system is a temporal sequence of (approximate) steady state configurations. Each member of such a quasi-steady sequence describes a snapshot of the system taken at a particular time during its evolution (Chapter 4).

An important aspect of slow evolution is that in a set of cases relevant for space and astrophysics the evolution leads to the formation of thin current sheets (TCS). That formation is spontaneous in the sense that the small length scale associated with thin current sheets is not present in the external driving forces. Thus we are dealing with spontaneous formation of structure.

As it will turn out, thin current sheets seem to play a crucial role in transitions from quiescence to activity (Parker, 1972; Priest, 1981; Parker, 1994). Parker has argued that braiding of coronal magnetic fields caused by foot-point motion leads to the formation of tangential discontinuities, which cause dynamic resistive dissipation even for extremely small dissipation. These processes are believed to play an important role in solar activity. Theoretical models (e.g., Birn *et al.*, 1998a) as well as observations (e.g., Kaufmann, 1987) have led to the conclusion that thin current sheets also play a central role in magnetospheric activity. This chapter concentrates on the formation of thin current sheets, their role in the transitions from quiescent to active states will be discussed in Part III.

As we have seen, a characteristic property of steady state solutions is the large freedom that they offer for the choice of parameters and free functions. For instance, the general form of asymptotic tail solutions (5.161) contains three free functions, $z_0(x_1)$, $\hat{p}(x_1)$ and $p(A)$. As discussed, part of that freedom allows us to impose boundary conditions. The remaining freedom provides the possibility to satisfy constraints, such as conservation laws.

For a reminder, let us look at mass conservation expressed by the continuity equation, which we repeat here for convenience,

$$\frac{\partial \rho}{\partial t} + \nabla \cdot (\rho \boldsymbol{v}) = 0 \,. \tag{8.1}$$

For the purpose of a snapshot of a quasi-static equilibrium sequence, this equation, to lowest order, is satisfied trivially as $\partial/\partial t$ and $\boldsymbol{v}$ vanish in that order. Equation (8.1), however, can no longer be ignored for a quasi-static sequence if we consider a large time interval, during which appreciable changes of density ρ occur. If we scale time variation and velocity with a parameter δ, such that $\partial \rho/\partial t = O(\delta)$, $v = O(\delta)$, the long time interval τ for appreciable changes is of order $\tau = O(1/\delta)$.

Integrating (8.1) over a domain D with surface ∂D, which has the outward pointing unit normal $\boldsymbol{n}$ and moves with velocity $\boldsymbol{u}$, one finds

$$\frac{\mathrm{d}}{\mathrm{d}t} \int_{\mathrm{D}} \rho \mathrm{d}^3 r + \int_{\partial \mathrm{D}} \rho \boldsymbol{n} \cdot (\boldsymbol{v} - \boldsymbol{u}) \mathrm{d}^2 r = 0 \,. \tag{8.2}$$

For vanishing outflow, represented by the surface integral, total mass is conserved. Here, mass conservation remains to be a nontrivial property even for arbitrarily small values of δ. As we will see below, for ideal MHD with appropriate boundary conditions mass conservation may be formulated separately for each magnetic flux tube.

For the full set of MHD quantities, one finds the following scaling for quasi-static evolution,

$$p, \boldsymbol{B}, \rho, \nabla = O(1) \tag{8.3}$$

$$\boldsymbol{j}, \boldsymbol{v}, \boldsymbol{E}, \frac{\partial}{\partial t} = O(\delta) \,. \tag{8.4}$$

The zeroth order quantities obey the static equilibrium equations.

In certain cases a conservation law can be replaced by connecting the system to a large reservoir, which fixes the value of a quantity such as pressure or density.

8.2 Perturbed Harris sheet

Consider a finite portion of a Harris sheet (Section 5.3.2) with flux function $A_0(z)$, enclosed in a rectangular box of dimensions $(0 \leq x \leq l, -s \leq z \leq s)$ in the x, z-plane (upper panel of Fig. 8.1). Note that the current density is bounded (see Fig. 5.5).

Now, consider a two-dimensional quasi-static deformation of that system following Hahm and Kulsrud (1985). This may be achieved by magnetic flux transfer through the boundary, described by changing the values of A on the unperturbed upper and lower boundaries. The perturbation vanishes on the lateral boundaries.

For simplicity, we assume the following symmetry properties with respect to the axis $(z = 0)$. Let A be symmetric, i.e., $A(x, -z) = A(x, z)$, which implies antisymmetry for B_x, and let v_x be symmetric and v_z antisymmetric. The present gauge implies $E_y = -\partial A / \partial t$ (for an explanation see Section 8.3.1). We set $v_y = 0$, $B_y = 0$. As we will see, it makes a qualitative difference whether or not Ohm's law in its ideal form (3.60) is imposed.

Let us first allow for a general nonideal form of Ohm's law, which allows $E \neq 0$ at the neutral sheet (case 1). In a first step, we consider only the upper half of the box $(z \geq 0)$. For the perturbed (half-box) problem we then have the following boundary conditions:

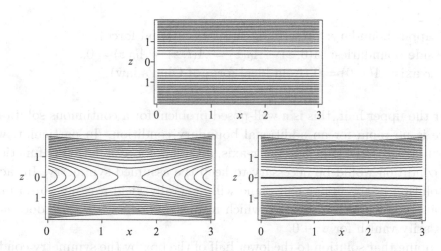

Fig. 8.1 Magnetic field lines of a Harris sheet without (top panel) and with boundary perturbation for box dimensions $l = 3$, $s = 2$. The nonideal continuous solution (lower left) shows changing magnetic topology. Imposing Ohm's law in ideal form (lower right) keeps magnetic topology but causes a singular current sheet (thick line).

case 1

upper boundary: prescribing $A(x, s)$ (external force),

side boundaries: $A(0, z) - A_0(z) = A(l, z) - A_0(z) = 0$,

x-axis: $\partial A/\partial z = 0$ for $z = 0$ (symmetry, assuming continuous B_x).

In a second step we map the solution to the lower half of the box, again using symmetry. Obviously, this generates a continuous solution. It has the structure shown in the lower left panel of Fig. 8.1. The perturbation has changed the magnetic topology in that the field lines connect to the boundary in a different way as compared to the unperturbed sheet.

Let us now impose ideal Ohm's law (3.60) (case 2). This case is qualitatively different from the previous case, because it requires that the electric field vanishes at the neutral sheet.

Ideal Ohm's law implies that the flux function A is a constant of the motion. This is a consequence of magnetic flux conservation and follows directly from the y-component of (3.60), which gives $\partial A/\partial t + \boldsymbol{v} \cdot \nabla A = 0$. Thus, the moving plasma sees a constant value of A. Since the symmetry requires that v_z vanishes on the axis, the plasma that was situated on the axis originally will stay on the axis. Since A is a constant of the motion, A remains zero on the axis as the perturbation is applied. This boundary condition replaces the vanishing of the normal derivative of case 1. Thus, for ideal MHD we find the boundary conditions:

case 2

upper boundary: prescribing $A(x, s)$ (external force),

side boundaries: $A(0, z) - A_0(z) = A(l, z) - A_0(z) = 0$,

x-axis: $A(x, 0) = 0$ (from ideal form of Ohm's law).

For the upper half, this is a well-posed problem for a continuous solution. There is no room for an additional boundary condition. In particular, we cannot impose $\partial A/\partial z = 0$ on the x-axis, although for a smooth flux function that condition would be necessary to keep the assumed symmetry. In fact, the solution in the upper half plane will generally yield a non-zero normal derivative as z approaches zero, which means that $B_x = -\partial A/\partial z$ does not necessarily vanish for $z \to 0$.

Mapping that solution to the lower half of the box by the symmetry condition $B_x(x, -z) = -B_x(x, z)$ results in a jump of the (tangential component of the) magnetic field at $z = 0$. Thus, the x-axis carries a tangential discontinuity. The absolute value of the sheet current is given by $|2B_{\lim}/\mu_0|$, where $B_{\lim}$ denotes the one-sided limit of $|B_x|$ for z approaching zero. Figure 8.1

illustrates such a solution in the lower right panel. The singular current sheet is marked by the thick horizontal line.

The solution either violates continuity (lower right panel of Fig. 8.1) or conservation of magnetic topology (lower left panel). If the conservation of magnetic topology is a built-in property of plasma dynamics (by Ohm's law in ideal form), a continuous solution is not available and a singular current sheet forms. If, on the other hand, Ohm's law contains a suitable nonideal term, the system can be expected to relax to a continuous solution with a change in magnetic topology. Note that this picture is consistent with Parker's general hypothesis (Parker, 1994) on the formation of tangential discontinuities.

Clearly, in a real plasma dissipative terms will become available when critical parameter conditions are reached. (This property is discussed in Chapter 9.) Thus, as the boundary perturbation is applied, a thin current sheet will build up. Beyond criticality, the dissipation will lead to a change of topology. Since the quasi-static evolution and the relaxation are different physical processes, they can be expected to occur on different time scales.

Although, strictly speaking, the state shown in the lower right panel of Fig. 8.1 will never be reached, the concept of such a state as a limiting case is valuable to understand the behaviour of real plasmas, where the current density becomes locally large.

Imposing inhomogeneous Dirichlet boundary conditions applied to the side boundaries (i.e., prescribing $A(0, z)$ and $A(l, z)$), also leads to singular current sheets even if the upper boundary remains unperturbed.

The current sheet formation in a perturbed Harris sheet was confirmed by a time-dependent treatment of the perturbation (Hahm and Kulsrud, 1985). It was shown that an Alfvén wave steepens at $z = 0$ because of the singularity in the Alfvén velocity. Nonlinear numerical computations have confirmed and extended these findings (Voge *et al.*, 1994; Rastätter *et al.*, 1994).

Example

A simple explicit example may illustrate the properties outlined above (Schindler and Birn, 1993). Consider a linear perturbation $a(x, z)$ of the flux function A_0 of a Harris sheet

$$A(x, z) = A_0(z) + a(x, z) \qquad (8.5)$$

with $A_0(z) = \ln(\cosh z)$. Inserting (8.5) into the Grad–Shafranov equation (5.78), we find after linearization and using dimensionless variables

$$-\Delta a = \frac{\mathrm{d}^2 p_0(A_0)}{\mathrm{d}A_0^2} a + \frac{\mathrm{d}p_1(A_0)}{\mathrm{d}A_0} , \qquad (8.6)$$

where the last term is due to a possible perturbation of the pressure function $p(A)$, which, however, is set to zero. This choice can be justified by the fact that all field lines intersect the vertical boundaries. They are assumed to connect to some region where the perturbation becomes negligible. There, the pressure $p(A)$ coincides with the unperturbed pressure $p_0(A)$. Since p is constant on field lines, $p(A)$ equals $p_0(A)$ everywhere. Thus, instead of determining the pressure from entropy conservation, we have simply used the notion of a pressure reservoir (for each flux tube separately). A corresponding argument can be applied to density.

Inserting the expression for the pressure function of the Harris sheet (5.120) into (8.6), we obtain (in dimensionless form)

$$\Delta a(x, z) + \frac{2}{\cosh^2 z} a(x, z) = 0 . \tag{8.7}$$

This equation can be solved in closed form using separation of variables and a Fourier representation of the x-dependence, incorporating the lateral boundary conditions,

$$a(x, s) = \sum_{n=1}^{\infty} t_n \sin(k_n x), \quad k_n = \pi n/l. \tag{8.8}$$

One finds

$$a(x, z) = \sum_{n=1}^{\infty} t_n \frac{v_n(z)}{v_n(s)} \sin(k_n x) . \tag{8.9}$$

The function $v_n(z)$ is different for the two cases,

case 1:

$$v_n(z) = \sinh(k_n z) \tanh(z) - k_n \cosh(k_n z) , \tag{8.10}$$

case 2:

$$v_n(z) = \cosh(k_n |z|) \tanh(|z|) - k_n \sinh(k_n |z|) . \tag{8.11}$$

The expression (8.9) with (8.10) and (8.11) have been used to compute the perturbed solutions of Fig. 8.1, where t_1 was set to 0.2, all other t_n vanishing.

The surface current density of the singular current sheet arising in case 2 is given by the jump of B_x on the axis (Section 5.1.6)

$$K(x) = -2 \sum_{n=1}^{\infty} \frac{t_n}{v_n(s)} \left(1 - k_n^2\right) \sin(k_n x) . \tag{8.12}$$

Note that the response $a(x, z)$ to a given boundary condition represented by the coefficients t_n may differ strongly, depending on parameter values. In the neighbourhood of a zero of one of the denominators $v_n(s)$ in (8.9) the perturbation formally becomes arbitrarily large and linear theory no longer gives satisfactory answers. We will return to this property later in connection with bifurcation theory. To emphasize this aspect the parameters of the case plotted in Fig. 8.1 are chosen such that the response to the boundary perturbation is rather strong. The wave number is $\pi/3$, the singularity formally would occur at wave number 0.92 for case 1 and 1.00 for case 2. The surface current density (8.12) remains bounded as the singularity is approached. Similar expressions have been derived for the case, where inhomogeneous Dirichlet conditions hold at the lateral boundaries.

We add the solution of (8.7) for the ideal MHD case, where Dirichlet conditions are prescribed at one of the lateral boundaries, say at $x = 0$, the other boundary moved to infinity. Then the boundary conditions are

case 3

> upper boundary: $a(x, s) = 0$,
> side boundaries: $a(0, z) = \sum_{n=1}^{\infty} r_n w_n(z)$, $a(l, z) \to 0$ for $l \to \infty$,
> x-axis: $a(x, 0) = 0$ (from ideal form of Ohm's law),

where

$$w_n(z) = \tanh z \cos(\kappa_n z) + \kappa_n \sin(\kappa_n z). \tag{8.13}$$

The solution is

$$a(x, z) = \sum_{n=1}^{\infty} r_n w_n(z) \, e^{-\kappa_n x}, \tag{8.14}$$

where the numbers κ_n are the solutions of

$$\tanh s + \kappa \tan(\kappa s) = 0. \tag{8.15}$$

As before, a singular sheet current forms on the x-axis with density

$$K(x) = -2 \sum_{n=1}^{\infty} r_n (1 + \kappa_n^2) \, e^{-\kappa_n x}. \tag{8.16}$$

There is no singularity in the expansion in this case.

One-dimensional adiabatic compression of a Harris sheet can also lead to thin or singular current sheets. Examples include cases with plasma losses or with expansion along the direction of the magnetic field. (For details see www.tp4.rub.de/~ks/tc.pdf.)

8.3 MHD tail model with flux transfer

As we have seen in the previous section, exploring the perturbed Harris sheet is a rather straightforward matter with clear-cut results on the formation of TCS. However, if the starting configuration is two- or three-dimensional and if the evolution is allowed to be nonlinear, the situation is far more complicated. Particularly, this applies to magnetic fields that, although becoming weak locally, do not go to zero. Then, as we will see, the resulting structure and, in particular, current sheet formation crucially depend on details of boundary and initial conditions. Notably, there exist regimes where the solution remains smooth with no indication of current sheet formation. In this section we illustrate these aspects for a magnetospheric tail model.

8.3.1 Model and procedure

Here we take a look at quasi-static sequences with translational invariance with respect to y.

We begin with a remark on the choice of the electromagnetic gauge. As in the static case, we will use a gauge where the electromagnetic potentials are independent of y, however, here we allow for time-dependence. In contrast to the static case, the time-dependence makes this gauge generally applicable. The y-component of the electric field does no longer have to be excluded, because it is generated by the time-dependence of the y-component of the vector potential, which coincides with the flux function A. Note that there is still the freedom of adding an arbitrary function of time to A.

The model consists of a sequence of states, each of which is an asymptotic magnetohydrostatic solution for small aspect ratios as derived in Section 5.4.1. Here is a brief outline of the model, which is described in detail elsewhere (Schindler and Birn, 1982).

To lowest order, and assuming symmetry with respect to $z = 0$, the flux function $A(x, z, t)$ is implicitly given by

$$z = \int_{\hat{A}(x,t)}^{A} \frac{\mathrm{d}A'}{\sqrt{2\mu_0(\hat{p}(x,t) - p(A',t))}}, \tag{8.17}$$

where $\hat{A}(x, t)$ is determined by $\hat{p}(x, t)$ and $p(A, t)$ through the equation

$$p(\hat{A}(x,t), t) = \hat{p}(x, t). \tag{8.18}$$

The dependence on x is weak in accordance with the assumption of a small aspect ratio.

Formally, (8.17) can be regarded a solution of local pressure balance

$$\frac{1}{2\mu_0}\left(\frac{\partial A(x,z,t)}{\partial z}\right)^2 + p(A,t) = \hat{p}(x,t), \tag{8.19}$$

where $p(A,t)$ and $\hat{p}(x,t)$ are arbitrary positive functions, which can be used to accommodate additional constraints. A given value of A, at time t, generates a magnetic field line in the x-interval where $\hat{p}(x,t) \geq p(A,t)$. The pressure balance results from the Grad–Shafranov equation (5.78) to lowest order in the aspect ratio.

Considering continuous fields, symmetry requires that $\partial A/\partial z$ vanishes for $z = 0$. In view of (8.18) and (8.19) this boundary condition may be expressed as $A(x,0,t) = \hat{A}(x,t)$.

One possible way of taking into account a boundary condition at the magnetopause is to specify the function $\hat{p}(x,t)$. This is because, typically, magnetic flux transfer to the magnetotail is associated with an increase of the magnetic field strength B_l in the tail lobes. By pressure balance (8.19), B_l is related to $\hat{p}$. Thus, specifying $B_l(x,t)$ determines $\hat{p}(x,t) = B_l(x,t)^2/(2\mu_0)$, if the lobe pressure is ignored as being small. This reasoning suggests choosing $\hat{p}(x,t)$ as an appropriate external signature driving the temporal evolution.

Fig. 8.2 shows a typical flux tube in the region considered ($x \geq 0$). In view of the strong convergence of field lines on the left the flux tube is treated as closed at $x = 0$.

Under the present conditions the ideal MHD model implies that mass conservation holds for each flux tube separately. This follows by applying (8.2) to an arbitrary flux tube of the type shown in Fig. 8.2. One might wonder how this is possible, as the velocity $\boldsymbol{u}$ is to be identified with the velocity of the bounding magnetic field lines, which, however, is not defined unambiguously. Here the ambiguity stems from the freedom of adding an

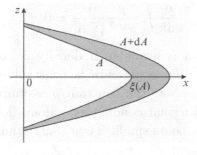

Fig. 8.2 Magnetic flux tube A of width $\mathrm{d}A$ in two dimensions. For conservation aspects the flux tube is considered to be closed at $x = 0$.

arbitrary function of time to a particular choice of the flux function. To show that mass conservation is not subject to this ambiguity (i.e., gauge invariant) let us replace $A(x, z)$ by $A(x, z) + g(t)$, where $g(t)$ is arbitrary. A magnetic field line is then identified by a fixed value of $A(x, z) + g(t)$ and the field line velocity $\boldsymbol{u}$ satisfies the equation

$$\frac{\partial A}{\partial t} + \frac{\partial g}{\partial t} + \boldsymbol{u} \cdot \nabla A = 0 . \tag{8.20}$$

A corresponding equation for the plasma velocity $\boldsymbol{v}$ follows from the y-component of Ohm's law (4.5),

$$\frac{\partial A}{\partial t} + \frac{\partial g}{\partial t} + \boldsymbol{v} \cdot \nabla A = 0 . \tag{8.21}$$

Ignoring the end sections of the flux tube boundary on the z-axis, the normal is $\boldsymbol{n} = \nabla A / |\nabla A|$ such that by (8.20) and (8.21) one finds $(\boldsymbol{v} - \boldsymbol{u}) \cdot \boldsymbol{n} = 0$, independent of the choice of $g(t)$. Thus, mass conservation holds for each flux tube separately, i.e.,

$$\frac{\mathrm{d}}{\mathrm{d}t} \int_A \rho \frac{\mathrm{d}s}{|\nabla A|} = 0 , \tag{8.22}$$

where the integral is a line integral along the field line section lying in the region $x \geq 0$. The denominator $|\nabla A|$ is the Jacobian of the coordinate transformation $(x, z) \rightarrow (A, s)$, where s is the arclength on field lines.

Correspondingly, an adiabatic law (4.6) holds for each flux tube also. This follows by combining (4.6) with (4.3) to give the adiabatic law in conservation form

$$\frac{\partial}{\partial t} \left(\frac{p}{\rho^{\gamma-1}} \right) + \nabla \cdot \left(\frac{p}{\rho^{\gamma-1}} \boldsymbol{v} \right) = 0 , \tag{8.23}$$

which can be handled in the same way as mass conservation. One finds

$$\frac{\mathrm{d}}{\mathrm{d}t} \int_A \frac{p}{\rho^{\gamma-1}} \frac{\mathrm{d}s}{|\nabla A|} = 0 . \tag{8.24}$$

A useful simplification is available if density is constant on field lines such that $\rho = \rho(A, t)$. Since pressure is constant on field lines in each magnetohydrostatic snapshot, the constant-ρ assumption can be justified by the notion of high thermal conductance along field lines, which keeps parallel temperature gradients small. Then (8.22) and (8.24) simplify as

$$\frac{\mathrm{d}}{\mathrm{d}t}(\rho V) = 0, \qquad \frac{\mathrm{d}}{\mathrm{d}t} \left(\frac{pV}{\rho^{\gamma-1}} \right) = 0 , \tag{8.25}$$

where V is the (differential) flux tube volume

$$V = \int_A \frac{\mathrm{d}s}{|\nabla A|} \, . \tag{8.26}$$

The volume of a flux tube of width $\mathrm{d}A$ is $V\mathrm{d}A$. (For simplicity, we keep the notion of a 'volume', although in the present two-dimensional geometry we are dealing with an area.)

Eliminating ρ between the two equations of (8.25) one finds the familiar thermodynamic adiabatic law, valid for each flux tube A,

$$\frac{\mathrm{d}}{\mathrm{d}t}\left(pV^\gamma\right) = 0 \, . \tag{8.27}$$

The adiabatic law (8.27) determines the pressure function $p(A, t)$ for a given driver $\hat{p}(x, t)$. To show this, we first write (8.26) in the form

$$V(A, t) = \int_0^{\xi(A,t)} \frac{\mathrm{d}x}{|B_x|}$$

$$= \int_0^{\xi(A,t)} \frac{\mathrm{d}x}{\sqrt{2\mu_0(\hat{p}(x, t) - p(A, t))}} \, . \tag{8.28}$$

Here $\xi(A, t)$ is the x-coordinate of the point (vertex) where at time t the field line A intersects the x-axis (see Fig. 8.2), and pressure balance (8.19) was used to replace B_x. Understanding V as a function of p and t and changing the integration variable from x to $\hat{p}$, one finds

$$V(p, t) = \frac{1}{\sqrt{2\mu_0}} \int_p^{\hat{p}(0,t)} \frac{\mathrm{d}\hat{p}}{f(\hat{p}, t)\sqrt{\hat{p} - p}} \, , \tag{8.29}$$

where $-\partial\hat{p}/\partial x$ is expressed as $f(\hat{p}, t)$.

By time-integrating (8.27) and using (8.29) one obtains

$$p^{1/\gamma} \int_p^{\hat{p}(0,t)} \frac{\mathrm{d}\hat{p}}{f(\hat{p}, t)\sqrt{\hat{p} - p}} = M(A) \, , \tag{8.30}$$

where $M(A)$ is determined by the initial state. Solving (8.30) for p gives the pressure function $p(A, t)$.

Inserting $p(A,t)$ together with the specified driver $\hat{p}(x,t)$ into (8.17) and solving that equation for A completes the construction of the quasi-static sequence $A(x,z,t)$ as a result of varying lobe field strength. The latter enters (8.30) through the function $f(\hat{p},t)$.

8.3.2 Selfsimilar solutions

Here we look at the time-evolution of tail solutions $A(x,z,t)$ satisfying (8.17) that can be understood simply by selfsimilar stretching or compression of an initial configuration $A_0(x,z)$. Accordingly, the flux function has the form

$$A(x,z,t) = A_0(\tilde{x},\tilde{z}), \qquad \tilde{x} = \Lambda(t)x, \; \tilde{z} = \Gamma(t)z, \tag{8.31}$$

with $\Lambda(0) = \Gamma(0) = 1$. Further, each physical quantity Q factorizes as

$$Q(x,z,t) = Q_1(t)Q_0(\tilde{x},\tilde{z}). \tag{8.32}$$

The first task is to consider the variables describing the quasi-static snapshots ($O(1)$ variables in the δ-expansion, see (8.3)) and to determine the corresponding functions Q_1 in terms of the driver function $\kappa(t) = \hat{p}_1(t)$. The second step addresses quantities of order δ, particularly v and E.

With (8.31) and (8.32) pressure balance (8.19) takes the form

$$\frac{1}{2\mu_0}\Gamma(t)^2 \left(\frac{\partial A_0}{\partial \tilde{z}}\right)^2 + p_1(t)\,p_0(A) = \hat{p}_1(t)\,\hat{p}_0(\tilde{x}) \,. \tag{8.33}$$

In view of the similarity condition (8.31), the time-dependence must drop out of (8.33), which gives

$$p_1 = \kappa, \quad \Gamma = \kappa^{1/2}. \tag{8.34}$$

The flux tube volume (8.28) factorizes as

$$\begin{aligned} V(A,t) &= \int_0^{\xi(A,t)} \frac{\mathrm{d}x}{|B_x|} \\ &= \frac{1}{\Lambda\Gamma} \int_0^{\tilde{\xi}(A)} \left|\frac{\partial A_0}{\partial \tilde{z}}\right|^{-1} \mathrm{d}\tilde{x} \end{aligned} \tag{8.35}$$

where $\tilde{\xi} = \Lambda\xi$. The second line of (8.35) yields

$$V_1 = \frac{1}{\Lambda\Gamma} \,. \tag{8.36}$$

The adiabatic law (8.27) gives

$$(p_1(t)p_0(A))^{1/\gamma} V_1(t)V_0(A) = M(A) \tag{8.37}$$

which implies

$$p_1^{1/\gamma} V_1 = 1. \tag{8.38}$$

The time factor of density is obtained from (8.22),

$$\rho_1 = \frac{1}{V_1}. \tag{8.39}$$

From (8.31), (8.34), (8.36), (8.38) and (8.39) one finds that for all snapshot quantities Q the time factors are power laws of the form $Q_1(t) = \kappa(t)^{m_Q}$. The powers m_Q are listed in Table 8.1.

Quantities of order δ depend on the time rate of change represented by the logarithmic derivative $\dot{\kappa}/\kappa$, where $\dot{\kappa} = d\kappa/dt$,

$$v_x = -R\frac{\dot{\kappa}}{\kappa} x \tag{8.40}$$

$$v_z = -\frac{1}{2}\frac{\dot{\kappa}}{\kappa} z \tag{8.41}$$

$$E_y = -\frac{\dot{\kappa}}{\kappa}\left(R\tilde{x}\frac{\partial A_0}{\partial \tilde{x}} + \frac{z}{2}\frac{\partial A_0}{\partial \tilde{z}}\right). \tag{8.42}$$

The present selfsimilar solutions require boundary conditions compatible with the assumed similarity conditions. This is a substantial constraint on possible modes of evolution of the tail configuration.

Other than cases discussed in later sections, selfsimilar evolution excludes a thin current sheet forming inside a main current sheet that does not contain a thin sheet initially. As any snapshot quantity, the current density j remains similar to the initial current density. The corresponding time factor κ (Table 8.1) is simply the result of lobe field strength scaling as $\kappa^{1/2}$ and characteristic length $L_z = 1/\Gamma$ as $\kappa^{-1/2}$.

The electric field shows substantial spatial variation. For $\gamma = 5/3$ one finds $R = 1/10$, such that in typical cases the second term in (8.42) dominates. Then E_y shows an approximate linear variation with z in lobe regions where $\partial A_0/\partial \tilde{z}$ is nearly constant.

Table 8.1 *The powers m_Q that control the time-dependence of the snapshot quantities Q, $R = 1/\gamma - 1/2$.*

Q	A	p	j	B_x	B_z	Λ	Γ	ρ	V
m_Q	0	1	1	$1/2$	R	R	$1/2$	$1/\gamma$	$-1/\gamma$

8.3.3 Steady state and pressure crisis

The theory of quasi-static evolution also applies to steady states with plasma flow velocities sufficiently small for the δ-scaling (see (8.3) and (8.4)) to be applicable. A non-vanishing component E_y can be included by keeping A time-dependent, where the assumption of a steady state electromagnetic field reduces $A(x, z, t)$ to the form

$$A(x, z, t) = A_s(x, z) - E_y t , \qquad (8.43)$$

$A_s(x, z)$ being the flux function of a static magnetic field configuration and E_y a constant of order δ. All snapshot states are identical in their physical properties.

Regarding the adiabatic law (8.30), two steady state properties lead to modifications. First, the right side M becomes independent of A because, unlike in the time-dependent case, all flux tubes experience the same time history and therefore contain the same entropy. Let the constant value of M be M_0. Second, it is no longer possible to assign a finite pressure to a flux tube with a vanishing volume. The adiabatic law $p^{1/\gamma}V = M_0$ implies that p diverges as $V \to 0$ such that $\hat{p}(0, t)$ has to be set to infinity. Although that singularity is unphysical, the fact that p becomes large as the volume becomes small is entirely realistic. So we simply exclude the immediate neighbourhood of the singularity from the discussion to have a realistic description.

Taking both modifications into account, the adiabatic law (8.30) takes the form

$$p^{1/\gamma} \int_p^\infty \frac{\mathrm{d}\hat{p}}{f(\hat{p})\sqrt{\hat{p} - p}} = M_0 . \qquad (8.44)$$

The equation (8.44) is to be read as a constraint on the function $\hat{p}(x)$. There is no longer any freedom left for specifying $\hat{p}(x)$ by a boundary condition. In fact, (8.44) is Abel's integral equation for the function $1/f(\hat{p})$. The solution is given by (Whittaker and Watson, 1973)

$$\frac{1}{f(\hat{p})} = -\frac{M_0}{\pi} \frac{\mathrm{d}}{\mathrm{d}\hat{p}} \int_{\hat{p}}^\infty \frac{\mathrm{d}p}{p^{1/\gamma}\sqrt{p - \hat{p}}} . \qquad (8.45)$$

Noting that $f = -\mathrm{d}\hat{p}/\mathrm{d}x$, this equation is readily integrated with respect to x, giving

$$x = \frac{M_0}{\pi} \int_{\hat{p}}^{\infty} \frac{\mathrm{d}p}{p^{1/\gamma}\sqrt{p-\hat{p}}}$$

$$= \frac{M_0}{\pi} \hat{p}^{1/2-1/\gamma} \int_{1}^{\infty} \frac{\mathrm{d}\zeta}{\zeta^{1/\gamma}\sqrt{\zeta-1}} \tag{8.46}$$

$$= C\hat{p}^{-R}, \tag{8.47}$$

where C is a constant and $R = 1/\gamma - 1/2$. The additive integration constant was set to zero such that the pressure singularity is placed at $x = 0$. We fix C by choosing $\hat{p} = p_L$ at $x = L$. Inverting (8.47) then gives

$$\frac{\hat{p}}{p_L} = \left(\frac{x}{L}\right)^{-1/R}. \tag{8.48}$$

Fig. 8.3 shows $Y = \hat{p}/p_L$ versus $X = x/L$ for several values of γ. A model curve, which is regarded as a reasonable fit to the observed pressure decay in the Earth's magnetotail, is shown also. The pressure singularity is excluded by using the interval (0.25, 1) as a representative region of the tail. Evidently, the steady state curves must be regarded as unrealistic.

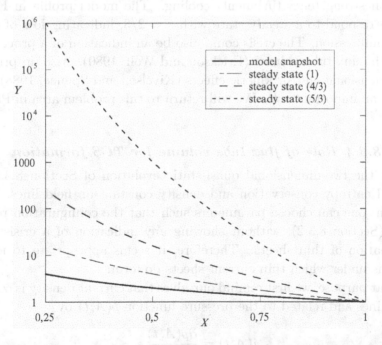

Fig. 8.3 Illustration of the pressure crisis. The dimensionless pressure $Y = \hat{p}/p_L$ is plotted as a function of $X = x/L$ under steady state conditions for $\gamma = 1, 4/3, 5/3$. The solid curve shows the function $Y = 1/X$, which corresponds to a realistic snapshot model of the Earth's magnetotail.

In particular, the classical ideal gas value $\gamma = 5/3$ is off by a factor greater than 10^5.

This property has been called the *pressure inconsistency, pressure balance inconsistency* or *pressure crisis*. It was first found by Erickson and Wolf (1980) using a somewhat different approach. The present picture, based on Schindler and Birn (1982), is fully self-consistent (in the limit of small aspect ratio).

The physical reason of the pressure crisis is simply the large change in the flux tube volume that the convecting plasma experiences. The pressure increases accordingly by a large factor. However, these large pressures are inconsistent with any realistic values. Pressure balance would require an equally large total (magnetic plus kinetic) pressure applied at the tail boundary, which is not available.

The pressure crisis has become an important issue of magnetospheric physics. The task is to find out in what way nature avoids this crisis. In principle, there are several possibilities. The tail could respond to flux transfer in a time-dependent fashion. Note that there is no indication of a crisis in the time-varying model of Section 8.3. Another possibility is a steady state with strong (e.g., turbulent) cooling. The model profile in Fig. 8.3 would correspond to a steady state with $\gamma = 2/3$, indicating loss of energy during compression. The crisis could also be an indication of a process that reduces the flux tube volume (Erickson and Wolf, 1980). Also, in principle, three-dimensional and/or kinetic effects (Kivelson and Spence, 1988) might also play an important role. We will return to this problem area in Part IV.

8.3.4 Role of flux tube volume for TCS formation

Consider the two-dimensional quasi-static evolution of Section 8.3.1 with mass and entropy conservation and density constant on field lines. As we have seen, one can choose parameters such that the configuration remains smooth (Section 8.3.2), without showing any indication of a crisis nor of the formation of thin sheets. Therefore, it seems appropriate to identify conditions under which thin current sheets do form.

For that purpose, we first remind ourselves that current density is constant on field lines and related to the pressure function $p(A, t)$ by

$$j(A, t) = \frac{\partial p(A, t)}{\partial A} .$$

(8.49)

Thus, a steep gradient of the pressure function results in the formation of a thin current sheet.

An alternative version of that condition is obtained by expressing the pressure function by the flux tube volume $V(A, t)$, using the adiabatic law (8.27) in the form

$$p(A, t) = p(A, 0) \left(\frac{V(A, 0)}{V(A, t)} \right)^{\gamma}, \qquad (8.50)$$

such that (8.49) takes the form

$$j(A, t) = j(A, 0) \left(\frac{V(A, 0)}{V(A, t)} \right)^{\gamma} + p(A, 0) \frac{\partial}{\partial A} \left[\left(\frac{V(A, 0)}{V(A, t)} \right)^{\gamma} \right]. \qquad (8.51)$$

Thus, a thin current sheet forms if one of the two terms in (8.51) or both become large. Either the flux tube volume itself or its gradient (as a function of A), or both, must increase strongly.

Although the discussion of the flux tube volume gives insight into the physical mechanism at work, it does not indicate whether or not a given external driving force causes a thin current sheet. To answer that question one has to relate the current density to the driver function $\hat{p}$. The following example illustrates how this is possible.

Example

The magnetospheric model described in Section 8.3.1 is well suited to construct an example of the formation of thin current sheets based on local properties of the flux tube volume. In fact, the inner magnetosphere has small flux tubes and the tail flux tubes are large with a rather rapid transition from one region to the other. In the present example, the rapid decrease of the flux tube volume toward the Earth is idealized by setting $V = 0$ at the earthward tail boundary, which is located at $x = 0$.

We use the gauge freedom (Section 8.3.1) to set $A(0, 0, t) = 0$, such that $\hat{A}(0, t) = 0$ and therefore $p(0, t) = \hat{p}(0, t)$ (see (8.18)).

Anticipating that the near-Earth region is particularly sensitive to current sheet formation, we evaluate the expression (8.29) for $V(A, t)$ under the condition that $p(A, t)$ differs from $\hat{p}(0, t)$ only slightly. Then, in the integral in (8.29) the square root in the denominator varies much faster than $f(\hat{p})$. Thus, taking $f(\hat{p})$ in front of the integral, we find the asymptotic expression

$$V(A, t) = -\sqrt{\frac{2}{\mu_0}} \frac{\sqrt{\hat{p}(0, t) - p(A, t)}}{\hat{p}'(0, t)}, \qquad (8.52)$$

where the prime symbol denotes differentiation with respect to x. The expression (8.52) gives

$$\frac{V(A,0)}{V(A,t)} = \sqrt{\frac{\hat{p}(0,0) - p(A,0)}{\hat{p}(0,t) - p(A,t)} \frac{\hat{p}'(0,t)}{\hat{p}'(0,0)}}$$

$$= \sqrt{\frac{j(0,0)}{j(0,t)} \frac{\hat{p}'(0,t)}{\hat{p}'(0,0)}} \,, \tag{8.53}$$

where the second line is the result of taking the limit $A \to 0$ (making use of l'Hôpital's rule) and using (8.49). The adiabatic law (8.50) then gives approximately

$$\frac{j(0,t)}{j(0,0)} = \left[\frac{\hat{p}(0,t)}{\hat{p}(0,0)}\right]^{-2/\gamma} \left[\frac{\hat{p}'(0,t)}{\hat{p}'(0,0)}\right]^2 . \tag{8.54}$$

For realistic values of γ, such as $\gamma = 5/3$, the second factor in (8.54) dominates. Thus, we see that thin current sheet formation is favoured by large gradients of the pressure function $\hat{p}(x)$.

The present model was first studied by Wiegelmann and Schindler (1995) for the special choice

$$\hat{p} = p_0 \frac{\tau^2}{1 + \tau^2 x/L_x}, \quad p(A,0) = p_0 \, e^{-2A/A_c}, \quad \gamma = 5/3, \tag{8.55}$$

where $\tau = 1 + t/t_c$. The lobe field strength B_l at $x = 0$ increases with time linearly. Note that (8.55) significantly breaks selfsimilarity (Section 8.3.2). It was found that $j(0,t)$ rises with time as $\tau^{28/5}$, which also follows from the general expression (8.54), specialized for (8.55).

It is of interest to consider this case also away from $A = 0$. Solving (8.50) for $p(A,t)$ by a power expansion with respect to $a = A/A_c$, valid up to order a^2, one finds

$$p(A,t) = \left[\tau^2 - 2\tau^{28/5} a + \left(\tfrac{88}{15} \tau^{46/5} - \tfrac{58}{15}\tau^{28/5}\right) a^2\right] p_0 \,, \tag{8.56}$$

giving current density and flux tube volume as

$$j(A,t) = \left[-2\tau^{28/5} + \left(\tfrac{176}{15} \tau^{46/5} - \tfrac{116}{15}\tau^{28/5}\right) a\right] \frac{p_0}{A_c} \tag{8.57}$$

$$V(A,t) = \left[2\tau^{-6/5}a^{1/2} + \left(\tfrac{29}{15}\tau^{-6/5} + \tfrac{12}{5}\tau^{12/5}\right) a^{3/2}\right] \frac{L}{\sqrt{\mu_0 p_0}}. \tag{8.58}$$

From (8.58) we can determine which of the terms in (8.51) is leading. One finds

$$j(A,0)\left(\frac{V(A,0)}{V(A,t)}\right)^{\gamma} = -2\tau^2\frac{p_0}{A_c}$$

$$p(A,0)\frac{\partial}{\partial A}\left[\left(\frac{V(A,0)}{V(A,t)}\right)^{\gamma}\right] = \left(2\tau^2 - 2\tau^{28/5}\right)\frac{p_0}{A_c},$$

(8.59)

which indicates that it is the gradient of the flux tube volume rather than the flux tube volume itself that gives rise to the strong increase of the current density.

For a prescribed value of B_l the z-integrated current is fixed. Therefore, the large current density near the origin can only occur in a thin sheet. A rough estimate for the sheet thickness is $d = 2|B_l/(\mu_0 j)|\nu$, where ν is the fraction of the total current flowing in the thin current sheet. Since $\nu \leq 1$ we get at $x = 0$

$$d \leq \frac{2}{|j(0,t)|}\sqrt{\frac{2\hat{p}(0,t)}{\mu_0}} = L_0\tau^{-23/5},$$

(8.60)

where L_0 is the thickness of the initial sheet at $\tau = 1$. Thus, the sheet shows pronounced thinning as the flux transfer proceeds.

8.4 TCS from boundary deformation

So far current sheet formation was discussed as a result of magnetic flux transfer, which was represented by the total pressure $\hat{p}(x,t)$. Here, we look at boundary deformation by external forces as an alternative driver.

8.4.1 The model

Again, the model is based on asymptotic tail theory as described in Section 5.4.1. The magnetic flux function $A(x,z)$ is given by (8.17). The construction of the temporal evolution, which is represented by the functions $p(A,t)$ and $\hat{p}(x,t)$, differs from that of Section 8.3.1 because a different boundary condition is applied (Birn and Schindler, 2002). The boundary condition prescribes the shape of a bounding magnetic field line.

For magnetohydrostatic equilibria, which here we are dealing with for fixed t, the corresponding theory was described in Section 5.4.2. There, it was shown that the boundary condition $z = a(x)$ for the shape of field line with pressure p_b determines the function $p(A)$. The procedure actually uses the boundary condition in the form of a prescribed function $a(\hat{p})$. As $\hat{p}(x)$

was also prescribed and assumed to be monotonic, the step from $a(x)$ to $a(\hat{p})$ was trivial.

Here, the situation is different in that $\hat{p}$ is not free to be chosen but will be determined from entropy conservation. Therefore, the boundary condition is directly formulated as prescribing the function $a(\hat{p}, t)$. The shape of the bounding field line in the form $a(x, t)$ is available after finding $\hat{p}(x, t)$.

In this way one can compute $p(A, t)$ using the procedure of Section 5.4.2. This is done by writing the solution (8.17) for the bounding field line in the form (Birn and Schindler, 2002)

$$a(\hat{p}, t) = \int_{p_b}^{\hat{p}} \left(-\frac{\partial A}{\partial p} \right) \frac{dp}{\sqrt{2\mu_0(\hat{p} - p)}} \tag{8.61}$$

where p_b is the pressure on the boundary. Abel-inversion (Whittaker and Watson, 1973) of (8.61) gives $\partial p / \partial A$ and integration with respect to p gives

$$A(p, t) = -\frac{\sqrt{2\mu_0}}{\pi} \int_{p_b}^{p} a(\hat{p}, t) \frac{d\hat{p}}{\sqrt{p - \hat{p}}} \ ; \tag{8.62}$$

solving (8.62) for p gives the pressure function $p(A, t)$.

It remains to determine $\hat{p}(x, t)$ from entropy conservation (8.27), which we write as

$$p(A, t) V(A, t)^\gamma = M(A) \ , \tag{8.63}$$

where $M(A)$ is determined by the initial configuration.

Using that $p(A, t)$ is known, and assuming monotonic dependence on A (see the discussion below), one conveniently replaces A by p as the running variable in (8.63). Solving for V one then finds V as a function of p and t,

$$V(p, t) = \left[\frac{M(A(p, t))}{p} \right]^{1/\gamma} . \tag{8.64}$$

The flux tube volume in the form of (8.29) is then understood as an equation for $f(\hat{p}, t)$. Again using Abel-inversion, one finds

$$\frac{1}{f(\hat{p}, t)} = -\frac{\sqrt{2\mu_0}}{\pi} \frac{\partial}{\partial \hat{p}} \int_{\hat{p}}^{\hat{p}(0, t)} V(p, t) \frac{dp}{\sqrt{p - \hat{p}}} \ . \tag{8.65}$$

As $f = -\partial \hat{p} / \partial x$ (see Section 8.3.1), (8.65) can be integrated with respect to x,

$$x(\hat{p}) = \frac{\sqrt{2\mu_0}}{\pi} \int_{\hat{p}}^{\hat{p}(0, t)} V(p, t) \frac{dp}{\sqrt{p - \hat{p}}} \ . \tag{8.66}$$

Solving (8.66) for $\hat{p}$ gives $\hat{p}(x,t)$. Having found $p(A,t)$ and $\hat{p}(x,t)$, the solution for the flux function $A(x,z,t)$ can now be obtained from (8.17).

The present procedure requires that $p(A,t)$ and $\hat{p}(x,t)$ are monotonic functions of A and x, respectively. This property is equivalent to the absence of neutral points and thus ensures the conservation of magnetic topology (Birn and Schindler, 2002). It is significant that $a(\hat{p},t)$ is not restricted to being monotonic in $\hat{p}$, so that non-monotonic deformations $a(x,t)$ are admitted.

Note that the current density $j(A,t) = \partial p(A,t)/\partial t$ is available already from the first step, i.e., from (8.62). That result already indicates whether or not a thin current sheet forms, manifested by a steep gradient of $p(A)$. Surprisingly, it turns out that it is not very difficult to find cases where at some stage during the temporal evolution an infinite slope occurs, corresponding to a singularity of the current density. Here is an example.

8.4.2 An example with loss of equilibrium

Following Birn and Schindler (2002), an initial state is chosen with the pressure function

$$p(A,0) = p_{\mathrm{b}} \exp(-2(A - A_{\mathrm{b}})/A_{\mathrm{c}}) , \qquad (8.67)$$

where the subscript b refers to the boundary. This choice corresponds to the boundary profile

$$a(\hat{p},0) = \frac{A_{\mathrm{c}}}{\sqrt{2\mu_0 p_{\mathrm{b}}}} \sqrt{\frac{\hat{p}}{p_{\mathrm{b}}}} \, \mathrm{arccosh}\sqrt{\frac{\hat{p}}{p_{\mathrm{b}}}} \, . \qquad (8.68)$$

A deformation is applied to this profile by choosing the boundary condition

$$a(\hat{p}) = \frac{A_{\mathrm{c}}}{\sqrt{2\mu_0 p_{\mathrm{b}}}} \sqrt{\frac{\hat{p}}{p_{\mathrm{b}}}} \, \mathrm{arccosh}\sqrt{\frac{\hat{p}}{p_{\mathrm{b}}}} \left[1 - \frac{a_1}{1 + ((\hat{p} - p_{\mathrm{m}})/\Delta p)^2}\right] . \qquad (8.69)$$

The term in the square brackets adds a local indentation of amplitude a_1, location p_{m} and width Δp. These parameters can be arbitrary functions of time.

After obtaining $p(A,t)$ from (8.62) one finds that, typically, for fixed p_{m} and Δp and a_1 increasing with time, $p(A,t)$ remains to be a monotonic function of A only for amplitudes a_1 below a critical value. At the critical amplitude the slope becomes infinite at some A-value, such that the current density assumes a singularity (Fig. 8.4).

For larger amplitudes the pressure function is no longer monotonic, indicating the presence of neutral points. Therefore, the present topology-conserving sequence that exists for subcritical amplitudes is terminated when the critical amplitude is reached. In that sense the singularity marks loss of equilibrium.

The next step is to compute $M(A)$ from the initial state, choosing $\hat{p}(x,0) = p_0/(1 + x/L)$. Then $\hat{p}(x,t)$ follows from (8.66). The magnetic field lines are found from (8.17). Fig. 8.5 shows the field lines for the three cases of Fig. 8.4. The effect of the thin current sheet is clearly seen in panels b and c. The indentations that cause the current singularity remain moderate (panels b and c of Fig. 8.5). The magnetic field remains bounded although the current density diverges.

Loss of equilibrium has been confirmed by more extended investigations (Birn and Schindler, 2002; Birn *et al.*, 2003) as well as by a 3D generalization (Birn *et al.*, 2004).

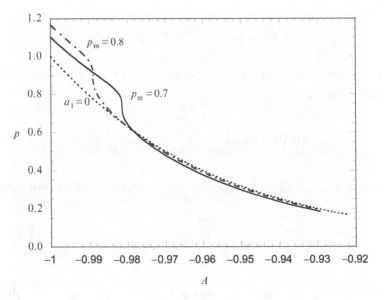

Fig. 8.4 The pressure function $p(A)$ obtained from the boundary condition (8.69) for the initial state ($a_1 = 0$) and for two cases ($p_\mathrm{m} = 0.7$ and $p_\mathrm{m} = 0.8$) at critical amplitudes; the width is fixed at $\Delta p = 0.1$, dimensionless variables are used. (Reproduced from Birn and Schindler (2002) by permission of the American Geophysical Union.)

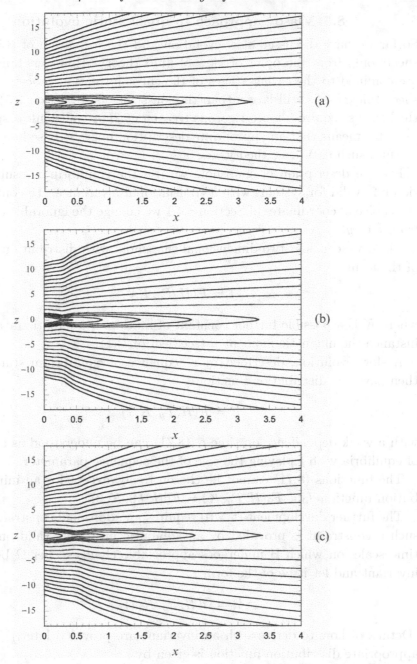

Fig. 8.5 Magnetic field lines corresponding to the three cases of Fig. 8.4: case (a) with $a_1 = 0$; cases (b) and (c) have critical amplitudes and $p_m = 0.8$ and $p_m = 0.7$, respectively. (Reproduced from Birn and Schindler (2002) by permission of the American Geophysical Union.)

8.5 Vlasov approach to quasi-static evolution

So far we have discussed slow evolution and the formation of thin current sheets only from the point of view of fluid theory. However, fluid theories are confined to the earlier stages of the sheet formation process, when the sheet thickness is still larger than the ion scales (see Section 3.5). Smaller thicknesses require à more rigorous treatment. For collisionless space plasmas this means that an appropriate description must be based on a particle picture, such as Vlasov theory.

For the description of snapshots we use the equilibrium results of Section 6.2, valid for systems with a translational invariance. To conform with the choice of coordinates of Section 8.4.1 we change the ignorable coordinate from z to y.

Thus, we consider two-dimensional snapshots with distribution functions of the form

$$f = F(H, P_y, K) \,, \tag{8.70}$$

where K is a possible further constant of motion or adiabatic invariant, for instance the magnetic moment μ (see Section 3.1).

A slow evolution, described by a sequence of equilibrium states, would then have the distribution function

$$f = F(H, P_y, K, t) \tag{8.71}$$

with a weak dependence on time t. (8.71) can be understood as a sequence of equilibria with t playing the role of the sequence parameter.

The functions (8.71) cannot be chosen freely, except for an initial distribution function $f_0 = F_0(H, P_y, K) = F(H, P_y, K, 0)$.

The further development ($t > 0$) requires an additional constraint. Often, such a constraint is provided by an adiabatic invariant, valid on the long time scale, on which H undergoes appreciable changes. Let Ω be such an invariant and let it be of the form

$$\Omega = \Omega(H, P_y, K, t) \,. \tag{8.72}$$

(Details of how to derive such an invariant are provided later.) Then, the appropriate distribution function is given by

$$f = G(\Omega, P_y, K) \,, \tag{8.73}$$

where we have assumed that K is invariant on the long time scale. Importantly, the distribution function (8.73) is expressed only by long-term constants of the motion. This makes (8.73) a valid Vlasov distribution function, albeit in an approximate sense. The function G is determined by F_0.

We illustrate this procedure for the case of chaotic motion.

8.5.1 Adiabatic sequences for chaotic particle motion

As discussed in Section 6.4, the particle motion in sufficiently thin sheets may become chaotic. Here we deal with the slow evolution in that regime. More precisely, we assume that the particles cover the hypersurface defined by their values of H and P_y in the sense of ergodic motion. Then, the equilibrium distribution function has the form $F(H, P_y)$ and slow evolution is described by a sequence of such states. With the additional assumption of quasi-neutrality, the equilibrium model of Section 6.2.2 applies to the individual snapshots. In addition, let us assume that the magnetic field strength is bounded away from zero.

It should be noted that distribution functions of the form $F(H, P_y)$ are not exclusively the consequence of chaotic particle motion (Section 6.4) but, conceivably, may also arise from the presence of a weak fluctuation field (Nötzel et al., 1985). Although the following explicitly addresses chaotic motion, the results can be expected to apply to cases with suitable fluctuation fields also.

Outline of the procedure

The main aim is to determine the snapshot state for an arbitrary time t. The present procedure is largely analogous to the MHD approach by Birn and Schindler (2002) as described in Section 8.4. We summarize that approach by breaking it up in a sequence of steps. This is useful because it will turn out that only one of these steps has to be modified in the Vlasov picture. Since time is fixed, t is suppressed in the arguments. We distinguish the following steps of the fluid theory.

Step 1: A boundary perturbation is chosen in the form $a(\hat{p})$. This leads to the pressure function $p(A)$:

$$a(\hat{p}) \rightarrow A(p) \rightarrow p(A). \tag{8.74}$$

Step 2: Using entropy conservation ($pV^{5/3}$ constant), one finds the differential flux tube volume:

$$p(A) \rightarrow V(p). \tag{8.75}$$

Step 3: The definition of V then gives the pressure $\hat{p}$:

$$V(p) \rightarrow x(\hat{p}) \rightarrow \hat{p}(x). \tag{8.76}$$

Step 4: The magnetic flux function $A(x, z)$ follows as

$$(p(A),\ \hat{p}(x)) \rightarrow A(x, z). \tag{8.77}$$

Step 5: The current density is

$$j(A) = \frac{\mathrm{d}p(A)}{\mathrm{d}A}, \quad j(x, z) = j(A(x, z)) . \tag{8.78}$$

The present kinetic approach allows us to proceed in a similar way. All steps, with the exception of step 2, make use only of the asymptotic solution of the Grad–Shafranov equation for small aspect ratios (Section 5.4.1) and of the expression for the flux tube volume. As shown in Section 6.2, the present assumptions lead to a Grad–Shafranov equation, too. The underlying property is that distribution functions of the form $F(H, P_y)$ give a pressure tensor that is isotropic in the plane perpendicular to the direction of translational invariance, which allows us to define a pressure function $p(A)$. Thus, steps 1, 3, 4, 5 can be taken over directly from the fluid theory.

In step 2 the fluid invariant, the entropy, has to be replaced by a suitable adiabatic invariant of the particle motion. Also, the electric potential appears as an additional unknown, which has to be determined, too.

The invariant

To establish the (kinetic) invariant for the long-term evolution, we first have to discuss the topological structure of the hypersurfaces of constant H (*energy surfaces*) in 4-dimensional phase space spanned by (x, z, P_x, P_z).

The Hamiltonian has the form (see (3.14))

$$H = \frac{1}{2m}(P_x^2 + P_z^2) + \psi(A, P_y, t) \tag{8.79}$$

with

$$\psi(A, P_y, t) = \frac{1}{2m}(P_y - qA)^2 + q\phi(A, t) , \tag{8.80}$$

where quasi-neutrality ensures that ϕ depends on x, z only through A. Note, however, that the A-dependence of ψ is not available explicitly before $\phi(A, t)$ is known. When dealing with ϕ directly, one should understand ψ as a function of A, ϕ, P_y, t.

For illustration, we begin with the simple case where the electric potential ϕ vanishes. In that case the energy surfaces are singly connected and nested in the sense that a given surface with $H = H_0$ encloses all surfaces with

$H < H_0$. These properties can be visualized by noticing that for $\phi = 0$ and a fixed value of P_y the Hamiltonian depends on A and $\tilde{P} = \sqrt{P_x^2 + P_z^2}$ only. In the $\tilde{P}, A$-plane the level curves of H are nested ellipses with centres at $\tilde{P} = 0$, $A = P_y/q$.

Under such conditions and with ergodic particle motion covering the entire H-surface, the volume of the domain enclosed by a particle's energy surface is an adiabatic invariant of that particle. This is well known in statistical mechanics, where the entropy of a microcanonical ensemble is derived from the invariance of phase space volume. A derivation matching the present circumstances is given in www.tp4.rub.de/~ks/tb.pdf. There, cases of non-vanishing ϕ are included. It turns out that the above arguments still apply, as long as (for fixed P_y and t) the function $\psi(A)$ has a single minimum ψ_{min} and varies monotonically with A elsewhere (Fig. 8.6). Then, the energy surfaces are still singly connected and nested in qualitatively the same way as in the simpler case where ϕ vanishes. Typically, this means that ϕ must not get too large. In any given application the assumption regarding ψ has to be checked explicitly.

In the present 4-dimensional phase space the invariant Ω is given by

$$\Omega(H, P_y, t) = \int S(H - H(x, z, P_x, P_z, P_y, t))\, dx\, dz\, dP_x\, dP_z \,, \qquad (8.81)$$

where S denotes the unit-step function.

After transforming to polar coordinates $\tilde{P}, \alpha$ in momentum subspace spanned by (P_x, P_z), then using $\eta = \tilde{P}^2/2m$ instead of $\tilde{P}$ as integration variable, and transforming x, z to A, s, where s is the arclength on field lines, we carry out the α and s integrations and obtain

$$\Omega(H, P_y, t) = 2\pi m \int S(H - \psi(A, P_y, t) - \eta)V(A, t)\, d\eta\, dA \,, \qquad (8.82)$$

where

$$V(A, t) = \int_A \frac{ds}{B(A, s, t)} \qquad (8.83)$$

is the differential flux tube volume, the integration being carried out along the entire field line with flux label A. Here we assume that field lines have finite length, as is typical for regions with closed magnetic flux.

For fixed P_y and t the step-function in (8.82) limits the integration to the region bounded by the curve $\psi(A)$ and the line $\eta = H$ in the A, η-plane (Fig. 8.6). Thus we find (after carrying out the η-integration)

$$\Omega(H, P_y, t) = 2\pi m \int_{A_1(H, P_y, t)}^{A_2(H, P_y, t)} (H - \psi(A, P_y, t))V(A, t)\, dA, \qquad (8.84)$$

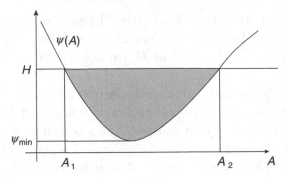

Fig. 8.6 The shaded area is the region where the integrand of the integral in (8.82) does not vanish, A_1 and A_2 are the limiting values of A appearing in (8.84).

where $A_1(H, P_y, t)$ and $A_2(H, P_y, t)$ are the solutions of the equation $H = \psi(A, P_y, t)$, solved for A (see Fig. 8.6).

We note that the derivative of Ω with respect to H is

$$\frac{\partial \Omega(H, P_y, t)}{\partial H} = 2\pi m D(H, P_y, t) \tag{8.85}$$

with

$$D(H, P_y, t) = \int_{A_1(H, P_y, t)}^{A_2(H, P_y, t)} V(A, t) \, dA , \tag{8.86}$$

which is positive. The function $D(H, P_y, t)$ simply represents the area of the region in the x, z-plane that is accessible to a particle with constants of the motion H, P_y at time t.

The distribution function

Suppose an adiabatic sequence of quasi-static equilibrium states, driven by slowly varying boundary conditions, starts out with an initial distribution function $F_0(H, P_y)$ and with potentials $A_0(x, z)$, $\phi_0(A_0)$. Then, specializing (8.73) to the present case with no additional invariant K, the function $G(\Omega, P_y)$ is obtained from the initial distribution function in the following way. Since the invariant Ω is monotonic in H (see (8.85) and (8.86)) the expression (8.84) can be solved for H,

$$H = H(\Omega, P_y, t) . \tag{8.87}$$

Choosing $t = 0$ in (8.87) one finds

$$G(\Omega, P_y) = F_0(H_0(\Omega, P_y), P_y) , \tag{8.88}$$

where $H_0(\Omega, P_y) = H(\Omega, P_y, 0)$. For an arbitrary state during the sequence the instantaneous distribution function is obtained as (see (8.73))

$$F(H, P_y, t) = G(\Omega(H, P_y, t), P_y) \, . \tag{8.89}$$

Kinetic version of step 2

As outlined above, we have to replace step 2 of the fluid theory by a corresponding kinetic version. Since the electric potential is an additional unknown, we look for a way to determine both $V(A)$ and $\phi(A)$.

The corresponding equations are the charge neutrality condition (vanishing of space charge σ, see (6.16)) and the definition of the pressure (6.17). With the time parameter t suppressed, these equations yield

$$
\begin{aligned}
\sigma(A, \phi) &= \sum q \int F \, d^3 v \\
&= \sum \frac{2\pi q}{m^2} \int_{q\phi}^{\infty} dH \int_{P_1}^{P_2} F(H, P_y) \, dP_y = 0 \tag{8.90} \\
p(A, \phi) &= \sum \int \frac{m}{2}(v_x^2 + v_z^2) F \, d^3 v \\
&= \sum \frac{2\pi}{m^2} \int_{q\phi}^{\infty} dH \int_{P_1}^{P_2} (H - \psi(A, \phi, P_y)) F(H, P_y) \, dP_y \tag{8.91}
\end{aligned}
$$

where $\sum$ sums over particle species and P_1, $P_2 \geq P_1$ are functions of A, ϕ, H, defined as the solutions of $H = \psi(A, \phi, P_y)$, solved for P_y; F is taken from (8.89). From (8.90) one determines $\phi(A)$ and after inserting that result into (8.91) one finds $V(A)$. This completes the new step 2.

Recovering the fluid approach for $m \to 0$

Ideally, a Vlasov description of slow evolution should be embedded in a fluid description in such a way that the Vlasov model approaches the fluid model in the limit where the intrinsic plasma scales are small compared to the macroscopic scales. A convenient way to look at that limit is to let the particle mass go to zero. Here we show that, indeed, we recover the fluid result in that limit.

In the derivation we deal with the more general case where we allow for the presence of one or more species that are treated exactly. This covers the situation where the electrons have reached the small-mass limit while the ions have not.

Let the contributions of a species that can be described in the small-mass limit to the sums in (8.90) and (8.91) be σ_1 and p_1, respectively.

To lowest order in m one finds the integration limits in (8.84) as

$$A_1 = \frac{P_y}{q} - \frac{1}{|q|}\sqrt{2m(H - \phi(P_y/q))}, \qquad A_2 = \frac{P_y}{q} + \frac{1}{|q|}\sqrt{2m(H - \phi(P_y/q))}\,.$$

$$(8.92)$$

Introducing $\xi = A - P_y/q$ as integration variable and noting that $\xi = O(\sqrt{m})$, and $(P_y - qA)^2/2m = q^2\xi^2/2m = O(1)$, one finds from (8.84) the invariant to lowest order in m,

$$\Omega(H, P_y) = \frac{4\pi}{3|q|}\left(2m(H - q\phi(P_y/q))\right)^{3/2}V(P_y/q)\,.$$

$$(8.93)$$

Solving for H and evaluating the result for $t = 0$ gives

$$H_0(\Omega, P_y) = \frac{1}{2m}\left(\frac{|q|\Omega}{4\pi V_0(P_y/q)}\right)^{2/3} + \phi_0(P_y/q)\,.$$

$$(8.94)$$

Inserting this expression into the initial distribution function and then using (8.93) to express Ω by H (according to (8.87)), one finds for an arbitrary instantaneous distribution function

$$F(H, P_y) = F_0\left(\left(\frac{V(P_y/q)}{V_0(P_y/q)}\right)^{2/3}(H - q\phi(P_y/q)) + q\phi_0(P_y/q), P_y\right)\,.$$

$$(8.95)$$

With that distribution function σ_1 and p_1 are readily computed from (8.90) and (8.91). To lowest order in m one finds

$$P_1(H, A, \phi) = qA - \sqrt{2m(H - q\phi)}, \quad P_2(H, A, \phi) = qA + \sqrt{2m(H - q\phi)}$$

$$(8.96)$$

$$\sigma_1(A, \phi) =$$
$$\frac{4\pi q}{m^2}\int_{q\phi}^{\infty}\sqrt{2m(H - q\phi)}F_0\left(\left(\frac{V(A)}{V_0(A)}\right)^{2/3}(H - q\phi) + q\phi_0(A), qA\right)\mathrm{d}H.$$

$$(8.97)$$

Introducing $h = (V(A)/V_0(A))^{2/3}(H - q\phi) + q\phi_0(A)$ as integration variable one finds

$$\sigma_1(A, \phi) = \frac{4\pi\sqrt{2}q}{m^{3/2}}\frac{V_0(A)}{V(A)}\int_{q\phi_0}^{\infty}\sqrt{h - q\phi_0(A)}F_0(h, qA)\,\mathrm{d}h\,.$$

$$(8.98)$$

Identifying

$$\sigma_{10} = \frac{4\pi\sqrt{2}q}{m^{3/2}} \int_{q\phi_0}^{\infty} \sqrt{h - q\phi_0(A)} F_0(h, qA) \, dh \qquad (8.99)$$

as the initial space charge density, one finally obtains

$$\sigma_1(A) = \sigma_{10}(A)\frac{V_0(A)}{V(A)} . \qquad (8.100)$$

This is the same result as that of a fluid description. The analogous behaviour is found for the pressure (8.91). The procedure is similar to that of σ_1. Regarding the additional factor $(H - q\phi)$, it is important to note that the term $(P_y - qA)^2/2m$ remains of order 1. Then to lowest order in m one finds

$$p_1(A) = p_{10}(A) \left(\frac{V_0(A)}{V(A)}\right)^{5/3} . \qquad (8.101)$$

Again, this is the behaviour of an (ideal) adiabatic fluid.

One might wonder why the present case, which formally is described by a Hamiltonian with two degrees of freedom, reproduces the exponent (5/3) of a gas with 3 degrees of freedom. The reason is that the energy represented by the 'potential' ψ is the kinetic energy associated with the velocity in the invariant direction and thus all 3 physical degrees of freedom are taken into account.

If the small-mass limit applies to all particle species, the kinetic version of step 2 reduces to the fluid limit. In this case the present kinetic model comprises the fluid theory as an asymptotic regime, without further assumptions being required.

A model based on gyro-centre motion with pitch angle scattering keeping the pressure tensor isotropic gives similar results. Particularly, the invariant (8.93) is exactly the same (Wolf, 1983; Garner *et al.*, 2003).

8.5.2 A kinetic snapshot containing a thin current sheet

Constructing an explicit kinetic equilibrium sequence is beyond the present scope. Instead, we will give an example of a kinetic snapshot, which is embedded in a corresponding fluid case. The latter is taken from the sequence discussed in Section 8.4, which illustrates the formation of a thin current sheet reaching a singularity of the current density at a critical amplitude of the external perturbation. Here we give an outline only, for details see Schindler and Birn (2002).

A two-species quasi-neutral plasma is considered that consists of electrons and protons. Anticipating that for thin current sheets the P_y-dependence of the distribution functions $F_s(H, P_y)$ plays a more important role than the H-dependence, we fix the H-dependence as exponential and leave the P_y-dependence open at this point to match it to the fluid case asymptotically for small ρ_i. Thus, we set

$$F_s(H, P_y) = C_s \exp\left(-\frac{H_s}{k_B T_s}\right) g_s(P_y). \qquad (8.102)$$

Dimensionless quantities are used, which are defined by the following normalizations. Coordinates x and z are normalized by a characteristic sheet thickness L (in cases of a double structure L is the thickness of the wider sheet), the thickness of the thin sheet, normalized by L, is denoted by d. Bulk velocities v_{xs} and v_{zs} are normalized by $\sqrt{k_B T_s / m_s}$, P_y by $eB_0 L$, where B_0 is a typical magnetic field strength outside the sheet, A by $B_0 L$, ϕ by $k_B T / e$, where $T = T_e + T_i$, H_s by $k_B T_s$, densities n_s by their value n_0 for vanishing A and ϕ, j_y by $B_0 / \mu_0 L$, and pressure by B_0^2 / μ_0. The constants C_s can be expressed by these normalization parameters. Dimensionless quantities are denoted by the same symbols as the original quantities, except that the subscript y at P_y and at j_y is dropped. Also we assume that parameters satisfy the relationship $n_0 k_B T = B_0^2 / 2\mu_0$, in accordance with pressure balance in the limit of a strictly one-dimensional sheet. The sign of particle charges is denoted by $\lambda_s = q_s / e$. Quasi-neutrality gives $n_e = n_i = n$.

The equilibrium theory involves the following dimensionless parameters,

$$\tau_e = \frac{T_e}{T}, \quad \tau_i = \frac{T_i}{T}, \quad \rho_i = \frac{\sqrt{m_i k_B T_i}}{eB_0 L}, \quad \rho_e = \frac{\sqrt{m_e k_B T_e}}{eB_0 L}, \qquad (8.103)$$

where, for convenience, we listed both τ_e and τ_i although their sum equals 1. Under these conditions, the distribution function (8.102) leads to densities

$$n_s = \exp\left(-\frac{\lambda_s}{\tau_s}\phi\right) \hat{n}_s(A), \qquad (8.104)$$

where

$$\hat{n}_s(A) = \frac{I_{2s}(A)}{I_{1s}} \qquad (8.105)$$

with

$$I_{1s} = \int_{-\infty}^{\infty} \exp\left(-\frac{P^2}{2\rho_s^2}\right) g_s(P)\, dP \tag{8.106}$$

$$I_{2s}(A) = \int_{-\infty}^{\infty} \exp\left(-\frac{(P-\lambda_s A)^2}{2\rho_s^2}\right) g_s(P)\, dP . \tag{8.107}$$

For details see Schindler and Birn (2002). Similarly, one finds for the current density

$$j_s = \frac{\lambda_s \tau_s}{2} \exp\left(-\frac{\lambda_s}{\tau_s}\phi\right) \frac{I_{3s}(A)}{I_{1s}} , \tag{8.108}$$

where

$$I_{3s}(A) = \int_{-\infty}^{\infty} \frac{(P-\lambda_s A)}{\rho_s^2} \exp\left(-\frac{(P-\lambda_s A)^2}{2\rho_s^2}\right) g_s(P)\, dP \tag{8.109}$$

and pressure becomes

$$p_s = \frac{n_s \tau_s}{2} . \tag{8.110}$$

The quasi-neutrality condition $n_e = n_i = n$ determines the electric potential ϕ

$$\phi(A) = \tau_e \tau_i \ln\left(\frac{\hat{n}_i(A)}{\hat{n}_e(A)}\right), \tag{8.111}$$

such that from (8.104) we find for n

$$n = \hat{n}_i^{\tau_i} \hat{n}_e^{\tau_e}. \tag{8.112}$$

Consider the case where ρ is small compared to 1 for both ions and electrons. In the limit $\rho_s \to 0$ the factor $\exp\left(-(P-\lambda_s A)^2/2\rho_s^2\right)$ localizes the integrands in (8.107) and (8.109) to the neighbourhood of $P = \lambda_s A$ and one obtains

$$n = g_i(A)^{\tau_i} g_e(-A)^{\tau_e} \tag{8.113}$$

$$\phi = \tau_e \tau_i \ln\left(\frac{g_i(A) g_e(0)}{g_e(-A) g_i(0)}\right) \tag{8.114}$$

$$j_e = -\frac{\tau_e}{2}\left(\frac{g_i(A)}{g_i(0)}\right)^{\tau_i}\left(\frac{g_e(-A)}{g_e(0)}\right)^{\tau_e}\frac{g_e'(-A)}{g_e(-A)} \tag{8.115}$$

$$j_i = \frac{\tau_i}{2}\left(\frac{g_i(A)}{g_i(0)}\right)^{\tau_i}\left(\frac{g_e(-A)}{g_e(0)}\right)^{\tau_e}\frac{g_i'(A)}{g_i(A)} . \tag{8.116}$$

For corrections taking small non-vanishing ρ into account, see Schindler and Birn (2002).

The special choice $g_e(-P) = g_i(P)$ corresponds to the same spatial distributions of the particle orbits. This property allows us to study thermal particle effects in thin current sheet configurations obtained from magnetohydrostatics. As an example we choose the equilibrium sequence discussed in Section 8.4. Figures 8.7 and 8.8 show properties of a member of that sequence corresponding to an amplitude of the boundary perturbation slightly below the critical amplitude ($a_1 = 0.52$, $p_m = 0.7$, $\Delta p = 0.2$), $g_e(-P) = g_i(P)$, a realistic value of m_i/m_e (ions being protons) and $T_i/T_e = 5$. Fig. 8.7

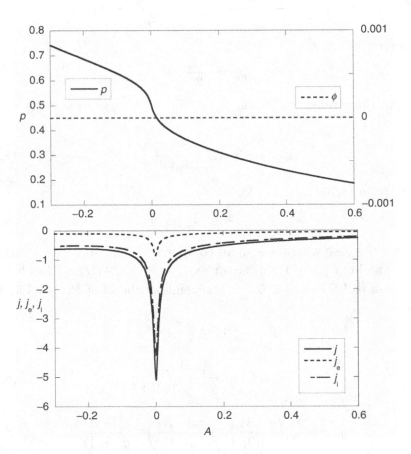

Fig. 8.7 Pressure p, electric potential ϕ and current densities j, j_e, j_i versus the flux function A for $\rho_i \to 0$. (Reproduced from Schindler and Birn (2002) by permission of the American Geophysical Union.)

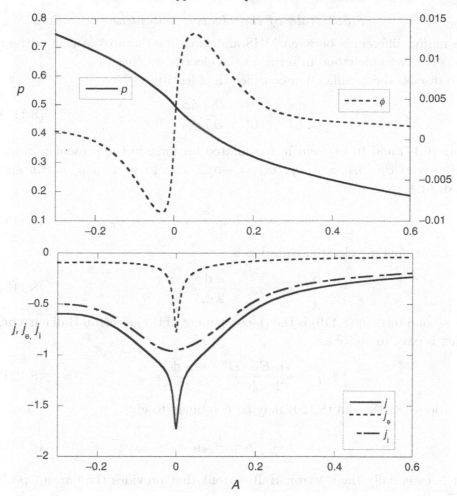

Fig. 8.8 Pressure p, electric potential ϕ and current densities j, j_e, j_i versus the flux function A for $\rho_i = 0.1$. (Reproduced from Schindler and Birn (2002) by permission of the American Geophysical Union.)

corresponds to the limit $\rho_i \to 0$. The maximum current density magnitude is about a factor of 10 larger than the background value with a rather sharp peak, with the corresponding change of slope of the pressure. As expected, the electric potential is vanishingly small and the contributions of electrons and ions to the total current density scale as their temperatures.

For $\rho_i = 1.0$ (Fig. 8.8) the ion current is smoothed considerably and even electron effects begin to become visible through a reduction of the maximum of $|j_e|$ (note the difference in scale). Nevertheless, even in this extreme case, there is still a current sheet with $|j|$ enhanced by about 50%.

8.5.3 *Role of the electric potential*

The major differences between MHS and kinetic structures of thin current sheets can be understood in terms of the electric potential ϕ.

To discuss the modification consider the identity

$$\frac{\mathrm{d}p_i}{\mathrm{d}A} = \frac{\partial p_i}{\partial A} + \frac{\partial p_i}{\partial \phi}\frac{\mathrm{d}\phi}{\mathrm{d}A}\ . \tag{8.117}$$

Using (6.17) and (6.18), which, formulated for ions, in the present normal-ization read $\partial p_i/\partial A = j_i$, $\partial p_i/\partial \phi = -n/2$, one finds with $p_i = \tau_i p$ and $j = \mathrm{d}p/\mathrm{d}A$, that

$$j_i = \tau_i j + \frac{n}{2}\frac{\mathrm{d}\phi}{\mathrm{d}A}\ . \tag{8.118}$$

The corresponding electron equation is

$$j_e = \tau_e j - \frac{n}{2}\frac{\mathrm{d}\phi}{\mathrm{d}A}\ . \tag{8.119}$$

The second term in (8.119) is the (y-component of the) electron Hall current, which is easy to verify as

$$\boldsymbol{j}_{eH} = -\frac{n}{2}\frac{\boldsymbol{E}\times\boldsymbol{B}}{B^2} = -\frac{n}{2}\frac{\mathrm{d}\phi}{\mathrm{d}A}\boldsymbol{e}_y\ . \tag{8.120}$$

Equations (8.118) and (8.120) may be combined to give

$$j = \frac{1}{\tau_i}j_i + \frac{1}{\tau_i}j_{eH}\ . \tag{8.121}$$

It is essentially the electron Hall current that provides the current peak in Fig. 8.8.

Under the present assumptions the electric potential would map along field lines to regions away from the thin current sheet. Possible consequences of this feature are discussed in Part IV.

8.6 Further aspects

Here we add a brief outline of further aspects relevant for slow evolution and/or thin current sheet formation.

8.6.1 *Potential fields*

For ignorable plasma pressure quasi-static magnetic fields are force-free fields and in the simplest case potential fields. Even in the latter case, singular current sheets can form by quasi-static evolution (Bungey and Priest, 1995). Here, we give a simple example. The arguments are similar to those used in the case of the perturbed Harris sheet in Section 8.2.

Suppose two antiparallel line dipoles of equal absolute strength oriented parallel to the y-axis start out with a large (formally infinite) distance between them and quasi-statically approach each other, moving along the x-axis. The plasma pressure is assumed to be negligible but the electrical conductivity to be large such that Ohm's law can be assumed to hold in its ideal form. As outlined earlier, this means that a gauge can be found so that the flux function $A(x, y, t)$ is a constant of motion (setting $g(t) = 0$ in (8.21)). Let the dipole positions be x_0 and $-x_0$, where x_0 is a positive function of time. Then, by symmetry, the plasma at $x = 0$ does not move along x, such that $A(0, y, t) = 0$ for all times.

As a first attempt to construct $A(x, y, t)$ one might try to simply add the two dipole fields. The top panel of Fig. 8.9 shows that case for $x_0 = 1$. However, one finds that $A(0, y, t) \neq 0$. In fact, noting that in suitable non-dimensional units the flux function A_D of a positive dipole (directed along the positive y-axis) located at $(x_0, 0)$ is given by

$$A_D(x, y, x_0) = \frac{x - x_0}{(x - x_0)^2 + y^2},$$
(8.122)

we find $A_D(0, y, x_0) - A_D(0, y, -x_0) = -2x_0/(x_0{}^2 + y^2)$, which does not vanish for $x_0 \neq 0$.

Therefore, the vacuum solution is not available in a highly conducting plasma, even if the pressure is negligible. The vacuum electric field is in conflict with ideal Ohm's law. Since the vacuum field of the dipoles is a unique solution of Laplace's equation, it is obvious that the solution of our plasma problem cannot be completely smooth. Again, we have to make a choice between field continuity and the ideal plasma behaviour.

To construct the solution, let us start with the vacuum case of two parallel dipoles (middle panel of Fig. 8.9). This solution is satisfactory for the plasma case in the left half-plane including $x = 0$, because $A_D(0, y, x_0) + A_D(0, y, -x_0) = 0$. In the right half-plane that function reproduces the dipole singularity at the correct location, however, for a dipole with the wrong sign. Thus, we simply have to change the sign in the right half-plane and obtain the wanted solution in the form

$$A(x, y, t) = \begin{cases} A_D(x, y, x_0) + A_D(x, y, -x_0), & x \geq 0 \\ -A_D(x, y, x_0) - A_D(x, y, -x_0), & x < 0. \end{cases}$$
(8.123)

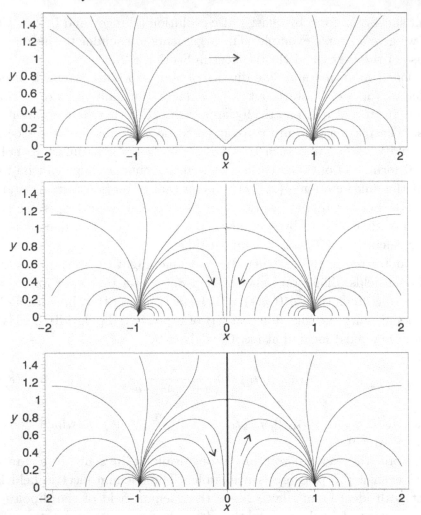

Fig. 8.9 Two line dipoles approaching each other. Their instantaneous locations are $x = \pm 1, y = 0$. The graphs are confined to $y > 0$. The top panel shows the vacuum solution for antiparallel dipoles. The middle panel gives the vacuum solution for parallel dipoles, which is used to construct the wanted solution (bottom panel) for antiparallel dipoles embedded in a highly conducting plasma. The latter shows a singular current sheet (thick line).

Since B_y shows a jump on the y-axis, that axis carries a tangential discontinuity (thick line in the bottom panel of Fig. 8.9). The surface current density has a z-component only, which is given by

$$K_z = 2 \frac{x_0{}^2 - y^2}{(x_0{}^2 + y^2)^2} \, . \tag{8.124}$$

K_z changes its sign at $y = \pm 1$ and the integrated surface current vanishes.

In general, two-dimensional potential fields can be constructed by using complex variables, because the real and imaginary parts of an analytical function satisfy Laplace's equation. For that purpose one defines a complex magnetic field

$$B_c(z) = B_y + iB_x, \quad z = x + iy .$$

(8.125)

The Cauchy–Riemann differential equations guarantee that $\nabla \cdot \boldsymbol{B} = 0$. A complex flux function is obtained by

$$A_c = -\int^z B_c \, dz ,$$

(8.126)

such that the physical flux function is given by

$$A(x, y) = \mathrm{Re}(A_c) .$$

(8.127)

For the dipole field (8.122) the complex formulation is

$$B_c = \frac{1}{(z - z_0)^2}, \quad A_c = \frac{1}{z - z_0} .$$

(8.128)

For suitable choices of B_c or A_c singular current sheets may arise from branch cuts (Biskamp, 1993; Bungey and Priest, 1995).

As an example let us choose

$$B_c = \sqrt{z^2 - L^2} ,$$

(8.129)

the corresponding field lines are shown in Fig. 8.10 for $L = 1$.

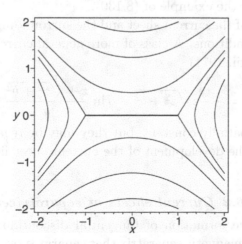

Fig. 8.10 Field lines of model (8.129), the thick line indicates the singular current sheet.

A singular current sheet extends on the x-axis for $-L < x < L$, where the sheet current density goes to zero at $x = \pm 1$. The end points are magnetic neutral points, but other than at x-points, where the magnetic field varies linearly with distance, here B goes to zero as the square root of distance. As the two separatrices and the current sheet form the letter Y, these neutral points are often referred to as y-type neutral points.

The complex flux function is given by

$$A_c = \frac{1}{2}\left(L^2 \ln(z + \sqrt{z^2 - L^2}) - z\sqrt{z^2 - L^2} \right). \qquad (8.130)$$

At large distances from the origin one finds asymptotically

$$A_c = -\frac{z^2}{2} + \frac{L^2}{2} \ln(z) \qquad (8.131)$$

which is the superposition of a current-free neutral point configuration (hyperbolic field lines) and the field of a line singularity at the origin. This is expected, because the current sheet singularity viewed from a large distance shrinks to a line singularity.

It has been suggested that current sheets of the form of Fig. 8.10 can arise from a pressure-free collapse of a hyperbolic x-line singularity (Dungey, 1953). However, it seems that a finite pressure might prevent the collapse from progressing (for a detailed discussion see Priest and Forbes (2000)).

A modified interpretation assumes that during a quasi-static evolution (with flux-conserving flow) the configuration instantaneously relaxes into a potential field (Longcope, 2001). Then, by adding magnetic flux, a current sheet is expected to form and its width L increases with the increasing flux, as can be seen from the example of (8.130).

The actual form of the current sheet and the surrounding field will depend on the boundary conditions. A class of more general current sheet fields has the form (Syrovatskii, 1971)

$$A_c = -\frac{\alpha}{2}z\sqrt{z^2 + b^2} - \beta \ln \frac{z + \sqrt{z^2 + b^2}}{b}, \qquad (8.132)$$

where α, β, b are spatially constant, but they may be regarded as functions of time describing the development of the current sheet field.

8.6.2 Current sheets at separatrices

Here we address the formation of tangential discontinuities in the MHD picture. Consider a magnetic separatrix that separates two regions from each other, in which some magnetic field and/or plasma properties are different.

A typical example is the separatrix between a planetary dipolar field and the interplanetary magnetic field. For simplicity, let us assume that the field lines on the magnetospheric side are all connected with the interior of the planet, while the interplanetary field lines have a much larger extent. (For the present argument their actual connection properties are irrelevant, what counts is the difference in connectivity of the field lines across the separatrix.) Clearly, in this situation the differential flux volume has a jump at the separatrix. Thus, by the previous discussion we can expect that a quasi-static evolution will lead to a singular current sheet, even if the initial fields are continuous.

For singular current sheets to occur it suffices that the thermodynamic properties of the plasmas on both sides of the separatrix are different. This difference leads to a jump in the plasma pressure and, by pressure balance, to a jump in the magnetic field and thus to a non-vanishing surface current density. The following simple example illustrates this process.

Consider the situation shown in Fig. 8.11, where the separatrix located at position x separates the regions 1 and 2 with two thermodynamically different MHD media. This system undergoes adiabatic compression by slow reduction of length L with the following properties:

initial conditions:

$$B_1 = B_2 = B_0, \quad p_1 = p_2 = p_0, \quad x = x_0, \qquad (8.133)$$

magnetic flux conservation for each region:

$$B_1 x = B_0 x_0, \quad B_2(L - x) = B_0(L_0 - x_0), \qquad (8.134)$$

thermodynamic constraints:

$$p_1 = p_0, \quad p_2(L - x)^\gamma = p_0(L_0 - x_0)^\gamma, \qquad (8.135)$$

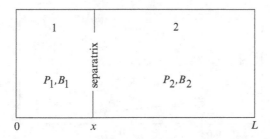

Fig. 8.11 Model configuration in which a singular current sheet forms at the separatrix located at position x and separating thermodynamically different regions 1 and 2 with pressures and magnetic field magnitudes p_1, B_1 and p_2, B_2, respectively; the magnetic field direction is perpendicular to the plane of the graph.

where the label '0' refers to initial values. Note that initially pressure and magnetic field are continuous. For region 1 it is assumed that the pressure remains constant. This may be due to large flux tubes, plasma mass or energy losses. For region 2 an adiabatic pressure law is assumed.

Pressure balance at the separatrix

$$p_1 + \frac{B_1^2}{2\mu_0} = p_2 + \frac{B_2^2}{2\mu_0} \tag{8.136}$$

together with the conditions (8.133), (8.134) and (8.135), determine the variables x, p_1, B_1, p_2, B_2 as functions of the box width L. The magnitude of the surface current density is

$$K = |B_2 - B_1|/\mu_0| . \tag{8.137}$$

Fig. 8.12 gives an example. It seems that this mechanism, among others, is able to explain the sharp boundary of a magnetosphere, i.e., its magnetopause.

More realistic cases reveal a relationship between separatrices and magnetic neutral points, a topic that will be taken up in Chapters 11 and 14. For a current sheet arising from the compression of a Harris sheet, see www.tp4.rub.de/~ks/tc.pdf.

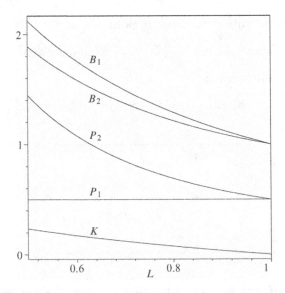

Fig. 8.12 The variables p_1, B_1, p_2, B_2 (quantities normalized by their initial values) and the magnitude of surface current density K (normalized by B_0/μ_0) as functions of the box width L for $x_0/L_0 = 2\mu_0 p_0/B_0^2 = 0.5$.

8.6.3 Quasi-static formation of tangential discontinuities?

We have seen that, in some cases, thin current sheets are represented by tangential discontinuities. In a mathematical context, where discontinuous fields are admitted, there is no reason to raise questions about representing current sheets as singularities. From a more realistic physical point of view, the following question arises. For realistic (continuous) fields the tangential discontinuity must be understood as an approximate description of a narrow sheet in which the magnetic field rapidly changes. For illustration let us choose a case where the discontinuity separates oppositely directed fields and there exists a plane where B vanishes. It was pointed out by Parker (1994) that under (quasi-)static conditions the inner structure of the sheet must be in equilibrium also, which means that pressure balance holds across the sheet. In particular, the pressure p_n at the neutral plane is determined by the total pressure $P = p + B^2/(2\mu_0)$ which is constant across the sheet, therefore $p_n = P$.

Except for 1D cases, the pressure and the magnetic field vary along the discontinuity, see Fig. 8.9 for an example. This means that P and therefore p_n vary along the discontinuity. In the absence of a magnetic field component normal to the sheet (a characteristic feature of a tangential discontinuity, see Section 3.9), there is no force to compensate the pressure gradient at the neutral plane. There is lack of equilibrium and plasma flow will appear, which transports plasma in the direction of $-\nabla p_n$. In the regions of higher p_n the pressure will be reduced so that the local sheet structure will be compressed. Under suitable boundary conditions this would continue as a runaway process and there will be regions where the sheet becomes arbitrarily thin. This is Parker's nonequilibrium argument (Parker, 1994).

A consequence of this phenomenon is rapid decay of current sheets by magnetic reconnection (see Chapter 11), which would be a powerful mechanism for dissipating magnetic energy. This would also mean that the quasi-static build-up of thin current sheets, as discussed in previous parts of this chapter, would turn into a dynamic phase (see also Zhang and Low (2002)).

However, this nonequilibrium argument, which may play an important role under suitable circumstances, does not apply to thin current sheets in general. The thin current sheets that were considered in Sections 8.3 and 8.4 are not affected by internal nonequilibrium conditions, although they can become arbitrarily thin (The loss of equilibrium of Section 8.4.2 has a different cause.) They contain a small normal magnetic field component B_n, which together with the large current density gives rise to a $j \times B$ force that balances the internal pressure gradient. It is true that in the limit of vanishing thickness B_n vanishes, but the current density becomes infinitely large so that their product remains finite. This follows from the fact that in the models described in Sections 8.3 and 8.4 the thin current sheets are inherent parts of quasi-static states.

This property may be relevant for a variety of self-consistent models of current sheets (see also Zhang and Low (2002)). For an explicit illustration, let us discuss a simple case of the type considered in Section 5.4.1, choosing

$$p(A) = \frac{1}{2}e^{-2A/\delta}, \quad \hat{p} = \frac{1}{1+x_1^2} \, . \tag{8.138}$$

The current density is given by

$$j_y = -\frac{2\hat{p}}{\delta \cosh^2(z\sqrt{2\hat{p}}/\delta)} \tag{8.139}$$

which in the limit $\delta \to 0$ assumes the singular form

$$j_y \to -2\sqrt{2\hat{p}}\delta(z) \tag{8.140}$$

consistent with a tangential discontinuity.

Fig. 8.13 shows the magnetic field lines and the current density for $\delta = 0.1$ (top graphs) and the field lines for the limit $\delta \to 0$ (bottom graph), illustrating the tangential discontinuity. For any value of $\delta \neq 0$ the configuration is in static equilibrium in the sense of the asymptotics of Section 5.4.1. In particular, there is force balance along the sheet. Fig. 8.13a indicates the internal structure that causes a B_z-component of order δ; the current density (Fig. 8.13b) is of order $1/\delta$; their product balances the pressure gradient, which is of order 1.

Even in that approach a quasi-static sequence can find its end in a singularity so that there is no equilibrium beyond that point, but that is a different matter (see Section 8.4.2).

Another possibility to avoid sheet nonequilibrium in thin sheets is to apply a gravity force to balance the internal pressure gradient (e.g., Zhang and Low, 2002).

8.6.4 Weak singularities

So far we have discussed singular current sheets mostly in the sense of a tangential discontinuity, where the current density has a δ-function singularity, leading to a non-vanishing surface current density and a step-function jump in the magnetic field. This, however, is not the only possibility of a

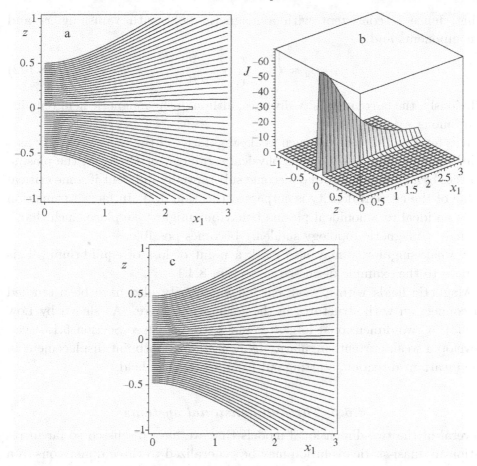

Fig. 8.13 Formation of a tangential discontinuity with pressure balance using the model of (8.138). Graph a shows the field lines for $\delta = 0.1$, graph b the corresponding current density, graph c gives the field lines in the limit $\delta \to 0$ where the current sheet shrinks to a tangential discontinuity (thick line).

current sheet, in which the current density diverges. It is also possible that the current density has an integrable singularity, with no surface current and with a continuous magnetic field.

We illustrate this possibility by the following simple one-dimensional example. Let the current density have a y-component only, which is given by

$$j_z = \begin{cases} 1, & x \le 0 \\ \frac{1}{\sqrt{x}}, & x > 0 \end{cases}. \tag{8.141}$$

This choice is consistent with a magnetic field with vanishing x- and z-components and

$$B_y = \begin{cases} x, & x \leq 0 \\ 2\sqrt{x}, & x > 0. \end{cases} \tag{8.142}$$

Obviously, the current density diverges, although the magnetic field remains continuous.

It is tempting to consider such weak singularity of the current density as a mathematical oddity with little physical significance. However, the plasma physics of a current sheet may become substantially different if some critical value of the current density is surpassed (Section 9.2). In fact, a transition from an ideal to a nonideal plasma behaviour might take place, such that a change of magnetic topology suddenly becomes possible.

A weak singularity may also mark a point of loss of equilibrium. This applies to the example discussed in Section 8.4.1.

Magnetic fields with a weak current singularity also have been studied in connection with structures in the solar atmosphere. As shown by Low (1993), a two-dimensional force-free magnetic field (see Section 5.1.5) can develop a weak current singularity by continuous footpoint displacement in the invariant direction, starting out from a potential field.

8.6.5 Three-dimensional systems

Several of the two-dimensional models that we have discussed so far in relation to quasi-static evolution may be generalized to three dimensions in a straightforward way. In particular, this applies to the flux-transfer model of Section 8.3. The snapshots may be taken from the approach described in Section 5.4.3. Accordingly, the magnetic field is expressed by Euler potentials α and β, and the adiabatic law is to be applied to every flux tube (α, β).

A corresponding model has been evaluated for magnetospheric conditions (Birn and Schindler, 1983). Several features that we encountered in the two-dimensional case persist in three dimensions. Examples are the existence of similarity solutions, strong shielding of an external electric field and the pressure crisis.

The formation of thin current sheets based on large gradients of the flux tube volume, as discussed in Section 8.3.4, in principle also applies to three-dimensional fields. This is readily illustrated for fields possessing Euler potentials. From (5.21) and the adiabatic law

$$p(\alpha, \beta) = p_0(\alpha, \beta) \left(\frac{V_0(\alpha, \beta)}{V(\alpha, \beta)} \right)^{\gamma} \tag{8.143}$$

one finds

$$\boldsymbol{j} \cdot \nabla \beta = \frac{\partial p_0}{\partial \alpha} \left(\frac{V_0}{V}\right)^{\gamma} + p_0 \frac{\partial}{\partial \alpha} \left(\frac{V_0}{V}\right)^{\gamma}. \tag{8.144}$$

Even if the initial pressure gradient and current density are not particularly large, a large gradient of V can soon lead to a large gradient of V_0/V, such that $\boldsymbol{j} \cdot \nabla \beta$ will become large. A corresponding argument applies to $\boldsymbol{j} \cdot \nabla \alpha$. Numerical simulations have confirmed the formation of thin current sheets in the evolution of three-dimensional fields under suitable conditions (e.g., Birn *et al.*, 2004).

Part III
Dynamics

In this part we turn to the description of dynamical plasma states. In view of our main objectives particular emphasis is placed on the question of how the sudden transition from a quiescent state to a fast evolving dynamical state may take place and how such transitions may be described quantitatively.

There are two major ways in which a physical system can go from a quiescent to a dynamical state.

In the first case the dynamical behaviour may be the obvious result of external forces and, in particular, the energy associated with the dynamical evolution is supplied from external sources. In other words, there is no release of previously stored energy. It is typical for this case that small external perturbations will have small internal effects.

The other possibility is that there is some amount of energy (usually called 'free energy') available in the original quiescent system, which leads to an enhancement of the dynamical reaction. In response to the external perturbation the system moves away from its original state and, because of the available free energy, the energy associated with the dynamic development is not bounded by the energy supplied by external action. In this way a small perturbation can have a large effect.

If the first case applies to a system for all relevant perturbations, that system is *stable*. If a system admits a perturbation realizing the second possibility, the system is *unstable*. (There are many everyday examples, e.g., a ball on a hilltop.)

Although these (rather fuzzy) notions of stability and instability can be quite valuable in qualitative arguments, they have to be put on a more rigorous basis before they become useful as quantitative tools. This will be the topic of Section 10.1.

Obviously, understanding activity requires a thorough discussion of the relevant instability processes. Therefore, a major fraction of this part will be devoted to instabilities (Chapter 10).

Even if the stability problem is solved, the consequences that a given instability has for an actual physical system are not easy to assess. Typically, the following difficulty arises. Although for a given system it is important to obtain as much stability information as possible, an unstable quiescent state remains a theoretical fiction; it does not exist in the real world, because a broad spectrum of external perturbations, which is always present, is likely to excite the unstable modes.

Analysing bifurcation properties will shed some light on the ways in which instabilities might become effective in real systems. In that context physical constraints and their sudden breakdown will be found to play an important role. (In the picture of a ball on a hilltop, a constraint could be realized by a human hand holding the ball and the breakdown by the release occurring when suddenly the hand opens.)

A set of constraints relevant for space plasma activity results from ideal Ohm's law (see Section 3.8). Their breakdown would require that at some stage of the slow evolution nonideal effects in Ohm's law become relevant. Therefore, we begin with a discussion of such nonideal effects.

9

Nonideal effects

For the dynamic processes that are of interest for activity of space plasmas an important issue is the deviation from the ideal form of Ohm's law (3.60).

Nonideal processes play an important role for stability properties and they are crucial for breaking constraints holding under ideal MHD conditions, among them magnetic flux and line conservation (see Section 3.8.2). Before addressing the consequences of such breakdown of conservation laws, we rewrite Ohm's law in a suitable form, called *generalized Ohm's law*, and discuss possible nonideal effects and the underlying physics.

9.1 Generalized Ohm's law

We start our discussion by considering a form of Ohm's law that is more general than (3.57). It is obtained by multiplying the electron kinetic equation (3.18) with $m_e \boldsymbol{w}$ and integrating over velocity space. The result is the momentum equation of the collisionless case (3.34), specified for electrons, with the additional collision term $M_e^{(bc)} = \int m_e \boldsymbol{w} \partial f_e / \partial t|_{coll} \, d^3 w$, arising from binary collisions,

$$m_e n_e \frac{\partial \boldsymbol{v}_e}{\partial t} + m_e n_e \boldsymbol{v}_e \cdot \nabla \boldsymbol{v}_e = -\nabla \cdot \mathcal{P}_e - e n_e (\boldsymbol{E} + \boldsymbol{v}_e \times \boldsymbol{B})$$

$$- m_e n_e \nabla \psi + \boldsymbol{M}_e^{(bc)}. \tag{9.1}$$

The superscript 'bc' stands for *binary collisions*. To rewrite (9.1) in the form of Ohm's law, we again make the assumption that there is only one ion species present and that the ions are singly charged. Then, making use of

185

the definitions of charge density σ, current density j and plasma bulk flow velocity v

$$\sigma \;=\; en_i + en_e \tag{9.2}$$

$$j \;=\; en_i v_i - en_e v_e \tag{9.3}$$

$$v \;=\; \frac{n_i m_i v_i + n_e m_e v_e}{n_i m_i + n_e m_e}, \tag{9.4}$$

we rewrite (9.1) in the form

$$\boldsymbol{E} + \boldsymbol{v} \times \boldsymbol{B} = -\frac{1}{en_e}\nabla \cdot \mathcal{P}_e + \frac{(\boldsymbol{j} - \sigma \boldsymbol{v}_e) \times \boldsymbol{B}}{en_e(1 + \frac{\sigma}{en_e} + \frac{m_e}{m_i})}$$

$$-\frac{m_e}{e}\left(\frac{\partial \boldsymbol{v}_e}{\partial t} + \boldsymbol{v}_e \cdot \nabla \boldsymbol{v}_e\right) - \frac{m_e}{e}\nabla \psi + \frac{\boldsymbol{M}_e^{(bc)}}{en_e}. \tag{9.5}$$

In this detailed version of generalized Ohm's law the nonideal effects are introduced by the right-hand side. From left to right the terms are referred to as the pressure term, the generalized Hall term, the electron inertia term, the gravity term and the resistive term based on particle collisions. If there is no magnetic field, typically the resistive term is parallel to j. One then writes

$$\frac{\boldsymbol{M}_e^{(bc)}}{en_e} = \eta \boldsymbol{j} \tag{9.6}$$

where η is the resistivity. The term $\eta \boldsymbol{j}$ is widely used also for non-vanishing magnetic fields, ignoring the fact that then $\boldsymbol{M}_e^{(bc)}$ and $\boldsymbol{j}$ are not strictly parallel (Spitzer, 1962; Braginskii, 1965).

For quasi-neutrality ($\sigma = 0, n_e = n_i = n$), $m_e/m_i \ll 1$ and the resistive term given by (9.6), (9.5) reduces to

$$\boldsymbol{E} + \boldsymbol{v} \times \boldsymbol{B} = -\frac{1}{en_e}\nabla \cdot \mathcal{P}_e + \frac{\boldsymbol{j} \times \boldsymbol{B}}{en_e} - \frac{m_e}{e}\left(\frac{\partial \boldsymbol{v}_e}{\partial t} + \boldsymbol{v}_e \cdot \nabla \boldsymbol{v}_e\right) + \eta \boldsymbol{j}, \tag{9.7}$$

where the inertia term was kept to cover cases where it cannot be ignored because v_e becomes large.

The pressure term is sometimes written as

$$\mathcal{P}_e = p_e \mathcal{I} + \mathcal{P}_e^{(V)}, \tag{9.8}$$

where

$$p = \frac{1}{3}\mathrm{Tr}(\mathcal{P}_e), \quad \mathcal{P}_e^{(V)} = \mathcal{P}_e - \frac{1}{3}\mathrm{Tr}(\mathcal{P}_e)\mathcal{I}. \tag{9.9}$$

Here, p_e is the scalar electron pressure and $\mathcal{P}_e^{(V)}$ has vanishing trace and describes the non-isotropic part of $\mathcal{P}_e$, which typically accounts for viscous interactions.

For assessing the role that resistivity plays under various circumstances, it is necessary to link η to plasma properties that are directly observable. This is done in the following section.

9.2 Resistivity

In a fully ionized gas resistivity is due to Coulomb collisions between electrons and ions. If the corresponding collision time is τ_{ei}, the resistivity η may be expressed as

$$\eta = \frac{m_e}{e^2 n_e \tau_{ei}} . \tag{9.10}$$

Here, τ_{ei} is understood as the relaxation time for electron–ion momentum transfer, such that (in the ion rest frame) $M_e^{(bc)} = -m_e n_e v_e / \tau_{ei}$, which combined with (9.6) and $j = -e n_e v_e$ gives (9.10).

An appropriate expression for the collision time τ_{ei} follows from an explicit treatment of electron–ion collisions. Since, as we will see, collisional effects are often extremely rare in space plasmas, we skip rigorous derivations here and just list a number of relevant definitions with brief explanations.

Resistivity in fully ionized plasmas is a classical topic of plasma physics (e.g., Spitzer, 1962; Trubnikov, 1965). Here we follow Spitzer (1962).

The resistivity is given by

$$\eta = \frac{\pi^{3/2}}{4\sqrt{2}} \frac{e^2 \sqrt{m_e}}{(4\pi\epsilon_0)^2 (k_B T_e)^{3/2}} \frac{\lambda_c}{\gamma_E} . \tag{9.11}$$

The factor $\gamma_E = 0.582$ corrects for the deviation of the electron distribution function from a shifted Maxwellian (Spitzer, 1962) and λ_c is the *Coulomb logarithm*. The latter takes care of the cumulative effect of distant electron–ion encounters with small deflections, adding up to an average deflection of $90°$. Typical values of λ_c lie between 10 and 30, indicating that collisions caused by distant encounters typically dominate.

In the presence of a magnetic field an error arises from ignoring the fact that a more rigorous approach would result in a resistivity tensor that is anisotropic (Spitzer, 1962). However, this correction is widely ignored as it does not seem to have a drastic influence on dynamical properties.

However, there are other restrictions to the validity of (9.11) which do play a crucial role. We mention three particularly important items.

1. The expression (9.11) for η is valid only for *weakly coupled* plasmas. These are plasmas for which two-particle correlations dominate, which requires $\Lambda_p \gg 1$ (Balescu, 1975; Ichimaru, 1973). Weakly-coupled plasmas are also called 'ideal'. We do not use that term here to avoid confusion with the regime of the ideal form of Ohm's law, where collisional effects are ignored completely.

2. The theory of Coulomb collisions as used here is valid only for plasmas with macroscopic length scales larger than the mean free path $l_e = v_e \tau_{ei}$ of the electrons. If this condition is violated and if the collision time exceeds the typical time a particle spends within the system (*life-time*), it is appropriate to set the resistivity to zero. In that case quantities such as (9.11) represent extrapolations and have no quantitative physical significance.

3. The linear relationship between E and j, implied by the existence of a resistivity, breaks down if E exceeds a critical value. Then, the collisional momentum transfer saturates and the force $\boldsymbol{M}_e^{(bc)}$ is unable to balance the electric force $-en_e\boldsymbol{E}$. In plasmas without magnetic field (or in the presence of a magnetic field and an electric field component parallel to $\boldsymbol{B}$) the saturation results in acceleration (*runaway*) of the electrons. The runaway effect occurs if the electron drift velocity exceeds the electron thermal velocity v_{te}, such that the critical electric field strength (*Dreicer field*) approximately is given by (Dreicer, 1960)

$$E_c = \frac{m_e v_e}{e\tau_{ei}} .\tag{9.12}$$

So far we have looked at Coulomb collisions from a local particle point of view. We now turn to the relevance of collisional resistivity for macroscopic dynamics. A strict assessment requires solving the resistive MHD equations of Section 3.3.3 for the system of interest. A rough estimate can be obtained from considering the induction equation and Ohm's law alone. Eliminating the electric field between these equations one finds

$$\frac{\partial \boldsymbol{B}}{\partial t} - \nabla \times (\boldsymbol{v} \times \boldsymbol{B}) + \nabla \times (\frac{\eta}{\mu_0}\nabla \times \boldsymbol{B}) = 0 \tag{9.13}$$

which has the form of a vector diffusion equation. Let us consider the resistive decay of an initial equilibrium, assuming that the diffusion process can be regarded as a slow evolution with static snapshots. Then, typically, the second term of (9.13) will not dominate the first one and a rough estimate of the diffusion time scale τ_D can be obtained by equating the first term to the third term,

$$\tau_D = \frac{L^2 \mu_0}{\eta}. \tag{9.14}$$

Here it was assumed that there is only one characteristic length scale, L, such that ∇ scales as $1/L$. Also, $\partial/\partial t$ was replaced by $1/\tau_D$. The time τ_D is the characteristic time scale for the evolution due to resistive dissipation alone.

A non-dimensional measure for the relevance of resistive diffusion is obtained by comparing τ_D with the time scale of a typical ideal MHD process. Choosing as that time scale the Alfvén time $\tau_A = L/v_A$, where $v_A = B/(\mu_0\rho)^{1/2}$ is the Alfvén velocity, leads to defining the Lundquist number

$$S = \frac{\tau_D}{\tau_A} = \frac{\mu_0 L v_A}{\eta}. \tag{9.15}$$

If S is large compared to 1, resistive effects can be neglected for phenomena occurring on the Alfvén time scale or faster.

When using S it should be kept in mind that the Alfvén velocity was used as the typical macroscopic signal velocity in the plasma. This applies to plasmas for which the ratio of kinetic to magnetic pressure

$$\beta_p = \frac{2\mu_0 p}{B^2} \tag{9.16}$$

is not large compared to 1. Otherwise, v_A must be replaced by a suitable MHD phase velocity. One uses S even for $\beta \gg 1$ by introducing an appropriate effective magnetic field. For the centre of a Harris sheet (Section 5.3.2) with $T_i > T_e$ the effective magnetic field is the field outside the sheet. (By pressure balance the ion thermal velocity in the centre is approximately equal to the external Alfvén velocity.)

Table 9.1 shows several quantities relevant for Coulomb collisions for three different plasma media. The cases 1–3 lie in the general parameter ranges of the interstellar medium (HII regions), the solar corona and the Earth's magnetosphere, respectively. In the latter case the values for the central plasma sheet were taken (see Table 2.1), except for the magnetic field, where the lobe value is the appropriate effective field for the high-β plasma sheet. Note that in view of strong spatial and/or temporal variability unique 'typical' parameter sets cannot be defined. Therefore, the numbers do not give more than rough indications.

Clearly, the values of the plasma parameter Λ_p and of the Lundquist number S are large compared to 1 in all three cases, so that the regime of weak coupling applies and resistivity can be ignored in the macroscopic dynamics on scale L.

In case 3 the electron mean free path l_e by far exceeds L. Also, considering the plasma sheet, the formal collision time of 5.1×10^7 s is much larger than the time a particle typically spends in the sheet (smaller than one day).

Table 9.1 *Quantities relevant for Coulomb collisions for three media,*
standing for HII-regions (case 1), the solar corona (case 2) and the
Earth's plasma sheet (case 3). Brackets indicate extrapolation.

	case 1	case 2	case 3
n/m^{-3}	5×10^9	1×10^{15}	2×10^5
T_e/K	1×10^4	2×10^6	1×10^7
L/m	1×10^{15}	3×10^7	2×10^7
B/T	1×10^{-9}	1×10^{-2}	2×10^{-8}
$\omega_\mathrm{p}/\mathrm{s}^{-1}$	4×10^6	2×10^9	3×10^4
Λ_p	2×10^7	1×10^8	1×10^{14}
l_e/L	4×10^{-9}	2×10^{-2}	(2×10^9)
$\tau_\mathrm{ei}/\mathrm{s}$	6×10^0	7×10^{-2}	(2×10^9)
$\eta/(\mathrm{Vm/A})$	1×10^{-3}	5×10^{-7}	(7×10^{-8})
S	3×10^{14}	5×10^{14}	(3×10^{14})

Therefore, collisional effects can be ignored altogether. As mentioned above, in that case parameters involving resistivity are extrapolations without precise quantitative meaning and therefore are put in brackets in the table. This feature does not necessarily apply to radiation belt populations, where the particle life-times can be much longer.

If instead of a single scale L (as assumed in the present discussion) a system possesses two or more different length scales, a more complicated situation arises. Then, resistive processes associated with a scale much smaller than the corresponding value given in Table 9.1 may still play a significant role.

9.3 Microturbulence

As we have seen above, electric fluctuations associated with the motion of discrete plasma particles are the cause of collisional resistivity. Here we consider the consequences of fluctuations in a collisionless plasma. Thus, we are dealing with the Vlasov regime (Section 3.2). Microturbulence has become a very large field and here we can touch on a few basic facts only. A number of examples, relevant for the present context, are discussed in some detail in www.tp4.rub.de/~ks/ta.pdf.

In the absence of collisions, fluctuations largely arise from plasma instabilities. Here we are interested in instabilities that lead to turbulence with length and time scales small compared with the corresponding macroscopic scales. Such (linear) instabilities and the associated (nonlinear) turbulence are referred to as *microinstabilities* and *microturbulence*, respectively. From a macroscopic point of view microturbulence, suitably averaged over small scales, may be the cause of *turbulent* or *collective transport*, analogous to collisional transport. Then the collective electron–ion momentum transfer causes a contribution $M_e^{(\text{turb})}/(en_e)$ to generalized Ohm's law (9.5) which is analogous to $M_e^{(\text{bc})}/(en_e)$ based on collisions. If $M_e^{(\text{turb})}$ is parallel to j, we may write $M_e^{(\text{turb})}/(en_e) = \eta_{\text{turb}} j$ where η_{turb} is the *turbulent* or *collective resistivity*. (We avoid the notion of 'anomalous resistivity' as widely used in laboratory plasma physics, because in space plasmas turbulent resistivity seems to play a more important role than its collisional counterpart.)

A common aspect of collisional and turbulent resistivity is the presence of space charge and associated electric field fluctuations, which lead to momentum exchange between electrons and ions.

Since the collective resistivity will depend on the amplitudes of the fluctuations, a quantitative description must come from a nonlinear theory. Nevertheless, the question of the excitation of the fluctuations also poses a linear stability problem. Thus, we will first discuss the linear stability problem and then turn to the nonlinear regime.

9.3.1 *Microinstabilities*

Microinstabilities require deviation from thermodynamic equilibrium, such as counterstreaming, velocity shear, pressure anisotropy or gradients. Here we give a brief overview, concentrating on aspects that are of particular relevance for our present purposes. For detailed descriptions we refer to the literature (e.g., Davidson, 1972; Mikhailovskii, 1974; Hasegawa, 1975; Melrose, 1986; Gary, 1993). See also www.tp4.rub.de/~ks/ta.pdf.

The standard linear stability analysis in Vlasov theory starts from the collisionless Boltzmann equation (3.26). This equation is linearized for small perturbations (f_{s1}, E_1, B_1) added to a given background state solution (subscript 0). In most analytical treatments the background is a spatially homogeneous steady state or it varies on large length scales. The latter case is particularly important for systems where the instability is caused by the inhomogeneity. We begin by assuming that the background is homogeneous.

The linearized version of (3.26) is

$$\frac{\partial f_{s1}}{\partial t} + w \cdot \frac{\partial f_{s1}}{\partial r} + \frac{q_s}{m_s}(E_0 + w \times B_0) \cdot \frac{\partial f_{s1}}{\partial w}$$

$$= -\frac{q_s}{m_s}(E_1 + w \times B_1) \cdot \frac{\partial f_{s0}}{\partial w} . \quad (9.17)$$

The left side of (9.17) can be interpreted as the total time-derivative of f_{s1} taken along particle orbits in the unperturbed electric and magnetic fields. Thus, a (formal) solution of (9.17) is obtained by integrating the right side along those unperturbed orbits,

$$f_{s1}(\boldsymbol{r}, \boldsymbol{w}, t) = f_{s1}^{\mathrm{in}}(\boldsymbol{r}_0, \boldsymbol{w}_0)$$

$$- \frac{q_s}{m_s} \int_{t_0}^{t} (\boldsymbol{E}_1(\boldsymbol{r}', t') + \boldsymbol{w}' \times \boldsymbol{B}_1(\boldsymbol{r}', t')) \cdot \frac{\partial f_{s0}(\boldsymbol{w}')}{\partial \boldsymbol{w}'} \mathrm{d}t' , \quad (9.18)$$

where $\boldsymbol{r}' = \boldsymbol{r}'(\boldsymbol{r}, \boldsymbol{w}, t, t')$ describes the particle orbit that assumes position and velocity $(\boldsymbol{r}', \boldsymbol{w}')$ at time t' and $(\boldsymbol{r}, \boldsymbol{w})$ at time t. The variables appearing in the argument of the initial perturbation f_{s1}^{in} are $\boldsymbol{r}_0 = \boldsymbol{r}'(\boldsymbol{r}, \boldsymbol{w}, t, t_0)$, $\boldsymbol{w}_0 = \boldsymbol{w}'(\boldsymbol{r}, \boldsymbol{w}, t, t_0)$.

The fact that the coefficients multiplying the perturbations in (9.17) can be regarded as independent of $\boldsymbol{r}$ and t suggests the introduction of exponential modes. This would mean

$$\boldsymbol{E}_1 = \hat{\boldsymbol{E}}_1(\boldsymbol{k}, \omega) \, \mathrm{e}^{\mathrm{i}\boldsymbol{k}\cdot\boldsymbol{r}-\mathrm{i}\omega t} \qquad (9.19)$$

for the electric field perturbation and corresponding expressions for $\boldsymbol{B}_1$ and f_{s1}.

It turns out, however, that in general the Vlasov theory possesses modes of the type (9.19) only asymptotically for large times elapsed after an initial perturbation is applied (see e.g., Ichimaru, 1973). This follows from an analysis using a combined Laplace–Fourier transform assuming sufficiently smooth initial conditions (e.g., Krall and Trivelpiece, 1973), where the Laplace transform appropriately takes into account causality. Then, a mode with $\gamma = \mathrm{Im}(\omega) > 0$ can be treated as adiabatically switched on, i.e., as growing from vanishing initial conditions at $t = -\infty$,

$$\hat{f}_{s1}(\boldsymbol{k}, \omega) =$$

$$- \frac{q_s}{m_s} \int_{-\infty}^{0} (\hat{\boldsymbol{E}}_1(\boldsymbol{k}, \omega) + \boldsymbol{w}' \times \hat{\boldsymbol{B}}_1(\boldsymbol{k}, \omega)) \cdot \frac{\partial f_{s0}(\boldsymbol{w}')}{\partial \boldsymbol{w}'} \mathrm{e}^{\mathrm{i}\boldsymbol{k}\cdot(\boldsymbol{r}'-\boldsymbol{r})-\mathrm{i}\omega\tau} \mathrm{d}\tau , \qquad (9.20)$$

where the integration variable has been changed from t' to $\tau = t' - t$. Modes with $\gamma \leq 0$ require analytical continuation.

The central problem is to find the dispersion relation, which determines how ω is related to the wave number $\boldsymbol{k}$. To obtain the dispersion relation, one expresses $\hat{\boldsymbol{B}}_1$ by $\hat{\boldsymbol{E}}_1$ via the induction equation (3.28),

$$\hat{\boldsymbol{B}}_1 = \frac{1}{\omega} \boldsymbol{k} \times \hat{\boldsymbol{E}}_1 , \qquad (9.21)$$

such that in view of (9.20) the perturbation amplitude $\hat{f}_{s1}$ becomes linear and homogeneous in the components of $\hat{E}_1$. This property is carried over to the components of the electrical current density amplitude

$$\hat{j}_1 = \sum_s \int q_s w \hat{f}_{s1} \mathrm{d}^3 w \,, \tag{9.22}$$

which may be expressed as

$$\hat{j}_1(k, \omega) = \mathcal{C}(k, \omega) \cdot \hat{E}_1(k, \omega) \,, \tag{9.23}$$

where the tensor $\mathcal{C}(k, \omega)$ is the sum of the species contributions $\mathcal{C}_s(k, \omega)$.

The electric charge density is directly obtained from the current density via charge conservation (3.33),

$$\hat{\sigma}_1(k, \omega) = \frac{1}{\omega} k \cdot \hat{j}_1(k, \omega) \,. \tag{9.24}$$

Again, σ_1 and j_1 are sums over the contributions of all particle species.

The dispersion relation is determined by introducing modes of the form (9.19) into Maxwell's equations (3.27)–(3.30). Making use of (9.23) and (9.24) one obtains

$$\mathcal{K}(k, \omega) \cdot \hat{E}_1(k, \omega) = 0 \,, \tag{9.25}$$

where

$$\mathcal{K}(k, \omega) = \frac{c^2}{\omega^2}(kk - k^2 \mathcal{I}) + \frac{\mathrm{i}}{\epsilon_0 \omega} \mathcal{C}(k, \omega) + \mathcal{I} \,, \tag{9.26}$$

with $\mathcal{I}$ denoting the unit tensor.

For a non-vanishing electric field to exist, (9.25) implies that the determinant of $\mathcal{K}$ vanishes,

$$\det(\mathcal{K}(k, \omega)) = 0 \,, \tag{9.27}$$

which is the dispersion relation. It determines the complex frequency ω for a given wave number k, which for the present initial value problem is real.

The dispersion relation usually has a number of discrete solution branches for $\omega(k)$. The sign of γ determines whether a mode is stable ($\gamma < 0$) or unstable ($\gamma > 0$).

For longitudinal modes, satisfying $\nabla \times E_1 = 0$, $\hat{E}_1$ is parallel to k, $\hat{B}_1$ vanishes, and E_1 can be derived from a scalar potential ϕ_1 with $\hat{E}_1 = -\mathrm{i}k\hat{\phi}_1$. In that case the dispersion relation may be simplified considerably. One finds

$$1 + \chi(k, \omega) = 0 \tag{9.28}$$

where

$$\chi(\mathbf{k}, \omega) = \sum_s \chi_s(\mathbf{k}, \omega) \text{ with } \chi_s(\mathbf{k}, \omega) = \frac{i}{\epsilon_0 \omega k^2} \mathbf{k} \cdot \mathcal{C}_s(\mathbf{k}, \omega) \cdot \mathbf{k} \qquad (9.29)$$

is the *electric susceptibility*. The contributions χ_s can be determined from the relationship

$$\hat{\sigma}_{s1} = -\epsilon_0 k^2 \chi_s \hat{\phi}_1 . \qquad (9.30)$$

In view of the existence of a potential, longitudinal modes are widely referred to as *electrostatic modes*. Note that in the present context this term does not necessarily imply time-independence.

The present procedure can be generalized to include cases where the equilibrium has a weak spatial dependence in the sense that the wave number k is large compared to $1/L$, where L is the characteristic scale length of the equilibrium. Then, one uses a two-scale expansion similar to the procedure of Section 5.4. To lowest order in $(kL)^{-1}$ one finds local dispersion relations by the same procedure as outlined above. The only difference is the appearance of quantities with a weak spatial dependence describing the inhomogeneity. A few explicit examples of dispersion relations can be found in www.tp4.rub.de/~ks/ta.pdf.

9.3.2 Collective transport

A linear instability will generate a field of non-thermal fluctuations. The amplitudes of these fluctuations are determined by nonlinear effects, such as particle trapping in the wave troughs, nonlinear Landau damping or mode-coupling. To determine the amplitudes is the most difficult part of the theory of microturbulence, and the most reliable results seem to come from large particle simulation studies.

A problem which under some simplifying conditions can be treated by analytical techniques is to compute the turbulent momentum transfer $\mathbf{M}_{\mathrm{e}}^{\mathrm{turb}}$ analogous to the corresponding binary collision term $\mathbf{M}_{\mathrm{e}}^{\mathrm{bc}}$ in (9.1) from a *given* field of microturbulence. Under further restricted circumstances this leads to the *turbulent* or *collective* resistivity.

Consider a plasma with a clear separation of scales between a large scale background configuration and small scale turbulence. Let us write any field quantity Y as $Y = \langle Y \rangle + \delta Y$, where $\langle \dots \rangle$ denotes the ensemble average taken over a suitable set of realizations of the turbulence, implying $\langle \delta Y \rangle = 0$. Averaged quantities are large scale quantities. In contrast to the linear theory, let us begin by considering the exact equations.

For simplicity, however, we assume that δB can be ignored. At least approximately, this can be expected to apply to turbulence arising from longitudinal instability modes of moderate amplitudes.

Averaging the Vlasov equation (3.26) for species s, one then finds

$$\frac{\partial \langle f_s \rangle}{\partial t} + \boldsymbol{w} \cdot \nabla \langle f_s \rangle + \frac{q_s}{m_s} (\langle \boldsymbol{E} \rangle + \boldsymbol{w} \times \langle \boldsymbol{B} \rangle) \cdot \frac{\partial \langle f_s \rangle}{\partial \boldsymbol{w}} = \frac{\partial \langle f_s \rangle}{\partial t} \bigg|_{\text{turb}}, \qquad (9.31)$$

where

$$\frac{\partial \langle f_s \rangle}{\partial t} \bigg|_{\text{turb}} = - \left\langle \frac{q_s}{m_s} \delta \boldsymbol{E} \cdot \frac{\partial \delta f_s}{\partial \boldsymbol{w}} \right\rangle \qquad (9.32)$$

plays the role of a 'collision' term based on the presence of collective fluctuations, analogous to the Coulomb collision term in (3.18).

The collision term can be expressed in terms of correlation coefficients. Since $\delta B = 0$, δE can be derived from a scalar potential $\delta \phi$, which satisfies *Poisson's law*

$$-\Delta \delta \phi = \frac{1}{\epsilon_0} \sum_{s'} q_{s'} \int \delta f_{s'} \, \mathrm{d}^3 w \qquad (9.33)$$

with the solution

$$\delta \phi(\boldsymbol{r}, t) = \frac{1}{4\pi\epsilon_0} \sum_{s'} q_{s'} \int \mathrm{d}^3 r' \, \mathrm{d}^3 w' \frac{1}{|\boldsymbol{r} - \boldsymbol{r}'|} \delta f_{s'}(\boldsymbol{r}', \boldsymbol{w}', t). \qquad (9.34)$$

Inserting $\delta \boldsymbol{E} = -\nabla \delta \phi$ with (9.34) into (9.32) one obtains

$$\frac{\partial \langle f_s \rangle}{\partial t} \bigg|_{\text{turb}} =$$

$$\sum_{s'} \frac{q_s q_{s'}}{4\pi\epsilon_0 m_s} \int \mathrm{d}^3 r' \, \mathrm{d}^3 w' \frac{\partial}{\partial \boldsymbol{r}} \frac{1}{|\boldsymbol{r} - \boldsymbol{r}'|} \cdot \frac{\partial}{\partial \boldsymbol{w}} \langle \delta f_s(\boldsymbol{r}, \boldsymbol{w}, t) \delta f_{s'}(\boldsymbol{r}', \boldsymbol{w}', t) \rangle. \qquad (9.35)$$

In the momentum equation of species s (see (3.34)) the collective collisions generate a friction force represented by the momentum transfer

$$\boldsymbol{M}_s^{(\text{turb})} = \int m_s \boldsymbol{w} \frac{\partial \langle f_s \rangle}{\partial t} \bigg|_{\text{turb}} \mathrm{d}^3 w$$

$$= \sum_{s'} \boldsymbol{M}_{ss'}^{(\text{turb})} \qquad (9.36)$$

where

$$\boldsymbol{M}_{ss'}^{(\text{turb})} = -\frac{q_s q_{s'}}{4\pi\epsilon_0} \int \mathrm{d}^3 r' \frac{\partial}{\partial \boldsymbol{r}} \frac{1}{|\boldsymbol{r} - \boldsymbol{r}'|} \langle \delta n_s(\boldsymbol{r}, t) \delta n_{s'}(\boldsymbol{r}', t) \rangle. \qquad (9.37)$$

This indicates that the momentum transfer due to collective interaction results from the inter-species correlation of density fluctuations. (The self-transfer $M_{ss}^{(\mathrm{turb})}$ vanishes, as shown in the quasi-linear approximation farther below.)

Since, locally, both the background state and the statistical properties of turbulence are regarded as homogeneous, near r the density correlation function depends on r and r' only through $a = r - r'$, with an additional parametric dependence on r (not shown) describing the large scale variation.

For the next step let us include temporal correlations, such that the correlation function is a function of a and $\tau = t - t'$,

$$\langle \delta n_s(r, t) \delta n_{s'}(r', t') \rangle = S_{ss'}(a, \tau) , \qquad (9.38)$$

and let us introduce its Fourier representation

$$S_{ss'}(a, \tau) = \int \frac{\mathrm{d}^3 k}{(2\pi)^{3/2}} \int \frac{\mathrm{d}\omega}{(2\pi)^{1/2}} \hat{S}_{ss'}(k, \omega) \, \mathrm{e}^{\mathrm{i}k \cdot a - \mathrm{i}\omega\tau} , \qquad (9.39)$$

where, regarding the factors involving powers of 2π, the symmetric version of the Fourier transform has been chosen.

With (9.39) the momentum transfer (9.37) becomes

$$M_{ss'}^{(\mathrm{turb})} = -\frac{q_s q_{s'}}{\epsilon_0} \int \frac{\mathrm{d}^3 k}{(2\pi)^{3/2}} \int \frac{\mathrm{d}\omega}{(2\pi)^{1/2}} \frac{k}{k^2} \, \mathrm{Im}[\hat{S}_{ss'}(k, \omega)] . \qquad (9.40)$$

In deriving (9.40) one uses that $S_{ss'}$ is a real quantity and that by $\partial/\partial r(1/a) = -\partial/\partial r'(1/a)$ the r differentiation can be moved to $S_{ss'}$ by integration by parts. In view of (9.37) t' was reset to t and the integration variable r' in (9.37) has been replaced by a, where, for reasons of causality, the radial integration is kept well-defined by introducing an arbitrarily small damping, so that

$$\int \frac{1}{a} \, \mathrm{e}^{\mathrm{i}k \cdot a} \, \mathrm{d}^3 a = \frac{4\pi}{k^2} . \qquad (9.41)$$

For details see the literature (e.g., Tange and Ichimaru, 1974).

It is important to realize that (9.40) is exact except for the separation of scales and the restriction to longitudinal waves.

A drawback exists in that the density correlation function is difficult to assess both theoretically and observationally. A more useful expression is provided by relating the density correlation function to the energy spectrum of the turbulence. However, such a formulation generally requires further restrictions.

Here we will briefly describe the *quasi-linear approach*, which uses density fluctuations that are small enough to be covered by linear theory (e.g., Sagdeev and Galeev, 1969). Then one identifies fluctuation quantities with the linear modes (e.g., δn_s with n_{1s}) and uses (9.30) (applied to each species separately) to obtain

$$
q_s q_{s'} \langle n_{s1}(r, t) n_{s'1}(r', t) \rangle = \langle \sigma_{s1}(r, t) \sigma_{s'1}(r', t) \rangle
$$

$$
= \epsilon_0^2 \int \frac{d^3 k}{(2\pi)^{3/2}} k^2 \chi_s(k) \chi_{s'}^*(k) G(k) e^{i k \cdot a} \tag{9.42}
$$

where $G(k)$ is the Fourier transform

$$
G(k) = \int \frac{d^3 a}{(2\pi)^{3/2}} \langle E_1(r, t) E_1(r', t) \rangle e^{-i k \cdot a} . \tag{9.43}
$$

In view of the restriction to longitudinal modes the electric field perturbation is represented by the component $E_1 = k \cdot E_1 / k$. Again, local homogeneity is assumed, so that $\langle E_1(r, t) E_1(r', t) \rangle$ locally depends on $a = r - r'$ only.

The fact that we are dealing with eigenmodes of the type of (9.19) can be taken into account in space-time Fourier transformed quantities by a factor $\delta(\omega - \omega(k))$, where $\omega(k)$ is the relevant solution of the dispersion relation (9.28). Integration with respect to ω then generates functions of k alone, in particular, $\chi_s(k) = \chi_s(k, \omega(k))$.

The function $G(k)$, introduced by (9.43) as the Fourier transform of the correlation function of the electric field fluctuations, may be identified as the spectral density of $\langle |E_1|^2 \rangle$,

$$
\langle |E_1|^2 \rangle = \int \frac{d^3 k}{(2\pi)^{3/2}} G(k). \tag{9.44}
$$

This suggests following common practice and rename $G(k)$ as $\langle |E_1|^2(k) \rangle$; for basic properties of correlation functions see e.g., Landau and Lifshitz (1963, vol. *Statistical Physics*).

Using (9.42) in (9.37) and applying the same techniques that led to (9.40) gives

$$
M_{s,s'}^{(\text{turb})} = -\epsilon_0 \int \frac{d^3 k}{(2\pi)^{3/2}} k \, \text{Im}[\chi_s(k) \chi_{s'}^*(k)] \langle |E_1|^2(k) \rangle , \tag{9.45}
$$

where it was used that $\langle |E_1|^2(k) \rangle$ is real.

The total momentum transfer to a given species from all others is found as

$$
M_s^{(\text{turb})} = \sum_{s'} M_{ss'}^{(\text{turb})}
$$

$$
= \epsilon_0 \int \frac{d^3 k}{(2\pi)^{3/2}} k \, \text{Im}[\chi_s(k)] \langle |E_1|^2(k) \rangle , \tag{9.46}
$$

using the dispersion relation (9.28).

The property

$$M_{ss'}^{(\text{turb})} + M_{s's}^{(\text{turb})} = 0 \tag{9.47}$$

ensures momentum conservation, in particular $M_{ss}^{(\text{turb})} = 0$ and $\sum_{ss'} M_{ss'}^{(\text{turb})} = 0$.

As $M_s^{(\text{turb})}$ is a force that appears in the momentum equation of species s, $M_{\text{e}}^{(\text{turb})}$ gives rise to a term $M_{\text{e}}^{(\text{turb})}/(en_{\text{e}})$ in generalized Ohm's law (9.5) taking the role of $M_{\text{e}}^{(\text{bc})}/(en_{\text{e}})$. We see from (9.37) that this term is due to collective density fluctuations and associated fluctuations of the electric field (see (9.43)).

For current driven instabilities it is typical that the fluctuation term $M_{\text{e}}^{(\text{turb})}$ is parallel to the current density j. Then this term formally can be written as a resistive term in (9.5), where then the resistivity is termed η_{turb}, the *turbulent resistivity*.

For a two-species plasma with singly charged ions with densities $n_{\text{e}} = n_{\text{i}} = n$, one finds

$$\frac{M_{\text{e}}^{(\text{turb})}}{en} = \eta_{\text{turb}}\, j \tag{9.48}$$

or

$$\eta_{\text{turb}} = \frac{1}{e^2 n^2 v_{\text{d}}^2} M_{\text{e}}^{(\text{turb})} \cdot v_{\text{d}} , \tag{9.49}$$

where v_{d} is the drift velocity defined in terms of the current density, $j = en v_{\text{d}}$.

The main difficulty in evaluating the k-integration in (9.46) for a given application is the lack of sufficient knowledge of the spectral density $\langle |E_1|^2(k,\omega)\rangle$. A useful estimate is available for cases where the spectrum sharply peaks at a value of $k = k_0$, where the growth rate maximizes (Sagdeev and Galeev, 1969). Then, with the help of (9.44) one finds

$$\eta_{\text{turb}} \approx \frac{2k_0 \cdot v_{\text{d}}}{e^2 n^2 v_{\text{d}}^2} \,\text{Im}[\chi_{\text{e}}(k_0)] W_{\text{f}} \tag{9.50}$$

where

$$W_{\text{f}} = \frac{\epsilon_0}{2} \langle |E_1|^2 \rangle \tag{9.51}$$

is the average electric energy density associated with the fluctuations. Although an exact value of W_{f} may not be available either, it may be easier to arrive at an estimate for a single value than for an entire spectrum. One may find estimates for W_{f} from observations, numerical simulations or from theoretical estimates.

Originally, the approximation (9.50) was derived for the ion-acoustic in-stability. Here, suppressing a numerical factor, one finds $k_0 \approx 1/\lambda_D$. For $v_{de} > v_{ia}$, up to a numerical factor, this gives (see e.g., Sagdeev and Galeev, 1969; Treumann and Baumjohann, 1997)

$$\eta_{turb} \approx \frac{1}{\epsilon_0 \omega_p} \frac{W_f}{n k_B T_e} \qquad (9.52)$$

where it is to be noted that $v_d = -v_{de}$, and $v_{ia} = \sqrt{k_B T_e/m_i}$ is the ion acoustic speed. The expression (9.52) is known as *Sagdeev's formula*. Inter-estingly, it gives a rough estimate also for Coulomb collisions if for W_f one inserts the energy density of the electric field fluctuations due to Coulomb interaction.

In view of the approximations made to obtain (9.50) or (9.52) it cannot be expected that these expressions are precisely valid in a quantitative sense. Nevertheless, these and similar expressions have proven useful for order-of-magnitude estimates and for the interpretation of simulation results.

9.3.3 LHD turbulence

Here we briefly discuss the collective momentum transfer due to lower-hybrid-drift (LHD) turbulence (Davidson and Gladd, 1975; Huba *et al.*, 1978). The LHD instability is of particular interest for space plasmas with $T_e \ll T_i$, where several other modes, such as the ion-acoustic mode, fail. The linear theory is outlined in Gary (1993), see also www.tp4.rub.de/~ks/ta.pdf.

We choose the explicit solution obtained for a two-species plasma with small $\beta_p = 2\mu_0 p/B^2$ and in the limit of vanishing electron temperature, where one finds for the electron susceptibility x_e, the real part of the fre-quency ω_r and the growth rate r,

$$\chi_e = \frac{\omega_{pe}^2}{\Omega_e^2} - \frac{\omega_i^2 v_d}{v_i^2 k \omega} \qquad (9.53)$$

$$\omega_r = \frac{k v_d}{1 + k^2 v_i^2/\omega_{lh}^2}, \quad \omega_{en} = \sqrt{\Omega_e \Omega_i} \qquad (9.54)$$

$$\gamma = \sqrt{\frac{\pi}{2}} \frac{k^3 v_d^2 v_i \omega_{lh}^4}{(\omega_{lh}^2 + k^2 v_i^2)^3} . \qquad (9.55)$$

Both v_d and k are directed perpendicular to the magnetic field. Small β_p ensures that the LHD modes are approximately longitudinal. For $\beta > 1$ the

LHD instability does not play a significant role. For turbulence based on this mode with $k \parallel j$ one finds an expression for the turbulent resistivity from (9.49) with (9.46) where χ_e is inserted from (9.53).

The approximation (9.50) is obtained with $k_0 = \omega_{lh}/v_i$ and corresponding values of ω_r and γ from (9.54) and (9.55). One finds

$$\eta_{\text{turb}} = \frac{1}{2\epsilon_0} \sqrt{\frac{\pi}{2} \frac{\omega_{lh}}{\Omega_e^2}} \left(\frac{v_d}{v_i}\right)^2 \frac{\epsilon_0 \langle |E_1|^2 \rangle /2}{n m_e v_d^2/2}. \tag{9.56}$$

In cases where $\epsilon_0 \langle |E_1|^2 \rangle /2 \approx n m_e v_d^2/2$ (Huba *et al.*, 1977) the expression (9.56) becomes fully explicit in terms of equilibrium quantities.

A fully satisfactory theoretical estimate of the nonlinear stages of the lower-hybrid modes does not seem to be available. Work by Drake *et al.* (1983) and Drake *et al.* (1984) shows that the main saturation process is mode coupling with energy transfer from growing long wavelength modes to damped short wavelength modes.

9.4 Non-turbulent kinetic effects

In the previous section we saw that, under suitable conditions, the effect of kinetic fluctuations can be incorporated into a fluid picture in the form of a collective resistivity. Here we turn to nonideal effects for which such a connection with fluid theory does not seem to exist. This applies to plasmas without a significant level of microturbulence. In those cases it is no longer appropriate to refer to the fluid form of Ohm's law (9.5) but rather to Vlasov theory directly. We can still use its exact moment equations to discuss the different possibilities of nonideal effects.

We begin with the momentum balance (3.34) of the electrons where we ignore the gravity term, which is legitimate for most space plasma phenomena because of the smallness of the electron mass. We write (3.34) for electrons in the form

$$\boldsymbol{E} + \boldsymbol{v}_e \times \boldsymbol{B} = -\frac{1}{en_e} \nabla \cdot \mathcal{P}_e - \frac{m_e}{e} \left(\frac{\partial \boldsymbol{v}_e}{\partial t} + \boldsymbol{v}_e \cdot \nabla \boldsymbol{v}_e\right), \tag{9.57}$$

where $\mathcal{P}_e$ denotes the electron pressure tensor (see (3.35)), the time-evolution of which is governed by (3.44). Note that we are not averaging over any fast processes and that gravity is neglected.

To break magnetic flux or line conservation, it is necessary that the right side of (9.57) is different from zero (see Section 3.8.2). There are essentially two different groups of effects, 'pressure tensor effects' based on the term

involving the electron pressure tensor and 'inertial effects' due to the inertial term (last term in (9.57)). For orientation let us look at a few simple
examples.

The first example deals with the significance of the off-diagonal components of the pressure tensor. Consider a two-dimensional geometry with
translational invariance with respect to the y-direction and the magnetic
field lying in the x, z-plane. Then, the y-component of (9.57) is of particular
interest,

$$E'_y = -\frac{1}{en_e}\left(\frac{\partial P_{e,xy}}{\partial x} + \frac{\partial P_{e,zy}}{\partial z}\right) - \frac{m_e}{e}\left(\frac{\partial v_{e,y}}{\partial t} + \boldsymbol{v}_e \cdot \nabla v_{e,y}\right) \qquad (9.58)$$

where $\boldsymbol{E}' = \boldsymbol{E} + \boldsymbol{v}_e \times \boldsymbol{B}$.

Obviously, in (9.58) the pressure tensor effects are due to off-diagonal
components of the electron pressure tensor. These components can be attributed to nongyrotropic effects. In fact, under the present assumptions
$P_{e,xy}$ and $P_{e,zy}$ vanish for the gyrotropic form of $\mathcal{P}_e$ given by (3.69). We will
return to this case later.

The second example addresses the relative weight of the pressure and inertial terms in (9.57). Consider a spatially homogeneous background plasma
with electrons at rest and $\boldsymbol{B} = 0$. The background is perturbed by a longitudinal linear wave mode. In that case, the terms in (9.57) are easily
computed, using linear Vlasov theory. We restrict our attention to the $\boldsymbol{k}$-
component of (9.57) and write that component as

$$E'_k = T^{\mathrm{pr}}_k + T^{\mathrm{in}}_k . \qquad (9.59)$$

To find the pressure term T^{pr}_k and the inertial term T^{in}_k the essential
quantity containing the necessary information is χ_e, to be obtained from
(9.30) specialized for electrons, which gives

$$n_{e1} = \frac{\epsilon_0 k^2 \chi_e}{e}\phi_1 . \qquad (9.60)$$

With this expression together with the particle conservation law (see (3.31))
for electrons

$$\omega n_{e1} - n_{e0}\boldsymbol{k} \cdot \boldsymbol{v}_{e1} = 0 \qquad (9.61)$$

one finds

$$T^{\mathrm{in}}_k = \frac{\mathrm{i}\,\omega^2 k}{\omega^2_{\mathrm{pe}}}\chi_e\,\phi_1 . \qquad (9.62)$$

Similarly, one may compute T_k^{pr} from the definition of the pressure tensor. However, an easier way is to use (9.59), which gives immediately

$$T_k^{\mathrm{pr}} = -\mathrm{i}k\phi_1 \left(1 + \frac{\omega^2}{\omega_{\mathrm{pe}}^2}\chi_{\mathrm{e}} \right). \tag{9.63}$$

To get an expression that indicates which of the two contributions in (9.59) dominates, let us evaluate the magnitude of their ratio,

$$
\begin{aligned}
R_0 &= \left| \frac{T_k^{\mathrm{in}}}{T_k^{\mathrm{pr}}} \right| \\
&= \left| \frac{\frac{\omega^2}{\omega_{\mathrm{pe}}^2}\chi_{\mathrm{e}}}{1 + \frac{\omega^2}{\omega_{\mathrm{pe}}^2}\chi_{\mathrm{e}}} \right|.
\end{aligned}
\tag{9.64}
$$

One can expect that R_0 tends to be larger for faster modes than for slower modes. In fact, for Langmuir waves with frequency (real part) $\sqrt{(\omega_{\mathrm{p}}^2 + 3k^2 v_{\mathrm{e}}^2)}$, valid for $k_{\mathrm{e}}/k \gg 1$, one finds $R_0 = k_{\mathrm{e}}^2/(3k^2) \gg 1$. Here, the frequency is of the order of the (electron) plasma frequency ω_{p}, indicating that the electron inertia is important. For ion acoustic modes, which are much slower, the ratio R_0 equals $m_{\mathrm{e}}/m_{\mathrm{i}}$, consistent with negligible electron inertia.

For fast modes, such as modes varying on time scales near $1/\omega_{\mathrm{p}}$, one would prefer a description in terms of average microturbulence, rather than the present point of view, where the variations are fully resolved. Thus, for modes that are slow enough for the present resolved picture to be appropriate, it is reasonable to expect that the pressure term in (9.57) dominates.

10

Selected macroinstabilities

A macroinstability is an instability that has a length scale comparable with
an equilibrium length. Since there is a considerable variety of macroinsta-
bilities, even a brief description of each instability would break the present
scope. So we concentrate on instabilities that seem to play an important
role in space plasma activity.

We begin with a brief discussion of stability concepts and then turn to
particular dynamical models and resulting instabilities.

Since changes of the magnetic topology are believed to play an important
role for activity phenomena, considerable room is given to instabilities that
involve such changes, covering both fluid and kinetic models.

An instability that changes magnetic topology is particularly relevant if
the system considered is stable with respect to topology-conserving instabilities. This motivates the inclusion of ideal MHD modes.

10.1 Stability concepts

In general terms, a stable steady state is characterized by its robustness
against external perturbations, while for an unstable steady state there exists at least one perturbation that leads to substantial changes, which in
some cases have dramatic consequences. Turning such qualitative state-
ments into quantitative notions requires operational definitions of stability
and instability. There are several possibilities of such definitions.

One line of approach is based on exponential modes as used in our discussion of microinstabilities (Section 9.3.1). This approach considers time-
dependence of the form $\exp(-i\omega t)$; stability corresponds to $\text{Im}(\omega) \leq 0$,
instability to $\text{Im}(\omega) > 0$.

This approach has the advantage that it does not only tell us whether a
given steady state is stable or unstable but it also provides the frequencies ω,

203

including damping or growth rates. A drawback is that in some cases the discrete exponential modes do not form a complete set of eigenmodes. Also, as we have seen in Section 9.3.1 for Vlasov systems, exponential modes exist only in a time-asymptotic sense. It seems, however, that these deficiencies do not cause serious problems in practical applications.

Lyapunov's theory (Zubov, 1964; Arnold, 1979; Holm *et al.*, 1985) concentrates on exact formal definitions and criteria of stability and instability. Perturbations are measured by distances defined in a metric function space, allowing for nonlinear perturbations.

A more physics-oriented approach (*energy approach*) counts a system as stable if a suitably selected test energy remains bounded by the energy supplied from external sources. For stability of static equilibria the test energy is simply the kinetic energy of bulk flow.

Interestingly, all these schemes largely agree in the resulting stability criteria. For instance, the familiar MHD energy principle (Hain *et al.*, 1957; Bernstein *et al.*, 1958) results in each case.

We prefer the energy approach, because it turns out to be useful not only for MHD but also for Vlasov systems. Predominantly, the energy approach provides us with criteria that are sufficient for stability. Necessity requires additional considerations.

10.2 Ideal MHD stability

Here we give a brief derivation of the MHD energy principle and discuss a number of special cases and applications in detail. The derivation should illustrate the usefulness of the energy approach, which later will also be applied to kinetic systems.

10.2.1 The energy principle

Let us consider an ideal MHD system with a spatial domain D, a state being symbolized by a suitable state vector g, and a solution of the MHD equations by $g(t)$ satisfying initial conditions $g(t_0)$ and the boundary conditions. The equilibrium state g_0 is magnetohydrostatic. For time $t \geq t_0$ the system is closed in the sense that the energy flux through the boundary vanishes. Then, equation (3.79), integrated over D, gives energy conservation

$$\frac{\mathrm{d}W(g(t))}{\mathrm{d}t} = 0, \quad W(g(t)) = \mathcal{T}(g(t)) + \mathcal{V}(g(t)), \tag{10.1}$$

where

$$T(g) = \int_D \frac{\rho v^2}{2} \mathrm{d}^3 r \qquad (10.2)$$

$$V(g) = \int_D \left(\frac{p}{\gamma - 1} + \frac{B^2}{2\mu_0} \right) \mathrm{d}^3 r . \qquad (10.3)$$

In deriving (10.1) Gauss's integral theorem was used.

Obviously, T is the kinetic energy of bulk motion, which is non-negative, and V, at least formally, takes the role of the potential energy.

The system is allowed to have energy exchange with the surroundings only for $t < t_0$, so that for $t \geq t_0$ it has a (constant) energy, which can be different from the equilibrium value $W(g_0)$. Let $g = g_0 + \Delta g$, $\Delta W(\Delta g) = W(g_0 + \Delta g) - W(g_0)$, $\Delta T(\Delta g) = T(g_0 + \Delta g) - T(g_0)$ and $\Delta V(\Delta g) = V(g_0 + \Delta g) - V(g_0)$, where the equilibrium is kept fixed. Here we have included $T(g_0)$ for a later (non-MHD) application.

From (10.1), evaluated at some time t and for the equilibrium, and taking the difference, one obtains

$$\Delta T(\Delta g(t)) = \Delta W(\Delta g(t)) - \Delta V(\Delta g(t)) \qquad (10.4)$$

where $\Delta W(\Delta g(t))$ is constant in time. We use the following stability definition:

An equilibrium is stable if for all t
$$\Delta T(\Delta g(t)) < \Delta W(\Delta g(t)) \quad \text{for all } \Delta g(t_0). \quad (10.5)$$

In other words, the kinetic energy can only draw from the energy that was transferred to the system in the period before t_0. There is no dynamic conversion of equilibrium energy into kinetic energy (of bulk motion).

From (10.4) it is obvious that a system is stable if

$$V(\Delta g) > 0 \quad \text{for all } \Delta g \neq 0. \qquad (10.6)$$

For comparison with Lyapunov-based stability notions it should be noted that ΔT is not necessarily a norm in state space. For $\Delta T = 0$ the plasma velocity v vanishes but a static perturbation (neighbouring equilibrium) is not excluded. However, in practice this subtle difference seems to be irrelevant. Also, we argue that the energy approach more directly corresponds to the intuitive notion of a large scale plasma instability. (Note that a static perturbation with vanishing kinetic energy requires infinite time.)

The criterion (10.6) is not subject to any amplitude restriction. But for the further analysis we turn to the linearized case, valid for small perturbations. (Note, however, that the perturbations can be renormalized once the equations have been linearized.)

The equations for the perturbations are obtained from (3.58)–(3.64), linearized in the perturbations (subscript '1') of the static equilibrium quantities (subscript '0') satisfying (5.1)–(5.3). In the absence of a gravity force we find

$$\frac{\partial \rho_1}{\partial t} + \nabla \cdot (\rho_0 \boldsymbol{v}_1) = 0 \tag{10.7}$$

$$\rho_0 \frac{\partial \boldsymbol{v}_1}{\partial t} = -\nabla p_1 + \boldsymbol{j}_1 \times \boldsymbol{B}_0 + \boldsymbol{j}_0 \times \boldsymbol{B}_1 \tag{10.8}$$

$$\boldsymbol{E}_1 + \boldsymbol{v}_1 \times \boldsymbol{B}_0 = 0 \tag{10.9}$$

$$\frac{\partial p_1}{\partial t} + \boldsymbol{v}_1 \cdot \nabla p_0 + \gamma p_0 \nabla \cdot \boldsymbol{v}_1 = 0 \tag{10.10}$$

$$\nabla \times \boldsymbol{E}_1 = -\frac{\partial \boldsymbol{B}_1}{\partial t} \tag{10.11}$$

$$\nabla \cdot \boldsymbol{B}_1 = 0 \tag{10.12}$$

$$\nabla \times \boldsymbol{B}_1 = \mu_0 \boldsymbol{j}_1 , \tag{10.13}$$

where (10.7) has been used to derive (10.10).

After eliminating $\boldsymbol{E}_1$ and $\boldsymbol{j}_1$ by (10.9) and (10.13), the equations (10.7), (10.10) and (10.11) can be used to express ρ_1, p_1 and $\boldsymbol{B}_1$ by the displacement vector $\boldsymbol{\xi}$, which describes the displacement of a fluid element from its equilibrium position $\boldsymbol{r}_0$ to its perturbed position $\boldsymbol{r}$ at time t,

$$\boldsymbol{r}(t) = \boldsymbol{r}_0 + \boldsymbol{\xi}(\boldsymbol{r}_0, t). \tag{10.14}$$

Although this is a Lagrangian concept, for the present linearized version we can replace $\boldsymbol{r}_0$ by $\boldsymbol{r}$ in the argument of $\boldsymbol{\xi}$ and retain the Eulerian point of view, keeping in mind that all perturbations vanish for $\boldsymbol{\xi} \equiv 0$ and that (10.14) implies that

$$\boldsymbol{v}_1(\boldsymbol{r}, t) = \frac{\partial \boldsymbol{\xi}(\boldsymbol{r}, t)}{\partial t} . \tag{10.15}$$

The required closure with respect to energy is achieved by imposing the boundary condition that $\boldsymbol{\xi}$ vanishes on the boundary. This implies that there is no plasma flow across the boundary and that, for a non-vanishing normal magnetic field component, there is no transport of the field line footpoints (*line-tying*). Also, there is no Poynting flux across the boundary. We will refer to this kind of boundary as a *closed boundary*, it is also known as *rigid-wall*.

After integration with respect to time we find

$$\rho_1 = -\nabla \cdot (\rho_0 \boldsymbol{\xi}) \,, \tag{10.16}$$

$$p_1 = -\gamma p_0 \nabla \cdot \boldsymbol{\xi} - \boldsymbol{\xi} \cdot \nabla p_0 \tag{10.17}$$

$$\boldsymbol{B}_1 = \boldsymbol{Q}, \qquad \boldsymbol{Q} = \nabla \times (\boldsymbol{\xi} \times \boldsymbol{B}_0) \,. \tag{10.18}$$

Using these expressions in (10.8) with (10.13) gives

$$\rho_0 \frac{\partial^2 \boldsymbol{\xi}}{\partial t^2} = \boldsymbol{F}(\boldsymbol{\xi}) \,, \tag{10.19}$$

where

$$\boldsymbol{F}(\boldsymbol{\xi}) = \nabla(\gamma p_0 \nabla \cdot \boldsymbol{\xi} + \boldsymbol{\xi} \cdot \nabla p_0) + \frac{1}{\mu_0}(\nabla \times \boldsymbol{Q}) \times \boldsymbol{B}_0 + \boldsymbol{j}_0 \times \boldsymbol{Q} \,. \tag{10.20}$$

Equation (10.19) has the form of an equation of motion with force $\boldsymbol{F}$.

We are now in a position to show that the first variation $\mathcal{V}_1$ of $\mathcal{V}$ vanishes at the equilibrium, as can be expected from a functional playing the role of a potential energy. That variation is obtained by linearization of (10.3) with respect to the perturbations,

$$
\begin{aligned}
\mathcal{V}_1 &= \int_D \left(\frac{\boldsymbol{B}_0 \cdot \boldsymbol{B}_1}{\mu_0} + \frac{p_1}{\gamma - 1} \right) \mathrm{d}^3 r \\
&= \int_D \boldsymbol{\xi} \cdot (\nabla p_0 - \boldsymbol{j}_0 \times \boldsymbol{B}_0) \mathrm{d}^3 r \\
&= 0 \,,
\end{aligned}
\tag{10.21}
$$

where (10.17) and (10.18) were used and two integrations by parts were carried out with the surface integrals vanishing because of the boundary condition. The last step in (10.21) makes use of the fact that the unperturbed quantities satisfy the force balance of magnetohydrostatics (5.6). Thus, $\mathcal{V}$ is stationary in equilibrium.

Continuing the search for $\mathcal{T}$ and $\mathcal{V}$ in the small perturbation limit we express these functionals as quadratic functionals of $\partial \boldsymbol{\xi}/\partial t$ and $\boldsymbol{\xi}$, denoted by $\mathcal{T}_2$ and $\mathcal{V}_2$. A convenient way to find these quantities is to derive energy balance from the linear equation of motion (10.19). Multiplying (10.19) by $\dot{\boldsymbol{\xi}}$ (short for $\partial \boldsymbol{\xi}/\partial t$) and integrating over the domain D one finds

$$\frac{\mathrm{d}}{\mathrm{d}t} \int_D \frac{1}{2} \rho_0 \dot{\boldsymbol{\xi}}^2 \mathrm{d}^3 r = \int_D \dot{\boldsymbol{\xi}} \cdot \boldsymbol{F}(\boldsymbol{\xi}) \mathrm{d}^3 r \,. \tag{10.22}$$

At this stage it is important that the force (10.20), understood as a linear operator acting on $\boldsymbol{\xi}$, is symmetric, such that

$$\int_D \boldsymbol{\xi}_1 \cdot \boldsymbol{F}(\boldsymbol{\xi}_2)\, \mathrm{d}^3r = \int_D \boldsymbol{\xi}_2 \cdot \boldsymbol{F}(\boldsymbol{\xi}_1)\, \mathrm{d}^3r \,. \tag{10.23}$$

This property holds for arbitrary smooth, real displacements $\boldsymbol{\xi}_1$ and $\boldsymbol{\xi}_2$ satisfying the boundary condition. For a derivation of (10.23) see, for instance, Appendix A of Freidberg (1987). Making use of (10.23) we write (10.22) in the form

$$\frac{\mathrm{d}}{\mathrm{d}t} \int_D \frac{1}{2}\rho_0 \dot{\boldsymbol{\xi}}^2 \mathrm{d}^3r \;=\; \frac{1}{2}\int_D \dot{\boldsymbol{\xi}} \cdot \boldsymbol{F}(\boldsymbol{\xi})\, \mathrm{d}^3r + \frac{1}{2}\int_D \boldsymbol{\xi} \cdot \boldsymbol{F}(\dot{\boldsymbol{\xi}})\, \mathrm{d}^3r \tag{10.24}$$

$$= \;\frac{1}{2}\frac{\mathrm{d}}{\mathrm{d}t}\int_D \boldsymbol{\xi} \cdot \boldsymbol{F}(\boldsymbol{\xi})\, \mathrm{d}^3r \,, \tag{10.25}$$

which, by integration with respect to time gives us second-order energy conservation in the form

$$W_2 = \mathcal{T}_2 + \mathcal{V}_2 \tag{10.26}$$

with

$$\mathcal{T}_2 = \frac{1}{2}\int_D \rho_0 \dot{\boldsymbol{\xi}}^2 \mathrm{d}^3r \tag{10.27}$$

$$\mathcal{V}_2 = -\frac{1}{2}\int_D \boldsymbol{\xi} \cdot \boldsymbol{F}(\boldsymbol{\xi})\mathrm{d}^3r \tag{10.28}$$

and W_2 being a constant.

Using (10.20) one finds, after integration by parts, the following explicit expression for $\mathcal{V}_2$,

$$\mathcal{V}_2 = \frac{1}{2}\int_D \left(\frac{Q^2}{\mu_0} + \gamma p_0 (\nabla \cdot \boldsymbol{\xi})^2 + \boldsymbol{j}_0 \cdot \boldsymbol{\xi} \times \boldsymbol{Q} + \nabla \cdot \boldsymbol{\xi}\,\boldsymbol{\xi} \cdot \nabla p_0\right) \mathrm{d}^3r \,. \tag{10.29}$$

For applications it is useful to note that the parallel component $\xi_{\parallel} = \boldsymbol{B}_0 \cdot \boldsymbol{\xi}/B_0$ drops out of the last two terms in the integrand of (10.29).

The MHD problem is now in a form that allows us to apply the local (i.e., small perturbation) version of the criterion (10.6). Thus, we can draw the conclusion that for an ideal MHD system it is sufficient for local stability that $\mathcal{V}_2 > 0$ holds for all perturbations $\boldsymbol{\xi}$ that do not vanish identically and that satisfy the boundary condition.

So stability can be assessed by finding the minimum of (10.29), which exists in typical applications. (There is no maximum, because for a sufficiently small spatial scale of the test displacements the positive expression

$Q^2/\mu_0 + \gamma p_0(\nabla \cdot \boldsymbol{\xi})^2$ in (10.29) dominates and becomes arbitrarily large compared with the other terms.)

For evaluating the minimum it is appropriate to introduce a normalization condition, such as

$$\frac{1}{2}\int_D \rho_0 \boldsymbol{\xi}^2 \mathrm{d}^3 r = c_0 \, , \tag{10.30}$$

where c_0 is a positive constant. (10.30) excludes the trivial displacement $(\boldsymbol{\xi} = 0)$ and keeps negative minima bounded.

For the remaining part of this section, let us use (10.30) and assume that $\mathcal{V}_2$ has a minimum $\mathcal{V}_{2,\mathrm{m}}$ with minimizing test displacement (minimizer) $\boldsymbol{\xi}_\mathrm{m}$. Then a system is stable if $\mathcal{V}_{2,\mathrm{m}} > 0$.

The minimizer satisfies the Euler–Lagrange equation associated with the functional $\mathcal{V}_2$ given by (10.28)

$$-\boldsymbol{F}(\boldsymbol{\xi}) = \lambda \rho_0 \boldsymbol{\xi} \, . \tag{10.31}$$

The derivation of (10.31) uses (10.23). The eigenvalue λ is the Lagrangian multiplier associated with the condition (10.30), it is real because of (10.23). The lowest eigenvalue has the same sign as the minimum of $\mathcal{V}_2$, so that $\lambda_\mathrm{m} > 0$ implies stability. The eigenfunction associated with λ_m is the minimizer.

A modification is necessary if an external gravity field is included. The corresponding generalization of (10.29) is

$$\mathcal{V}_2 = \frac{1}{2}\int_D \left(\frac{|Q|^2}{\mu_0} + \boldsymbol{j}_0 \cdot \boldsymbol{\xi}^* \times \boldsymbol{Q} + \gamma p_0 |\nabla \cdot \boldsymbol{\xi}|^2 \right.$$
$$\left. + \nabla \cdot \boldsymbol{\xi}^* \, \boldsymbol{\xi} \cdot \nabla p_0 - \boldsymbol{\xi}^* \cdot \nabla \psi_0 \, \nabla \cdot (\rho_0 \boldsymbol{\xi}) \right) \mathrm{d}^3 r \, , \tag{10.32}$$

where $\psi(\boldsymbol{r})$ is the gravity potential.

We have written (10.32) in a form that demonstrates the changes that occur if one uses complex displacements (the definition of the dot product left unchanged). Note that the restriction to real displacements does not imply a loss of generality. Nevertheless, a complex formulation often is chosen for convenience.

In analogy to (10.23) for real $\boldsymbol{\xi}$, complex displacements are Hermitean, i.e., they have the property

$$\int_D \boldsymbol{\xi}_1^* \cdot \boldsymbol{F}(\boldsymbol{\xi}_2) \, \mathrm{d}^3 r = \int_D \boldsymbol{\xi}_2 \cdot \boldsymbol{F}(\boldsymbol{\xi}_1^*) \, \mathrm{d}^3 r \, , \tag{10.33}$$

where the star denotes the conjugate complex quantity. Equation (10.33) ensures real eigenvalues for complex displacements.

For reasons of comparison let us briefly discuss what one obtains by choosing the exponential time-dependence in (10.19). Using the complex formulation one finds

$$-\boldsymbol{F}(\boldsymbol{\xi}) = \omega^2 \rho_0 \boldsymbol{\xi} \,, \tag{10.34}$$

where ω^2 takes the role of the eigenvalue. Since (10.31) and (10.34) have the same eigenvalue spectrum we obtain full agreement for stable systems. For both stability notions it is sufficient for stability that all eigenvalues are positive.

The energy approach does not provide criteria for instability. In view of the agreement for stable systems we adopt the exponential-mode criterion, giving instability if there exists a negative eigenvalue.

An approach by Laval *et al.* (1965) avoids the incertitude regarding exponential modes by using the energy approach for stability and by giving an explicit proof of unbounded growth for the cases where $\mathcal{V}_2$ assumes a negative value. The result is fully consistent with the present approach.

We can summarize the results on MHD stability by the criterion:

It is necessary and sufficient for local stability of an ideal MHD system that the functional $\mathcal{V}_2(g)$ is positive for all perturbations g satisfying the boundary conditions. The system is unstable if there exists a perturbation for which $\mathcal{V}_2$ is negative. (10.35)

This formulation does not address states with vanishing minimum of $\mathcal{V}_2$ (marginal states); they form a separate category. They often can be ignored as being of zero measure.

10.2.2 Properties and specializations

Here we discuss a number of qualitative properties of the MHD energy functional $\mathcal{V}_2$ and specialize it for several choices of invariance. In each case it is assumed that the displacement vector $\boldsymbol{\xi}$ vanishes on the boundary of the domain D considered.

Role of pressure gradient and parallel current

Pressure gradients in combination with the curvature of magnetic field lines as well as the presence of a component of the electric current density parallel to the magnetic field play a particularly important role for MHD instabilities. There is a form of the MHD variational principle that exhibits these features explicitly. For a discussion of this form we ignore gravity, which, however, will be included in one of the cases considered farther below.

Then, for real $\boldsymbol{\xi}$, the variational functional is given by (10.29), which can be rewritten in the form

$$\mathcal{V}_2 = \frac{1}{2} \int_D \left(\frac{Q_\perp^2}{\mu_0} + \frac{B_0^2}{\mu_0} (\nabla \cdot \boldsymbol{\xi}_\perp + 2\boldsymbol{\xi}_\perp \cdot \boldsymbol{\kappa})^2 + \gamma p_0 (\nabla \cdot \boldsymbol{\xi})^2 \right.$$

$$\left. - 2\boldsymbol{\xi}_\perp \cdot \boldsymbol{\kappa} \boldsymbol{\xi}_\perp \cdot \nabla p_0 - j_{0\parallel} (\boldsymbol{\xi}_\perp \times \boldsymbol{b}_0) \cdot \boldsymbol{Q}_\perp \right) \mathrm{d}^3 r \ . \quad (10.36)$$

Here, the subscripts $\perp$, $\parallel$ denote the components perpendicular and parallel to the magnetic field, e.g., $\boldsymbol{\xi}_\perp = \boldsymbol{b}_0 \times (\boldsymbol{\xi} \times \boldsymbol{b}_0)$ and $j_{0\parallel} = \boldsymbol{b}_0 \cdot \boldsymbol{j}_0$, $\boldsymbol{b}_0$ being the unit vector in the direction of $\boldsymbol{B}_0$, and $\boldsymbol{\kappa}$ is the field line curvature vector

$$\boldsymbol{\kappa} = \boldsymbol{b}_0 \cdot \nabla \boldsymbol{b}_0 \ . \quad (10.37)$$

For a detailed derivation of (10.36) see Freidberg (1987).

All but one of the terms in (10.36) contain only $\boldsymbol{\xi}_\perp$, the parallel component of the displacement enters only through $(\nabla \cdot \boldsymbol{\xi})^2$. This property allows for an explicit minimization with respect to $\xi_\parallel$, generating a functional that contains $\boldsymbol{\xi}_\perp$ only. A complicating feature of this procedure is that a spatial average appears in the integrand of the resulting functional (see Section 10.2.5).

The first three terms in (10.36) are non-negative and thus stabilizing. Instabilities can arise only from the fourth and fifth terms containing a combination of curvature and pressure gradient and parallel current density, respectively. Instability requires that the sum of these terms becomes sufficiently negative to overcome the other terms. Therefore, depending on the dominating term in (10.36), one distinguishes between instabilities based on pressure gradients and curvature and instabilities based on parallel currents. The former require that in a sufficiently large region $\boldsymbol{\xi}_\perp \cdot \boldsymbol{\kappa}$ and $\boldsymbol{\xi}_\perp \cdot \nabla p_0$ have the same sign. Particularly favourable for instability are regions where $\boldsymbol{\kappa}$ and ∇p_0 are parallel. This is relevant for the ballooning mode discussed in Section 10.2.5.

Inspection of (10.36) also indicates that an MHD instability is a macroinstability, i.e., an instability that possesses a spatial scale that is comparable with or larger than the smallest equilibrium scale. This can be seen by observing that perturbations $\boldsymbol{\xi}$ with spatial scales much smaller than the equilibrium scales in all directions are stable. As the positive-definite terms involve two derivatives of $\boldsymbol{\xi}$ and the last two terms contain only one derivative, $\mathcal{V}_2$ will be positive for sufficiently small perturbation scales. If the perturbation has different scales in different directions, it turns out that at least one of the scales must be comparable with or larger than the smallest equilibrium scale.

Significance of the steady state potential T for stability

In Chapter 7 we have seen that under fairly general conditions the steady state potential T acts as a Lagrangian for steady states in the sense that the steady state field equations can be obtained from the condition that the first variation of the integral

$$\mathcal{U} = \int_D T \, d^3 r \qquad (10.38)$$

vanishes.

Here we raise the question whether, in the present context of ideal MHD, the second variation of $\mathcal{U}$ provides information on stability properties.

As in the steady state discussion we assume that the magnetic field possesses Euler potentials α, β. Then, for ideal MHD the potential T is given by (7.9), such that $\mathcal{U}$ has the form

$$\mathcal{U} = \int_D \left(\frac{(\nabla\alpha \times \nabla\beta)^2}{2\mu_0} - p(\alpha, \beta) \right) d^3 r \ . \qquad (10.39)$$

The aim is to compare $\mathcal{U}_2$, the second variation of $\mathcal{U}$, with $\mathcal{V}_2$. To obtain a meaningful comparison we must establish a relationship between the displacement vector $\boldsymbol{\xi}$ and the perturbations $\delta\alpha$ and $\delta\beta$ of the Euler potentials. This relationship is provided by the frozen-in condition of ideal MHD, which implies that potentials α and β can be found which are constants of the motion, such that in linearized form

$$\frac{\partial \delta\alpha}{\partial t} + \boldsymbol{v} \cdot \nabla\alpha_0 = 0, \quad \frac{\partial \delta\beta}{\partial t} + \boldsymbol{v} \cdot \nabla\beta_0 = 0 \ . \qquad (10.40)$$

Integration of (10.40) with respect to time, observing the appropriate initial condition that perturbations vanish for $\boldsymbol{\xi} = 0$, gives

$$\delta\alpha = -\boldsymbol{\xi} \cdot \nabla\alpha_0, \quad \delta\beta = -\boldsymbol{\xi} \cdot \nabla\beta_0 \ . \qquad (10.41)$$

Taking into account that the first variation of $\mathcal{U}$ vanishes, we find the second variation as

$$\mathcal{U}_2 = \frac{1}{2} \int_D \left(\frac{1}{\mu_0} (\nabla\delta\alpha \times \nabla\beta_0 + \nabla\alpha_0 \times \nabla\delta\beta)^2 + \frac{2}{\mu_0} \boldsymbol{B}_0 \cdot \nabla\delta\alpha \times \nabla\delta\beta \right.$$

$$\left. - \frac{\partial^2 p_0}{\partial \alpha_0^2} \delta\alpha^2 - 2\frac{\partial^2 p_0}{\partial \alpha_0 \partial \beta_0} \delta\alpha\delta\beta - \frac{\partial^2 p_0}{\partial \beta_0^2} \delta\beta^2 \right) d^3 r \ , \quad (10.42)$$

where $\boldsymbol{B}_0 = \nabla\alpha_0 \times \nabla\beta_0$ and $\delta\alpha$ and $\delta\beta$ have to be inserted from (10.41).

After appropriate manipulation of $\mathcal{U}_2$ including integration by parts, the comparison with $\mathcal{V}_2$ gives

$$\mathcal{V}_2 = \mathcal{U}_2 + \int_D \gamma p_0 (\nabla \cdot \boldsymbol{\xi})^2 \mathrm{d}^3 r \qquad (10.43)$$

consistent with the fact that $\mathcal{U}$ and $\mathcal{V}$ coincide for the (formal) choice of $\gamma = 0$. As the additional term on the right side of (10.43) is non-negative, $\mathcal{U}_2 \geq 0$ for all $\boldsymbol{\xi}$ vanishing on the boundary is sufficient for stability. Necessity cannot be expected because the adiabatic constraint has not been incorporated into the formulation of $\mathcal{U}_2$.

Thus we can conclude that for ideal MHD the notion of the steady state potential T is useful not only for formulating the steady state field equations but also for assessing stability properties. It will be interesting to find out in future studies whether the same is true for the other plasma models discussed in Chapter 7.

3D perturbations of 2D equilibrium

Let $\partial/\partial y = 0$ for the equilibrium quantities while the perturbations are kept fully three-dimensional. If an external gravity force is included, the equilibrium is described by $\hat{T}$ as given by (7.27), specialized for systems with translational invariance,

$$\hat{T}(A, B, \psi) = \frac{B^2}{2\mu_0} - P(A, \psi) . \qquad (10.44)$$

Note that this description requires that ∇A, ∇B and $\nabla \psi$ do not lie in a plane everywhere. A case violating that condition will be discussed later. The pressure p is written as $P(A, \psi)$, and density satisfies

$$\rho = -\partial P(A, \psi)/\partial \psi . \qquad (10.45)$$

The assumed symmetry allows us to rewrite (10.32) in the form (Schindler *et al.*, 1983)

$$\mathcal{V}_2 = \frac{1}{2\mu_0} \int \left\{ \left| \frac{\boldsymbol{B} \cdot \nabla A_1}{B_\mathrm{p}} \right|^2 + V_\mathrm{c} |A_1|^2 + |B_y \nabla \cdot \boldsymbol{\xi}_\mathrm{p} - \boldsymbol{B}_\mathrm{p} \cdot \nabla \xi_y|^2 \right.$$

$$+ \left| \frac{\nabla A \cdot \nabla A_1 + J A_1}{B_\mathrm{p}} - \frac{\partial}{\partial y}(B_\mathrm{p} \xi_y - B_y \xi_{\mathrm{p}\|}) \right|^2$$

$$+ \mu_0 \left(\gamma p + \frac{\rho^2}{\partial \rho / \partial \psi} \right) |\nabla \cdot \boldsymbol{\xi}|^2 - \mu_0 \frac{\partial \rho}{\partial \psi} \left| \boldsymbol{\xi} \cdot \nabla \psi + \frac{\rho}{\partial \rho / \partial \psi} \nabla \cdot \boldsymbol{\xi} \right|^2 \right\} \mathrm{d}^3 r ,$$

$$(10.46)$$

where

$$A_1 = -\boldsymbol{\xi} \cdot \nabla A, \qquad (10.47)$$

$$V_c = \nabla \cdot \frac{J\nabla A}{B_p^2} - \frac{J^2}{B_p^2} - \frac{\partial J}{\partial A} \qquad (10.48)$$

and

$$J = \mu_0 j_y = \mu_0 \frac{\partial}{\partial A}\left(p + \frac{B_y^2}{2\mu_0}\right) \qquad (10.49)$$

(see (5.80) noting that the ignorable coordinate has changed from z to y). As in Section 5.2.1, $\boldsymbol{B}_p$ is the poloidal part of the magnetic field, i.e., perpendicular to the invariant direction. The component of ξ perpendicular to $\boldsymbol{B}_p$ is represented by A_1, the component parallel to $\boldsymbol{B}_p$ by $\xi_{p\|}$. Note that the subscript '0' has been dropped for all equilibrium quantities. An alternative form of (10.46) was given by de Bruyne and Hood (1989).

Using A_1 as a test function the following precaution has to be taken. Since $\boldsymbol{\xi}$ by definition is unrestricted (except for continuity and boundary conditions), (10.47) implies that A_1 is subject to the constraint

$$A_1 = 0 \quad \text{where} \quad B_p = 0 . \qquad (10.50)$$

In the absence of gravity (10.46) reduces to

$$V_2 = \frac{1}{2\mu_0} \int \left\{ \left|\frac{\boldsymbol{B} \cdot \nabla A_1}{B_p}\right|^2 + V_c |A_1|^2 + |B_y \nabla \cdot \boldsymbol{\xi}_p - \boldsymbol{B}_p \cdot \nabla \xi_y|^2 \right.$$
$$\left. + \left|\frac{\nabla A \cdot \nabla A_1 + J A_1}{B_p} - \frac{\partial}{\partial y}(B_p \xi_y - B_y \xi_{p\|})\right|^2 + \mu_0 \gamma p |\nabla \cdot \boldsymbol{\xi}|^2 \right\} \mathrm{d}^3 r . \qquad (10.51)$$

In (10.46) one can distinguish two different effects that in principle might give rise to instability. The second term of the integrand shows that suitable current distributions might cause instability. This term combines the effect of pressure-gradient/curvature with that of parallel currents. The terms involving the gravity potential ψ in (10.46) can cause instability for suitable density distributions.

3D perturbations of 1D equilibrium

Here we consider the equilibria of Section 5.3.1 with a magnetic field of the form

$$\boldsymbol{B}_p = B_x(z)\boldsymbol{e}_x, \quad B_y = B_y(z) , \qquad (10.52)$$

there is no external gravity force so that (10.51) applies. So, in principle, the V_c term is the only term that could give rise to instability. However, with the choice (10.52), one finds that V_c vanishes, such that V_2 is non-negative for all perturbations. The equilibria are stable with respect to arbitrary ideal MHD modes. In particular, this applies to the specializations discussed in Section 5.3.1, the rotating field and the Harris sheet.

We add that for the present simple equilibria one can rewrite V_2 in the form

$$V_2 = \frac{1}{2\mu_0} \int \left\{ |\boldsymbol{B} \cdot \nabla \boldsymbol{\xi} - \boldsymbol{B} \nabla \cdot \boldsymbol{\xi}|^2 + \gamma p |\nabla \cdot \boldsymbol{\xi}|^2 \right\} \mathrm{d}^3 r . \tag{10.53}$$

In the first term of the integrand $\boldsymbol{\xi}$ can be replaced by $\boldsymbol{\xi}_\perp$. Clearly, the expression (10.53) can also be found directly from the general form of (10.32), ignoring gravity, but the procedure is rather cumbersome.

It is worthwhile pointing out that of the two destabilizing terms in (10.36) only the curvature term vanishes, because all field lines are straight lines. The $j_\parallel$ term does not necessarily vanish, but, as (10.53) implies, that term cannot lead to negative values of V_2 for the present equilibria.

2D perturbations of 2D equilibrium

Let us now assume two-dimensionality in the sense that translational invariance with respect to the y-direction is imposed on both the equilibrium and perturbed states. Here, for convenience, the term 'stability' is used in the same restricted sense, applying to two-dimensional modes only. It suffices to limit the spatial integrations to the x, z-plane.

Then, after integration by parts of the term $|A_1|^2 \nabla \cdot (J \nabla A / B_\mathrm{p}^2)$, which needs (10.50), one finds that (10.46) assumes the form

$$\begin{aligned}
V_2 = \frac{1}{2\mu_0} \int \Bigg\{ & |\nabla A_1|^2 - \frac{\mathrm{d}J}{\mathrm{d}A} |A_1|^2 + |B_y \nabla \cdot \boldsymbol{\xi}_\mathrm{p} - \boldsymbol{B}_\mathrm{p} \cdot \nabla \xi_y|^2 \\
& + \gamma \mu_0 p \left(1 + \frac{\rho^2}{\gamma p \partial \rho / \partial \psi} \right) |\nabla \cdot \boldsymbol{\xi}|^2 \\
& - \mu_0 \frac{\partial \rho}{\partial \psi} \left| \boldsymbol{\xi} \cdot \nabla \psi + \frac{\rho}{\partial \rho / \partial \psi} \nabla \cdot \boldsymbol{\xi} \right|^2 \Bigg\} \mathrm{d}^2 r . \tag{10.54}
\end{aligned}$$

Let us first look at the effect of gravity. It is sufficient for the absence of gravity-associated instabilities that

$$1 + \frac{\rho^2}{\gamma p \partial \rho / \partial \psi} \geq 0 . \tag{10.55}$$

In order to see the physical significance of (10.55), let us discuss the limiting case,

$$\frac{\partial \rho}{\partial \psi} = -\frac{\rho^2}{\gamma p} .$$ (10.56)

With the help of (10.45), the equation (10.56) is easily shown to be solved by $p/\rho^\gamma = c(A)$, where $c(A)$ is an integration constant with respect to ψ. Thus, the limiting case corresponds to an adiabatic pressure profile on each field line. This is in agreement with the well-known limiting case of the convection instability (Lang, 1974), which we obtain here for each field line separately.

In the absence of gravity the functional (10.54) reduces to

$$\mathcal{V}_2 = \frac{1}{2\mu_0} \int \left\{ |\nabla A_1|^2 - \frac{\mathrm{d}J}{\mathrm{d}A}|A_1|^2 + |B_y \nabla \cdot \boldsymbol{\xi}_\mathrm{p} - \boldsymbol{B}_\mathrm{p} \cdot \nabla \xi_y|^2 \right.$$
$$\left. + \gamma \mu_0 p |\nabla \cdot \boldsymbol{\xi}|^2 \right\} \mathrm{d}^2 r .$$ (10.57)

From (10.57) one can obtain the following stability result. If (after suitable rotation of the coordinate system in the x, z-plane) there exists a non-vanishing Cartesian component of the poloidal magnetic field $\boldsymbol{B}_\mathrm{p}$, the functional (10.57) is positive definite, implying stability.

For a proof let us consider the functional

$$F_2 = \frac{1}{2\mu_0} \int \left\{ |\nabla A_1|^2 - \frac{\mathrm{d}J}{\mathrm{d}A}|A_1|^2 \right\} \mathrm{d}^2 r .$$ (10.58)

In view of (10.57) positive-definiteness of F_2 is sufficient for stability for the present class of 2D perturbations. Arranging the coordinate system such that the non-vanishing component of $\boldsymbol{B}_\mathrm{p}$ is the x-component and differentiating the Grad–Shafranov equation of the equilibrium (see (5.78) but the invariant direction adjusted to y) with respect to z gives

$$-\Delta B_x = \frac{\mathrm{d}J}{\mathrm{d}A} B_x .$$ (10.59)

Using (10.59) the functional (10.58) can be written as

$$F_2 = \frac{1}{2\mu_0} \int \left\{ |\nabla A_1|^2 + \frac{\Delta B_x}{B_x}|A_1|^2 \right\} \mathrm{d}^2 r .$$ (10.60)

Now let us substitute $A_1 = \eta B_x$, which after integration by parts gives

$$F_2 = \frac{1}{2\mu_0} \int B_x^2 |\nabla \eta|^2 \mathrm{d}^2 r ,$$ (10.61)

which is positive definite for unrestricted η. As B_p does not vanish the restriction (10.50) does not apply and A_1 is unrestricted. The condition $B_x \neq 0$ assures that property of A_1 for unrestricted functions η. This completes the proof.

For reasons of reference we add that (10.58) is the second variation of

$$F = \int \left(\frac{(\nabla A)^2}{2\mu_0} - p(A) \right) \mathrm{d}^2 r, \tag{10.62}$$

the first variation vanishes (Grad, 1964).

For one-dimensional sheets of the form (10.52) the same argument proves stability for arbitrary $B_x(z)$. At points where B_x vanishes the substitution $A_1 = \eta B_x$ generates a constraint on A_1 that is equivalent to (10.50), and therefore admissible. For the present class of 2D perturbations this result confirms the stability of fields of the form (10.52) shown above for 3D perturbations.

Remark on the significance of the constraint (10.50)

Here we return to the constraint (10.50) and illustrate its significance by a simple example, the Harris sheet introduced in Section 5.3.2, subject to 2D perturbations. Using the same dimensionless formulation as in Section 5.3.2, we obtain

$$F_2 = \frac{1}{2} \int \left\{ |\nabla A_1|^2 - \frac{2}{\cosh^2(z)} |A_1|^2 \right\} \mathrm{d}x\,\mathrm{d}z . \tag{10.63}$$

Here, without loss of generality, modes of the form

$$A_1(x, z) = \sqrt{k/2\pi} \exp(ikx) a(z) \tag{10.64}$$

are chosen and the x-integration is extended over the period $2\pi/k$.

First, let us minimize this functional without the constraint (10.50) by solving the corresponding Euler–Lagrange equation. As that equation is linear-homogeneous in A_1 we are free to choose a normalization condition, which we set to $\int |A_1|^2/2 \,\mathrm{d}x\,\mathrm{d}z = 1$ which implies $\int_{-\infty}^{\infty} |a|^2/2 \,\mathrm{d}z = 1$. Introducing a corresponding Lagrangian multiplier λ we find the Euler–Lagrange equation in the form of the eigenvalue equation

$$-a'' - \frac{2}{\cosh^2(z)} a = \Lambda a \tag{10.65}$$

which is a one-dimensional eigenvalue problem of the Sturm–Liouville type (Morse and Feshbach, 1953) with the eigenvalue $\Lambda = \lambda - k^2$. The lowest eigenvalue λ equals the minimum value of F_2.

Equation (10.65) has a solution

$$\Lambda_0 = -1, \quad a_0 = \frac{1}{\cosh(z)}. \tag{10.66}$$

This solution satisfies the boundary condition $(a(\pm\infty) = 0)$ and has no zeros and therefore $\Lambda = -1$ is the lowest eigenvalue. As k is arbitrary, the choice $k \to 0$ leads to the lowest eigenvalue $\lambda = -1$. Thus, the minimum of F_2 is negative.

If we take the constraint (10.50) into account, it is clear that the solution (10.66) has to be removed because it violates the constraint (at $z = 0$). This removal is realized by setting $A_1 = B_x \eta$ with η being continuous. However, as shown above (see (10.61)), there is no η that makes F_2 negative. So, the Harris sheet is stabilized by the frozen-in constraint.

For later reference we add here that this argument implies that Λ_0 is the only negative eigenvalue of (10.65). If there was a second negative eigenvalue, the corresponding eigenfunction would be antisymmetric with respect to z and therefore it would satisfy the constraint (10.50), implying that a function η would exist with the corresponding F_2 being negative. This contradicts (10.61).

The case $B_y = 0$

For $B_y = 0$ the functional (10.57), minimized with respect to ξ_y, reduces to

$$\mathcal{V}_2 = \frac{1}{2} \int \left(\frac{|\nabla A_1|^2}{\mu_0} - \frac{\mathrm{d}j_0}{\mathrm{d}A_0} |A_1|^2 + \gamma p_0 |\nabla \cdot \boldsymbol{\xi}|^2 \right) \mathrm{d}x\,\mathrm{d}z, \tag{10.67}$$

subject to (10.50). Minimizing with respect to the component $\xi_{\parallel} = \boldsymbol{\xi} \cdot \boldsymbol{B_0}/B_0$ one obtains a functional of A_1 alone. The corresponding Euler–Lagrange equation is $\boldsymbol{B_0} \cdot \nabla(\nabla \cdot \boldsymbol{\xi}) = 0$, such that

$$\nabla \cdot \boldsymbol{\xi} = g(A_0) \tag{10.68}$$

where $g(A_0)$ is determined in the following way. Noting that the perpendicular component of $\boldsymbol{\xi}$ is $-A_1 \nabla A_0/B_0^2$, one writes (10.68) as

$$B_0 \frac{\partial}{\partial s_0} \left(\frac{\xi_{\parallel}}{B_0} \right) - \nabla \cdot \left(\frac{A_1 \nabla A_0}{B_0^2} \right) = g(A_0). \tag{10.69}$$

Dividing (10.69) by B_0, integrating with respect to s_0 and taking into account that $\xi_{\parallel}$ vanishes on the boundary, one finds

$$-\int \nabla \cdot \left(\frac{A_1 \nabla A_0}{B_0^2} \right) \frac{\mathrm{d}s_0}{B_0} = V(A_0) g(A_0) \tag{10.70}$$

where $V(A_0)$ is the differential flux tube volume $\int \mathrm{d}s_0/B_0$. Noting that $\int \cdots \mathrm{d}s_0/B_0 = (\mathrm{d}/\mathrm{d}A_0) \int_{\mathrm{D}(A_0)} \cdots \mathrm{d}x \, \mathrm{d}z$, where the domain D_{A_0} is bounded by the field line A_0 and by the boundary, and applying Gauss's theorem one obtains $g(A_0) = -Q_1(A_0)$, where

$$Q_1 = \frac{1}{V(A_0)} \frac{\mathrm{d}}{\mathrm{d}A_0} \int_{A_0} A_1 \frac{\mathrm{d}s_0}{B_0}. \tag{10.71}$$

Inserting $g(A_0)$ for $\nabla \cdot \boldsymbol{\xi}$ in (10.67) one obtains the minimum of $\mathcal{V}_2$ in the form

$$\mathcal{V}_2 = \frac{1}{2} \int \left(\frac{|\nabla A_1|^2}{\mu_0} - \frac{\mathrm{d}j_0}{\mathrm{d}A_0} |A_1|^2 + \gamma p_0 |Q_1|^2 \right) \mathrm{d}x \, \mathrm{d}z , \tag{10.72}$$

subject to (10.50). This functional will be recovered later in the discussion of Vlasov plasmas.

If the equilibrium is one-dimensional with $\boldsymbol{B} = B_x(z)\boldsymbol{e}_x$ and infinitely extended along x, one chooses perturbations of the form (10.64) which gives $Q_1 = 0$, so that

$$\mathcal{V}_2 = \frac{1}{2} \int \left(\frac{|\nabla A_1|^2}{\mu_0} - \frac{\mathrm{d}j_0}{\mathrm{d}A_0} |A_1|^2 \right) \mathrm{d}x \, \mathrm{d}z , \tag{10.73}$$

subject to (10.50).

10.2.3 Rayleigh–Taylor instability

Here, we return to the case of two-dimensional perturbations of a one-dimensional equilibrium, however, taking gravity into account. Suppose that the equilibrium magnetic field has a y-component only, that B_y, pressure p and density ρ depend on z only, and that the gravity force points into the negative z-direction such that $\psi = gz$ where g is a constant gravity acceleration. Note that in this case the gradients of A, B and ψ are coaxial such that the equilibrium representation (10.44), (10.45) breaks down. Therefore, we treat this degenerate case separately. For simplicity, we confine the discussion to two-dimensional ($\partial/\partial y = 0$) displacements $\boldsymbol{\xi}$, which are incompressible (i.e., $\nabla \cdot \boldsymbol{\xi} = 0$), and lie in the poloidal plane ($\xi_y = 0$). Under these conditions the equilibrium is characterized by the equation

$$\frac{\mathrm{d}}{\mathrm{d}z} \left(p(z) + \frac{B_y(z)^2}{2\mu_0} \right) + \rho(z)g = 0 \tag{10.74}$$

and the general expression (10.32) for $\mathcal{V}_2$ reduces to

$$\mathcal{V}_2 = -\frac{g}{2} \int \rho' |\xi_z|^2 \mathrm{d}^3 r, \quad \rho' = \frac{\mathrm{d}\rho}{\mathrm{d}z}, \tag{10.75}$$

where all but the last term in (10.32) have cancelled out. We conclude that there is instability for $\rho' > 0$, i.e., for heavier fluid layers located above lighter fluid layers. Perturbations will grow by lowering the centre of mass of the fluid, which thereby gains potential energy.

Since both the magnetic field perturbation $Q = -\xi_z B_y'(z) e_y$ and the equilibrium magnetic field have y-components only, the perturbation simply consists of rearrangement of straight magnetic field lines together with the plasma. This perturbation can be described as an 'interchange' of plasma and magnetic flux leading to a net gain of gravitational potential energy. Field line curvature can have a similar effect.

10.2.4 *Kink instability of plasma columns*

Here we briefly consider a particular instability of a cylindrical plasma column. For concreteness let us assume that the column has a radius r_0 and that it is surrounded by a vacuum region ($r > r_0$). The equilibrium is static and is independent of φ and z, using cylindrical coordinates r, φ, z. The invariance properties suggest modes of the form $f(r, \varphi, z) = \hat{f}(r) \exp(im\varphi + ikz)$ for any perturbation quantity f. It turns out that the modes with $m = \pm 1$ are most difficult to stabilize. These are the *kink* modes. They lead to helical deformations of the column (Fig. 10.1). The physical reason is that the azimuthal magnetic field is deformed such that magnetic forces enhance the perturbation (Fig. 10.1). The instability is absent if there is no longitudinal current so that the azimuthal magnetic field vanishes. Another stable configuration is a column with a sufficiently small length L_z together with a finite longitudinal magnetic field. Details can be found in the plasma literature (e.g., Freidberg, 1987). Here we just add a remark on the mechanism.

The instability is based on a resonance between the helices defined by the perturbation with wave number $k = (m/r) e_\varphi + k e_z$ and the helix traced by a field line. The resonance condition is given by $k \cdot B = 0$. As this is a local condition, depending on the radius r, one has to identify the most unstable helix. Consider the case where the resonance is in the vacuum region (*external kink mode*). As the external B_z-component is constant and $B_\varphi \propto 1/r$ the external resonance condition becomes independent of r, so that one can formulate it at $r = r_0$. Thus, using that $kL_z \geq 2\pi$, one finds that for kink modes resonance is absent if and only if

$$\left| \frac{B_\varphi(r_0)}{B_z} \right| < \frac{2\pi r_0}{L_z} . \tag{10.76}$$

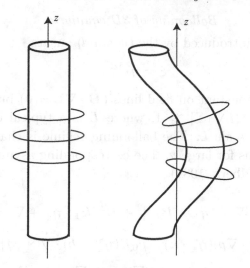

Fig. 10.1 The kink instability leads to a helical deformation (right) of a cylindrical plasma column (left). The field lines of the azimuthal magnetic field become denser on the inner side causing a net magnetic force, so that an initial helical perturbation grows.

Expressed in terms of the longitudinal current I_z, the limiting current for a stable column is given by

$$I_z = \frac{(2\pi r_0)^2 B_z}{\mu_0 L_z} \qquad (10.77)$$

which is known as the *Kruskal–Shafranov limit.*

For the prototype field of a force-free column (Gold and Hoyle, 1960)

$$B_\varphi = B_0 \frac{r/a}{1 + (r/a)^2}, \quad B_z = B_0 \frac{1}{1 + (r/a)^2} \qquad (10.78)$$

a rough estimate is obtained from (10.77) by identifying a with r_0, which gives $L_z/a = 2\pi$ as the stability limit. A rigorous approach (Hood and Priest, 1981) for a closed boundary gave a limit of 2.49π. (See also Einaudi and van Hoven, 1983 and An, 1984.)

10.2.5 The ballooning limit

Here we turn to instabilities with 3D perturbations that are strongly localized in a direction perpendicular to B. Such modes are generally termed *ballooning modes*, they are based on the pressure-gradient/curvature term in (10.36). We begin with the general case of 3D equilibria and then specialize for two dimensions including examples. Gravity is ignored.

Ballooning of 3D equilibria

The localization is introduced by the (*eikonal*) ansatz

$$\boldsymbol{\xi} = \boldsymbol{\eta}\, e^{i\Sigma} . \tag{10.79}$$

The function Σ is constant on field lines ($\boldsymbol{B} \cdot \nabla\Sigma = 0$) but varies strongly across $\boldsymbol{B}$, i.e., $\nu = |L\nabla_\perp\Sigma| \gg 1$, where L is a typical equilibrium scale length; $\boldsymbol{\eta}$ varies on scale L. The ballooning regime is characterized by asymptotic expressions for large ν. The corresponding asymptotic form of $\mathcal{V}_2$ is obtained as (Freidberg, 1987)

$$
\begin{aligned}
\mathcal{V}_2 = \frac{1}{2\mu_0} \int d^3r \Big[&|\nabla \times (\boldsymbol{\eta}_\perp \times \boldsymbol{B})_\perp|^2 + B^2 |i\boldsymbol{k}_\perp \cdot \boldsymbol{\eta}_\perp + \nabla \cdot \boldsymbol{\eta}_\perp + 2\boldsymbol{\kappa} \cdot \boldsymbol{\eta}_\perp|^2 \\
&- 2\mu_0(\boldsymbol{\eta}_\perp \cdot \nabla p)(\boldsymbol{\eta}_\perp^* \cdot \boldsymbol{\kappa}) - \mu_0 j_\parallel (\boldsymbol{\eta}_\perp^* \times \boldsymbol{b}) \cdot \nabla \times (\boldsymbol{\eta}_\perp \times \boldsymbol{B})_\perp \\
&+ \mu_0\gamma p |\nabla \cdot \boldsymbol{\eta}_\parallel + \nabla \cdot \boldsymbol{\eta}_\perp + i\boldsymbol{k}_\perp \cdot \boldsymbol{\eta}_\perp|^2 \Big] , \tag{10.80}
\end{aligned}
$$

where $\boldsymbol{k}_\perp$ equals $\nabla\Sigma$. In (10.80) we have added the compressibility term $\gamma p |\nabla \cdot \boldsymbol{\xi}|^2$ (the last term), which was left out by Freidberg, who concentrated on sufficient stability criteria. By using a power expansion $\boldsymbol{\eta}_\perp = \boldsymbol{\eta}_{\perp 0} + \boldsymbol{\eta}_{\perp 1} + \cdots$ with respect to $1/\nu$ and by minimizing $\mathcal{V}_2$ with respect to $\boldsymbol{\eta}_{\perp 0}$, $\boldsymbol{\eta}_{\perp 1}$ and $\boldsymbol{\eta}_\parallel$ one finds that the stability problem reduces to finding the minimum of the one-dimensional functional (Schindler and Birn, 2004)

$$
w = \frac{1}{2\mu_0} \int \Big[k_\perp^2 |\boldsymbol{b} \cdot \nabla X|^2 - \frac{2\mu_0}{B^2}(\boldsymbol{b} \times \boldsymbol{k}_\perp) \cdot \nabla p \,(\boldsymbol{b} \times \boldsymbol{k}_\perp) \cdot \boldsymbol{\kappa}\, |X|^2
$$
$$
+ \frac{1}{q} \Big| \overline{2(\boldsymbol{b} \times \boldsymbol{k}_\perp) \cdot \boldsymbol{\kappa}\, X/B} \Big|^2 \Big] \frac{ds}{B} , \tag{10.81}
$$

where s is the distance along field lines and X is defined through the minimizing $\boldsymbol{\eta}_{\perp 0}$, which is given by $X\boldsymbol{b} \times \boldsymbol{k}_\perp/B$. The bar denotes the average

$$\overline{(\cdots)} = \frac{\int (\cdots) ds/B}{\int ds/B} , \tag{10.82}$$

where the integral is extended over the entire field line section inside the domain under consideration, and

$$q = \frac{1}{\mu_0\gamma p} + \frac{1}{B^2} . \tag{10.83}$$

The last term in (10.81) stems from the compressibility term in (10.80). To avoid a rather involved discussion of the role of field singularities, it is assumed that $B \neq 0$ and that all field lines reach the boundary. However, singularities will be included in the discussion of the two-dimensional case.

Assuming that $\boldsymbol{B}$ can be represented by Euler potentials as $\boldsymbol{B} = \nabla\alpha \times \nabla\beta$ (Section 5.1.2) one can arrange the potentials such that Σ depends on β alone.

Defining $X_1 = d\Sigma/d\beta X$ one obtains

$$w = \frac{1}{2\mu_0} \int \left[(\nabla\beta)^2 |\mathbf{b} \cdot \nabla X_1|^2 - \frac{2\mu_0}{B^2} (\mathbf{b} \times \nabla\beta) \cdot \nabla p \, (\mathbf{b} \times \nabla\beta) \cdot \boldsymbol{\kappa} \, |X_1|^2 \right.$$
$$\left. + \frac{1}{q} \left| 2(\mathbf{b} \times \nabla\beta) \cdot \boldsymbol{\kappa} \, X_1/B \right|^2 \right] \frac{ds}{B} . \quad (10.84)$$

The expression (10.84) is put into a more compact form by expressing $\boldsymbol{\kappa}$ by its covariant components (there is no ∇s-component)

$$\boldsymbol{\kappa} = \kappa_\alpha \nabla\alpha + \kappa_\beta \nabla\beta \quad (10.85)$$

and by introducing the *curvature potential* U_c

$$U_c = -2\mu_0 \frac{\partial p}{\partial\alpha} \kappa_\alpha . \quad (10.86)$$

With these expressions (10.84) can be written as

$$w = \frac{1}{2\mu_0} \int \left[(\nabla\beta)^2 \left| \frac{\partial X_1}{\partial s} \right|^2 + U_c |X_1|^2 + \frac{1}{\overline{q} \, (\mu_0 \partial p/\partial\alpha)^2} \left| \overline{U_c X_1} \right|^2 \right] \frac{ds}{B} . \quad (10.87)$$

Instability can occur only through the term $U_c |X_1|^2$.

As the test function space has been restricted by (10.79) and $\nu \gg 1$ the functionals (10.81)–(10.87) can be used only to find instability. Therefore, it is not a severe additional drawback that the choice of $\boldsymbol{k}_\perp$ in (10.81) (or of $\nabla\beta$ in (10.84)) is not unique. It suffices to find a particular $\boldsymbol{k}_\perp$ (or $\nabla\beta$) that leads to an unstable field line. As seen below, for symmetric equilibria an appropriate choice is to have $\nabla\beta$ point in the invariant direction.

Rotational invariance

As a first symmetric case let us choose a system which in cylindrical coordinates r, φ, z is characterized by $\partial/\partial\varphi = 0$ and $B_\varphi = 0$. Then one can set $\alpha = r A_\varphi(r, z)$, where A_φ is the φ-component of the vector potential, and $\beta = \varphi$ (see Section 5.2.3). It will be seen below that this choice of β is the most efficient one.

For such systems (10.87) becomes

$$w = \frac{1}{2\mu_0} \int \left[\frac{1}{r^2} \left| \frac{\partial X_1}{\partial s} \right|^2 + U_{c,\text{rot}} |X_1|^2 + \frac{1}{\overline{q} \, (\mu_0 p')^2} \left| \overline{U_{c,\text{rot}} X_1} \right|^2 \right] \frac{ds}{B} \quad (10.88)$$

where now

$$U_{\mathrm{c,rot}} = -\frac{2\mu_0 \boldsymbol{\kappa} \cdot \nabla p}{r^2 B^2}$$

$$= -\frac{2\mu_0 p'}{B^2}\left(\mu_0 p' + \frac{\nabla B \cdot \nabla \alpha}{r^2 B}\right) \tag{10.89}$$

and $p' = \mathrm{d}p(\alpha)/\mathrm{d}\alpha$.

An expression equivalent to (10.88) was obtained by Bernstein *et al.* (1958), who also showed that negative values of (10.88) are not only sufficient but also necessary for instability. This confirms that the invariant direction is the most efficient direction of $\nabla \beta$.

Interchange mode

In one of their applications of (10.88) Bernstein *et al.* (1958) assumed longitudinal periodicity of equilibrium and perturbations in such a way that functions X_1 that are constant along field lines can be admitted as test functions. (The energy principle remains valid, although the boundary conditions differ from the closed boundary case considered here.) Also, it is assumed that the field lines pass through the entire period (Fig. 10.2). For constant X_1 one finds from (10.88)

$$w = \frac{|X_1|^2 V}{2\mu_0}\left(\overline{U_{\mathrm{c,rot}}} + \frac{\overline{U_{\mathrm{c,rot}}}^2}{\overline{q}\,(\mu_0 p')^2}\right), \tag{10.90}$$

where

$$V = \int \frac{\mathrm{d}s}{B} \tag{10.91}$$

is the differential flux volume.

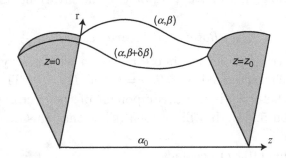

Fig. 10.2 The integration domain over which the identity (10.94) is integrated. The equilibrium quantities are periodic in z with period z_0; the upper boundary is a surface generated by the field lines (α, β') with $\beta \le \beta' \le \beta + \delta\beta$.

The expression (10.90) can be put in a more useful form by employing the relationship

$$\overline{V_c U_{\text{c,rot}}} + \overline{q} \left(\mu_0 p'\right)^2 = \frac{\mu_0 p'^2}{\gamma} \frac{\mathrm{d}S}{\mathrm{d}p}, \tag{10.92}$$

where

$$S = \ln\left(pV^\gamma\right) \tag{10.93}$$

is the flux tube entropy.

The relationship (10.92) is found from the identity

$$\nabla \cdot \left(\frac{P' \nabla \alpha}{r^2 B^2}\right) = P'' + U_{\text{c,rot}} + \frac{P'^2}{B^2}, \tag{10.94}$$

where $P = \mu_0 p$. Integrating (10.94) over the domain shown in Fig. 10.2 and applying Gauss's theorem, taking into account the assumed periodicity, one finds for infinitesimal $\delta\beta$

$$\delta\beta P' V = \delta\beta \int_{\alpha_0}^{\alpha} \mathrm{d}\alpha \int \frac{\mathrm{d}s}{B} \left(P'' + V_c U_{\text{c,rot}} + \frac{P'^2}{B^2}\right). \tag{10.95}$$

Differentiating (10.95) with respect to α and using (10.83) one finds (10.92).

Making use of (10.92) in (10.90) one finds

$$w = \frac{|X_1|^2 V}{2\mu_0^2 \gamma \overline{q}} \overline{U_{\text{c,rot}}} \frac{\mathrm{d}S}{\mathrm{d}p}, \tag{10.96}$$

implying instability for $\overline{U_{\text{c,rot}}} \, \mathrm{d}S/\mathrm{d}p < 0$. This criterion is equivalent to the condition $\Lambda < 0$ of Bernstein *et al.* (1958) (see their equation (6.27)).

In important applications (interchange mode) there exist field lines with $\overline{U_{\text{c,rot}<0}}$. In view of (10.89) (first line) this means that field line sections with the higher pressure on their concave side dominate. Then (10.96) implies instability if $\mathrm{d}S/\mathrm{d}p > 0$.

It should be noted that, because of the assumed periodicity, the interchange mode has rather limited applicability. In particular, the interchange mode does not exist for a closed boundary as it does not allow us to set X_1 to a constant. Nevertheless, as shown in the following section the sign of $\mathrm{d}S/\mathrm{d}p$ remains relevant for the stability of a particular class of configurations where the perturbations satisfy $\boldsymbol{\xi} = 0$ on the boundary.

Translational invariance

Let us now choose equilibria that are independent of the Cartesian y-component with $B_y = 0$ (see Section 5.2.1). Accordingly, we set $\alpha = A(x,z)$, where $A(x,z)$ is the y-component of the vector potential, and

$\beta = y$. The equilibria have the symmetry $A(x, -z) = A(x, z)$. Then, setting $\Sigma = \Sigma(y)$ and identifying $X_1 \exp(i\Sigma(y))$ with A_1' we find that (10.81) reduces to

$$w = \frac{1}{2\mu_0} \int \left(\left| \frac{\partial A_1}{\partial s} \right|^2 + V_c |A_1|^2 + \frac{1}{J^2 \bar{q}} |\overline{V_c A_1}|^2 \right) \frac{ds}{B}, \qquad (10.97)$$

where $J = \mu_0 j_y$ and U_c from (10.86) specializes to V_c as given by (10.48), which can be written as

$$V_c = -\frac{2\mu_0}{B^2} \boldsymbol{\kappa} \cdot \nabla p. \qquad (10.98)$$

For periodic systems one finds a relationship analogous to (10.92), which reads

$$\overline{V_c} + J^2 \bar{q} = \frac{J^2}{\mu_0 \gamma} \frac{dS}{dp} \qquad (10.99)$$

and leads to the interchange criterion, which again states instability for $V_c dS/dp > 0$.

As in the case of rotational invariance, positive-definiteness of (10.97) is necessary and sufficient for MHD stability. Here, the following simple explanation is available. The necessity is clear from the restriction of the test-function space to ballooning modes. The sufficiency can be seen by considering (10.51) with $B_y = 0$. Omitting the positive term $|\boldsymbol{B}_p \cdot \nabla \xi_y|^2$ results in an expression which implies stability if it is positive-definite. By minimizing that expression with respect to ξ_y and $\xi_\parallel$ one finds (10.97). Interestingly, the positive term that was left out also disappears in the ballooning limit. In fact, that is the only change that the ballooning limit causes on (10.51) with $B_y = 0$, so that the same expression arises in both cases. Thus, we conclude that under the present conditions (2D equilibrium with $B_y = 0$ and closed boundary), positive-definiteness of (10.97) is necessary and sufficient for stability with respect to arbitrary ideal MHD modes.

MHD stability of magnetotail equilibria

Here, this criterion is used to address stability of magnetotail equilibria with a closed boundary. The implied line-tying of magnetospheric field lines at the near-Earth boundary qualitatively represents a highly-conducting ionosphere, which is a reasonable approximation when the timescale for the evolution of the instability is short compared to resistive diffusion times at the ionosphere. Parallel flow is inhibited by the strong near-Earth magnetic field strength.

The plasma sheet is a potential candidate for a ballooning instability based on the strong field line curvature at the vertex of closed field lines (i.e., field lines passing through the centre of the plasma sheet) with $\kappa \cdot \nabla p > 0$. However, there are also stabilizing effects, such as the plasma compressibility, which also becomes large in the centre of the plasma sheet. Also, a constant background pressure stabilizes, as it does not contribute to the destabilizing pressure gradient but increases the compressibility effect via the quantity q. A detailed study of these effects was carried out by Schindler and Birn (2004).

In that work the stability was determined by numerical minimizations of (10.97) for several equilibrium models, considering strongly stretched field lines, typical for magnetotail configurations. Accordingly, the aspect ratio $\epsilon = L_z/L_x$ is chosen small compared to 1 (see Section 5.4.1).

Before giving a brief description of the results, for later application we raise the question how the relationship (10.99) is modified by the boundary condition (Fig. 10.3). As shown in Schindler and Birn (2004), a procedure analogous to the corresponding treatment for rotational invariance described above leads to the relationship

$$\overline{V_c} + J^2 \overline{q} = \frac{J^2}{\mu_0 \gamma} Q \,, \tag{10.100}$$

where

$$Q = \frac{\mathrm{d}S}{\mathrm{d}p} + c_b \tag{10.101}$$

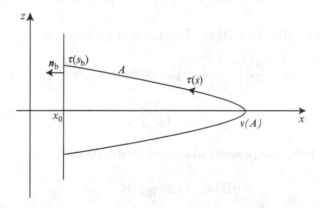

Fig. 10.3 Qualitative shape of a closed magnetic field line A in the plasma sheet with a conveniently defined running coordinate $\tau(s)$. The earthward boundary is located at $x = x_0$, the vertex at $x = v(A)$.

with

$$c_b = \frac{\gamma}{JV} \left[\frac{1}{B^2} \frac{n_b \cdot \nabla A}{|n_b \cdot B|} \right]_{s_b}, \tag{10.102}$$

s_b denoting the location on field line A where that field line crosses the boundary, which has an outward-pointing normal n_b (Fig. 10.3). In the present 2D case, integrations $\int \ldots ds$ are carried out on the field line sections in the half-plane $z > 0$.

The expression (10.100) with (10.101) shows that the boundary condition generates an additional term c_b. However, as shown in Schindler and Birn (2004), for strongly stretched fields c_b typically is a small correction (see also Lee and Wolf (1992)). If at the boundary B is considered as of order 1, one finds $V = O(1/\epsilon)$, $n_b \cdot \nabla A = O(\epsilon)$ and $n_b \cdot B = O(1)$ so that $c_b = O(\epsilon^2)$. So, in an approximate sense, Q equals dS/dp and (10.100) approximates (10.99). Accordingly, we will see that the entropy derivative dS/dp plays an important role also for the stability of the present systems with closed boundary.

As the first example, let us choose an equilibrium represented by the linear version of the Grad–Shafranov equation (5.136) including a constant background pressure.

Expressed in non-dimensional variables (formally setting $\mu_0 = 1$, see Section 5.3.4) the magnetic flux function is given by

$$A = -\frac{2}{\pi} \cos(\frac{\pi}{2}z) \, e^{-x_1} \quad \text{for } 0 \leq z \leq 1, \tag{10.103}$$

$$A = \frac{1}{\epsilon} \sin(\epsilon(z-1)) \, e^{-x_1} \quad \text{for } z > 1, \tag{10.104}$$

with $x_1 = \epsilon x$ and $A(-z) = A(z)$. The plasma pressure is

$$p = \frac{A^2}{2} \left(\frac{\pi^2}{4} - \epsilon^2 \right) + p_0 \quad \text{for } 0 \leq z \leq 1, \tag{10.105}$$

$$p = p_0 \quad \text{for } z > 1. \tag{10.106}$$

The magnetic field components obtained from (10.103) are

$$B_x = -\sin(\frac{\pi}{2}z) \, e^{-x_1} \tag{10.107}$$

$$B_z = \frac{2\epsilon}{\pi} \cos(\frac{\pi}{2}z) \, e^{-x_1} = -\epsilon A \,. \tag{10.108}$$

A set of magnetic field lines is shown in Fig. 10.4. The equilibrium parameters are ϵ and the pressure ratio $K = p_0/p_m$, where $p_m = (1 - 4\epsilon^2/\pi^2)/2 + p_0$ is the maximum pressure in the domain considered. The closed field lines are parameterized by v_1, the x_1-coordinate of the vertex (Fig. 10.3).

For symmetric modes the most interesting feature is the presence of an unstable region embedded in an extended region of stability (Fig. 10.5). In the figure the points are the result of numerical minimizations, iterated for vanishing minimum of (10.97). The smooth curve corresponds to the

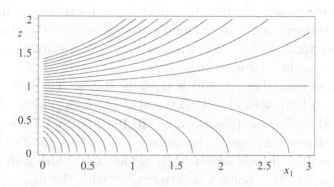

Fig. 10.4 Field lines of the model (10.103)–(10.106). The left boundary is placed at $x_1 = 0$. (Reproduced from Schindler and Birn (2004) by permission of the American Geophysical Union.)

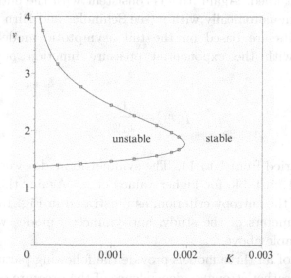

Fig. 10.5 Stability diagram with $\epsilon = 0.1$ of the model (10.103)–(10.106). (Reproduced from Schindler and Birn (2004) by permission of the American Geophysical Union.)

marginal entropy criterion $dS/dp = 0$. The differences are smaller than the resolution of the graph. An explanation of this surprising result is provided below.

Note that a rather small amount of background pressure suffices for complete stabilization.

A typical unstable symmetric mode is sharply peaked at the vertex, where the combination of pressure gradient and strong curvature provides the free energy for the instability. As the antisymmetric mode is forced to vanish at the vertex, it explores the region of large curvature less efficiently, which stabilizes. Correspondingly, all antisymmetric modes that were investigated in a scan of parameter space were found stable, in spite of the absence of the compressibility term. (This is consistent with an exact stability criterion for antisymmetric modes discussed below.)

The region of open flux ($A > 0$) is trivially stable. There is no pressure gradient (see 10.106), such that $V_c = 0$ and w becomes positive definite.

In the closed field region (described by (10.103), (10.105)) the equilibrium has a rather strong (exponential) pressure decay along the tail axis. This seems to destabilize in comparison with models that show a more gradual pressure variation. This point was investigated with the help of the model (5.133), where the pressure on the x_1-axis is given by $(1 + 1/(2\sqrt{x_1}))^2/2$, showing a rather moderate decrease compared with the exponential law of the first model. Numerical minimizations gave stability for the entire parameter set studied. Again, this is consistent with the entropy criterion, as S decreases monotonically with p (see Schindler and Birn (2004)).

Further results are based on the tail asymptotic model discussed in Section 5.4.1 with the exponential pressure function $p(A)$ and $\hat{p}(x_1)$ chosen as

$$\hat{p}(x_1) = \frac{1}{x_1{}^n} \tag{10.109}$$

where n was varied from 1 to 14. The symmetric modes were found stable for $n < 10$ and unstable for higher values of n. Again, this behaviour is consistent with the entropy criterion, as illustrated in Fig. 10.6.

For the parameters of the study, antisymmetric modes were stable for $n < 6$ and unstable above.

The findings of all three models provide the following picture. Instability occurs only for rather strong x_1-dependence of the pressure on the tail axis, such as $p \propto \exp(-x_1)$ or $\propto x_1^{-n}$, $n > 10$. For symmetric modes rather small background pressures lead to complete stabilization. Within the numerical

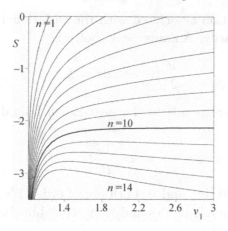

Fig. 10.6 Entropy S is plotted as a function of the vertex location v_1 for the tail asymptotic model with (10.109). S increases monotonically with v_1 for $n < 10$. (Reproduced from Schindler and Birn (2004) by permission of the American Geophysical Union.)

accuracy, stability of symmetric modes was found to be consistent with the entropy criterion in the form that the modes are stable if and only if $dS/dp < 0$.

In contrast to intuitive expectation, the large field line curvature in the central plasma sheet does not necessarily lead to ballooning instability. Note, however, that this result was established only for strongly stretched configurations; less stretched equilibria may well behave differently (Roux *et al.*, 1991; Cheng and Zaharia, 2004).

The ballooning stability of tail configurations has been studied by a number of other authors (Miura *et al.*, 1989; Lee and Wolf, 1992; Hurricane, 1997; Bhattacharjee *et al.*, 1998; Miura, 2000; Cheng and Zaharia, 2004). In most cases their results are largely consistent with the findings described above. A discrepancy with Miura (2000) is likely to be the result of his neglecting the compressibility term (for details see Schindler and Birn (2004)).

The central role of the entropy criterion, as for instance demonstrated by Fig. 10.5, can be understood from analytic considerations (Schindler and Birn, 2004). Here is a brief outline of the arguments. The discussion is confined to the region of closed field lines of 2D equilibria with higher pressure on the concave side of the field lines, so that $\boldsymbol{\kappa} \cdot \nabla p > 0$, implying that the curvature potential V_c is negative. The running coordinate along field lines is chosen as $\tau = \int_0^s B \, ds$, and the perturbation A_1 is written as $\alpha(\tau)$. Also, one uses the fact that the norm of the perturbation employed to obtain the

Euler–Lagrange equation associated with the variational expression (10.97) can be chosen arbitrarily. A convenient choice is

$$\frac{1}{2} \int_0^{\tau_b} \alpha^2 \phi \, d\tau = 1, \qquad (10.110)$$

where $\tau_b = \tau(s_b)$ corresponds to the near-Earth boundary (see Fig. 10.3) and

$$\phi = -\frac{V_c}{B^2}. \qquad (10.111)$$

The length of the field line and τ_b both are of order $1/\epsilon$.

Dealing with symmetric modes and using the abbreviations

$$D = \frac{1}{J^2 \int q \, d\tau / B^2}, \qquad Y = \int_0^{\tau_b} \phi \alpha \, d\tau, \qquad (10.112)$$

the Euler–Lagrange equation for the minimizer of (10.97) assumes the form

$$\ddot{\alpha} + (1 + \lambda)\phi\alpha = D\phi Y, \qquad (10.113)$$

where the dot denotes differentiation with respect to τ. The boundary conditions are $\dot{\alpha}(0) = 0, \alpha(\tau_b) = 0$.

Also consider the associated homogeneous eigenvalue problem

$$\ddot{\eta} + \sigma\phi\eta = 0, \qquad \dot{\eta}(0) = 0, \eta(\tau_b) = 0. \qquad (10.114)$$

In contrast to (10.113), the problem (10.114) is a Sturm–Liouville eigenvalue problem (e.g., Morse and Feshbach, 1953) with its characteristic ordering properties of eigenvalues and zeros of eigenfunctions (well known from one-dimensional Schrödinger problems).

The equation (10.113) may be solved in two steps (Hurricane, 1997). First, one sets $Y = 1$ and finds a solution of (10.113) for arbitrary values of λ, except for the values $\lambda = \sigma_\nu - 1$, where the solution has singularities. Then, the eigenvalues follow from the equation

$$Y(\lambda) = 1. \qquad (10.115)$$

It is convenient to express the solution of (10.113) with $Y = 1$ in terms of the solution $u(\tau, \lambda)$ of the homogeneous equation

$$\ddot{u} + (1 + \lambda)\phi u = 0 \qquad (10.116)$$

understood as an initial value problem with initial conditions $u(0, \lambda) = 1, \dot{u}(0, \lambda) = 0$ (not to be confused with the eigenvalue problem (10.114)).

Then, the solution of (10.113) with $Y = 1$ is

$$\alpha(\tau, \lambda) = \frac{D}{1 + \lambda} \left(1 - \frac{u(\tau, \lambda)}{u_b(\lambda)} \right) \qquad (10.117)$$

with $u_b(\lambda) = u(\tau_b, \lambda)$. With the help of (10.116), integrated from 0 to τ_b, this gives

$$Y(\lambda) = \frac{DZ}{1 + \lambda} + \frac{D}{(1 + \lambda)^2} \frac{\dot{u}_b(\lambda)}{u_b(\lambda)} \qquad (10.118)$$

where $Z = \int_0^{\tau_b} \phi \, d\tau$ and $\dot{u}_b(\lambda) = \dot{u}(\tau_b, \lambda)$.

It turns out that, for small ϵ, (10.118) assumes a rather simple structure. If ϵ is sufficiently small, the second term of (10.118) is significant only in small intervals of λ containing a singularity at $\lambda_\nu = \sigma_\nu - 1$, (corresponding to vanishing u_b). Outside those regions $Y(\lambda)$ can therefore be approximated by the first term,

$$Y_0(\lambda) = \frac{DZ}{1 + \lambda} \qquad (10.119)$$

which gives an eigenvalue

$$\lambda_Q = DZ - 1. \qquad (10.120)$$

With (10.100) λ_Q can be written as

$$\lambda_Q = -\frac{Q}{\mu_0 \gamma \bar{q}}. \qquad (10.121)$$

Here, some exceptional cases are ignored that are physically insignificant and can be avoided by small changes of parameters.

As $Y_0(\lambda)$ is a monotonic function, all eigenvalues other than (10.121) must be associated with the singularities of (10.118) arising from zeros of u_b.

Fig. 10.7 gives an example of the function $Y(\lambda)$, evaluated for the asymptotic tail model with (10.109). The figure also indicates that, outside the singularities, the expression (10.119) (smooth curves) is a good approximation. The graph includes the singularity at $\lambda = \sigma_2 - 1$.

The narrow structure of the singular parts of $Y(\lambda)$ and other details can be understood in terms of the following properties (for derivations see Schindler and Birn (2004)):

$$0 < \sigma_0 = O(\epsilon^3), \quad 0 < \sigma_2 = O(1), \quad \lambda_{\min} \geq \sigma_0 - 1$$

$$\lambda_Q = O(\epsilon), \quad \delta\lambda_n = O(\epsilon^3), \quad n = 0, 2, \ldots \qquad (10.122)$$

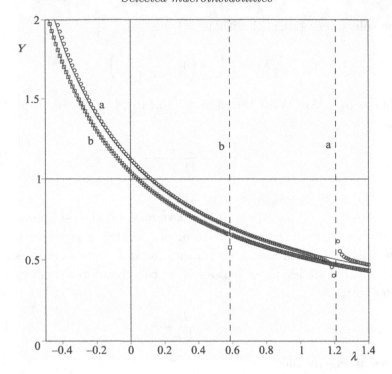

Fig. 10.7 The function $Y(\lambda)$ for the tail-asymptotic model with (10.109) for $n = 2$, $x_{10} = 1$, $v_1 = 2$ and $\epsilon = 0.3$ (curve a) and $\epsilon = 0.1$ (curve b). The smooth curves correspond to (10.119). The dashed vertical lines indicate the positions of the singularities at $\lambda = \sigma_2 - 1$. (Reproduced from Schindler and Birn (2004) by permission of the American Geophysical Union.)

where λ_{min} is the minimum eigenvalue of (10.113) and $\delta\lambda$ characterizes the width of the thin structures of $Y(\lambda)$ associated with the singularities. These properties immediately lead to the following result:

$$\lambda_{\mathrm{min}} = \min(\lambda_Q, \sigma_2 - 1)\,, \tag{10.123}$$

where a small difference of order ϵ^3 has been ignored.

Thus, the stability problem is reduced to determining whether σ_2 is smaller or larger than $\lambda_Q + 1$. Due to the Sturm–Liouville properties of (10.114) this can be done by simply plotting $u(\tau, \lambda_Q)$. If that curve has not more than one internal $(0 < \tau < \tau_b)$ zero, then $\lambda_{\mathrm{min}} = \lambda_Q$; if it has two or more internal zeros, then $\lambda_{\mathrm{min}} = \sigma_2 - 1$. For sufficiently small ϵ it suffices to investigate $u(\tau, 0)$ instead of $u(\tau\,\lambda_Q)$. For the three models discussed above it was established that $\lambda_{\mathrm{min}} = \lambda_Q$. For the first model, this explains the excellent fit of the minimization results by the entropy criterion in Fig. 10.5.

These results suggest that under the present equilibrium properties, $Q < 0$, or approximately $dS/dp < 0$, is necessary and sufficient for the stability with respect to arbitrary symmetric MHD modes, an equally simple and powerful criterion.

One might wonder how it is possible that stability turns out to be determined by Q, which (other than, e.g., V_c) is not particularly sensitive to the strong curvature at the vertex. The explanation is that the dominant contributions from curvature and from compressibility cancel each other. This follows from (10.120) which can be written as

$$\lambda_Q = -\frac{1}{J^2\bar{q}}(\bar{V}_c + J^2\bar{q}). \tag{10.124}$$

Both $\bar{V}_c$ and $J^2\bar{q}$ are of order $1/\epsilon$, the leading terms being $-\overline{(J^2/B^2)}$ and $\overline{(J^2/B^2)}$, which cancel out.

So far the discussion was confined to symmetric modes. Regarding antisymmetric modes, there is an interesting sufficient stability criterion that can be stated as follows: Let B_z be symmetric and B_x antisymmetric with respect to z and consider a field line A that is closed. If

$$\left\{ B_x B_z \neq 0, \ B_x B_z \frac{\partial^2 B_z^2}{\partial A \partial z} \leq 0 \right\} \quad \text{for} \quad z > 0 \tag{10.125}$$

on that field line, then the field line is stable with respect to antisymmetric modes. A proof of this criterion is given in Schindler and Birn (2004). This criterion directly proves stability of antisymmetric modes of the model (10.103), where B_z is constant on field lines.

10.3 The resistive tearing instability

For the purposes of this book, instabilities that involve a change of magnetic topology are of particular interest. This is because such instabilities allow for an efficient release of previously stored magnetic energy. A simple potential candidate for showing an instability that changes the magnetic topology is a one-dimensional current sheet with a reversal of the magnetic field direction. As discussed above (Section 10.2.2), in ideal MHD such a configuration is stabilized by the frozen-in constraint. The purpose of this section is to relax that constraint by allowing for a (small) resistivity. As we will see, this in fact leads to an instability that changes magnetic topology. This instability, the *resistive tearing mode* (Furth *et al.*, 1963), in many ways is the prototype of topology-changing instabilities, and therefore it is discussed in detail. (An analogous kinetic process will be discussed farther below.)

The resistive tearing instability is described by the equations of resistive MHD (Section 3.3.3), specialized for the present purpose.

10.3.1 The model equations

Let us assume translational invariance with respect to the y-direction and let the poloidal magnetic field (x- and z-components) be derived from a flux function $A(x, z, t)$, which gives the magnetic field of the form

$$\boldsymbol{B} = \nabla A(x, z, t) \times \boldsymbol{e}_y + B_y(x, z, t)\boldsymbol{e}_y \; . \tag{10.126}$$

The electromagnetic gauge is fixed such that the electric potential has no y-dependence implying $E_y = -\partial A/\partial t$.

For simplicity, we follow the standard procedure in assuming that the velocity field $\boldsymbol{v}$ is divergence-free. This assumption tends to be applicable to processes that are slow compared with the propagation of ideal MHD waves, which is the case for the tearing mode. Thus, in analogy with (10.126) we write $\boldsymbol{v}$ in the form

$$\boldsymbol{v} = \nabla D(x, z, t) \times \boldsymbol{e}_y + v_y(x, z, t)\boldsymbol{e}_y \; , \tag{10.127}$$

where the stream function $D = D(x, z)$ generates the poloidal velocity $\boldsymbol{v}_p$.

Under these conditions, the resistive MHD equations (see Section 3.3.3) assume the form

$$\frac{\partial \rho}{\partial t} + [\rho, D] = 0 \tag{10.128}$$

$$\nabla \cdot \left(\rho \nabla \frac{\partial D}{\partial t} \right) - [D, \rho \Delta D] - \frac{1}{2}[\rho, v_p^2] + \frac{1}{\mu_0}[A, \Delta A] = 0 \tag{10.129}$$

$$\frac{\partial A}{\partial t} + [A, D] - \frac{\eta}{\mu_0}\Delta A = 0 \tag{10.130}$$

$$\rho\frac{\partial v_y}{\partial t} + \rho[v_y, D] + \frac{1}{\mu_0}[A, B_y] = 0 \tag{10.131}$$

$$\frac{\partial B_y}{\partial t} + [B_y, D] + [A, v_y] - \nabla \cdot \left(\frac{\eta}{\mu_0}\nabla B_y \right) = 0 \; . \tag{10.132}$$

Here, the square brackets signify

$$[f, g] = \frac{\partial f}{\partial z}\frac{\partial g}{\partial x} - \frac{\partial f}{\partial x}\frac{\partial g}{\partial z} \tag{10.133}$$

for arbitrary functions $f(x, z)$, $g(x, z)$.

The equations (10.128), (10.130) and (10.131) express mass conservation and the y-components of Ohm's law and of momentum balance, respectively.

Equations (10.129) and (10.132) are generated by applying the operator $e_y \cdot \nabla \times$ to the poloidal components of the momentum equation and Ohm's law to eliminate the gradients of the pressure and of the electric potential.

It is an interesting property of (10.128)–(10.132) that the equations (10.128)–(10.130) do not involve v_y and B_y, so that, in a first step, they can be solved for ρ, A, D without considering the others. When a solution is found, v_y and B_y can then be obtained from (10.130) and (10.131) in a second step. Here we will deal mainly with the first step, which already gives us the stability properties.

Here we consider the limit of small resistivity η. To lowest order in η the static version of that system becomes

$$[A_0, \Delta A_0] = 0 \tag{10.134}$$

$$[A_0, B_{y0}] = 0. \tag{10.135}$$

These equations represent the Grad–Shafranov equation (5.78) in a differentiated form.

Let us choose a one-dimensional static state $A_0(z), B_{y0}(z)$ satisfying (10.134) and (10.135) subject to small two-dimensional perturbations depending on x, z, t.

For the static state a Harris sheet is chosen (Section 5.3.2) with B_{y0} superimposed. Accordingly, we set

$$A_0 = -\ln \cosh(z) \tag{10.136}$$

such that

$$B_{x0}(z) = \tanh(z); \tag{10.137}$$

B_{y0} is arbitrary except that pressure balance must be satisfied. The density ρ is set to a constant ρ_0, so that in the following the continuity equation (10.128) can be ignored.

In (10.136) and (10.137) and in the following non-dimensional quantities are used with B normalized by the magnitude B_0 of the asymptotic component B_x outside the Harris sheet, x and z by the sheet width L, A by $B_0 L$, D by $v_A L$, and v_y by v_A, where $v_A = B_0/\sqrt{\mu_0 \rho_0}$ is the Alfvén velocity.

The perturbations imposed on the static state $A_0, B_{y0}, D_0 = 0, v_{y0} = 0$ are denoted by ψ, b_y, θ, v_y, respectively. Anticipating the symmetry of the tearing mode we choose ψ symmetric and θ antisymmetric with respect to z.

In the equations for the perturbations the resistivity has to be taken into account. This is because under the present conditions the unstable mode develops a thin boundary layer centred at $z = 0$, in which gradients of the perturbations become large so that the terms involving η cannot be neglected. The presence of the boundary layer calls for a singular perturbation treatment (e.g., Eckhaus, 1973), applying different scaling prescriptions to the region outside the layer (*external region*) and to the layer itself (*internal region*). Since there is no such boundary layer in the background state, the present asymptotics, in leading order, treats that state correctly as an equilibrium, ignoring slow resistive diffusion.

The next step is to linearize (10.129) and (10.130) for small perturbations, which gives

$$\Delta \frac{\partial \theta}{\partial t} + [\psi, \Delta A_0] + [A_0, \Delta \psi] = 0 \tag{10.138}$$

$$\frac{\partial \psi}{\partial t} + [A_0, \theta] - \frac{1}{S}\Delta \psi = 0 \tag{10.139}$$

with

$$S = \frac{\mu_0 L v_{\mathrm{A}}}{\eta} \tag{10.140}$$

being the Lundquist number. For simplicity, η is treated as a constant parameter. (Since resistivity is ignored in the external region the assumption of constant resistivity is actually used in the internal region only.) The present regime of small η is characterized by $S \gg 1$.

Choosing modes of the form

$$\psi(x, z, t) = \hat{\psi}(z)\,\mathrm{e}^{\mathrm{i}kx + \gamma t} \tag{10.141}$$

with a corresponding expression for θ, (10.138) and (10.139) give two ordinary differential equations for $\hat{\psi}$ and $\hat{\theta}$,

$$\gamma \hat{\theta}'' - k^2 \gamma \hat{\theta} + \mathrm{i}k B_{x0}'' \hat{\psi} + \mathrm{i}k^3 B_{x0} \hat{\psi} - \mathrm{i}k B_{x0} \hat{\psi}'' = 0 \tag{10.142}$$

$$\gamma \hat{\psi} - \mathrm{i}k B_{x0} \hat{\theta} + \frac{1}{S}k^2 \hat{\psi} - \frac{1}{S}\hat{\psi}'' = 0 \,. \tag{10.143}$$

To solve these equations one has to anticipate the ordering of the terms. The present regime is characterized by

$$\gamma^2 \ll k^2 \lesssim 1 \tag{10.144}$$

$$|\gamma| \gg \frac{1}{S} \,. \tag{10.145}$$

Here, the first inequality implies that γ is much smaller than 1, which means that the time variation is slow compared with the Alfvén time scale (unity in the present non-dimensional form). On the other hand, (10.145) says that the mode is fast compared to resistive diffusion of the current sheet, which would take place on the time scale S. Thus, we expect that the tearing time scale lies between the Alfvén scale and the diffusion scale.

The equations (10.142) and (10.143) are solved analytically by a singular perturbation method for external and internal regions separately. The dispersion relation $\gamma(k)$ is then obtained by a matching procedure. Here we follow the work of Janicke (1980) and Otto (1991). It turns out that it is sufficient to consider $k \geq 0$.

10.3.2 External solution

In the external region gradients are scaled as of order 1. In view of the symmetry of $\hat{\psi}$, it suffices to confine the discussion to the half plane $z \geq 0$.

From Ohm's law (10.143) one concludes that $\hat{\theta}$ is of the same order as $\gamma\hat{\psi}/(kB_{x0})$. Using that in (10.142) one finds to lowest order

$$\hat{\psi}'' + \left(\frac{2}{\cosh^2 z} - k^2\right)\hat{\psi} = 0 . \tag{10.146}$$

The general solution of this equation is

$$\hat{\psi} = K_1 e^{-kz}(\tanh z + k) + K_2 e^{kz}(\tanh z - k) \tag{10.147}$$

where K_1 and K_2 are arbitrary constants. Since we are looking for an instability, $\hat{\psi}$ must decay to zero for large values of z, implying $K_2 = 0$. Thus the external solution is

$$\hat{\psi}_e = K_1 e^{-kz}(\tanh z + k) . \tag{10.148}$$

For later reference we compute

$$\frac{\hat{\psi}_e'(0)}{\hat{\psi}_e(0)} = \frac{1 - k^2}{k} . \tag{10.149}$$

10.3.3 Internal solution

The internal region is a thin layer around $z = 0$ of width $\epsilon \ll 1$, where the value of ϵ is to be determined later; derivatives of the perturbations scale as $1/\epsilon$. To achieve that scaling formally, one introduces a new coordinate $\zeta = z/\epsilon$. In the layer ζ as well as derivatives of the perturbations with respect to ζ are of order unity. The equilibrium magnetic field, on the other hand, remains a smooth function of z. Inside the layer $B_{x0}(z) = B_{x0}(\epsilon\zeta)$ can be represented by the leading term of a Taylor expansion with respect to ϵ. In this way one finds the following leading contributions to (10.142) and (10.143)

$$\gamma\hat{\theta}'' - ikz\hat{\psi}'' = 0 \tag{10.150}$$

$$\gamma\hat{\psi} - ikz\hat{\theta} - \frac{1}{S}\hat{\psi}'' = 0 . \tag{10.151}$$

Since the only purpose of the new coordinate was to compare the scaling of the different terms, in (10.150) and (10.151) we have returned to expressing the variables in terms of the coordinate z. The final formulation is obtained by integrating (10.150) once and eliminating θ by (10.151), which gives

$$z\hat{\psi}''' - \hat{\psi}'' - (\chi^2 z^3 + \lambda\chi z)\hat{\psi}' + (\lambda\chi + \chi^2 z^2)\hat{\psi} = \chi^2 z^2 \tilde{c} , \tag{10.152}$$

where

$$\chi = \frac{kS^{1/2}}{\gamma^{1/2}} \tag{10.153}$$

$$\lambda = \frac{\gamma^{3/2}S^{1/2}}{k} \tag{10.154}$$

and $\tilde{c}$ being an integration constant. The general solution of (10.152) is available in the following form

$$\hat{\psi}_i(z) = \tilde{c} + c_0 z + c_1 z \int_0^{\chi z^2} m_1(w)dw - c_2 z \int_{\chi z^2}^{\infty} m_2(w)dw$$

$$- c_p z \int_{\chi z^2}^{\infty} m_p(w)dw , \tag{10.155}$$

where

$$m_1(w) = e^{-w/2}M\left(\frac{\lambda+5}{4}, \frac{5}{2}, w\right) \tag{10.156}$$

$$m_2(w) = e^{-w/2}U\left(\frac{\lambda+5}{4}, \frac{5}{2}, w\right) \tag{10.157}$$

$$m_p(w) = m_2(w)\int_0^w m_1(w')dw' + m_1(w)\int_w^{\infty} m_2(w')dw' \tag{10.158}$$

$$c_p = \frac{\lambda\sqrt{\chi}\,\Gamma(\frac{\lambda+5}{4})}{8\Gamma(\frac{5}{2})}\tilde{c} \tag{10.159}$$

with M and U denoting the confluent hypergeometric functions, also called *Kummer* functions (Abramowitz and Stegun, 1965). In (10.155) the four integration constants c_0, c_1, c_2 and $\tilde{c}$ are subject to three boundary conditions. The solution should remain bounded for $z \to \infty$, which requires $c_1 = 0$.

The conditions that $\hat{\psi}_i$ be an even function of z yields $\hat{\psi}_i'(0) = 0$ and $\hat{\psi}_i'''(0) = 0$. The latter boundary conditions allow us to express c_0 and c_2 by $\tilde{c}$, which remains unspecified, because we are dealing with a linear problem. (The same applies to K_1 in (10.148).) For later reference we compute

$$\lim_{z \to \infty} \frac{\hat{\psi}_i'(z)}{\hat{\psi}_i(z)} = \frac{\pi \lambda \sqrt{\chi}\, \Gamma\left(\frac{\lambda+3}{4}\right)}{(1-\lambda^2)\Gamma\left(\frac{\lambda+1}{4}\right)} . \tag{10.160}$$

The explicit handling of the boundary conditions and the derivation of (10.160) is rather involved and must be omitted here. The details can be found in the literature (Janicke, 1980; Otto, 1987, 1991). (A simpler approximate procedure is indicated below.)

10.3.4 The dispersion relation

In the transition region between the external and the internal regions solutions (10.148) and (10.155) do not smoothly match for arbitrary values of k and γ. For the present case, singular perturbation theory provides the following condition for a smooth transition from one solution to the other (Eckhaus, 1973)

$$\frac{\hat{\psi}_e'(0)}{\hat{\psi}_e(0)} = \lim_{z \to \infty} \frac{\hat{\psi}_i'(z)}{\hat{\psi}_i(z)} . \tag{10.161}$$

Strictly speaking, this is a necessary condition. For sufficiency one needs the existence of an intermediate scaling for which both expansions coincide. This is satisfied in the present case (Janicke, 1980).

With (10.149) and (10.160) the matching condition (10.161) assumes the form

$$\frac{\pi \lambda \sqrt{\chi}\, \Gamma\left(\frac{\lambda+3}{4}\right)}{(1-\lambda^2)\Gamma\left(\frac{\lambda+1}{4}\right)} = \frac{1-k^2}{k} . \tag{10.162}$$

This equation provides the desired relationship between wave number k and growth rate γ, in other words, (10.162) is the dispersion relation of the resistive tearing mode.

The thickness δz of the resistive layer is obtained from the internal solution (10.155), the z-scale of which is determined by its depending on $z^2\chi$. Writing that expression as $(z/\delta z)^2$ one finds for the characteristic length δz (which also may be identified with ϵ)

$$\delta z = 1/\sqrt{\chi}$$
$$= \frac{\gamma^{1/4}}{k^{1/2}S^{1/4}} . \tag{10.163}$$

Analysing the dispersion relation numerically for a given value of S one finds that within the present regime there is a range of k-values with $0 \leq k \leq k_m$ where for each k there is a unique positive solution for the growth rate γ. For a more detailed discussion it is useful to introduce the growth rate parameter

$$Q = \gamma \sqrt{S} \tag{10.164}$$

such that $\lambda = Q^{3/2}/(kS^{1/4})$. Then the dispersion relation (10.162) becomes

$$\frac{\pi Q^2 \Gamma \left(\frac{\lambda+3}{4}\right)}{\sqrt{\lambda}(1-\lambda^2)\Gamma \left(\frac{\lambda+1}{4}\right)} = 1 - k^2 . \tag{10.165}$$

This equation is particularly useful for $k \ll 1$, where, approximately, Q becomes a function of λ only. This function assumes its maximum Q_0 at $\lambda = \lambda_0$, with

$$\lambda_0 = 0.362 \tag{10.166}$$
$$Q_0 = 0.623 . \tag{10.167}$$

Using the definitions (10.154) and (10.164) one finds the associated values of k and γ

$$k_0 = \frac{1.358}{S^{1/4}} \tag{10.168}$$

$$\gamma_0 = \frac{0.623}{S^{1/2}} , \tag{10.169}$$

valid asymptotically for $S \gg 1$. The associated thickness of the resistive layer is obtained from (10.163) as

$$\delta z_0 = \frac{0.762}{S^{1/4}} . \tag{10.170}$$

Fig. 10.8 shows the dispersion relation in the form $Q(k)$ for several values of S and Fig. 10.9 depicts the field lines of a tearing-perturbed state with its characteristic chain of magnetic islands. The islands form because the attraction of parallel currents dominates the counteracting field deformation. The non-vanishing resistivity breaks the frozen-in constraint thereby allowing the change of magnetic topology.

For the range $1/S^{1/4} \ll k \leq 1$ (to the right of the maximum of the corresponding curve in Fig. 10.8) λ becomes small compared to 1. Using this in (10.165) one finds

$$\gamma = 0.953(1 - k^2)^{4/5} \frac{1}{S^{3/5}k^{2/5}} . \tag{10.171}$$

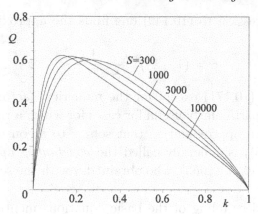

Fig. 10.8 The dispersion relation $Q(k)$ of the resistive tearing mode for four values of the Lundquist number S.

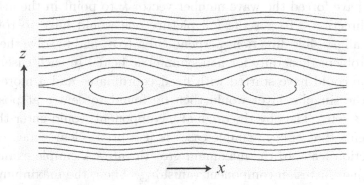

Fig. 10.9 The magnetic field lines of a Harris sheet subject to a tearing perturbation. The Lundquist number S is chosen large such that the internal region is not resolved. The amplitude of the perturbation is 0.25 and the wave number $k = 0.5$.

Disregarding the precise value of the numerical factor, this approximation has a simple interpretation. Approximating the first order current density in the internal region by a constant, one can set $\psi''(\delta z) \approx \psi''(0)$. This implies $\psi'(\delta z) \approx \psi''(0)\delta z$ and

$$\psi(\delta z) \approx \psi(0)\left(1 + \frac{\psi''(0)}{\psi(0)}\frac{\delta z^2}{2}\right) \tag{10.172}$$

$$\approx \psi(0)\left(1 + \frac{\lambda}{2}\right) \approx \psi(0)\,, \tag{10.173}$$

where (10.151) at $z = 0$ and (10.163) were used. Thus, one obtains

$$\frac{\psi'(\delta z)}{\psi(\delta z)} = \frac{\gamma^{5/4}S^{3/4}}{k^{1/2}}\,. \tag{10.174}$$

Equating this expression to (10.149) one finds the approximate dispersion relation

$$\gamma = (1 - k^2)^{4/5} \frac{1}{S^{3/5} k^{2/5}} , \tag{10.175}$$

which agrees with (10.171) except for the numerical factor. This approximate procedure is particularly useful for cases for which a rigorous matching is not available. An approximation that sets ψ to a constant in the inner region (see (10.173)) is generally called the *constant ψ approximation*, introduced by Furth *et al.* (1963), who obtained γ with the scaling of (10.171) in this way.

Because of the decoupling of the basic equations mentioned above, the growth rate given by (10.165) does not depend on B_{y0}. However, this does not mean that B_y completely drops out of the stability problem. The reason is that we have forced the wave number vector $\boldsymbol{k}$ to point in the arbitrarily chosen x-direction. To get the full picture one would have to rotate the x-axis into all possible directions perpendicular to $\boldsymbol{e}_z$ and to solve the present stability problem (generally for non-symmetric fields) in each case. (Alternatively, we could have started with fixed coordinates and a more general $\boldsymbol{k}$-vector, the singular layer would then be found to centre at positions z where $\boldsymbol{k} \cdot \boldsymbol{B} = 0$.) Thereby the original y-component would enter the problem. It seems that only in simple cases is it a priori clear which choice of the x-direction leads to the maximum growth rate. A simple example is a field with one Cartesian component vanishing. Then, the maximum growth rate corresponds to placing $\boldsymbol{e}_x$ perpendicular to that direction. This means that for the Harris sheet with no component along the invariant direction the maximum growth rate is available from the dispersion relation (10.162), which was derived with the appropriate choice of $\boldsymbol{e}_x$.

10.3.5 *Effect of an embedded thin current sheet*

Here we consider the case where the equilibrium structure contains a thin current sheet. Since the analysis closely follows the procedure described in the previous section, we give only a brief outline. (For details see Schindler and Birn (1999).)

We choose a magnetic field of the form

$$\boldsymbol{B}_{x0} = -\frac{\tanh(z) + \kappa \tanh(\frac{z}{\nu})}{1 + \kappa} \boldsymbol{e}_x . \tag{10.176}$$

Here, $\boldsymbol{B}_{x0}$ again is normalized by the asymptotic field magnitude B_0, and z by the width L of the broader sheet. The superimposed thin sheet, in non-dimensional form, has the width $\nu < 1$. The parameter κ controls the strength of the thin current sheet; the Harris sheet is recovered for $\kappa = 0$. The maximum magnitude of the current density j_y, normalized by $B_0/(\mu_0 L)$, is given by

$$j_{\mathrm{m}} = \frac{1 + \kappa/\nu}{1 + \kappa} . \tag{10.177}$$

Figure 10.10 shows the current density profiles $j_y(z)$ for several choices of κ and ν.

The plasma pressure p is given by pressure balance $p + B^2/2 = 1/2$. Following the procedure described in the previous section one finds the dispersion relation in the form (Schindler and Birn, 1999)

$$\frac{\pi \tilde{Q}^2}{k\sqrt{\tilde{\lambda}}(1 - \tilde{\lambda}^2)} \frac{\Gamma\left(\frac{\tilde{\lambda}+3}{4}\right)}{\Gamma\left(\frac{\tilde{\lambda}+1}{4}\right)} = g(k) , \tag{10.178}$$

where $\tilde{Q} = \gamma\sqrt{\tilde{S}}/j_{\mathrm{m}}$ with $\tilde{S} = v_A \mu_0 L^*/\eta$, $L^* = L/j_{\mathrm{m}}$ and $\tilde{\lambda} = \tilde{Q}^{3/2}/(k\tilde{S}^{1/4})$, and $g(k)$ denotes $\hat{\psi}'/\hat{\psi}$ computed from the external solution for $z \to 0$. In the external region ψ satisfies

$$\hat{\psi}'' - \left(\frac{B_{x0}''(z)}{B_{x0}(z)} - k^2\right)\hat{\psi} = 0 , \tag{10.179}$$

which generalizes (10.146). By obtaining $g(k)$ from (10.179) and solving (10.178) numerically one finds the dispersion relation explicitly in the

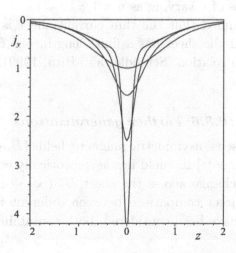

Fig. 10.10 Current density $j_y(z)$ of the magnetic field model (10.176) for the Harris sheet ($\kappa = 0$) and for $\kappa = 0.6$ with ν chosen as 0.4, 0.2, and 0.1, corresponding to increasing j_{m} (from Schindler and Birn (1999) by permission of the American Geophysical Union).

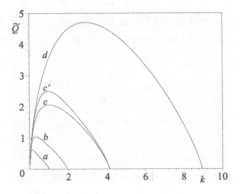

Fig. 10.11 Dispersion relation of the resistive tearing instability for the field model (10.176). The parameters are chosen as $\kappa = 0$ (Harris sheet, curve a), as $\kappa = 0.6$ and $\nu = 0.4$ (curve b), as $\kappa = 0.6$ and $\nu = 0.2$ (curve c), and as $\kappa = 0.6$ and $\nu = 0.1$ (curve d). The curve c' corresponds to $\kappa \to \infty$, $\nu = 0.243$, representing a Harris sheet of width 0.243, chosen such that the points of marginal stability are the same as in case c. (Reproduced from Schindler and Birn (1999) by permission of the American Geophysical Union.)

form $\tilde{Q}(k)$. Several examples are shown in Fig. 10.11. It is evident that the presence of the thin current sheet significantly enlarges the unstable wave number range and increases the maximum growth rate. Qualitatively, the shape of the curves resembles the dispersion relation of a Harris sheet with an appropriately chosen sheet width, as illustrated by a comparison between curves c and c'. In view of the definition of $\tilde{S}$, the same value of $\tilde{S}$ corresponds to values of η varying as $\eta \propto 1/j_{\mathrm{m}}$.

Here we have assumed that the thin current sheet is of infinite extent in the x-direction. If the sheet has a finite length, one can expect a more structured dispersion relation (Schindler and Birn, 1999).

10.3.6 Further generalizations

First let us briefly discuss asymmetric magnetic fields $(B_{x0}(-z) \neq -B_{x0}(z))$, assuming that for large $|z|$ the field has asymptotically constant values and that the total field change across the sheet $B_{x0}(\infty) - B_{x0}(-\infty)$ is set to 2, which allows a direct comparison between different field models. The following two cases have been considered, both containing the Harris sheet as a special case.

A constant field in the x-direction is superimposed on the Harris sheet field (Biskamp, 1982). In that case the asymmetry has a stabilizing effect, the growth rate decreases. This can be attributed to the fact that the singular layer moves to positions of smaller current density compared with the Harris sheet.

An alternative model uses different Harris solutions for positive and negative values of z (Birk and Otto, 1991). Here the asymmetry results from different asymptotic fields for $z = \pm\infty$ and different sheet widths necessary for achieving smooth matching at $z = 0$. In this case the asymmetry increases the growth rate strongly. In contrast to the previous model, here the singular layer remains at the position of the current density maximum, which even increases.

A generalization that is particularly interesting for magnetospheric tail applications adds the effect of a small magnetic field component normal to the sheet, as it arises from a weak two-dimensionality of the equilibrium (Section 5.4.1) with aspect ratio ϵ. For constant resistivity with $S \gg 1$ one finds that the dispersion relation (10.178) is valid as long as $\epsilon \ll 1/\sqrt{S}$, although in that regime the normal component can already have a significant effect on the mode (Janicke, 1980; Otto, 1991). Larger values of the normal component have a significant stabilizing influence (Paris, 1987).

Cases with general resistivity models were studied by Otto (1991). The resistivity was treated as an arbitrary function of the current density, mass density and temperature. Losses through thermal radiation were also taken into account. It was shown that a resistivity depending on current density can lead to a considerable enhancement of the growth rate compared to constant resistivity.

10.3.7 Nonlinear tearing

The nonlinear development of the tearing instability has been studied by both theoretical and numerical simulation techniques (e.g., White *et al.*, 1977; Biskamp and Welter, 1977). An important result is that the growth of the tearing islands undergoes a strong saturation process. Tearing saturation has been estimated by taking (weak) nonlinearities into account. However, if the domain considered is large enough, a pair of islands will coalesce to form one larger island, such that growth may continue in that fashion. For details see Biskamp (1993).

Certain two-dimensional configurations are not subject to severe saturation. This applies to two- and three-dimensional models that were developed for solar flares and magnetospheric substorms (e.g., Birn, 1980; Otto, 1991; Forbes and Priest, 1983; Mikić *et al.*, 1988), based on current sheet configurations with a non-vanishing normal component B_n. It should be noted that in this case linear theory breaks down before the point where a neutral

line forms and magnetic topology is changed. It takes a nonlinear perturbation to achieve a local compensation of the equilibrium component B_n.
Fig. 10.12 illustrates the magnetic field structure as it develops with time
in a stretched two-dimensional equilibrium discussed in Section 5.4.1. Here
a linear tearing mode is extrapolated to a finite, however still small, amplitude. The main feature is the characteristic island structure bounded
by a field line that intersects itself in an x-type neutral point. This type
of island is usually referred to as a *plasmoid*. Fig. 11.12 shows a corresponding figure from a simulation study by Otto (1987), which confirms
the formation of a plasmoid under nonlinear resistive dynamics. Depending on the details of the configuration, more than one plasmoid may form.

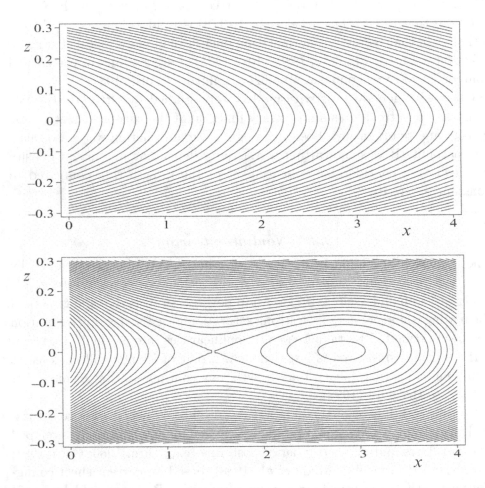

Fig. 10.12 Expected structure of a generalized tearing mode of a two-dimensional
equilibrium, obtained by extrapolation of a linear mode.

We will return to plasmoid dynamics in the context of magnetic reconnection (Chapter 11).

10.4 Collisionless tearing

Instabilities with signatures typical for tearing modes can also exist in collisionless current sheets. Two limiting cases are of particular interest.

In the first case, a sufficiently strong level of microturbulence, such as considered in Section 9.3, leads to a tearing mode that can be described by the resistive fluid picture outlined in the previous section. An expression for the resistivity (such as (9.56)) would be obtained from an appropriate turbulence theory. For resistive tearing with examples of collective resistivity see Birk and Otto (1991).

The other limiting possibility is a non-turbulent version of tearing based on Landau resonance (Section 9.3.1). This process, which is usually called *collisionless tearing*, is addressed in the remaining part of this section.

A useful benchmark problem of collisionless tearing consists in choosing singly charged ions and setting the mass ratio m_i/m_e to 1. In that case the electric coupling, which in realistic cases arises from the differences in gyro-scales, vanishes. Although this approach does not directly cover any real space applications, it has proven quite useful. By comparing results of more realistic treatments with equal-mass results, one gains insight into the role of the electric coupling. As we will see, in some cases electric coupling can have dramatic effects.

10.4.1 Harris sheet with equal masses

Here we consider the Harris sheet (see Sections 5.3.2, 6.2.2) and set $m_i = m_e = m$ and $q_i = -q_e = e$. The tearing mode is selected by choosing perturbations independent of y and A_1 symmetric with respect to z.

For the Harris sheet the equilibrium distribution function for particle species s has the form (Section 6.2.2)

$$f_{s0} = F_s(H_{s0}, P) = C_s \, e^{-\alpha_s P - \beta_s H_{s0}} \qquad (10.180)$$

where C_s, if expressed by the maximum density n_0, is

$$C_s = n_0 \left(\frac{m\beta_s}{2\pi} \right)^{3/2} e^{-m\alpha_s^2/(2\beta_s)} \qquad (10.181)$$

and H_{s0} is the equilibrium Hamiltonian and P the y-component of the canonical momentum. (P does not carry labels, because P is used as an independent variable.) The drift velocity v_{ys} is given by $-\alpha_s/\beta_s$ and the temperature T_s by $1/(k_B\beta_s)$. Charge neutrality requires $q_i\alpha_i = q_e\alpha_e$ (see (6.38)) such that we can set $\alpha_i = -\alpha_e = \alpha$.

Particle and current densities have the form

$$n = \frac{n_0}{\cosh^2(z/L)} \tag{10.182}$$

$$j_{y0} = -\sum_s \frac{e\alpha n_0}{\beta_s} e^{-e\alpha A_0} = -\frac{e\alpha}{\beta} \frac{n_0}{\cosh^2(z/L)} \tag{10.183}$$

with $L = (2\beta/(\mu_0\alpha^2 e^2 n_0))^{1/2}$ and $1/\beta = 1/\beta_e + 1/\beta_i$. There is no equilibrium B_y-component and the magnetic flux function A_0 is given by

$$A_0 = \frac{2}{e\alpha} \ln\cosh\left(\frac{z}{L}\right). \tag{10.184}$$

The growth rate γ is computed from the linearized Vlasov equation. A crucial parameter is d/L, where $d = \sqrt{2Lr_g}$ is the width of the region of exotic orbits (Section 6.4), r_g being the gyro-radius in the asymptotic magnetic field B_0, and L is the width of the current sheet. Here we consider the cases $d/L \ll 1$ and $d/L \approx 1$, where approximate expressions for γ are available.

As $\partial/\partial y = 0$ holds for both equilibrium and perturbations, the canonical momentum component P is a constant of motion. Therefore, the particle distribution functions, if written as $f(x, z, w_x, w_z, P, t)$, satisfy a Vlasov equation in the 4-dimensional phase space (x, z, w_x, w_z), i.e., an equation of the form (see (3.17))

$$\frac{df}{dt} = \frac{\partial f}{\partial t} + \frac{\partial f}{\partial x}\frac{dx}{dt} + \frac{\partial f}{\partial z}\frac{dz}{dt} + \frac{\partial f}{\partial w_x}\frac{dw_x}{dt} + \frac{\partial f}{\partial w_z}\frac{dw_z}{dt} = 0. \tag{10.185}$$

Such an equation holds for each of the two species. Using $E_x = E_z = 0$, which is consistent with the assumed exact neutrality, and inserting the equations of single particle dynamics into (10.185) one finds for each particle species s

$$\frac{df_s}{dt} = \frac{\partial f_s}{\partial t} + w_x\frac{\partial f_s}{\partial x} + w_z\frac{\partial f_s}{\partial z} + \frac{q_s}{m}\left(\frac{P - q_s A}{m}B_z - w_z B_y\right)\frac{\partial f_s}{\partial w_x}$$

$$+ \frac{q_s}{m}\left(w_x B_y - \frac{P - q_s A}{m}B_x\right)\frac{\partial f_s}{\partial w_z} = 0 \tag{10.186}$$

where w_y has been replaced by $(P - q_s A)/m$. By linearizing (10.186) and using (10.180), we find

$$
\frac{\mathrm{d}f_{s1}}{\mathrm{d}t} = -q_s \frac{\partial F_s}{\partial H_{s0}} \frac{P - q_s A_0}{m} (w_x B_{z1} - w_z B_{x1}) - \frac{q_s^2}{m} \frac{\partial F_s}{\partial H_{s0}} B_{x0} \, w_z A_1
$$

$$
= -\frac{\mathrm{d}}{\mathrm{d}t} \left(\frac{q_s}{m} \frac{\partial F_s}{\partial H_{s0}} (P - q_s A_0) A_1 \right) + \frac{q_s}{m} \frac{\partial F_s}{\partial H_{s0}} (P - q_s A_0) \frac{\partial A_1}{\partial t} .
$$

$$(10.187)$$

Here $\mathrm{d}/\mathrm{d}t$ is taken along unperturbed orbits. Integrating (10.187) gives

$$
f_{s1} = -\frac{q_s}{m} \frac{\partial F_s}{\partial H_{s0}} (P - q_s A_0) A_1 + \frac{q_s}{m} \frac{\partial F_s}{\partial H_{s0}} \int_{-\infty}^{t} (P - q_s A_0)' \left(\frac{\partial A_1}{\partial t} \right)' \mathrm{d}t',
$$

$$(10.188)$$

where the prime symbol refers to the unperturbed orbit. (Setting the lower limit of the integral to $-\infty$ is explained in Section 9.3.1.)

Let us now specialize for $d \ll L$. In the region of gyrotropic orbits $(z > d)$ the time-average of $w_y = (P - q_s A_0)/m$ is the gradient-B drift velocity (Section 3.1), which is of the order of $v_t r_g/L$, where v_t is the thermal velocity. Therefore, the integral term in (10.188) can be neglected. Then f_{s1} becomes the quasi-static perturbation, in direct analogy to the resistive case discussed in the previous section.

In the internal region particles move rather freely along the y-direction such that the average value of w_y is of order v_t and the integral term has to be kept. A useful approximation treats w_y as a constant by taking $(P - q_s A_0)/m$ in front of the integral.

We consider modes of the form $A_1 = \hat{A}_1(z) \exp(\gamma t + ikx)$. Noting that $x' = x + w_x(t' - t)$ with w_x constant, and anticipating that $\hat{A}_1(z)$ has negligible variation inside the internal region (constant ψ-approximation of the previous section) we obtain for the internal distribution function

$$
f_{s1} = -\frac{q_s}{m} \frac{\partial F_s}{\partial H_{s0}} (P - q_s A_0) A_1 + \gamma \frac{q_s}{m} \frac{\partial F_s}{\partial H_{s0}} \frac{P - q_s A_0}{\gamma + ikw_x} A_1. \qquad (10.189)
$$

With these results the perturbed current density

$$
j_{y1} = \sum_s q_s \int \frac{P - q_s A_0}{m} f_{s1} \, \mathrm{d}\tau - \sum_s \frac{q_s^2}{m} A_1 \int F_s \, \mathrm{d}\tau, \qquad (10.190)
$$

where $d\tau$ stands for $dw_x dw_z dP/m$, becomes (e.g., Galeev, 1984)

$$
\dot{j}_{y1} = \begin{cases} \frac{\partial j_{y0}}{\partial A_0} A_1 + \gamma A_1 \sum_s \frac{q_s^2}{m^2} \int \frac{(P - q_s A_0)^2}{\gamma + ikw_x} \frac{\partial F_s}{\partial H_{s0}} d\tau & z < d \\ \frac{\partial j_{y0}}{\partial A_0} A_1 & z > d. \end{cases} \tag{10.191}
$$

The external solution of $\nabla \times \boldsymbol{B}_1 = \mu_0 j_{y1}$ is the same as in the resistive case, such that, in (dimensional) analogy to (10.149), we find

$$
\left[\frac{1}{\hat{A}_1} \frac{d\hat{A}_1}{dz} \right]_{z=0} = \frac{1 - k^2 L^2}{kL^2}. \tag{10.192}
$$

Evaluating the integral in (10.191) gives with (10.180)

$$
\int \frac{(P - q_s A_0)^2}{\gamma + ikw_x} \frac{\partial F_s}{\partial H_{s0}} d\tau = -\sqrt{\frac{\pi \beta_s}{2}} \frac{m^{3/2} n_0}{k \cosh^2(z/L)} \left(1 + 2\frac{r_g^2}{L^2} \right). \tag{10.193}
$$

Here, the plasma dispersion function, which arises from the w_x-integration, was taken in the limit of small γ. For the internal region this allows us to write

$$
j_{y1} = \frac{2A_1}{\mu_0 L^2 \cosh^2(z/L)} \left(1 - \gamma M \left(1 - 2\frac{r_g^2}{L^2} \right) \right), \quad M = \frac{\sqrt{\pi} L^2}{2 r_g^2 k v_t}. \tag{10.194}
$$

The amplitude $\hat{A}_1$ is obtained by solving the equation (y-component of $\nabla \times \boldsymbol{B}_1 = \mu_0 \boldsymbol{j}_1$)

$$
\frac{d^2 \hat{A}_1}{dz^2} - \left(k^2 - \frac{2 - 2\gamma M(1 + 2r_g^2/L^2)}{L^2 \cosh^2(L/z)} \right) \hat{A}_1 = 0. \tag{10.195}
$$

The dispersion relation is obtained by applying the simplified procedure that proved useful in the resistive case. Anticipating that γM is of order $\sqrt{L/d}$, a boundary layer expansion with respect to $\sqrt{d/L}$, which uses the variable $\theta = z/\sqrt{d/L}$, shows that to lowest order in $\sqrt{d/L}$ and for $k^2 L^2 < 1$ (10.195) becomes

$$
\frac{d^2 \hat{A}_1}{dz^2} = \frac{2\gamma M}{L^2} \hat{A}_1(0). \tag{10.196}
$$

Solving that equation with the symmetry condition $d\hat{A}_1/dz = 0$ at $z = 0$, and equating $(d\hat{A}_1/dz)/\hat{A}_1$ of the external and internal solutions at $z = d$, evaluated to lowest order, one finds the growth rate

$$
\gamma = c_0 \Omega_0 \left(\frac{r_g}{L} \right)^{5/2} (1 - k^2 L^2) \tag{10.197}
$$

where c_0 is a numerical factor of order 1. The expansion that led to (10.197) formally breaks down for kL near 0 and 1. Nevertheless, $kL = 1$ gives the correct value ($\gamma = 0$).

Several authors (Laval *et al.*, 1966; Schindler, 1974) dealt with the growth rate γ, or its maximum value, using somewhat different approaches. The results are consistent with (10.197), see also Galeev (1984).

Next, we study the case of a thin current sheet, where $d \approx L$, following Pritchett *et al.* (1991). In that case the entire sheet can be regarded as filled with particles performing exotic orbits. Therefore, it seems reasonable to generate a model equation by extending the validity of (10.195) over the entire sheet. This equation must be solved without the help of a boundary layer expansion.

Fortunately, (10.195) has a relevant analytical solution. Setting $A_1 = \cosh^{-r}(z/L)$ one finds a solution satisfying the boundary conditions for $r = kL$ and $k^2L^2 + kL - 2 - 2\gamma M(1 + 2r_g^2/L^2) = 0$. The latter condition gives the growth rate as (see Pritchett *et al.*, 1991)

$$\gamma = \frac{\Omega_0}{\sqrt{\pi}}\left(\frac{r_g}{L}\right)^3 \frac{kL(2 + kL)(1 - kL)}{1 + 2r_g^2/L^2}. \tag{10.198}$$

As expected, the growth rate of the thin sheet (equation (10.198) with $L \approx r_g$) is considerably larger than that of the wide sheet (equation (10.197) with $L \gg r_g$).

An improvement of (10.198) for arbitrary values of r_g/L, however available only in an implicit form, was obtained by Brittnacher *et al.* (1995).

10.4.2 Realistic particle masses

For $m_i \neq m_e$ the perturbation can no longer be described by a single function. The different particle masses lead to charge separation effects and therefore to electric fields in the x, z-plane. Also, B_y no longer vanishes. Numerical solution of the linearized Vlasov equations for realistic mass ratios were carried out by Hoshino (1987) and Daughton (1999). Fig. 10.13 shows the growth rate as a function of r_{gi}/L for a fixed value of kL. The growth rate for $m_i/m_e = 1836$ is reduced by a factor of 1.5–2 in comparison with the $m_i/m_e = 1$ result. This seems to reflect a general trend, which is also confirmed by particle simulations (Pritchett, 1994).

For further discussions of the properties of the collisionless tearing mode we now turn to a variation method, described in the following section.

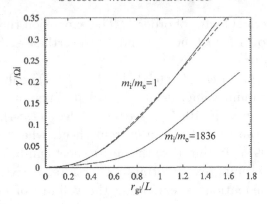

Fig. 10.13 Growth rate of the collisionless tearing mode as a function of r_{gi}/L for $T_i = T_e$ and $kL = 0.5$ after Daughton (1999). The solid lines correspond to numerical results, the dashed line to the analytical expression of Brittnacher *et al.* (1995), consistent with (10.198) for r_{gi}/L near 1. (With permission from William Daughton. Copyright 1999, AIP.)

10.5 Variational principle for 2D collisionless plasmas

For two-dimensional Vlasov plasmas stability properties can be expressed in terms of a nonlinear variational principle (Schindler and Goldstein, 1983), which uses the energy approach that was described in Section 10.2.1. As in the MHD case, for the present purpose it is sufficient to work with the version valid for linear perturbations. That version may be obtained in the small amplitude limit of the exact variational principle (see Schindler and Goldstein, 1983), however, it is more readily obtained by a more direct method.

The method was originally developed for one-dimensional equilibria with two-dimensional perturbations (Schindler, 1966). An equivalent approach is due to Laval *et al.* (1966). Later, the formalism was generalized for applications to two-dimensional equilibria (Schindler *et al.*, 1973; Lembège and Pellat, 1982). Since collisionless tearing of two-dimensional configurations has been playing an important role in the discussion of the onset of magnetospheric activity, we will give a rather detailed account of the associated variational principle.

10.5.1 Formulation

The equilibrium is chosen to be two-dimensional as in Section 6.2; however, in view of magnetospheric application the invariant direction is changed to e_y. As in Section 10.4.1, the equilibrium distribution function of particle species s is of the form

$$f_{s0} = F_s(H_{s0}, P). \tag{10.199}$$

(Again, P is treated as an independent variable and therefore has no label.) Distribution functions f_s are normalized such that species s has number density $n_s = \int f_s d\tau_s$ with $d\tau_s = dw_x dw_z d(P/m_s)$. We define $F'_s = \partial F_s/\partial H_{s0}$ and choose $F'_s < 0$. The consideration is restricted to quasi-neutral plasmas (Section 6.2.2).

Again, $A = A(x, z, t)$ is the magnetic flux function of the poloidal magnetic field (projection onto the x, z-plane). The equilibrium magnetic field $\boldsymbol{B}_0$ is assumed to be poloidal, i.e., $B_{y0} = 0$. However, the perturbations generally involve a non-vanishing B_y-component. The equilibrium flux function A_0 satisfies a Grad–Shafranov equation of the form (6.24).

The boundary conditions for $t > 0$ correspond to energy conservation, i.e., there is no energy flux across the boundary. Thus, as in the MHD case, we can use energy conservation as the starting point. Details are given in Appendix 2; here we summarize and discuss the results. The perturbations are treated as real quantities, but a brief prescription of how to generate the corresponding results for complex perturbations is added (Section 10.5.4).

The energy conservation law is obtained from (3.40) by integration over the spatial domain,

$$W = \sum_s \int \left(\frac{m_s}{2}(w_x^2 + w_z^2) + \frac{1}{2m_s}(P - q_s A)^2 \right) f_s \, d\Omega_s$$

$$+ \frac{1}{2\mu_0} \int B^2 \, dx \, dz . \tag{10.200}$$

Here W is a constant measuring the energy of the system after an external perturbation has been supplied (at $t \leq 0$) and $d\Omega_s$ stands for $dx \, dz \, d\tau_s$. In the expression (10.200) electric energy density is ignored, which is consistent with quasi-neutrality. Gravity is ignored for simplicity.

There are two ways of taking into account the dynamical constraints (Schindler et al., 1973). One can either use the explicit solution expressed by a time-integral such as (10.188), or incorporate a set of corresponding constraints. Here we follow the latter path.

As shown in Appendix 2, incorporation of the conservation of P and of Liouville's theorem allows us to show that the first-order contribution to W

vanishes and that the second-order energy expression can be written in the form

$$
W_2 = -\frac{1}{2} \sum_s \int \frac{1}{F_s'} \left(f_{s1} + \frac{q_s}{m_s}(P - q_s A_0) A_1 F_s' \right)^2 d\Omega_s
$$

$$
+ \frac{1}{2} \int \left(\frac{(\nabla A_1)^2}{\mu_0} - \frac{\partial j_0}{\partial A_0} A_1^2 + \frac{B_{y1}^2}{\mu_0} \right) dx\, dz \;, \quad (10.201)
$$

where in view of (6.13) we have used

$$
\frac{\partial j_0}{\partial A_0} = -\sum_s \int \left(\frac{q_s^2}{m_s^2}(P - q_s A_0)^2 F_s' + \frac{q_s^2}{m_s} F_s \right) d\tau_s \;. \quad (10.202)
$$

Due to the energy closure W_2 is constant for $t > 0$. To apply the formalism of Section 10.2.1, the next step is to define the quantities T_2 and V_2. The latter is obtained by minimizing (10.201) with respect to (smooth) perturbations f_{s1} and B_{y1}. The minimum defines V_2 as a functional of A_1, where A_1 is a smooth test function which is arbitrary except for the boundary condition. Then $T_2 = W_2 - V_2$, which is non-negative by construction. With these definitions we can apply (10.6), which states stability if

$$
V_2(A_1) > 0 \quad \text{for all } A_1. \quad (10.203)
$$

The quality of the resulting variational principle (minimization of V with respect to A_1) depends on the constraints that are taken into account in the minimization of W_2 with respect to f_{s1} and B_{y1}. These constraints reflect the fact that in the actual dynamics the perturbations f_{s1}, B_{y1}, A_1 and ϕ_1 are coupled to each other. There is no general prescription for the choice of the constraints in a given case other than their consistency with dynamical properties. Since constraints can only raise the minimum, the above stability criterion is valid independent of the set of constraints that are incorporated. Yet, working with too few constraints or imposing no constraint at all causes the danger that for a system that is actually stable the potential $V_2(A_1)$ may assume negative values. Then, stability cannot be demonstrated and the procedure becomes useless. On the other hand, working with all relevant constraints might lead to rather complicated procedures. We discuss two examples.

No constraints

In the absence of any constraints on the choice of f_{s1} the minimization of W_2 simply consists of setting the terms involving f_{s1} and B_{y1} in (10.201) to zero. Note that both terms are non-negative, the first term because of $F'_s < 0$. Thus we find

$$V_2 = \frac{1}{2} \int \left(\frac{(\nabla A_1)^2}{\mu_0} - \frac{\partial j_0}{\partial A_0} A_1^2 \right) \mathrm{d}x\,\mathrm{d}z \,. \tag{10.204}$$

By construction, the property (10.203) would imply stability. In practice, however, this is of limited use, because in many cases of interest the minimum is negative, so that no stability result is available (see the discussion of one-dimensional equilibria below).

Phase space condition and quasi-neutrality

Here the constraints are Liouville's theorem and the quasi-neutrality condition. For simplification let us introduce the abbreviation

$$\psi_s = \frac{1}{2m_s}(P - q_s A)^2 + q_s \phi \,. \tag{10.205}$$

As shown in Appendix 2, the constraints can be expressed as

$$\langle f_{s1} \rangle_s = 0 \tag{10.206}$$

$$\sum_s \int q_s f_{s1} \mathrm{d}\tau_s = 0 \tag{10.207}$$

where

$$\langle \ldots \rangle_s = \frac{\int_{\psi_{s0} \leq H_{s0}} \ldots \mathrm{d}x\,\mathrm{d}z}{\int_{\psi_{s0} \leq H_{s0}} \mathrm{d}x\,\mathrm{d}z} \,. \tag{10.208}$$

Note that these constraints are exact for quasi-neutral two-dimensional Vlasov plasmas. So adiabatic particle motion is covered as well as non-adiabatic motion.

Minimizing (10.201) under the constraints (10.206) and (10.207) gives (see Appendix 2)

$$V_2 = \frac{1}{2} \int \left(\frac{(\nabla A_1)^2}{\mu_0} - \frac{\partial j_0}{\partial A_0} A_1^2 + 2\frac{\partial \sigma_0}{\partial A_0} A_1 \phi_1 + \frac{\partial \sigma_0}{\partial \phi_0} \phi_1^2 \right) \mathrm{d}x\,\mathrm{d}z$$

$$- \frac{1}{2} \sum_s \int F'_s \langle \psi_{s1} \rangle^2 \, \mathrm{d}\Omega_s \,, \tag{10.209}$$

where for each A_1 the potential ϕ_1 is determined by the quasi-neutrality condition (10.207), which takes the form

$$\sum_s \int q_s F'_s(\psi_{s1} - \langle \psi_{s1} \rangle_s)\mathrm{d}\tau_s = 0 \,. \tag{10.210}$$

Making use of a Neumann series, Kiessling and Krallmann (1998) have solved (10.210) for ϕ_1 explicitly in the form $\phi_1(A_1)$.

The derivatives of σ_0 and j_0 contained in (10.209) are obtained from (6.12) and (6.13), one finds

$$\frac{\partial \sigma_0}{\partial \phi_0} = \sum_s \int q_s^2 F_s' \, \mathrm{d}\tau \tag{10.211}$$

$$\frac{\partial \sigma_0}{\partial A_0} = -\sum_s \int \frac{q_s^2}{m_s}(P - q_s A_0) F_s' \, \mathrm{d}\tau \tag{10.212}$$

and $\partial j_0/\partial A_0$ is given by (10.202).

We have addressed the derivation of $\mathcal{V}_2$ as a minimization, anticipating that the variation procedure gives a minimum of W_2 rather than a maximum or a saddle point. It is easily confirmed that under the present conditions a minimum actually occurs (see Appendix 2).

The expression (10.209) has been derived also by different methods (Schindler *et al.*, 1973; Schindler and Goldstein, 1983).

It is possible to write $\mathcal{V}_2$ in a form in which the effects of A_1 and ϕ_1 separate (Goldstein and Schindler, 1982)

$$\mathcal{V}_2 = \frac{1}{2} \int \left(\frac{(\nabla A_1)^2}{\mu_0} - \frac{\partial j_0}{\partial A_0} A_1^2 \right) \mathrm{d}x \, \mathrm{d}z$$

$$- \frac{1}{2} \sum_s \int F_s' \frac{q_s^2}{m_s^2} \left\langle (P - q_s A_0) A_1 \right\rangle^2 \mathrm{d}\Omega_s$$

$$- \frac{1}{2} \sum_s \int F_s' q_s^2 \left(\phi_1 - \langle \phi_1 \rangle \right)^2 \mathrm{d}\Omega_s . \tag{10.213}$$

A comparison between (10.204) and (10.209) shows that the last two terms are due to the imposed constraints. It is obvious that (in view $F_s' < 0$) these terms are non-negative, i.e., stabilizing.

Another form of $\mathcal{V}_2$ allows a more direct comparison with the corresponding ideal MHD result,

$$\mathcal{V}_2 = \frac{1}{2} \int \left(\frac{(\nabla A_1)^2}{\mu_0} - \frac{\mathrm{d}j_0}{\mathrm{d}A_0} A_1^2 - \frac{\partial \sigma_0}{\partial \phi_0} \left[\!\left[\left(\langle \Psi_{s1} \rangle - [\![\langle \Psi_{s1} \rangle]\!] \right)^2 \right]\!\right] \right) \mathrm{d}x \, \mathrm{d}z.$$

$$\tag{10.214}$$

Here and later on, the following notation is used:

$$\llbracket \dots \rrbracket = \frac{\sum_s \int \dots q_s^2 F_s' \, d\tau_s}{\sum_s \int q_s^2 F_s' \, d\tau_s} , \qquad (10.215)$$

$$\llbracket \dots \rrbracket_s = \frac{\int \dots q_s^2 F_s' \, d\tau_s}{\int q_s^2 F_s' \, d\tau_s} \qquad (10.216)$$

and $\Psi_{s1} = \psi_{s1}/q_s$.

The quasi-neutrality constraint (10.210) assumes the form

$$\llbracket \langle \Psi_{s1} \rangle \rrbracket = \llbracket \Psi_{s1} \rrbracket . \qquad (10.217)$$

As in the MHD case of (10.57), in (10.214) the total derivative dj_0/dA_0 appears, which is related to the partial derivative appearing in (10.213) by the relationship

$$\frac{dj_0}{dA_0} = \frac{\partial j_0}{\partial A_0} + \frac{\partial j_0}{\partial \phi_0} \frac{d\phi_0}{dA_0}$$

$$= \frac{\partial j_0}{\partial A_0} + \left(\frac{\partial \sigma_0}{\partial A_0} \right)^2 / \frac{\partial \sigma_0}{\partial \phi_0} , \qquad (10.218)$$

where (6.14) was used and

$$\frac{d\phi_0}{dA_0} = -\frac{\partial \sigma_0}{\partial A_0} / \frac{\partial \sigma_0}{\partial \phi_0} \qquad (10.219)$$

resulting from quasi-neutrality of the equilibrium.

In comparing collisionless results with MHD it is essential to take into account that in MHD the additional idealness constraint (10.50) must be imposed. In the presence of neutral points or sheets that can make a significant difference (see the corresponding remark in Section 10.2.2).

Using their solution of (10.210) mentioned above, Kiessling and Krallmann (1998) eliminated ϕ_1 from $\mathcal{V}_2$ and obtained a functional of A_1 alone of the form (see Appendix 2)

$$\mathcal{V}_2 = \frac{1}{2} \int \left(\frac{(\nabla A_1)^2}{\mu_0} - \frac{\partial j_0}{\partial A_0} A_1^2 \right) dx \, dz$$

$$+ \frac{1}{2} \int \int A_1(\boldsymbol{r}) K(\boldsymbol{r}, \boldsymbol{r}') A_1(\boldsymbol{r}') d^2 r \, d^2 r' , \qquad (10.220)$$

where they expressed the kernel K explicitly in terms of the solution of (10.210).

One-dimensional equilibrium

Consider one-dimensional equilibria with $\boldsymbol{B} = B_x(z)\boldsymbol{e}_x$ and infinitely extended in the x-direction. Then the perturbations can be assumed to be of the form (10.64) (using complex perturbations, see Section 10.5.4). Then all averages $[\![\cdots]\!]$ vanish and one finds

$$\mathcal{V}_2 = \frac{1}{2} \int \left(\frac{(\nabla A_1)^2}{\mu_0} - \frac{\mathrm{d}j_0}{\mathrm{d}A_0} A_1^2 \right) \mathrm{d}x \, \mathrm{d}z \ , \tag{10.221}$$

which coincides with the corresponding MHD result (10.73), except for the absence of the MHD constraint (10.50). As discussed in Section 10.2.2 without that constraint the minimum of the functional (10.214) is negative. This is consistent with the instabilities found in Section 10.4.

10.5.2 The limit of small electron mass

Here we return to 2D equilibria. Because of the large value of the mass ratio m_i/m_e, which lies near 1836 for a proton/electron plasma, the dynamics of the electrons often differs appreciably from the ion dynamics. Here, we consider a case which is motivated by the regime of adiabatic electrons. A straightforward procedure would introduce the additional adiabatic invariants (Section 3.1) as constraints in the minimization of W_2 and then consider the limit of small electron mass, keeping the electron temperature fixed. It turns out, however, that a useful stability criterion can be found in a simpler way by going to the small electron mass limit even without taking adiabatic invariants into account. Again, this possibility is based on the fact that the incorporation of additional constraints can only stabilize.

So let us go back to $\mathcal{V}_2$ as given by (10.214) and take the limit $m_e \to 0$, with fixed electron temperature. For simplicity, a single ion species with $q_i = e$ (e.g., protons) is considered. The distribution functions are kept general, except for $F_s' < 0$.

The limit is carried out in Appendix 2. With the definition

$$a_s = q_s^2 \int F_s' \, \mathrm{d}\tau_s \tag{10.222}$$

it is shown that for small electron mass (10.214) assumes the limiting form

$$\mathcal{V}_2 = \frac{1}{2} \int \left(\frac{(\nabla A_1)^2}{\mu_0} - \frac{\mathrm{d}j_0}{\mathrm{d}A_0} A_1^2 + |a_i| [\![(\langle \Psi_{i1} \rangle_i - [\![\langle \Psi_{s1} \rangle_i]\!]_i)^2]\!]_i \right.$$
$$\left. + n_0 Q_1^2 \left(\frac{5}{3} k_B T_e + \frac{e^2 n_0}{|a_i|} \right) \right) \mathrm{d}x \, \mathrm{d}z \ . \tag{10.223}$$

Here, $n_0 = n_{e0} = n_{i0}$ in view of quasi-neutrality and

$$Q_1(A) = \frac{1}{V(A)} \frac{\mathrm{d}}{\mathrm{d}A} \int_A A_1 \frac{\mathrm{d}s_0}{B_0} \tag{10.224}$$

is the expression that we already encountered in the discussion of two-dimensional MHD processes (see (10.71)) with $V = \int \mathrm{d}s_0/B_0$ being the differential flux tube volume and T_e the kinetic electron temperature

$$k_B T_e = \frac{2}{3n_0} \int \frac{m_e}{2} (\boldsymbol{w} - \boldsymbol{v})^2 F_e \mathrm{d}\tau_e , \tag{10.225}$$

evaluated to lowest order in m_e.

In the case of exponential distribution functions (10.223) assumes the form

$$\mathcal{V}_2 = \frac{1}{2} \int \left(\frac{(\nabla A_1)^2}{\mu_0} - \frac{\mathrm{d}j_0}{\mathrm{d}A_0} A_1^2 + |a_i| \left[\!\left[(\langle \Psi_{i1} \rangle_i - [\![\langle \Psi_{s1} \rangle_i]\!]_i)^2 \right]\!\right]_i \right.$$
$$\left. + n_0 Q_1^2 \left(\frac{5}{3} k_B T_e + k_B T_i \right) \right) \mathrm{d}x \, \mathrm{d}z . \tag{10.226}$$

It is also of interest to assess the validity of the formal limit $m_e \to 0$ by carrying the expansion one step further. One finds that typical corrections are of the order

$$\delta_{\mathrm{cor}} = \frac{m_e k_B T_e}{e^2 A^{*2}} \tag{10.227}$$

where A^* represents a characteristic scale for variation of ϕ_0 or of $\int A_1 \mathrm{d}s_0/B_0$ with A_0 (see Appendix 2).

10.5.3 A lower bound of W_2

Useful stability information can also be obtained by determining a lower bound of W_2, using Schwarz's inequality. This method has been playing a significant role in the literature (Pellat *et al.*, 1991; Quest *et al.*, 1996). As shown in Appendix 2, one finds the inequality

$$W_2 \geq \frac{1}{2} \int \left(\frac{(\nabla A_1)^2}{\mu_0} - \frac{\partial j_0}{\partial A_0} A_1^2 + \sum_s \frac{q_s^2 \tilde{n}_{s1}^2}{|a_s|} \right) \mathrm{d}x \, \mathrm{d}z , \tag{10.228}$$

where $\tilde{n}_{s1}$ is a modified density perturbation defined by

$$\tilde{n}_{s1} = \int \tilde{f}_{s1} \mathrm{d}\tau , \tag{10.229}$$

with

$$\tilde{f}_{s1} = f_{s1} + \frac{q_s}{m_s} (P - q_s A_0) A_1 F_s' . \tag{10.230}$$

Here again it is interesting to find the electron contribution in the small-mass limit. This cannot be done in the fully systematic way that was used above, because by applying the inequality dynamical information that would have been required was lost. However, one can evaluate the small-mass limit of $\overline{\tilde{n}_{e1}}$ in a separate procedure. A corresponding derivation in Appendix 2 gives

$$\overline{\tilde{n}_{e1}} = n_{e0}(A_0)Q_1(A_0) \,. \tag{10.231}$$

(Equivalent derivations are due to Pellat *et al.* (1991) and Quest *et al.* (1996).) Inserting (10.231) into (10.228) gives

$$W_2 \geq \frac{1}{2} \int \left(\frac{(\nabla A_1)^2}{\mu_0} - \frac{\partial j_0}{\partial A_0} A_1^2 + \sum_{\text{ions}} \frac{q_s^2 \overline{\tilde{n}_{s1}}^2}{|a_s|} + \frac{e^2}{|a_e|} n_{e0}^2 Q_1^2 \right) \mathrm{d}x \, \mathrm{d}z \,. \tag{10.232}$$

Specializing again for a single ion species with $q_i = e$, and, in addition, choosing exponential distribution functions and the frame where ϕ_0 vanishes, one finds that $\overline{\hat{n}_{i1}} = \overline{\hat{n}_{e1}}$, so that (10.232) becomes

$$W_2 \geq \frac{1}{2} \int \left(\frac{(\nabla A_1)^2}{\mu_0} - \frac{\mathrm{d}j_0}{\mathrm{d}A_0} A_1^2 + n_0 k_B T_0 Q_1^2 \right) \mathrm{d}x \, \mathrm{d}z \,, \tag{10.233}$$

with $T_0 = T_e + T_i$. Equation (10.233) was first derived by Pellat *et al.* (1991). As expected, (10.233) is consistent with (10.226). (Note that by (10.218) $-\mathrm{d}j_0/\mathrm{d}A_0 \geq -\partial j_0/\partial A_0$.)

The functional (10.233), more precisely its complex version given below, resembles the corresponding MHD functional (10.72). Note, however, that, unlike the Vlasov version (10.233), in the MHD case the perturbation A_1 is subject to the constraint (10.50). As discussed in Section 10.2, this difference can have an essential influence on stability. For $B_p \neq 0$ under the present assumptions (notably $m_e \to 0$) MHD stability, evaluated for $\gamma = 1$, implies Vlasov stability.

10.5.4 Complex perturbations

So far, without loss of generality, we have assumed that perturbations such as A_1 have real values. If for convenience one prefers to use complex-valued perturbations, the analysis is easily adjusted. For quadratic functionals of a single perturbation containing only squares of linear expressions, the squares simply are to be replaced by absolute squares. For example, (10.233) for complex A_1 becomes

$$W_2 \geq \frac{1}{2} \int \left(\frac{|\nabla A_1|^2}{\mu_0} - \frac{\mathrm{d}j_0}{\mathrm{d}A_0} |A_1|^2 + n_0 k_B T_0 |Q_1|^2 \right) \mathrm{d}x \, \mathrm{d}z . \qquad (10.234)$$

The same applies to other functionals such as (10.214), (10.223), (10.226) or (10.232).

10.5.5 *Effect of a normal magnetic field component in the Earth's magnetotail*

For many realistic current sheet configurations deviations from the one-dimensional Harris-type equilibrium can become important. In particular, this is the case for the presence of a small magnetic field component B_n normal to the sheet, as it exists in the weakly two-dimensional configurations discussed in Section 5.4.1. (Note that the Grad–Shafranov theory described there in the framework of MHD also applies to the kinetic equilibria discussed here, as explained in Section 6.2.2.) Here we consider the plasma sheet in the Earth's magnetotail, following work by Lembège and Pellat (1982), Pellat *et al.* (1991) and Quest *et al.* (1996).

Unless B_n gets extremely small, the electrons can still be described as adiabatic, such that the small-mass regime applies. The quantitative condition is that the correction (10.227) is negligible compared to 1.

In that regime the electrons give rise to a strong stabilizing effect, at least in the WKB regime (familiar from optics and quantum mechanics), where the modes have x-scales of the order of the equilibrium sheet width L_z. (As before, the x-axis points along and the z-axis perpendicular to the sheet and all observables are independent of y.) We give a brief description of the argument, starting from the lower bound (10.234).

The task is to identify a condition for that lower bound to be positive. The sheet equilibrium is a strongly stretched tail equilibrium, as discussed in Section 5.4.1. Accordingly, one chooses an equilibrium flux function $A_0(x, z)$, which depends on the x-coordinate only weakly, such that the equilibrium length scales L_x and L_z satisfy $L_z \ll L_x$. Let $B_n = B_z(x, 0)$ and B_0 the magnetic field strength outside the sheet and $b = |B_n/B_0|$, then $b \ll 1$.

Rather than carrying out an exact minimization we follow Lembège and Pellat (1982) and assume a plausible test function for A_1. Restricting the perturbation to the WKB regime allows us to choose the test modes of the form

$$A_1 = \hat{A}_1(z) \mathrm{e}^{\mathrm{i}kx} \qquad (10.235)$$

where an additional weak x-dependence of the wave number k and of the amplitude $\hat{A}_1$ is not shown explicitly.

With regard to Q_1 (given by (10.224)) it turns out that the main contribution to the integral $I = \int A_1 ds/B$ stems from a $|z|$-region with $|z|/L_z = O(\sqrt{b})$. For the evaluation of Q_1 this motivates choosing a simplified local approximation to the magnetic field structure with parabolic field lines:

$$A_0 = B_n x - \frac{B_0 z^2}{2L_z} . \tag{10.236}$$

Furthermore, changing the integration variable in I from s to z and treating $\hat{A}_1$ as approximately constant over the relevant integration interval one finds

$$I = \frac{\hat{A}_1(0)}{B_n} e^{\mathrm{i}\frac{kA_0}{B_n}} \int_{-\infty}^{\infty} e^{\mathrm{i}\frac{kz^2}{2L_z b}} \, dz , \tag{10.237}$$

where x in the exponent was expressed by A_0 and z using (10.236). The weak x-dependence of equilibrium quantities is ignored in the integration and a convenient way to ensure convergence is adding $-rz^2$ with $r > 0$ in the exponent and taking the limit $r \to 0$ of the resulting expression. One obtains

$$I = (1 \pm \mathrm{i})\frac{\hat{A}_1(0)}{B_n}\sqrt{\frac{\pi L b}{k}} e^{\mathrm{i}\frac{kA_0}{B_n}} , \tag{10.238}$$

where the sign in the first bracket is determined by the sign of B_n. With the help of (10.224) one finds from (10.234)

$$W_2 \geq \frac{1}{2} \int \left(\frac{|\nabla A_1|^2}{\mu_0} - \frac{\partial j_0}{\partial A_0}|A_1|^2 + \frac{\pi k L_z}{b}\frac{1}{V^2 B_n^2}\frac{|\hat{A}_0(0)|^2}{\mu_0} \right) dx\, dz , \tag{10.239}$$

where the local pressure balance has been used to eliminate $n_0 k_B T_0$.

Using (10.66) and assuming monotonic decrease of $|\hat{A}_1(z)|^2$ with $|z|$ increasing from zero, one finds from (10.239)

$$W_2 > \int \frac{1}{2\mu_0 L_z^2} \left(-1 + \frac{\pi k L_z}{b} \left(\frac{L_z}{V B_n} \right)^2 \right) |A_1(z)|^2 \, dx\, dz . \tag{10.240}$$

The ratio $L_z/(V B_n)$ is of order 1. For orientation, let us use the model (10.236) for $|z| \leq 1$, although the parabolic model is of only qualitative significance outside the thin region at the centre of the sheet. Then $L_z/(V B_n) = 1/2$, such that (10.239) gives stability for

$$kL_z > \frac{4}{\pi}b . \tag{10.241}$$

This stability criterion was first derived by Lembège and Pellat (1982).

For this criterion to be valid it is necessary that the correction δ_{cor} given by (10.227) is negligible. From (10.238) one finds $A^* = B_{\text{n}}/k$ such that the condition that the correction does not significantly alter (10.241) is

$$\frac{r_{\text{ge0}}}{L_z} \ll c_0 \frac{b}{kL_z}, \tag{10.242}$$

where r_{ge0} is the electron gyroradius with respect to B_0; note that kL_z is of order 1 in view of the WKB condition. Electron stabilization of the plasma sheet plays an important role in the substorm cycle (see Part IV).

It remains to address the physical mechanism that leads to the strong electron stabilization in the adiabatic regime. One way is to look at the current density associated with the minimizing electron distribution function f_{e1} of the form (see Appendix 2)

$$f_{\text{e1}}^{(m)} = F_{\text{e}}'(\psi_{\text{e1}} - \langle \psi_{\text{e1}} \rangle_{\text{e}}) . \tag{10.243}$$

One finds

$$j_{\text{e1}} = \frac{\mathrm{d}j_0}{\mathrm{d}A_0} A_1 + \left(\frac{5}{3} n_0 k_{\text{B}} T_{\text{e}} + n_0 k_{\text{B}} T_{\text{i}} \right) \frac{\mathrm{d}Q_1}{\mathrm{d}A_0} \tag{10.244}$$

where the second term, resulting from $\langle \psi_{\text{e1}} \rangle_{\text{e}}$, is shown only in leading order in $1/b$. The associated contribution to the second order energy integral is found as

$$\int (-j_{\text{e1}} A_1) \, \mathrm{d}x \, \mathrm{d}z = \int \left(-\frac{\mathrm{d}j_{\text{e0}}}{\mathrm{d}A_0} A_1^2 + n_0 Q_1^2 \left(\frac{5}{3} k_{\text{B}} T_{\text{e}} + k_{\text{B}} T_{\text{i}} \right) \right) \mathrm{d}x \, \mathrm{d}z , \tag{10.245}$$

after integration by parts with respect to A_0 and, again, considering the second term only in leading order. The comparison with (10.226) shows that the first term in (10.244) gives the electron contribution to the driving term and that the second term generates the strong stabilization through the factor Q_1. So, the second part of (10.244) is responsible for the stabilization.

The physical interpretation differs for the two stabilizing contributions in (10.245). To discuss the part depending on T_{i} let us look at the limit of small electron temperature of the exponential model with $\phi_0 = 0$. Then, the stabilizing current of (10.244) can be identified with the electron Hall current density j_H, based on the $\boldsymbol{E} \times \boldsymbol{B}$ drift (3.6) of the electrons where the electric field results from the quasi-neutrality condition. One finds

$$j_\mathrm{H} = -en_0 \frac{\boldsymbol{E} \times \boldsymbol{B}}{B^2} \cdot \boldsymbol{e}_y \qquad (10.246)$$

$$= \frac{en_0}{B^2} \boldsymbol{E} \cdot \nabla A , \qquad (10.247)$$

and to first order

$$j_\mathrm{H1} = -en_0 \frac{\mathrm{d}\varphi_1(A_0)}{\mathrm{d}A_0} \qquad (10.248)$$

where $\varphi_1(A_0)$ denotes the leading part of ϕ_1. Evaluating φ_1 with the help of the quasi-neutrality condition, which couples the ions into the electron current, one finds (for details see Appendix 2)

$$j_\mathrm{H1} = n_0 k_\mathrm{B} T_\mathrm{i} \frac{\mathrm{d}Q_1}{\mathrm{d}A_0} \qquad (10.249)$$

which confirms that the stabilizing part of the electron current (10.244) for $T_\mathrm{e} \ll T_\mathrm{i}$ indeed reduces to the electron Hall current.

The term of (10.245) depending on T_e is best interpreted as resulting from compressibility. It suffices to point at the fact that this term agrees with the corresponding MHD term if γ is set to 5/3 (see (10.72), there written for complex perturbations). In fact, the present electron description with P being the only constant of the motion, with m_e going to zero and with isotropy of the equilibrium distribution function (in the frame moving with the bulk velocity) essentially behaves like a fluid with an adiabatic index of 5/3. We have encountered that fact already in Section 6.3.

In the magnetospheric literature most authors emphasize the stabilizing effect of electron compressibility. However, in (10.245) compressibility has the relative weight $(5T_\mathrm{e}/3)/(5T_\mathrm{e}/3 + T_\mathrm{i})$, which becomes small for $T_\mathrm{e} \ll T_\mathrm{i}$, so that in this case the Hall effect dominates.

10.5.6 Ion tearing

Here we briefly return to the case of equal masses as introduced in Section 10.4.1 with exponential distribution functions. Let us add the condition $T_\mathrm{e} = T_\mathrm{i}$. Then there is no charge separation in the poloidal plane and the current densities are equal ($j_\mathrm{e} = j_\mathrm{i}$). Correspondingly, the poloidal electric field vanishes and (10.214) becomes

$$\mathcal{V}_2 = \frac{1}{2} \int \left(\frac{(\nabla A_1)^2}{\mu_0} - \frac{\mathrm{d}j_0}{\mathrm{d}A_0} A_1^2 - 2a_\mathrm{i} \left[\!\left[\langle \Psi_\mathrm{i1} \rangle_\mathrm{i}^2 \right]\!\right]_\mathrm{i} \right) \mathrm{d}x \, \mathrm{d}z . \qquad (10.250)$$

If the particles are strongly non-adiabatic (r_{g0}/L of order 1 or larger) the compressibility term is no longer large. If it becomes sufficiently small for an instability to occur, one refers to it as a *pure ion tearing mode*.

In the opposite case, represented by the small-mass limit, one finds that the compressibility term shown in (10.223) for the electrons applies to both particle species,

$$\mathcal{V}_2 = \frac{1}{2} \int \left(\frac{(\nabla A_1)^2}{\mu_0} - \frac{\mathrm{d}j_0}{\mathrm{d}A_0} A_1^2 + \frac{5}{3} n_0 k_{\mathrm{B}} T_0 Q_1^2 \right) \mathrm{d}x \,\mathrm{d}z \qquad (10.251)$$

with $T_0 = T_{\mathrm{e}} + T_{\mathrm{i}}$. Again, this expression coincides with the MHD case (10.72). Note that the assumed smallness of r_{gn} excludes magnetic neutral points so that the constraint (10.50) does not apply.

10.5.7 Remarks on boundary conditions and validity

The present approach is based on conservation of energy and of integrals of the form $\int G(f_s, P)\mathrm{d}\Omega_s$ (see Appendix 2). In addition to the equations of Vlasov theory this requires the vanishing of certain surface integrals. For the present quasi-neutral case with translational invariance, and with equilibrium distribution functions of the form (10.199), the conditions $A = A_0$, $f_s = f_{s0}$, satisfied on the boundary in the x, z-plane, guarantee that the surface integrals vanish. There is no boundary condition for ϕ, consistent with the quasi-neutrality condition, which does not involve any differential operator acting on ϕ.

From these boundary conditions one would expect that the minimizer f_{s1}^{m} has to vanish on the boundary, which, however, is not the case. Still, that does not cause a serious difficulty. As the variational expression (10.201) does not involve a derivative of f_{s1}, the minimizer f_{s1}^{m} can be approximated arbitrarily well by functions that do satisfy the boundary condition. In other words, $\mathcal{V}_2$ is not a minimum but an infimum on the test function space. As this difference does not seem to have significant physical consequences, we ignore this subtlety and keep using the terminology that would apply to a minimum.

It is important to note that in deriving the expressions (10.209) or (10.214) the only dynamical constraints that were imposed (within the framework of Vlasov theory with prescribed boundary conditions) were the constancy of P_y and quasi-neutrality. Here we repeat that applying any further constraint can only raise the minimum of W_2. The following implications are particularly worthwhile mentioning.

Pitch-angle diffusion of adiabatic electrons is not an additional destabilizing mechanism, as long as the fluctuation fields are subject to the assumed translational invariance and the scattering is a Vlasov process satisfying the present boundary conditions. Then the scattering process is included in the dynamics allowable in the derivation of $\mathcal{V}_2$. Pitch-angle diffusion would simply destroy the adiabatic invariance of the electron motion. However, since the adiabatic invariants were not imposed on the dynamics in the present approach, the minimum cannot be lowered by their violation. There had been discrepancies about this point in the literature until it was clarified by Brittnacher *et al.* (1994) and Pellat *et al.* (1991) that pitch-angle diffusion is not an efficient destabilizer.

Also it has been argued that the presence of transient electrons would have a destabilizing effect (Sitnov *et al.*, 1998), so that the criterion (10.241) would be relaxed considerably. Again, systems with transient and nontransient electron populations are included in the present approach, as long as P is conserved and $F'_e < 0$. This would lead one to argue that, in contrast to the results of Sitnov *et al.* (1998), criteria such as (10.241) cannot be relaxed by taking transient particles into account. It would be desirable for future work to identify the origin of this discrepancy.

11

Magnetic reconnection

The tearing instability, which was a dominant topic in the previous chapter, is an important example of magnetic reconnection. In the present chapter we will approach magnetic reconnection from a more general point of view and include steady state and three-dimensional processes.

11.1 Introduction

Magnetic reconnection can be regarded as the process that resolves the following dilemma. On sufficiently large length and time scales a plasma behaves approximately as an ideal fluid. The main reason is that in this regime generalized Ohm's law (9.7) reduces to its ideal limit (3.60). As a consequence, the magnetic field is frozen into the plasma motion, which sets severe limitations to the accessible dynamical states (Section 3.8).

Already in the early stages of space exploration it became clear that these limitations cannot be reconciled with observations. In particular, space plasma activity seemed to involve the conversion of large amounts of magnetic energy into kinetic energy of bulk plasma and random particle motion and of high energy particle populations. Such conversion, however, appeared to be strongly inhibited, if not ruled out, by the frozen-in condition.

A particularly clear manifestation of that property is the stabilization of one-dimensional current sheets by the constraint (10.50) that was discussed in Section 10.2.2. As seen there, that constraint excludes the change of the topological structure of the magnetic field. Breaking that constraint by introducing a resistivity led to instability. So, already that simple example indicated the central role that nonideal processes play in the present context.

When one began to look for processes that were able to break the frozen-in property in an effective way the following question arose. How can nonideal processes become efficient if all relevant terms of generalized Ohm's law (9.7) appear to be negligible? There is an obvious answer: With the exception of the term containing $\partial \boldsymbol{v}_e / \partial t$ (which is negligible unless the time scale is as short as $1/\omega_{pe}$, see Section 9.4) these terms are negligible only on large overall length scales. If a local structure developed on a much smaller scale, nonideal processes could become important in that structure. A significant example is the occurrence of a singular layer in the tearing mode (Section 10.3).

There is a second task that reconnection has to fulfill. Nothing would be gained if the effect of the local nonidealness remained local. It would be of little interest if the effect was limited to plasma passing through a small nonideal region. So it is necessary that the local nonideal process, to a sufficient extent, leads to consequences on larger scales. Such consequences, for instance, may concern energy transfer or the violation of magnetic line conservation (see Section 3.8.2). This is the second property of magnetic reconnection.

With both properties satisfied, magnetic reconnection enables plasmas and fields of different origin to mix and large scale plasma structures to transform magnetic to kinetic energy of the plasma particles in an efficient way. According to major lines of present thinking, this is what happens in solar flares or magnetospheric substorms, and possibly in many other plasma processes in the universe.

Note that instead of starting from a clear-cut definition of reconnection we have formulated requirements in qualitative terms. This reflects the present situation. Although several attempts have been made to define reconnection more precisely, none of the suggested notions has proven to be fully satisfactory.

It seems reasonable to begin with considering the simplest geometry that allows for efficient conversion of magnetic to kinetic energy with the frozen-in condition violated. At first sight, one might think that in a one-dimensional configuration with magnetic field reversal (e.g., a Harris sheet) a constant electric field imposed along the direction of the electric current might lead to a steady state energy conversion process, because the Poynting vector would transport electromagnetic energy toward the neutral sheet from both sides. But away from the neutral sheet, where the plasma may be treated as ideal, the $\boldsymbol{E} \times \boldsymbol{B}$ motion would lead to an inward flow of the plasma also. By mass conservation the plasma would pile up inside the sheet, which excludes a steady state process. Even in a time-dependent system the piling-up of the plasma would generate forces that would soon stop the incoming flow. Therefore, reconnection requires a plasma flow with at least two spatial dimensions. We will see to what extent this applies to the magnetic field as well.

The first attempt to deal with processes that today we would group under magnetic reconnection is generally attributed to Giovanelli (1946), who emphasized the importance of a weak field region for efficient particle acceleration. Beginning in the late 1950s, several authors, notably Sweet (1958), Parker (1963b) and Petschek (1964), developed the first fluid models in the context of energy conversion in solar flares; Dungey (1961) addressed the interaction between the magnetized interplanetary medium and the Earth's magnetosphere. Several aspects of these pioneering contributions will briefly be discussed in the following section. Today, magnetic reconnection is an established field of research with many papers appearing every year dealing with various facets of this fascinating phenomenon. Numerous details that go beyond the present rather concise description are given in monographs especially devoted to magnetic reconnection, such as Biskamp (2000) or Priest and Forbes (2000).

11.2 Two-dimensional fluid models

In this section we describe two-dimensional configurations by resistive MHD (Section 3.3.3). The physical quantities are independent of the (Cartesian) y-coordinate and the discussion is focussed on steady states.

11.2.1 Basic configuration and properties

The basic configuration is shown in Fig. 11.1. The magnetic field B and the plasma velocity v are assumed to lie in the x, z-plane, while the electric field has a non-vanishing y-component. As in the case of resistive tearing the plasma is treated as incompressible with a constant density ρ_0, and the resistivity η is kept constant unless stated otherwise. The plasma is almost ideal, such that the Lundquist number (9.15), evaluated with the global length scale L, is much larger than 1.

To obtain an effect on the large scale, one assumes a stagnation flow with velocity v (big arrows) and a magnetic field structure with a neutral point at the origin, such that the x-components of the magnetic fields in the upper and lower inflow regions are oppositely directed. (Viewed three-dimensionally, instead of the neutral point, a *neutral line* extends along the y-axis.)

The local nonidealness is realized by the presence of a *diffusion region*, centred at the origin and characterized by length scales δ and Δ with

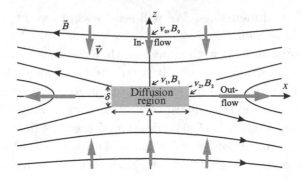

Fig. 11.1 Qualitative pattern of two-dimensional reconnection with v and B subject to the following symmetry relations: $v_x(-x, z) = -v_x(x, z)$, $v_x(x, -z) = v_x(x, z)$, $v_z(-x, z) = v_z(x, z)$, $v_z(x, -z) = -v_z(x, z)$, $v_y = 0$, $B_x(-x, z) = B_x(x, z)$, $B_x(x, -z) = -B_x(x, z)$, $B_z(-x, z) = -B_z(x, z)$, $B_z(x, -z) = B_z(x, z)$, $B_y = 0$.

$L \geq \Delta \gg \delta$. There, the plasma and magnetic fields can decouple effectively, so that the frozen-in property of ideal MHD is broken.

The fields may not be smooth everywhere, narrow regions with strong spatial variation that may be idealized as discontinuities (Section 3.9) may be present. In the case of shocks the associated nonidealness must also be taken into account. For the occurrence of shocks in an otherwise incompressible medium see Section 3.9. (Shocks are not indicated in Fig. 11.1, but Fig. 11.6 shows a configuration with a slow shock.)

The term *diffusion region* suggests the presence of a diffusion process. This notion results from the structure of the induction equation with the electric field inserted from (3.67),

$$\frac{\partial \boldsymbol{B}}{\partial t} = \nabla \times (\boldsymbol{v} \times \boldsymbol{B}) + \frac{\eta}{\mu_0} \nabla^2 \boldsymbol{B} , \qquad (11.1)$$

which shows that the evolution of the magnetic field is determined by advection (first term on the right side) and diffusion (second term). Considerable diffusion can take place if a length scale becomes small. Advection without diffusion leads to the frozen-in motion (Section 3.8). Magnetic reconnection is based on the interplay between both processes.

That the requirement of a large-scale effect is satisfied for the configuration of Fig. 11.1 can be seen as follows. Under the present steady state assumption $\nabla \times \boldsymbol{E}$ vanishes, which in two dimensions implies that E_y is a constant, say $-E_0$, with $E_0 > 0$ in the configuration of Fig. 11.1. The presence of the diffusion region allows for a non-vanishing value of E_0. Under ideal conditions (with $\eta \boldsymbol{j}$ negligible) the y-component of Ohm's law (3.67) would require $E_y = 0$ at the neutral point, such that E_0 would have to vanish and with it the plasma flow perpendicular to $\boldsymbol{B}$ in the external region everywhere. So, the existence of the large-scale stagnation flow is directly tied to the presence of a localized diffusion region.

In some of the models to be discussed in this chapter the diffusion region is a sheet formally extending along the entire x-axis (or an analogous coordinate axis in other geometries). The dynamics of this limiting case is largely referred to as *annihilation*.

Under the present assumptions the resistive MHD equations (Section 3.3.3) become

$$\rho_0 \boldsymbol{v} \cdot \nabla \boldsymbol{v} = -\nabla p + \boldsymbol{j} \times \boldsymbol{B} \tag{11.2}$$

$$\boldsymbol{E} + \boldsymbol{v} \times \boldsymbol{B} = \eta \boldsymbol{j} \tag{11.3}$$

$$\nabla \cdot \boldsymbol{v} = 0 \tag{11.4}$$

$$\nabla \times \boldsymbol{B} = \mu_0 \boldsymbol{j} \tag{11.5}$$

$$\nabla \cdot \boldsymbol{B} = 0. \tag{11.6}$$

For several purposes, such as the search for analytical solutions or numerical studies, it is convenient to represent both $\boldsymbol{v}$ and $\boldsymbol{B}$ by flux functions $D(x, z)$ and $A(x, z, t)$, respectively. Here A has the form $A'(x, z) + E_0 t$, so that E_y is generated by $-\partial A/\partial t$. Then, from (11.2)–(11.6) one finds the equations

$$[\Delta D, D] = [\Delta A, A] \tag{11.7}$$

$$M_{\mathrm{A}} + [A, D] = \frac{1}{S_0} \Delta A, \tag{11.8}$$

which are written in non-dimensional form, such that A is normalized by $B_0 L$, the stream function D by $v_{\mathrm{A}_0} L$ and coordinates by L. Here $v_{\mathrm{A}_0} = B_0/\sqrt{\mu_0 \rho_0}$ is the Alfvén velocity in the outer inflow region and M_{A} is given by

$$M_{\mathrm{A}} = \frac{E_0}{v_{\mathrm{A}_0} B_0}. \tag{11.9}$$

The bracket-symbol is defined in (10.133) and $S_0 = \mu_0 v_{\mathrm{A}_0} L/\eta$. The equations (11.2)–(11.6) and (11.7)–(11.8) are equivalent, except that in (11.7) the pressure gradient has been eliminated by taking the curl. The equations (11.7) and (11.8) can also be obtained from (10.128)–(10.132) when tailored to the present conditions.

Magnetic reconnection as described here is irreversible. This follows from the irreversibility of the general resistive MHD equations and is also obvious from their present specialization (11.7) and (11.8). For $\eta = 0$ these equations would be invariant against inversion of the direction of the plasma velocity and of the electric field leaving $\boldsymbol{B}$ and $\boldsymbol{j}$ unchanged, reflecting the reversibility of the ideal equations. For non-vanishing resistivity this symmetry is broken, the system becomes irreversible. A direct indication of the irreversibility is the fact that at the neutral point (11.3) reduces to $E_y = \eta j_y$, which in the present geometry implies $\boldsymbol{E} \cdot \boldsymbol{j} = \eta j^2 \geq 0$, so that $\boldsymbol{E}$ cannot be reversed with $\boldsymbol{j}$ unchanged. It is an interesting question whether irreversibility is a general property of magnetic reconnection.

Solutions of (11.7) and (11.8) have three dimensionless parameters, two of them are M_A and S_0; the third does not appear in (11.7)–(11.8) because it is associated with the pressure, which has been eliminated. The parameter M_A is called *reconnection rate*. As ideal Ohm's law implies $E_0 = v_0 B_0$, M_A equals the inflow Alfvén Mach number v_0/v_{A0}. The definition (11.9) can also be understood as a non-dimensional representation of the electric field strength E_0, which is often used as the dimensional form of the reconnection rate. It directly measures the rate at which magnetic flux conservation with respect to the plasma velocity is violated by the presence of the diffusion region. To illustrate that property, let us consider the rectangle S_r (Fig. 11.2) located in the plane $z = 0$; line (c) is located in the ideal region. To the rectangle S_r let us apply the balance of magnetic flux (see (3.80))

$$\frac{\mathrm{d}}{\mathrm{d}t} \int_{S_r} \boldsymbol{n} \cdot \boldsymbol{B} \, \mathrm{d}x \, \mathrm{d}y = -\oint_{\partial S_r} (\boldsymbol{E} + \boldsymbol{v} \times \boldsymbol{B}) \cdot \mathrm{d}\boldsymbol{s}$$

$$= -\int_{-1/2}^{1/2} E_y \, \mathrm{d}y$$

$$= E_0 \,, \tag{11.10}$$

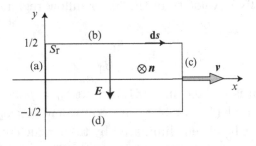

Fig. 11.2 Breakdown of magnetic flux conservation in the outflow region. The magnetic flux considered is connected with the rectangle S_r, whose sides move with the local plasma velocity. The normal $\boldsymbol{n}$ points in the negative z-direction. The line (c) is located in the ideal outflow region.

where ∂S_r is the boundary of S_r consisting of the lines (a),(b),(c),(d). Note that only line (a) contributes to the line integral in (11.10). On the inflow side there is a corresponding decrease of magnetic flux.

It is an important property of the present model (Fig. 11.1) that v violates line conservation (Section 3.8.2) and that this violation affects the ideal region. Two plasma elements that initially are situated on the same field line above the separatrix but on different sides of the z-axis (Fig. 11.3), by their perpendicular velocity $E \times B/B^2$, end up on different field lines in the outflow region. Note that the plasma elements cannot cross the z-axis because of symmetry (Fig. 11.1). As time proceeds, more and more field lines appear between the outflowing elements (although the corresponding magnetic flux is zero for the assumed symmetry). Eventually, the elements get separated by distances on the large scale L, even if their original distance was arbitrarily small. This gives the required large-scale effect.

Let us now consider magnetic field lines. In an ideal MHD system, the field lines can be regarded as being transported by the fluid velocity. (The condition (3.90) is satisfied for $w = v, \Lambda = 0$.) Applying this picture to the ideal inflow region the question arises, what happens to the field lines after they become connected to the nonideal region? Intuitively, they re-connect at the neutral point. What is the corresponding formal property? The answer is that there is no smooth transport of field lines. In formal terms this is manifested by the nonexistence of a smooth velocity field w

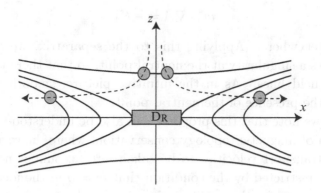

Fig. 11.3 Breakdown of magnetic line conservation in the ideal region, illustrated for two plasma elements. The full lines are magnetic field lines, broken lines the trajectories of the plasma elements. The elements start in the inflow region on the same field line and then move to different field lines. The outflowing elements become separated by an increasing number of field lines. D_R is the diffusion region. The presence of shocks (Fig. 11.6) would not affect the qualitative conclusion.

that coincides with the plasma velocity in the ideal region and satisfies the transport condition (see Section 3.8.2)

$$\nabla \times (\boldsymbol{w}^* \times \boldsymbol{B}) = \lambda \boldsymbol{B} \,, \tag{11.11}$$

where the * symbol indicates the condition that the transport velocity coincides with $\boldsymbol{v}$ in the ideal region.

It is instructive to give a formal proof of this property in addition to intuitive arguments such as given above. We show that the assumption of a smooth solution $\boldsymbol{w}^*$ leads to a contradiction. First we note that the divergence of (11.11) implies that λ is constant on field lines. As all field lines connect to the ideal region where $\lambda = 0$, λ must vanish everywhere. Further, let us write

$$\nabla \times (\boldsymbol{w}^* \times \boldsymbol{B}) = \nabla w_y^* \times \nabla A - \nabla(\boldsymbol{w}^* \cdot \nabla A) \times \boldsymbol{e}_y \tag{11.12}$$

where we have used $\boldsymbol{B} = \nabla A \times \boldsymbol{e}_y$.

Noting that $B_y = 0$ and that ∇A vanishes only at an isolated point and treating derivatives of $\boldsymbol{B}$ and $\boldsymbol{w}^*$ as continuous (Section 3.8.2), this implies

$$\boldsymbol{B} \cdot \nabla(\boldsymbol{w}^* \cdot \nabla A) = 0 \,. \tag{11.13}$$

In the ideal region $\boldsymbol{v}$ satisfies ideal Ohm's law, which has the y-component $\boldsymbol{v} \cdot \nabla A = E_y = -E_0$. Thus, the identification of $\boldsymbol{w}^*$ with $\boldsymbol{v}$ in the ideal region requires that $\boldsymbol{w}^* \cdot \nabla A = -E_0$ in that region. Now, (11.13) implies that $\boldsymbol{w}^* \cdot \nabla A$ is constant on the entire field lines. This means that

$$\boldsymbol{w}^* \cdot \nabla A = -E_0 \tag{11.14}$$

must hold everywhere. Applying this to the separatrix, we see that $\boldsymbol{w}^*$ necessarily has a singularity at the neutral point, so that there is no smooth transport of field lines. As in the intuitive picture, smooth transport is excluded by the presence of the neutral point.

In passing we note that this property can also be understood as a variant of breakdown of magnetic topology conservation, where, however, the class of (smooth) transport velocities to be taken into account is not arbitrary, but has to be restricted by the condition that $\boldsymbol{w} = \boldsymbol{v}$ in the ideal region.

As we have seen, for the configuration discussed here, the presence of a separatrix plays a crucial role in the breakdown of line conservation. Vasyliunas (1975) tied his notion of *magnetic merging* to plasma motion across separatrices. Unlike merging, the present criterion, based on breakdown of line conservation, can be generalized to include configurations without separatrix (Section 11.5).

For obtaining quantitative examples of reconnection one must solve the nonlinear partial differential equations (11.7) and (11.8). Although they might appear as not too complicated, because of their nonlinearity only very few analytical solutions have been found. We will start with discussing a class of annihilation solutions. Then we turn to more general configurations, where drastic simplifications and assumptions are the price for obtaining explicit results. Where necessary, the analytical results are complemented by results of simulations.

11.2.2 Exact solutions: Magnetic annihilation

We begin by confirming that one-dimensional solutions with A and D depending on only one coordinate, say on z, are useless for reconnection purposes. In that case all []-brackets in (11.7) and (11.8) vanish, implying that there is neither advection nor diffusion, which contrasts the picture of reconnection as an interplay between the two processes.

It turns out that the requirement of at least two spatial dimensions primarily applies to the plasma flow rather than to the magnetic field. So it is natural to look for solutions of (11.7) and (11.8) with a one-dimensional magnetic field, say $\boldsymbol{B} = B_x(z)\boldsymbol{e}_x$, so that A does not depend on x. (By the vanishing of B_z the magnetic field is qualitatively different from that of Fig. 11.1, so that those arguments of the previous section that refer to a separatrix are not applicable.) Under the present simplifications (11.7) and (11.8) reduce to

$$[\Delta D, D] = 0 \tag{11.15}$$

$$M_A + \frac{\partial A}{\partial z}\frac{\partial D}{\partial x} - \frac{1}{S_0}\frac{\partial^2 A}{\partial z^2} = 0 . \tag{11.16}$$

In view of (11.16) $\partial D/\partial x$ does not depend on x. Luckily, a simple solution meeting this requirement can be found as

$$D = -xz , \tag{11.17}$$

such that (11.16) becomes (Sonnerup and Priest, 1975)

$$\frac{\partial^2 A}{\partial \zeta^2} + \zeta\frac{\partial A}{\partial \zeta} - M_A = 0 , \tag{11.18}$$

where $\zeta = \sqrt{S_0}z$.

For $B_x(0) = 0$ (11.18) is solved by

$$\frac{\partial A}{\partial \zeta} = \sqrt{2}M_A \, \mathrm{daw}(\zeta/\sqrt{2}) , \tag{11.19}$$

where

$$\mathrm{daw}(u) = \mathrm{e}^{-u^2} \int_0^u \mathrm{e}^{t^2} \, \mathrm{d}t \qquad (11.20)$$

is the Dawson function (Abramowitz and Stegun, 1965). The solution obtained from (11.19) is illustrated in Fig. 11.4.

The current density is concentrated in a region where ζ is of order 1 and z of order $1/\sqrt{S_0}$. This region represents the diffusion region, which here extends over the entire x-range.

As, for a moment, we have abandoned the neutral-point configuration of Fig. 11.1, magnetic flux and line conservation have to be reconsidered. For the plasma velocity one finds

$$\nabla \times (\boldsymbol{E} + \boldsymbol{v} \times \boldsymbol{B}) = \lambda \boldsymbol{B} \,, \qquad (11.21)$$

where

$$\lambda = -\frac{\eta}{B_x} \frac{\mathrm{d}j}{\mathrm{d}z}, \qquad (11.22)$$

which remains bounded for $z \to 0$. By (3.87) and (3.81) this means that the plasma velocity is line-conserving but not flux-conserving. The former is clear from the fact that v_z depends on z only, so that for the assumed 1D field structure all plasma elements that share a field line at one time keep sharing the field line at all times.

The plasma pressure p imposes a limitation of validity (Litvinenko *et al.*, 1996). It is determined by (11.2), which gives

$$p = p_0 - \frac{\rho v^2}{2} - \frac{B^2}{2\mu_0} \,, \qquad (11.23)$$

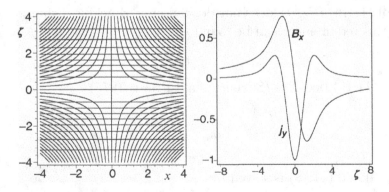

Fig. 11.4 Streamlines and (horizontal) magnetic field lines (left) and j_y and B_x in suitable normalization (right) of the Sonnerup–Priest solution (11.17), (11.19).

where p_0 is the pressure at the origin. For a given value of p_0 the pressure formally becomes negative outside a finite region containing the origin, so that the solution is valid only inside that region. Here, (11.23) puts a strong limitation on the flow velocity

$$\frac{\rho v^2}{2} \leq p_0 . \tag{11.24}$$

Although formally there is unlimited acceleration by the $\boldsymbol{j} \times \boldsymbol{B}$ force, the actual energy conversion is rather limited.

As we have seen, the present annihilation configuration, characterized by a one-dimensional magnetic field, is qualitatively different from the reconnection configuration of Fig. 11.1.

The term *reconnective annihilation* is used for generalizations of annihilation, where the magnetic field is two-dimensional with a point singularity but the current density is still extended over the entire domain. Corresponding solutions were presented by Craig and Henton (1995), who showed that (11.7) and (11.8) are solved by

$$D = \frac{\beta}{\alpha}g(z) + \alpha xz + \frac{1}{2}\gamma z^2 \tag{11.25}$$

$$A = g(z) + \beta xz , \tag{11.26}$$

where $g(z)$ is a suitable solution of the ordinary differential equation

$$\frac{1}{S_0}g'' - \frac{\alpha^2 - \beta^2}{\alpha}zg' = M_A - \beta\gamma z^2 \tag{11.27}$$

which can be solved analytically. The example of Fig. 11.5 clearly shows a two-dimensional magnetic field structure with a point singularity. The

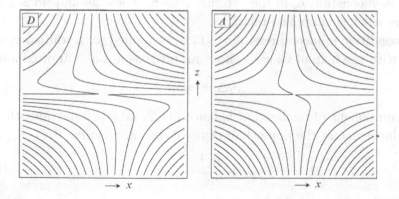

Fig. 11.5 Solutions D and A of the Craig–Henton model equations (11.25) and (11.26) for $\beta = 2$, $\alpha = 1$, $\gamma = 0$, $M_A = 0.1$, $S_0 = 100$.

separatrix extends into the ideal region. Therefore, one finds that, other than in the annihilation case, there is no smooth field line transport with $\boldsymbol{w} = \boldsymbol{v}$ in the ideal region, which implies violation of line and flux conservation for the plasma velocity. The diffusion region is still a one-dimensional current sheet. The positive pressure requirement still restricts the domain. On the x-axis (11.24) remains unchanged.

Equations (11.26), (11.25) reduce to the Sonnerup–Priest annihilation case by the choice $\alpha = -1$, $\beta = \gamma = 0$. Priest *et al.* (2000) succeeded in generalizing the choice (11.26) and (11.25), still reducing the problem to a set of ordinary differential equations.

A similar model was also formulated and solved in curvilinear coordinates, which substantially enlarged the class of configurations covered by this technique (Tassi *et al.*, 2002, 2003).

11.2.3 Simplified picture of reconnection

Our next step is to consider more general cases with two-dimensional magnetic fields, which have a structure as shown in Fig. 11.1. This is possible, however, only at the expense of drastic simplifications. Quantities in the outer inflow region will be characterized by their magnitudes at the point where the positive z-axis crosses the boundary, and are labelled by the subscript zero. The subscripts '1' and '2' refer to the centre inflow and outflow points on the boundary of the diffusion region and the subscript 'nl' is used for quantities on the neutral line. The symmetry properties are listed in the figure caption.

Further, let us assume that derivatives with respect to x are small compared with derivatives with respect to z and that $|B_z| \ll B_0$. Pressure is treated as a constant p_0 in the external region. These assumptions allow us to derive a set of simplified relations.

The condition of incompressibility (11.4) is applied to the diffusion region, which, with the help of Gauss's theorem, gives as a rough approximation

$$v_1 \Delta = v_2 \delta . \tag{11.28}$$

The z-component of momentum balance (11.2) at $x = 0$ provides the pressure balance across the upper half of the diffusion region

$$p_1 + \frac{B_1{}^2}{2\mu_0} = p_{\text{nl}} , \tag{11.29}$$

and the x-component of momentum balance (11.2) evaluated at $z = 0$, ignoring the small contribution from B_z, gives

$$\frac{\rho_2 v_2{}^2}{2} + p_2 = p_{\mathrm{nl}} \, . \tag{11.30}$$

From Ohm's law (11.3) applied to the outer and inner inflow points, to the outflow point and to the origin, one finds

$$E_0 = v_0 B_0 = v_1 B_1 = v_2 B_2 = \eta j_{\mathrm{nl}} \tag{11.31}$$

and *Ampère's law* (11.5) (replacing differentials by differences) gives

$$j_{\mathrm{nl}} = \frac{B_1}{\mu_0 \delta} \, . \tag{11.32}$$

By taking into account that $\rho_2 = \rho_1 = \rho_0$ and $p_1 = p_2 = p_0$ and by eliminating p_{nl}, j_{nl} and E_0 one obtains four equations, which we understand as determining v_1, v_2, B_2 and δ as functions of B_1 and Δ. One finds

$$v_1 = v_{\mathrm{A1}} \sqrt{\frac{L}{\Delta}} \frac{1}{\sqrt{S_1}} \tag{11.33}$$

$$v_2 = v_{\mathrm{A1}} \tag{11.34}$$

$$B_2 = B_1 \sqrt{\frac{L}{\Delta}} \frac{1}{\sqrt{S_1}} \tag{11.35}$$

$$\delta = L \sqrt{\frac{\Delta}{L}} \frac{1}{\sqrt{S_1}} \, , \tag{11.36}$$

where $S_1 = \mu_0 v_{\mathrm{A1}} L / \eta$.

That an equation determining Δ is missing reflects the fact that a full solution of the problem was not achieved. In the classical models of Sweet (1958) and Parker (1963b) and of Petschek (1964) that gap was filled by additional assumptions. A brief description of these models follows, for details see Vasyliunas (1975).

11.2.4 Sweet–Parker and Petschek models

In the Sweet–Parker model the diffusion region is a thin extended structure such that Δ becomes of the order of L. For simplicity, let us set $\Delta = L$. The external region is assumed to be largely homogeneous such that approximately B_1 can be identified with B_0 and one can set $S = S_1 = S_0$ and $M_{\mathrm{A}} = M_{\mathrm{A1}}$. Under these conditions, (11.33) gives the reconnection rate as

$$M_{\mathrm{A}} = \frac{1}{\sqrt{S}} \, . \tag{11.37}$$

This is the well-known Sweet–Parker reconnection rate. In stellar atmospheres and space plasmas S_0 usually assumes large values (Table 9.1) so that the Sweet–Parker reconnection rate generally is regarded as too low to be relevant.

Petschek's model includes a shock wave in each quadrant (Fig. 11.6). Note that the magnetic field is deflected toward the normal direction of the shock, so that the shock is of the slow mode type (Section 3.9). Although a full self-consistent incorporation of the shock dynamics is outside the scope of the model, Petschek (1964) succeeded in deriving a lower limit for Δ given by

$$\Delta > L\frac{\pi}{8}\frac{(\ln S)^2}{S}. \tag{11.38}$$

If, as in the Sweet–Parker model, one ignores the distinction between outer and inner inflow region, one finds from (11.33)

$$M_{\mathrm{A}} < \frac{\pi}{8}\frac{1}{\ln S}. \tag{11.39}$$

This upper limit is generally known as the *Petschek reconnection rate*. For $S \gg 1$ it is considerably larger than the Sweet–Parker rate.

In contrast to the annihilation solutions described above, neither the Sweet–Parker model nor the Petschek model represent a complete solution of the MHD equation. Nevertheless these models have been extremely valuable for further developments in reconnection theory.

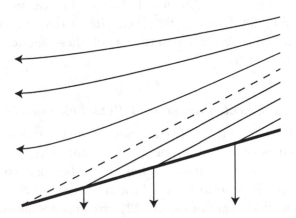

Fig. 11.6 Petschek's magnetic field configuration. The figure corresponds to the quadrant of positive x and z of Fig. 11.1 with the diffusion region contracted to a point; the thin lines are magnetic field lines, the broken line indicates the separatrix. A slow shock wave (thick line) leads to the characteristic weak magnetic field in the outflow region.

11.2.5 Modifications

The models by Sweet–Parker and Petschek triggered a considerable search for improvements. Here a brief, although incomplete, account is given of that development in a non-chronological order. (For more complete presentations see Priest and Forbes (2000) and Biskamp (2000).)

The identification of the upper limit given by (11.39) with the reconnection rate had the advantage that a rate faster than the slow Sweet–Parker rate became available. The disadvantage was that the two models gave conflicting answers to the same problem, which has caused a great deal of debate.

Several investigations produced results in favour of the Sweet–Parker regime. Fig. 11.7 shows, for example, a result Biskamp (1986) obtained by solving the equations (11.7) and (11.8) numerically for a particular set of boundary conditions. The solutions demonstrate that a Sweet–Parker current sheet rapidly develops for increasing S_0.

Kulsrud (2001) emphasized that the Petschek rate is only an upper limit and that the actual rate must come from a more detailed analysis. He managed to extract an extra condition from the steady state model equations. It states that the B_z-component in the outflow region to remain in a steady state requires that the supply of B_z from the diffusion region must balance the frozen-in down-sweeping of B_z in the outflowing plasma. As Kulsrud (2001) demonstrated, that condition demands that Δ/L must be of order 1, so that the Petschek reconnection rate would roughly agree with that of

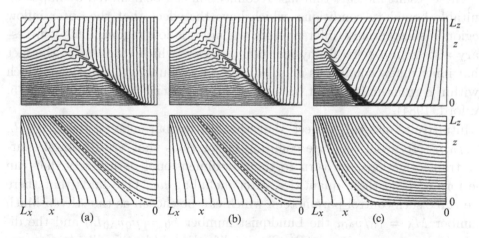

Fig. 11.7 Numerical solutions of (11.7) and (11.8) (Biskamp, 1986). The upper graphs show level curves of the stream function D, the lower graphs of the flux function A. In cases (b) and (c) the Lundquist number of case (a) is increased by factors of 2 and 4, respectively. (With permission from D. Biskamp.)

Sweet–Parker. In this picture the estimate (11.39) is formally correct, but the actual value is far below the upper limit. This is an interesting reconciliation of the two results. Yet, the resulting reconnection rate is still much too small to be of interest for typical applications.

Naturally, the search for measures that could enhance the reconnection rate has been a central topic. In other words, the aim is to identify regimes of *fast reconnection*, where *fast* means faster than the Sweet–Parker process.

As indicated in Fig. 11.6 the magnetic field strength of the Petschek model decreases as the origin is approached along the z-axis. In his detailed analysis, placing earlier work (e.g., Sonnerup, 1970; Yeh and Axford, 1970) in a new perspective, Vasyliunas (1975) pointed out that this effect reduces the maximum reconnection rate as compared with the opposite behaviour. This is based on the fact that a pressure decrease along the z-axis toward the origin in the inflow structure corresponds to a fast mode MHD expansion wave. By the same argument one expects an increase of the reconnection rate in the presence of a slow mode expansion. As an expansion shock does not exist (Section 3.9), this argument favours boundary conditions that generate an extended expansion wave structure.

Priest and Forbes (1986) presented a model in which they analysed the effect of slow and fast expansion in a systematic way. The Petschek model and the slow wave structure discussed above appeared as special cases. For sufficiently strong expansion they saw a flux pile-up phenomenon in the inner inflow region, which can cause a substantial increase of the reconnection rate.

These results indicate that fast reconnection can be achieved by imposing suitable boundary conditions, which have the effect that the outer inflow region is substantially different from the inner inflow region. This is necessary also from a general physical point of view which emphasizes the fact that inner inflow conditions must adjust to the boundary conditions, which (within certain limits) must be allowed to be prescribed (Vasyliunas, 1975; Axford, 1984).

Intuitively, one can expect that the control parameters are $v_0, B_0, L,$ ρ_0, p_0, η. To avoid separate counting of solutions that arise from similarity transformations, we turn to the non-dimensional parameters that can be formed from these quantities, after including μ_0. One finds that there are three such control parameters, which can be chosen as the Alfvén Mach number $M_A = v_0/v_{A0}$, the Lundquist number $S_0 = \mu_0 v_{A0} L \eta$ and the dimensionless pressure $2\mu_0 p_0/B_0^2$. To simplify the problem for illustration, we here follow the earlier practice to ignore the effect of the pressure. Then it must be possible to express normalized versions of all relevant quantities by

M_A and S_0. However, in view of the lack of information regarding Δ the parameter Δ/L appears in addition. Using (11.31) one finds from (11.33)–(11.36)

$$\frac{v_1}{v_{A0}} = \frac{B_2}{B_0} = \left(\frac{M_A}{S_0}\frac{L}{\Delta}\right)^{1/3} \tag{11.40}$$

$$\frac{v_2}{v_{A0}} = \frac{B_1}{B_0} = \left(M_A^2 S_0 \frac{\Delta}{L}\right)^{1/3} \tag{11.41}$$

$$\frac{\delta}{L} = \left(\frac{1}{M_A S_0^2}\frac{\Delta}{L}\right)^{1/3} . \tag{11.42}$$

Note that the Sweet–Parker regime is recovered by insisting on $B_1/B_0 = 1$, which for $\Delta/L = 1$ requires $M_A = 1/\sqrt{S_0}$.

It is of course still possible to choose boundary conditions in a way that the reconnection is slow. According to Priest and Forbes (1992a) this was the case in the simulations by Biskamp (1986). In fact, from his simulations he concluded that Δ/L scales as $M_A^4 S_0^2$, so that (11.40)–(11.42) give the scaling relations (see also Biskamp, 2000)

$$\frac{v_1}{v_{A0}} = \frac{B_2}{B_0} \sim \frac{1}{M_A S_0} \tag{11.43}$$

$$\frac{v_2}{v_{A0}} = \frac{B_1}{B_0} \sim M_A^2 S_0 \tag{11.44}$$

$$\frac{\delta}{L} \sim M_A , \tag{11.45}$$

which indicates the adjustment of the inner inflow conditions to the boundary conditions. However, the maximum reconnection rate, reached for $\Delta/L \approx 1$, still scales as $1/\sqrt{S_0}$, as in the Sweet–Parker case. Lee and Fu (1986) obtained scaling laws different from Biskamp's. Priest and Forbes (2000) argue that, again, the difference is due to different choices of boundary conditions.

The majority of the studies described so far use constant resistivity, the diffusion region being characterized by an increase of the current density alone. A further degree of freedom which influences the reconnection rate is the occurrence of a localized resistivity, which could be based on microturbulence causing collective transport (Section 9.3.2). For the Petschek model Biskamp (2000) argued that the lack of proper matching between external and diffusion regions for the classical constant resistivity case would be less serious for localized resistivity, probably due to the additional possibility of adjustment, so that fast Petschek-type reconnection might exist under these circumstances. Kulsrud (2001) finds that for the Petschek model in the presence of a collective resistivity, suitably depending on the current density (Section 9.3.2), the balance of B_z can be achieved in the regime $\Delta \ll L$, which would correspond to fast reconnection. These results are also consistent with simulations by Yan *et al.* (1992). Fast reconnection was studied also in a simulation by Ugai (1992) for different resistivity models.

The analytical models discussed above are based on crude approximations or they are focussed on particular aspects. As yet there is no fully satisfactory analytical model of steady state reconnection. An important step in that direction was made by Jamitzky and Scholer (1995), who treated configurations in the neighbourhood of annihilation structures by a boundary layer analysis for comparatively general boundary conditions.

The outflow ends of the diffusion region seem to have a complicated structure. The simulations by Biskamp (1993) have provided evidence indicating that a finite-length Sweet–Parker current sheet can have a structure similar to that of Syrovatskii's model given by (8.132).

11.2.6 Non-symmetric configurations

So far we have dealt with symmetric configurations (Fig. 11.1). Before considering more general cases let us tentatively superimpose a constant B_y-component on an otherwise planar field configuration. ($B_y e_y$ is called the *guide field*.) At first sight one might be misled to argue that this would not affect the structure of the solution in the x, z-plane. It is true that the general incompressible MHD equations with translational invariance in the y-direction (10.128)–(10.132) have exactly that property. However, the argument fails in the presence of shock waves. A characteristic feature of shocks is coplanarity (Section 3.9). It is easily seen that the superposition of a B_y-component would ruin that property. This does not contradict the general property of the resistive MHD equations mentioned above, because shocks by definition are compressive. (For discontinuities in an incompressible medium see the remark at the end of Section 3.9.) So, in the presence of shocks the poloidal structure of the configuration must change. The solution to this problem consists in the addition of discontinuities that are not subject to coplanarity. In the presence of a non-vanishing normal component B_n the only candidates are rotational discontinuities, which rotate the magnetic field in a required direction, so that the coplanarity of a subsequent shock wave is no longer a problem.

Heyn *et al.* (1988) have considered more general cases. They generalized Petschek's approach by studying the corresponding non-symmetric *Riemann problem*. The latter considers the decay of a plane tangential discontinuity into a system of MHD discontinuities and waves after introducing a finite B_n. The sequence of the discontinuities and waves is determined by the propagation speeds. In a Petschek-like reconnection configuration all discontinuities and waves originate from the diffusion region, which is assumed to be concentrated at the origin. The result is a wedge-shaped structure on each side of the diffusion region. If the z-components of v and B are assumed to remain small, the wedge remains narrow. For that case Heyn *et al.* (1988) find the steady state configuration illustrated in Fig. 11.8. The *reconnection layer* is the region bounded by the two rotational discontinuities.

The rotational discontinuities change the orientation of the magnetic field in the inflow regions so that the slow shock (upper part) or the *slow expansion wave* (lower part) can bring out changes in the magnetic field strength in the required coplanar geometry. The contact discontinuity provides a density jump required to match both inflow sides. The occurrence of a slow expansion wave instead of a slow shock on either side of the contact discontinuity depends on the inflow conditions on both sides. At the transition from a slow shock to a slow expansion wave the slow structure disappears. Fast shocks or fast expansion waves are not involved under the present circumstances. For a hybrid simulation of a non-symmetric reconnection layer see Nakamura and Scholer (2000).

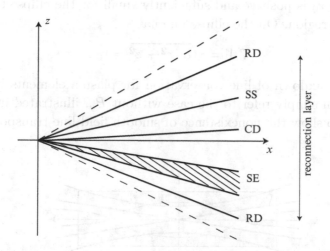

Fig. 11.8 A Petschek-like configuration for the general non-symmetric case after Heyn *et al.* (1988). The diffusion region, concentrated at the origin, is the source of two rotational discontinuities (RD) and either a slow shock (SS), shown in the upper part, or alternatively a slow expansion wave (SE), shown in the lower part, and a contact discontinuity (CD). The broken lines indicate the separatrices. For illustration, the slopes of the lines shown in the figure are exaggerated.

11.2.7 Reconnection with a guide field

The approach described in the previous section considered only the ideal region. Although a complete analytical approach is not available, here we concentrate on a particular problem that arises in the diffusion region when a guide field is superimposed.

The problem is that the guide field causes the neutral point, which is a dominant feature of the reconnection models without guide field, to disappear. On the one hand, one might expect that without a neutral point reconnection cannot operate, on the other hand, one might argue for physical continuity as a small guide field is superimposed. As we will see, continuity wins. We will show that the breakdown of line conservation is unaffected by the guide field.

So, let us consider a steady state configuration with translational invariance with respect to y but with a non-vanishing guide field ($B_y \neq 0$). The poloidal magnetic field ($B_x e_x + B_z e_z$) has a hyperbolic neutral point located at the origin (Fig. 11.9), surrounded by a diffusion region and let E_y be different from zero. Near the neutral point the flux function of the poloidal field has the form

$$A = \frac{1}{2}(\alpha x^2 - z^2)\,, \tag{11.46}$$

where α is a constant with $0 < \alpha < 1$. An ellipse $\alpha^2 x^2 + z^2 = r_0^2$ is placed at $y = 0$, where r_0 is positive and sufficiently small for the ellipse to lie inside the diffusion region. On the ellipse one has

$$|\nabla A| = \sqrt{\alpha^2 x^2 + z^2} = r_0\,. \tag{11.47}$$

For the breakdown of line conservation for plasma elements in the ideal region we can simply refer to the case without B_y, illustrated in Fig. 11.3. It remains to show the nonexistence of smooth field line transport. For the

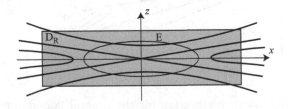

Fig. 11.9 Poloidal magnetic field structure with a hyperbolic neutral point at the origin and diffusion region D_R. A non-vanishing B_y-component (guide field) is superimposed (not shown). Field lines crossing the ellipse E (located at $y = 0$) connect the diffusion region with the ideal region.

purpose of an indirect proof, let us assume that there is a smooth velocity w^* and a smooth scalar field λ satisfying (Section 3.8.2)

$$\nabla \times (w^* \times B) = \lambda B \,, \tag{11.48}$$

with $w^* = v$ and $\lambda = 0$ in the ideal region. As all field lines, except the field line coinciding with the y-axis, connect to the ideal region, the property $B \cdot \nabla \lambda = 0$ together with continuity of λ gives $\lambda = 0$ everywhere. Scalar multiplication of (11.48) with ∇A yields $B \cdot \nabla (w^* \cdot \nabla A) = 0$, so that $w^* \cdot \nabla A$ is constant on all field lines. All field lines intersecting the ellipse E indicated in Fig. 11.9 connect D_R with the ideal region, so that $w^* \cdot \nabla A = E_y$ on those field lines.

Let θ be the angle between w^* and ∇A. Then, with the help of (11.47), we find that on the ellipse

$$|w^*| = \frac{|E_y|}{r_0 |\cos \theta|} \geq \frac{|E_y|}{r_0} \tag{11.49}$$

holds. This means that $|w|$ is unbounded as r_0 is chosen arbitrarily small. This contradicts the assumed existence of a smooth velocity field w^*. Note that the argument is similar to that used in the case $B_y = 0$ (Section 11.2.1). The only difference is that here we had to deal with the fact that there is one field line that stays inside the diffusion region.

We conclude that in two-dimensional steady state configuration with non-vanishing guide field as defined above, breakdown of magnetic line conservation occurs in qualitatively the same way as without B_y, so that large-scale effects occur. The two conditions (Section 11.1) for reconnection are satisfied.

11.2.8 Energy conversion

For a discussion of energy conversion let us return to the case $B_y = 0$. In fluid theories the energy balance is obtained by scalar multiplication of the momentum equation with the flow velocity v (see Section 3.8). Accordingly, one finds from (11.2)

$$v \cdot \nabla \left(\frac{1}{2} \rho v^2 + p \right) = v \cdot j \times B \tag{11.50}$$

and, using (11.3)–(11.6), energy balance takes the form

$$\nabla \cdot \left(\frac{1}{2} \rho v^2 v + pv + \frac{1}{\mu_0} E \times B \right) + \eta j^2 = 0, \tag{11.51}$$

or, with the help of Gauss's theorem,

$$\oint_{\partial d} \left(\frac{1}{2}\rho v^2 \boldsymbol{v} + p\boldsymbol{v} + \frac{1}{\mu_0}\boldsymbol{E} \times \boldsymbol{B} \right) \cdot \boldsymbol{n}\,\mathrm{d}l = -\int_d \eta j^2\,\mathrm{d}x\,\mathrm{d}z \qquad (11.52)$$

for an arbitrary domain d with boundary ∂d and normal $\boldsymbol{n}$ in the x, z-plane.

In comparison with the steady state version of the energy balance (3.79) of ideal MHD, (11.52) includes ohmic dissipation, but there is no internal energy reservoir as a result of incompressibility. (In the resistive compressible case (3.68) would ensure that ohmic heat goes into internal energy.)

Applying (11.52) to the diffusion region, one finds that in view of the assumption $p_1 = p_2$ and (11.28) the pressure term drops out and the leading term for large S_1 is

$$\int_{\mathrm{in}} \frac{1}{2}\rho v^2 |\boldsymbol{v}\cdot\boldsymbol{n}|\,\mathrm{d}l = \int_{\mathrm{out}} \frac{1}{\mu_0}\boldsymbol{E}\times\boldsymbol{B}\cdot\boldsymbol{n}\,\mathrm{d}l + \int_d \eta j^2\,\mathrm{d}x\,\mathrm{d}z \;, \qquad (11.53)$$

where all three terms are of the same order of magnitude. Thus, the incoming energy flow separates into two parts of equal order, one of which is ohmically dissipated, the other is transferred into energy of directed flow. So, in the diffusion region a substantial transfer from magnetic to kinetic energy takes place. In a more refined description the ohmic heating would lead to an increase of pressure.

In cases where the diffusion region is not extended over the entire system $(\Delta/L < 1)$ the energy conversion taking place inside the diffusion region concerns only a fraction of the plasma elements. A globally relevant effect would require energy conversion in the ideal region as well. Here details depend on the structure of the external solution. An obvious general requirement is a non-vanishing current density $\boldsymbol{j}$, as the transfer generally is provided by the work of the $\boldsymbol{j} \times \boldsymbol{B}$ force (see (11.50)). The current may be either extended or localized in thin structures, such as the slow shock waves in the case of Petschek's model or a current sheet that develops along the magnetic separatrix (Biskamp, 1986).

11.2.9 Layers of parallel flow

Configurations as shown in Fig. 11.1 emphasize the velocity components $\boldsymbol{v}_\perp$ perpendicular to the magnetic field. Here we address parallel flows and their possible concentration in layers.

Let us consider the magnetic field as being prescribed. Then, using the model (11.2)–(11.6), one can determine the perpendicular velocity from the

y-component of (11.3) and the parallel velocity from (11.4). In the ideal region these equations reduce to the simple system

$$v_\perp \cdot \nabla A = E_y \tag{11.54}$$

$$\nabla \cdot v = 0 , \tag{11.55}$$

where E_y is a constant.

Writing v as $v_\parallel B/B + v_\perp$, one finds from (11.54) and (11.55)

$$v_\perp = \frac{E_y}{B^2} \nabla A \tag{11.56}$$

$$\frac{v_\parallel}{B} = \left(\frac{v_\parallel}{B}\right)_0 - E_y \left(\frac{\partial V(A,s)}{\partial A} + \frac{\nabla A \cdot \nabla s}{B^3} - \left(\frac{\nabla A \cdot \nabla s}{B^3}\right)_0 \right) , \tag{11.57}$$

where s is the distance along field lines and the subscript zero refers to an arbitrary point on the field line. $V(A, s)$ is the flux tube volume

$$V(A,s) = \int_{s_0}^{s} \frac{\mathrm{d}s'}{B(A,s')} . \tag{11.58}$$

The derivation of (11.57) is analogous to that of the parallel current (5.42) when specialized for two-dimensional fields.

A layer of pronounced parallel flow can develop if $\partial V/\partial A$ becomes large in a limited interval of magnetic flux. The lower sketch of Fig. 11.10 illustrates

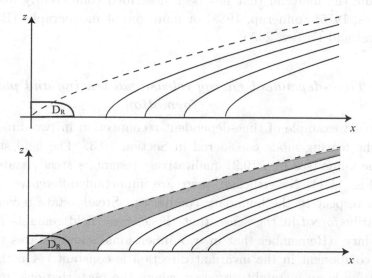

Fig. 11.10 A layer of parallel flow (shaded area in the lower graph) arises from a rapid change of the flux tube volume in the ideal region outside the diffusion region D_R. In the absence of such change no layer forms (upper graph).

this effect. The frozen-in flow, after crossing the separatrix from above, is squeezed into narrower flux tubes. Incompressibility requires adjustment by enhanced parallel flow. The effect is particularly strong in the flux tubes just below the separatrix. Formally, $V(A, s)$ diverges at the separatrix in an x-type neutral point geometry. Although the diffusion region has a regularizing effect, the parallel flow is still large in a layer just below the separatrix. Qualitatively this effect may also be present on the upper side of the separatrix in Fig. 11.10; however, it is less efficient because of the wider angle formed by the separatrices on the inflow side.

The presence of discontinuities would modify this simple picture; however, as the qualitative geometrical argument would still apply, layers of fast parallel flow might still form for a sufficiently fast decrease of the flux tube volume in the outflow region. If the diffusion region is a finite-length Sweet–Parker current sheet, parallel velocity layers form as long as there is a separatrix connecting to a neutral point, which, for instance, applies to a Syrovatskii layer.

Strong parallel flows that seem to be consistent with this picture have been observed in simulations (e.g., Fig. 11.7). Also, this effect has been applied to the plasma sheet in the Earth's magnetotail to explain the observed boundary layer flow (Schindler and Birn, 1987).

There are many more details available in the literature about steady state reconnection. We do not pursue this topic further here. There is no point in repeating the material that has been described competently in articles (Vasyliunas, 1975; Sonnerup, 1988) or more recent monographs (Biskamp, 2000; Priest and Forbes, 2000).

11.2.10 Time-dependent energy release via tearing and plasmoid formation

An important example of time-dependent reconnection in two-dimensional fields is the tearing mode considered in Section 10.3. The field structure around the x-line (see Fig. 10.9) qualitatively resembles steady state reconnection (Fig. 11.1). Nevertheless, there are important differences, particularly with respect to the boundary conditions. Steady state reconnection is necessarily *forced* in the sense that the electric field remains finite at the boundary. (Remember that in two-dimensional steady states the electric field component in the invariant direction is constant.) On the other hand, tearing is an instability process, where the perturbations, including the electric field component $-\partial A/\partial t$ in the invariant direction, vanish at the boundary. In that sense tearing is a *spontaneous* process. Note that this does not exclude that external forces drive the system towards the point of onset of the instability, which usually has dynamical properties distinctly different from the slow driver. The situation is similar to phase transitions in equilibrium thermodynamics, where quasi-static changes drive the system toward the point at which the phase transition takes place spontaneously. In some cases, processes can be regarded as spontaneous even if the boundary is partially open. For instance, some of the energy released by the instability may flow out of the system through the open parts of the boundary.

The tearing mode seems to be an important ingredient in the unloading of energy that originally was accumulated in stretched magnetic field configurations. This process involves the formation of a plasmoid (or several plasmoids). Plasmoids were introduced in Section 10.3.7, see Fig. 10.12. Here we consider that process in a configuration which contains an x-type neutral point already before the plasmoid starts forming.

Let us return to the case where the field is y-invariant with B_y set to zero. Suppose energy has been accumulated in a region of closed flux bounded by two separatrix surfaces that intersect in a neutral line as illustrated qualitatively in panel a of Fig. 11.11. The energy input can occur in different ways. One possibility is energy flow through the left boundary, which may be relevant for the solar corona. Another input mode is reconnection at the neutral line, which is widely believed to happen in the Earth's magnetotail.

If the left boundary is completely closed, the crucial role of reconnection for the supply of mass, energy and magnetic flux is best illustrated by considering the absence of reconnection. For illustration let us assume a plasma described by ideal MHD. Then, the flux function $A(x, z, t)$ describing the (time-dependent) magnetic field in the x, z-plane is constant on the separatrix by the following argument. For points $\xi(t), \eta(t)$ on the separatrix the value of the flux function $A_s(t)$ is given by

$$A(\xi(t), \eta(t), t) = A_s(t) . \tag{11.59}$$

Differentiating that expression with respect to t, specializing for the neutral point and applying Ohm's law ($\partial A/\partial t = 0$ at the neutral point) gives $dA_s/dt = 0$. So, the neutral point stays on the same field line, field lines being identified by their value of A.

This property immediately excludes the transfer of magnetic flux across the separatrix. The closed magnetic flux is given by $A_s - A_b$, where A_b is the flux function of the field line just touching the left boundary. By the closure condition, A_b is a constant, so that the constancy of A_s fixes the closed flux to a constant.

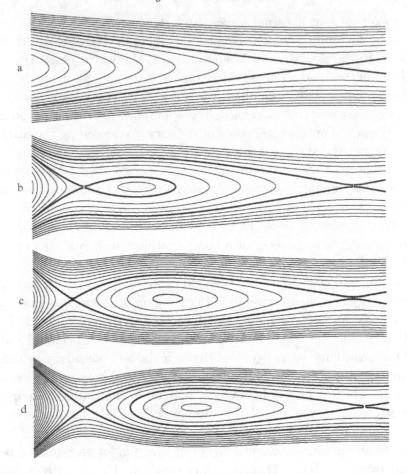

Fig. 11.11 Qualitative sketch of plasmoid formation and ejection in a stretched magnetic field configuration.

It is equally simple to show that there is no mass flux across the separatrix. Differentiating (11.59) for an arbitrary point on the separatrix gives

$$v_s \cdot \nabla A + \frac{\partial A}{\partial t} = 0 \,. \qquad (11.60)$$

Comparing with Ohm's law formulated for the same point gives

$$(v - v_s) \cdot \nabla A = 0 \,, \qquad (11.61)$$

so that there is no bulk flow across the separatrix.

A similar result is obtained for the energy. However, here the situation is complicated by the fact that, although energy flow through the separatrix is excluded, some energy can still be supplied by compression. To see this let us integrate the energy conservation law (3.79) over the closed flux region allowing for the boundary, the separatrix, to move with its velocity v_s. One finds

$$\frac{d}{dt} \int \left(\frac{\rho v^2}{2} + \frac{p}{\gamma - 1} + \frac{B^2}{2\mu_0} \right) d^3 r$$

$$= -\int \left(\frac{\rho v^2}{2} + \frac{p}{\gamma - 1} + \frac{B^2}{2\mu_0} \right) (v - v_s) \cdot n \, dl - \int \left(p + \frac{B^2}{2\mu_0} \right) v \cdot n \, dl \,,$$

$$(11.62)$$

where $n = \pm \nabla A / |\nabla A|$, the sign to be chosen so that n points outward, and the line integrals are to be evaluated on the separatrices; there is no contribution from the left boundary. The energy flux term (the first term on the right) vanishes, so that energy can be changed only by the compression work (second term).

In contrast to ideal MHD dynamics, magnetic reconnection, implying $\partial A / \partial t \neq 0$ and therefore $v - v_s \neq 0$ on the separatrix, would allow for increasing magnetic flux, and for mass and energy flux across the separatrix.

Since the energy cannot grow indefinitely, an unloading process must be at work also. In the case of resistive MHD reconnection, its irreversibility would exclude a simple reversal of the loading process as long as $|B_x|$ (outside the current sheet) dominates over $|B_z|$. Then the direction of the current density j_y is conserved and, by $E \cdot j > 0$ at the neutral line, the direction of E_y also. Reversal of reconnection would require a change of both the electric field direction and the current direction, which means that local steepening must cause a situation where $|\partial B_z / \partial x|$ becomes comparable with $|\partial B_x / \partial z|$, which does not occur in strongly stretched current sheet configurations. So, for stretched configurations the question arises of how the unloading takes place. As we will see, the unloading is based on a different reconnection process.

Since in the present stretched configuration the existing neutral line can only be used for input, a new neutral line forms inside the closed flux region. The standard way of achieving this is plasmoid formation via tearing (Section 10.3.7). Here the tearing mode picture is applicable as long as the tearing mode grows fast compared with the temporal changes occurring on large scales. This leads to the situation shown in panel b of Fig. 11.11. Reconnection at the inner (new) neutral line reduces the magnetic flux between the two separatrices while reconnection at the distant (seen from the left boundary) neutral line adds to that flux. But in view of the fast tearing process the net effect is a flux reduction. In other words, the two separatrices merge into one (panel c). Further reconnection, with the inner process still dominating, leads to a new topological structure (panel d). The plasmoid has left the closed flux region, which is relaxed, as indicated by the reduced stretching. The new neutral line now forms the end point of the closed flux region. Typically, in the final step (not shown in the figure) the plasmoid moves out of the system (to the right in panel d), driven by the outflow of continuing reconnection and/or by forces acting on the plasmoid.

The figure 11.11 does not represent a self-consistent solution of the reconnection problem. It is based on a sequence of flux functions that depends continuously on a parameter. So, the figure has only qualitative significance, but it illustrates the existence of a smooth transition between the field configurations of panels a and d.

The process of Fig. 11.11 has proven to be a basic key to the understanding of the unloading of closed flux regions stressed by energy input. In that scenario the build-up continues until a current sheet has formed that is sufficiently pronounced for a tearing instability to start. So, it is the tearing mode that initiates the unloading process.

This scenario can be tested by comparison with numerical simulation. A corresponding set of resistive MHD results obtained by Otto (1987) is shown in Fig. 11.12. The plasmoid forms spontaneously in the stretched configuration of panel a. The topological evolution is consistent with the one of Fig. 11.11.

Plasmoid formation and ejection occur in a similar form also when the outer neutral line is moved to infinity (Birn, 1980). For further 2D simulations with similar objectives see, e.g., Forbes and Priest (1983), Ugai (1982), Otto (2001) or Wiechen *et al.* (1997). Corresponding 3D simulations are discussed farther below.

Most empirical models addressing solar or magnetospheric activity involve plasmoids in one way or the other (see Part IV). Loss processes limiting plasma confinement in laboratory experiments have a similar topological evolution, although the geometry is different (Biskamp, 2000).

Regarding the reconnection process itself, time-dependent simulations have revealed several important nonlinear phenomena. The reconnection layer, generated by a tearing mode growing into its nonlinear regime, may become tearing unstable itself. This phenomenon is known as *secondary tearing*.

Two magnetic islands, such as realized by the cat's eye solution (5.131), can become unstable against *coalescence* into a single island (Biskamp and

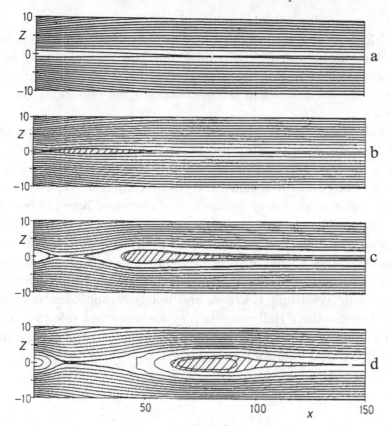

Fig. 11.12 Magnetic field lines of a resistive MHD simulation of plasmoid formation and ejection from Otto (1987), the resistivity is constant with $S = 100$. (With permission from A. Otto.)

Welter, 1980). There is an intermediate step with a thin current sheet forming between the two islands, driven by the attractive forces between the island currents. Reconnection at this current sheet initiates the merging of the two islands into one.

11.3 Kinetic reconnection in 2D collisionless plasmas

For collisionless plasmas the resistive MHD model has been useful for general qualitative orientation. For a detailed description of collisionless reconnection more refined models are necessary. The additional degrees of freedom, which these models typically offer, give rise to a picture that is more complex than the fluid picture considered so far.

Consider a two-dimensional collisionless quasi-neutral electron–proton plasma with the symmetry indicated in Fig. 11.1. The reconnection process typically develops its own B_y-component, but here we exclude a guide field from external sources unless stated otherwise. Note that structures on the Debye scale are not covered in this section (see the remark at the end of Section 3.2).

Both analytical estimates and simulations have led to the concept of a structured reconnection region (Shay *et al.*, 1998; Hesse *et al.*, 2001). The structure is shown in Fig. 11.13. The region in the centre is characterized by nongyrotropic electrons; it is called the *electron diffusion region* or *electron dissipation region* or *region of unmagnetized electrons*, here abbreviated as *electron region*. The electron region is embedded in the region where the ions are nongyrotropic (*ion region*). This region is larger than the electron region because larger magnetic fields are required to make the (heavier) ions gyrotropic. The part of the ion region that lies outside the electron region, characterized by gyrotropic electrons and nongyrotropic ions, is called the *Hall zone*, because there Hall currents play an important role (Section 3.6). Outside the ion region both species are gyrotropic and the plasma has the behaviour of an ideal fluid such as ideal MHD.

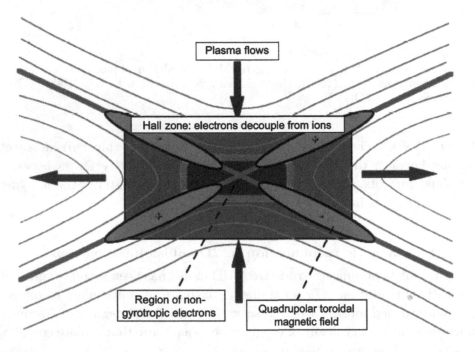

Fig. 11.13 The structured reconnection region of collisionless reconnection after Hesse *et al.* (2001) (by permission of the American Geophysical Union).

11.3.1 The Hall zone

The basic significance of the Hall zone lies in the fact that the ions are decoupled from the magnetic field while the electrons are not (e.g., Hesse *et al.*, 2001). Quasi-neutrality is established by a poloidal electric field. In a useful simplified picture the current is dominated by the $E \times B$ drift of the electrons, so that $j \approx -en\boldsymbol{v}_e$. The poloidal current density causes a B_y-component with a characteristic quadrupolar structure (Fig. 11.13), which in the present simplified picture can be seen as follows. The poloidal component of $\mu_0 \boldsymbol{j} = \nabla \times \boldsymbol{B}$ gives

$$\nabla B_y = \mu_0 \boldsymbol{e}_y \times \boldsymbol{j}_\mathrm{p} = -\mu_0 en \boldsymbol{e}_y \times \boldsymbol{v}_e . \tag{11.63}$$

Assuming a standard stagnation point pattern for $\boldsymbol{v}_e$, one finds that on any closed curve in the inner part of the Hall zone B_y has minima and maxima (Fig. 11.14), consistent with Fig. 11.13.

Although the ions can be considered as being essentially accelerated by the electric field, the momentum balance for the plasma as a whole has no electric force, so that, by self-consistency, the relevant electromagnetic force is the $\boldsymbol{j} \times \boldsymbol{B}$ force.

Applying mass continuity to the ion region, we find the same equation as obtained in the MHD case, (11.28), so that

$$v_1 = \frac{\delta_\mathrm{i}}{\Delta_\mathrm{i}} v_2 , \tag{11.64}$$

where the additional subscript i refers to the ion region. Again, as in the MHD case, by order of magnitude, $v_2 = v_{\mathrm{A}1}$, so that again (setting $\Delta_\mathrm{i} = \Delta$)

$$M_{\mathrm{A}1} = \frac{\delta_\mathrm{i}}{\Delta} . \tag{11.65}$$

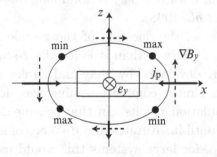

Fig. 11.14 Explanation of the quadrupolar structure of B_y. By (11.63) the expected structure of the poloidal current density $\boldsymbol{j}_\mathrm{p} = -en\boldsymbol{v}_{\mathrm{ep}}$ generates gradients of B_y, which lead to minima (min) and maxima (max) of B_y on closed curves enclosing the electron region (rectangle).

In the resistive MHD case the problem of slow reconnection arises from the small value of δ, which was found to be proportional to $1/\sqrt{S_1}$. Here the situation is much more favourable for fast reconnection because δ_i can be much larger than δ in the resistive case.

Let us estimate δ_i. In the nongyrotropic region the particles perform the exotic orbits described in Section 6.4. Assuming that E_z does not play the dominant role and that B_x increases linearly with z, one finds that an ion moving with $v_z = v$ at $z = 0$ is turned around by the magnetic field B_x at $z = d_i$, where d_i is the formal gyroradius evaluated at $z = d_i$. So one finds

$$\delta_i = \frac{m_i v}{e B_1} . \tag{11.66}$$

Frequently, δ_i is found to be of the order of c/ω_{pi} in apparent contrast to (11.66). However, if v in (11.66) is chosen as the Alfvén velocity v_{A1}, then δ_i does reduce to c/ω_{pi}. In fact, simulations have indicated that the ions during their inflow are strongly accelerated by a localized electric field E_z, forcing the ions to follow the electrons (Shay *et al.*, 1998). By this acceleration the ions reach a velocity of the order of v_{A1}, which would explain a sheet width of order c/ω_{pi}. (An equivalent picture has the ions oscillate in an electric potential well with scale c/ω_{pi}, which indicates that E_z plays a finite role but does not dominate the magnetic effect.)

It has been argued that the reason for $\delta_i = c/\omega_{pi}$ is the fact that c/ω_{pi} is the scale of the Hall physics. The scale length introduced by the Hall effect, e.g., obtained by equating the $\boldsymbol{v} \times \boldsymbol{B}$ term with the Hall term in Ohm's law, is $B/\mu_0 e n v$. Again, if v is of the order of the Alfvén velocity, this length becomes c/ω_{pi}.

Although, in principle, the ion thermal velocity would be a candidate for v in (11.66) also, simulation results have indicated that the ion temperature of the inflowing plasma does not play a dominant role for the structure of the ion region (Shay *et al.*, 1998).

As in the resistive case, the question of the x-scale Δ of the ion region is much more difficult to answer than it is for the z-scale. So it is remarkable that in the picture developed by Shay and Drake and coworkers (Shay *et al.*, 1998, 1999) an almost complete scaling was derived from analytical estimates and simulation results. In that scheme Δ increases with the macroscopic length L until it saturates at a distance of about $10\,c/\omega_{pi}$, when v_2 approaches v_{A1}. So, for large systems this would mean that by (11.65) the reconnection rate, measured by the M_{A1}, becomes a universal constant near 0.1. For sufficiently large L Petschek-like shocks form downstream of the separatrices. That the Rankine–Hugoniot relations (Section 3.9) are not satisfied quantitatively, has been attributed to the interaction of the shocked ion population with the ion beam ejected from the diffusion region.

As already mentioned, the simplest model of the ion region ignores the ion dynamics and uses Ohm's law in the form $E + v_e \times B = 0$. Then electron MHD applies (Section 3.6) and the currents are Hall currents. Accordingly, spatial structures are interpreted in terms of the quadratic dispersion of the whistler wave, characteristic for electron MHD. This interpretation was also applied to more refined modelling (Shay *et al.*, 1998). But is has been argued also, that ion kinetic effects can give fast reconnection even when whistler dispersion is switched off artificially (Karimabadi *et al.*, 2004).

Even if the scaling described above should turn out not to cover the entire parameter space of two-dimensional collisionless reconnection (Bhattacharjee *et al.*, 2005), it forms a valuable basis to build upon in future work.

Although the Hall zone dynamics plays an important role, it should be noted that without the presence of the electron region reconnection would not work, because the reconnection electric field E_y would not be balanced at the origin. Thus, it is of basic physical interest to address the electron region also.

11.3.2 The electron region

Let us assume that for the electrons it is their thermal velocity that determines the geometry of the nongyrotropic electron region. So, in analogy with (11.66) we obtain

$$\delta_e = \frac{m_e v_{te}}{e B_x(0, \delta_e)}, \quad \Delta_e = \frac{m_e v_{te}}{e B_z(\Delta_e, 0)}, \qquad (11.67)$$

where the brackets give the coordinates in the x, z-plane, here restricted to the first quadrant with $B_x(0, \delta_e)$ and $B_z(\Delta_e, 0)$ chosen positive. When needed, the implicit equations (11.67) are easily solved for δ_e and Δ_e, assuming linear variation of the magnetic field components.

From the discussion of Section 9.4 it follows that in the electron region the possible sources of nonidealness are electron nongyrotropy and electron inertia (see (9.58)). A simple estimate, based on waves, also indicated that inertia effects do not dominate if the relevant wave frequency is smaller than ω_{pe}. Adopting this criterion for the present problem it seems that we have to compare the characteristic time scale of the electron motion with $1/\omega_{pe}$. The (shortest) relevant length scale being δ_e, the timescale becomes $\tau = \delta_e/v_{te}$. This gives

$$\omega_{pe}\tau = \omega_{pe}\frac{\delta_e}{v_{te}} = \frac{\delta_e}{\lambda_D} \gg 1, \qquad (11.68)$$

because we are dealing with a quasi-neutral plasma on scales larger than the Debye length λ_D. Although applying the wave criterion to the reconnection problem is rather intuitive, it is still interesting to see that it favours thermal effects, here represented by the off-diagonal components of the pressure tensor. More importantly, this view is consistent with simulation results (Horiuchi and Sato, 1994; Hesse *et al.*, 1999; Ricci *et al.*, 2002). So, in the following let us concentrate on electron nongyrotropy. First, there is no B_y externally superimposed; later such a (guide) field will be included.

In neutral fluids the off-diagonal pressure tensor components describe viscous effects arising from velocity gradients. Here, at least formally, a similar situation arises. Let us consider the equation for the evolution of the pressure tensor (3.44), which provides useful expressions for the off-diagonal components (Kuznetsova *et al.*, 1998). For instance, one finds on the x-axis (upper index x)

$$P_{exy}^x =$$
$$- \left[\frac{P_{exx}}{\Omega_{ez}} \frac{\partial v_{ex}}{\partial x} + \frac{1}{2\Omega_{ez}} \left(P_{exx} \nabla \cdot \boldsymbol{v}_e + \left(\frac{\partial}{\partial t} + v_{ex} \frac{\partial}{\partial x} \right) P_{exx} + \nabla \cdot \mathcal{Q}_e|_{xx} \right) \right]^x$$

(11.69)

where the symmetry of Fig. 11.1 was assumed.

Except for the heat flux, the terms on the right side of equation (11.69) are linearly related to first order spatial derivatives of the flow velocity. As the interaction is not necessarily of genuinely viscous nature, Kuznetsova *et al.* (1998) speak of a *quasi-viscous* interaction. Although the first term in the square brackets of (11.69) alone was found useful for the interpretation of simulation results (Kuznetsova *et al.*, 1998), more detailed considerations have indicated that a fully satisfactory picture requires taking into account a contribution from the remaining terms (Hesse, 2005). Apart from details it seems clear that the off-diagonal pressure tensor components play an important role in supporting the reconnection electric field.

Let us add a brief remark on the case in which a guide field B_y is superimposed, which, in contrast to the quadrupolar B_y structure, is generated by external currents. A guide field is *strong* if $|B_y| \gtrsim |B_x(0,\delta)|$, so that the motion of a thermal electron would become significantly affected by the guide field. In that regime a relevant scale length is the electron gyroradius with respect to B_y, which becomes $\lesssim \delta_e$. Simulations (Hesse *et al.*, 2004) have indicated that on that scale electron pressure effects are still large enough to support the reconnection electric field, but heat flux is required to sustain the off-diagonal components of $\mathcal{P}_e$. The symmetry of the poloidal components of the magnetic field that exists for $B_y = 0$ (Fig. 11.1) is broken. Alternative suggestions for the physics of the electron region employ three-dimensional effects (Section 11.4).

In magnetospheric plasmas with small values of β, Scudder and Mozer (2005) found strong, highly localized electric field enhancements (EFE) perpendicular to B, leading to an effective demagnetization of the electrons. The authors suggest that this phenomenon plays an important role in supporting the reconnection electric field via electron pressure anisotropy.

11.3.3 Comparative simulation studies

The results of 2D reconnection discussed so far suggest that fast reconnection processes crucially depend on the presence of an ion region. The ion region seems to lead to a fast reconnection rate, essentially independent of the mechanism by which the frozen-in constraint is broken in the electron region. As discussed, there are also indications suggesting that the length Δ of the ion region (in the outflow direction) remains microscopic, i.e., for sufficiently large macroscopic length scales L it becomes independent of L. In models without an ion region, such as resistive MHD, the reconnection rate appears to depend on the electron dissipation process. In resistive MHD with constant resistivity a Sweet–Parker-type current sheet may form and the reconnection will be slow.

If this scenario was true, rates and other major reconnection features could be obtained from any model, provided it contains an ion region, such as Hall-MHD. This would mean a dramatic advantage, considering the cost of full particle simulations.

This aspect has been a major motive for performing studies in which a given reconnection problem is to be solved using a variety of different plasma models. The *GEM reconnection challenge* (see Birn *et al.* (2001) and references therein) compared results of full particle codes, hybrid codes with and without electron mass and the off-diagonal electron pressure tensor, Hall-MHD with and without electron inertia and MHD codes. The basic configuration was a thin Harris sheet with a width of $0.5 \, c/\omega_{pi}$ and a 20% background density and, for the models that include the electron mass, the mass ratio m_i/m_e was set to 25. The sheet is subject to an initial perturbation, which forms a magnetic island. The reconnection rate is given by the time derivative of the magnetic flux between the x- and o-points. The results well confirmed the expected properties described above. Fig. 11.15 shows the reconnected flux for four runs with different simulation models.

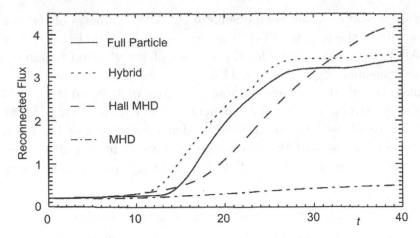

Fig. 11.15 GEM reconnection challenge: The reconnected magnetic flux versus time for four different simulation models (from Birn *et al.* (2001) by permission of the American Geophysical Union).

The models that include an ion region show very similar slopes over a wide time interval. The reconnection is fast with a rate of the order of $0.2\,v_{A0}B_0$, where v_{A0} is the Alfvén speed corresponding to the maximum magnetic field strength B_0 and sheet centre density. The agreement of reconnection rates is contrasted by substantial differences in the geometrical structure of the electron region and several other aspects. The resistive MHD case, with a constant resistivity and Lundquist number of 0.005, shows a distinctively smaller rate. The expected Sweet–Parker sheet forms for sufficiently small resistivities. MHD models achieved fast reconnection with localized resistivity, the maximum of S_0 being of order 1.

Although the linear phase, dominated by the tearing mode, shows considerable differences in the growth rates (manifested by the time shifts of the curves in Fig. 11.15), these differences do not develop further in the nonlinear regime.

The GEM challenge did not address the question of how the thin current sheet that was present in the start configuration was formed. Therefore, the current sheet formation was included in a second cooperative study, which grew out of a workshop held in 2004 at the Isaac Newton Institute at Cambridge, England, and has become known as the *Newton challenge*. Here the initial sheet was wider by a factor of 4 than in the GEM challenge and the reconnection process was initiated by a temporally limited, spatially varying inflow of magnetic flux. The forced thinning initiates the reconnection process.

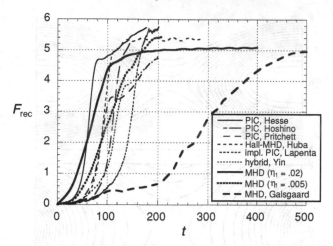

Fig. 11.16 Newton reconnection challenge. The reconnected magnetic flux versus time for four different simulation models; here η_1 is the Lundquist number of Birn's MHD simulations, reproduced from Birn *et al.* (2005) by permission of the American Geophysical Union.

Again, the particle studies and the Hall-MHD studies show very similar reconnection rates (see Fig. 11.16). The final amount of reconnected flux is also about the same with one exception, which has been explained as being caused by a somewhat smaller initial deformation. (Preliminary results suggest that both the reconnection rate and the final reconnected flux are influenced by the imposed perturbation.) Again, the MHD rates depend on the resistivity, although the deviations from the other rates are smaller than in the GEM case.

The final configurations are qualitatively similar in all cases, an example is shown in the upper panel of Fig. 11.17. Interestingly, the final state resembles an equilibrium solution (lower panel) obtained by solving the Grad–Shafranov equation for the boundary conditions reached after the flux inflow was shut off and for fixed pressure function $p(A)$ (see Section 6.2.2). This problem generally has more than one solution, the figure shows the one that minimizes free energy (Chapter 12). A substantial amount of the initial magnetic energy has been released.

11.4 Inclusion of the third dimension

The two-dimensional models as discussed so far apply to reconnection occurring in three-dimensional space only if the y-dependence is small and if the extent of the reconnection region along the y-direction is large enough for

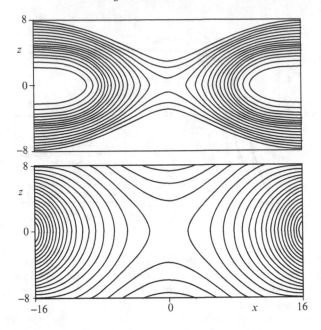

Fig. 11.17 Newton reconnection challenge: the final state; the upper panel shows the final state of a particle simulation by M. Hesse, the lower panel the equilibrium solution with the lowest (free) energy (reproduced from Birn *et al.* (2005) by permission of the American Geophysical Union).

edge effects to be negligible. In spite of these restrictions, two-dimensional reconnection research has proven to be a necessary intermediate step in the approach to the enormously complicated world of three-dimensional reconnection.

For a given property the inclusion of the third dimension may or may not lead to qualitatively new effects. Here we will give a few selected examples. In some cases the structure becomes genuinely three-dimensional without an obvious 2D analogy.

11.4.1 *Role of microinstabilities*

Typical current driven microinstabilities (Section 9.3.1) have a wave vector component in the current direction and therefore are excluded in two-dimensional reconnection studies. So, how is the picture changed by including the third dimension? The existing studies convey the impression that there is no rich ensemble of microinstabilities leading to a completely different picture over a broad range of parameters, but microinstabilities do seem to play an important role under a variety of particular circumstances.

At present a complete description of the role of microinstabilities is not yet available, so that we can only convey general impressions resulting from analytical studies and simulations. The following brief assessment is largely guided by a recent more detailed survey by Büchner and Daughton (2006) on this subject, emphasizing the role of microinstabilities in the plasma sheet of the Earth's magnetotail.

Microinstabilities can become important for magnetic reconnection in a direct way by providing a collective nonidealness that supports the reconnection electric field in Ohm's law (Section 9.3.2). Another possibility is that an unstable mode influences reconnection indirectly through nonlinear interaction with the reconnection dynamics.

In the following we comment on a number of instabilities that are considered as prime candidates for being relevant for the onset of reconnection.

Let us begin with the ion-acoustic instability. For the standard case of a two-component plasma with shifted Maxwellian distribution functions the ion-acoustic instability can be excited only in the regime where $T_e > T_i$, otherwise the instability is inhibited by strong Landau damping (Section 9.3.2). As magnetospheric studies are focussed on current sheets with $T_e < T_i$, ion acoustic modes do not seem to play a significant role in corresponding reconnection processes. Note that in the plasma sheet in the Earth's magnetotail the electron temperature is smaller than the ion temperature by a factor of about 5 (see Table 2.1). In principle, strong deviations from the Maxwellian might reduce the damping, although the relevance of this effect seems unclear. It is also an open question whether the conclusions regarding Landau damping also apply to strong ion acoustic turbulence, where the quasi-linear approach, on which they are based, is no longer valid.

Different forms of the *kink instabilities* exist in the MHD and Vlasov regimes. The ideal MHD kink mode typically occurs in cylindrical plasma columns, leading to a wavy structure of the column along its axes (Section 10.2.4). Highly stretched planar configurations such as the magnetospheric plasma sheet are MHD stable (Section 10.2.5). Observed oscillations (Sergeev *et al.*, 2004) are therefore likely to be of non-MHD nature. Their role in reconnection, however, is unclear. Vlasov versions are based on the relative speed between ions and electrons (*drift-kink mode*) or on ion–ion interaction (*ion–ion kink mode*) in the presence of a cold background plasma. This case is considered to be relevant for the magnetosphere in view of the hot plasma sheet and the cold lobe plasma. For the drift-kink mode, although showing significant effects in simulation studies with artificial ion–electron mass ratios near and below 100, the growth rate is drastically reduced in the regime of realistic mass ratios (Daughton, 1999).

From full particle simulations Karimabadi *et al.* (2003, 2004) found that in the presence of a 20% background density the ion–ion mode grows to significant amplitudes, but the kink and tearing modes seem to coexist without significant interaction. Correspondingly, during the final stages the current sheet reconnection process is similar to the corresponding two-dimensional case. A wavy structure can also arise from a Kelvin–Helmholtz instability in the nonlinear stage of the reconnection dynamics (Lapenta *et al.*, 2003).

The lower-hybrid drift instability is exceptional in that it is not subject to strong damping for $T_e < T_i$. However, it is sensitive to β ($= 2\mu_0 p/B^2$), such that the instability is inhibited for $\beta > 1$ (Section 9.3.2). Considering that in the centre of the magnetotail plasma sheet β typically is in the range of 10–100, this instability does not exist in the centre, i.e., in the region where it would be required if it was to support the reconnection electric field via a corresponding collective resistivity (Section 9.3.2). This has led to a general scepticism regarding the role of this instability as a reconnection-relevant source of collective resistivity under magnetotail conditions. Long-wavelength perturbations seem to penetrate the centre of the high-beta current sheet only for extremely thin sheets with thicknesses substantially below r_{gi} (evaluated with the external magnetic field).

However, it has been suggested also that there is an indirect way in which the lower-hybrid drift mode might influence reconnection. Even if the mode is localized to the edge of the current sheet (where $\beta < 1$), nonlinear interaction might modify the current distribution and lead to anisotropic electron heating, which in turn may enhance the effect of the tearing mode (Daughton *et al.*, 2004).

It should be noted that the large-β argument does not necessarily apply to configurations with a sufficiently strong guide field. There, the lower-hybrid drift instability has a good chance to play a significant role in current sheet reconnection (e.g., Silin and Büchner, 2005). However, there are also cases where, after some initial adjustments, in a fully three-dimensional system with a strong guide-field, reconnection develops to an almost translationally invariant state, where magnetic perturbations are aligned primarily along the main current flow direction (Hesse *et al.*, 2005b).

Further instabilities that have been suggested to be of interest for reconnection are *modified two-stream instabilities* (Lui, 2004) and modes that are symmetric with respect to the central plane of the current sheet, such as *sausage* modes. However, more work seems required to assess their part in reconnection processes. In particular, it seems important to take into account the non-locality of kinetic modes in thin current sheets (Büchner and Daughton, 2006).

11.4.2 3D plasmoid formation

For magnetospheric purposes the plasmoid formation described in Section 11.2.10 has been generalized to three dimensions. In their 3D resistive MHD computation Birn and Hones (1981) took into account a cross-tail flaring, however keeping $B_y = 0$. In the central plane ($y = 0$) the reconnection process showed a striking similarity to the two-dimensional case (Fig. 11.12). A near-Earth neutral line forms, large plasma flows develop and a rapidly growing plasmoid moves tailward. The shape of the neutral lines, however, is more complex than in the 2D case.

This trend becomes more pronounced when a non-vanishing B_y-component is taken into account. Using a 3D MHS equilibrium based on the technique described in Section 5.4.3, resistive MHD computations by Hesse and Birn (1991b) confirmed the plasmoid formation process. Fig. 11.18 shows the magnetic field structure in the x, z-plane. The plasmoid receives its tailward momentum mainly from the inflow of moving plasma through its boundary.

The magnetic field structure becomes much more involved than in cases with vanishing B_y, the reason being that the magnetic field lines inside the plasmoid become helical and leave the plasmoid at its ends, there connecting to field lines with earthward or tailward orientation. Fig. 11.19 illustrates three types of field line connections; *closed* lines have both sides, *half-open* lines one side and *open* lines no side connected with the Earth.

Determining the field line type by tracking the lines from their crossing points in the plane $z = 0$, Hesse and Birn (1991b) found that the connectivity changes rather rapidly with the crossing location (Fig. 11.20). For $B_y \to 0$ the number of crossings of a plasmoid field line becomes large and the scale of the connectivity variation in the x, z-plane becomes smaller and smaller, tending to a chaotic behaviour.

The involved magnetic field structure has raised questions about the role of topology in 3D reconnection (Hughes and Sibeck, 1987). We return to this problem in Section 11.5.

11.4.3 Flux linkage

During the nonlinear evolution of 3D fields new reconnection configurations may form that do not have direct two-dimensional counterparts. A typical example is linkage of magnetic flux tubes. Figure 11.21 gives an example from a resistive MHD simulation (Otto, 1995). The figure shows a snapshot of selected flux tubes. Reconnection releases the tension in the linked

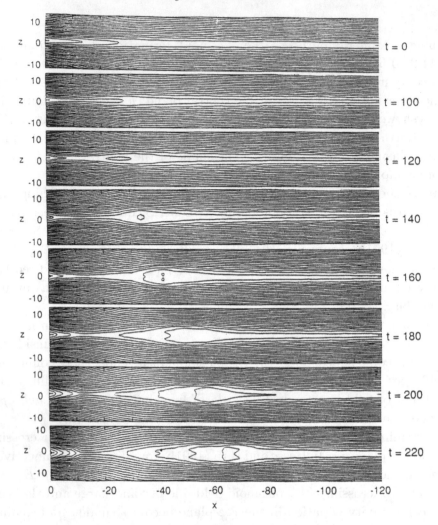

Fig. 11.18 The magnetic field structure in the central plane $y = 0$ of a 3D resistive MHD simulation (Hesse and Birn, 1991b). As B_y does not vanish, the lines are not field lines, they are based on B_x and B_z in the central plane (reproduced from Hesse and Birn (1991b) by permission of the American Geophysical Union).

flux tubes. The reconnection site is characterized by a strong electric field component parallel to $\boldsymbol{B}$. Linkage of flux tubes can arise from the interaction of multiple reconnection patches on an originally plane sheet separating sheared magnetic fields. This is meant to apply to the magnetopause of the Earth's magnetosphere, where the interplanetary magnetic field comes in close contact with the geomagnetic field.

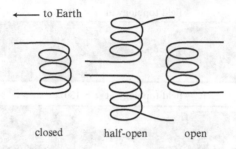

Fig. 11.19 External connection of magnetic field lines of a plasmoid (helical field line sections) in the Earth's magnetotail for $B_y \neq 0$.

11.4.4 Magnetic flux emergence

On the Sun, reconnection of magnetic fields of different origin occurs, where new flux emerges into the corona from below (Heyvaerts *et al.*, 1977). Fig. 11.22 is a sketch of a corresponding planar configuration. Galsgaard *et al.* (2005) studied flux emergence by a 3D MHD simulation. Fig. 11.23 illustrates some of the features observed after onset of the reconnection. A concentrated current sheet with the shape of an arch is formed in the contact region. The reconnection is accompanied by considerable plasma heating. It generates high-speed plasma outflows, which propagate as jets along the ambient magnetic field lines. The reconnection causes significant changes in the field connectivity.

11.5 Kinematics of 3D reconnection

Important properties of magnetic reconnection can adequately be discussed in terms of a *kinematic* description, which takes into account only the homogeneous subset of Maxwell's equations and Ohm's law

$$\nabla \times \boldsymbol{E} = -\frac{\partial \boldsymbol{B}}{\partial t} \tag{11.70}$$

$$\nabla \cdot \boldsymbol{B} = 0 \tag{11.71}$$

$$\boldsymbol{E} + \boldsymbol{v} \times \boldsymbol{B} = \boldsymbol{R} , \tag{11.72}$$

where $\boldsymbol{R}$ is the unspecified nonideal term of Ohm's law. Clearly, any general kinematic result will hold in any plasma model that includes the kinematic equations (11.70)–(11.72) as a subset of the model equations. Important examples of kinematic properties are the validity or the breakdown of magnetic flux and line conservation (Section 3.8.2).

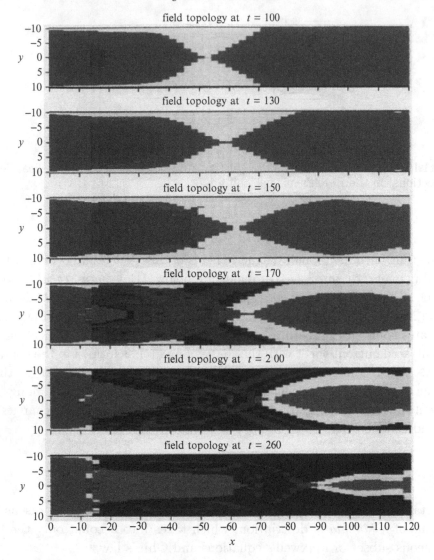

Fig. 11.20 Connectivity of field lines in a 3D resistive MHD magnetotail model
(Hesse and Birn, 1991b). The different connection types for plasmoid field lines
with multiple crossing points (Fig. 11.19) and non-plasmoid field lines (one crossing
point) are indicated by shading differences. For details one should consult the
original colour figure. (Reproduced from Hesse and Birn (1991b) by permission of
the American Geophysical Union.)

In this section we discuss reconnection from a general 3D kinematic point
of view. As the starting point a consideration of magnetic line conservation will demonstrate that two basic classes of configurations have to be
distinguished.

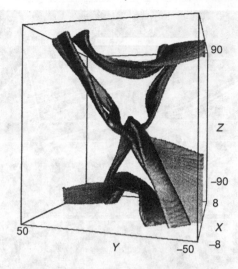

Fig. 11.21 Magnetic reconnection of linked flux tubes from a resistive MHD simulation (from Otto (1995) by permission of the American Geophysical Union).

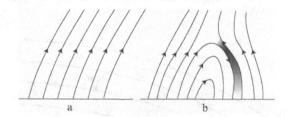

Fig. 11.22 A simple planar sketch of magnetic flux emergence into the corona. Magnetic field lines are shown before (a) and after (b) new magnetic flux has emerged. As a potential site of reconnection a current sheet (shaded) forms where antiparallel fields are pushed against each other.

Let us assume that a region of nonidealness is contained in a singly connected compact region D_R, outside of which $\mathbf{R}$ vanishes (Fig. 11.24). All field lines connect to the ideal region. The criterion for a smooth transport velocity $\mathbf{w}^*$ to be line conserving is (see Section 3.8.2)

$$\nabla \times (\mathbf{E} + \mathbf{w}^* \times \mathbf{B}) = \lambda \mathbf{B}, \tag{11.73}$$

where λ is a smooth scalar field. (Remember that the * symbol signifies coincidence with $\mathbf{v}$ in the ideal region.) Taking the divergence of (11.73) gives $\mathbf{B} \cdot \nabla \lambda = 0$, so that λ is constant on field lines. Since in the present case all field lines connect to the ideal region, λ must vanish everywhere, so that the criterion (11.73) becomes

$$\nabla \times (\mathbf{E} + \mathbf{w}^* \times \mathbf{B}) = 0. \tag{11.74}$$

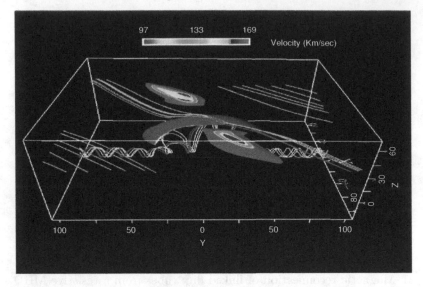

Fig. 11.23 Magnetic reconnection caused by magnetic flux emergence into the corona (Galsgaard *et al.*, 2005); features are explained in the text. (Reproduced by permission of the AAS.)

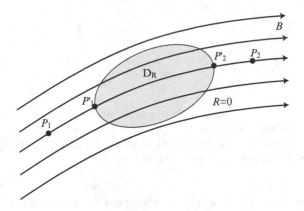

Fig. 11.24 In the present 3D kinematic picture the nonidealness is contained in a bounded singly-connected region D_R; the marked points lie on the same field line, P_1, P_2 in the ideal region on opposite sides of D_R and P'_1, P'_2 on the boundary.

For orientation, let us first consider the case in which

$$E_\parallel = \boldsymbol{E} \cdot \boldsymbol{B}/B = 0 \text{ and } B \neq 0 \text{ everywhere in the system.} \qquad (11.75)$$

This implies that $\boldsymbol{R} \cdot \boldsymbol{B} = 0$, so that $\boldsymbol{R}$ can be written as $\boldsymbol{B} \times (\boldsymbol{R} \times \boldsymbol{B})/B^2$. Then (11.74) is satisfied with the choice

$$\boldsymbol{w}^* = \boldsymbol{v} + \frac{\boldsymbol{R} \times \boldsymbol{B}}{B^2} . \qquad (11.76)$$

This transport velocity is smooth because with $B \neq 0$ there is no singularity; also (11.76) satisfies the requirement that $\boldsymbol{w}^* = \boldsymbol{v}$ in the ideal region where $R = 0$. So, under the conditions (11.75) the plasma velocity is line-conserving. There is deviation from the frozen-in motion in the ideal region, but there is no large-scale effect that could be associated with a reconnection process.

This leaves the following regimes as possible candidates for reconnection:

(a) $B \neq 0$ everywhere and there exist points with $E_\parallel \neq 0$; $\qquad$ (11.77)

(b) there exist points with $B = 0$. $\qquad$ (11.78)

These cases will be discussed in subsequent sections. The general framework that includes both candidates has been termed *general magnetic reconnection, GMR* (Schindler *et al.*, 1988; Hesse and Schindler, 1988).

11.5.1 Finite-B reconnection

For the discussion of case (11.77) we largely follow Hesse and Schindler (1988). For finite B this approach defines a general class of reconnection-relevant kinematic processes and it offers a simple expression for the reconnection rate in terms of the electric field component parallel to $\boldsymbol{B}$. As in 2D cases, breaking of magnetic line conservation is the central property (Axford, 1984).

Basic aspects of the approach

Let $\boldsymbol{B}$ be represented by Euler potentials, $\boldsymbol{B} = \nabla \alpha \times \nabla \beta$. This choice requires the absence of recurrent field lines (see Section 5.1.2). Using the vector potential $\boldsymbol{A} = \alpha \nabla \beta$, one can solve (11.70) by setting (Stern, 1970)

$$E = -\frac{\partial A}{\partial t} - \nabla \phi \qquad (11.79)$$

$$= -\frac{\partial \alpha}{\partial t} \nabla \beta + \frac{\partial \beta}{\partial t} \nabla \alpha - \nabla \psi , \qquad (11.80)$$

where $\psi = \phi + \alpha \, \partial \beta / \partial t$. Using α, β, s as coordinates, where s denotes the arclength along field lines, the three covariant components of Ohm's law (11.72) yield

$$\frac{d\alpha}{dt} = -\frac{\partial \psi}{\partial \beta} - R_\beta \qquad (11.81)$$

$$\frac{d\beta}{dt} = \frac{\partial \psi}{\partial \alpha} + R_\alpha \qquad (11.82)$$

$$-\frac{\partial \psi}{\partial s} = R_s = E_\parallel , \qquad (11.83)$$

where $d/dt = \partial/\partial t + \boldsymbol{v} \cdot \nabla$ and $\boldsymbol{R} = R_\alpha \nabla \alpha + R_\beta \nabla \beta + R_s \nabla s$.

Let us integrate (11.83) at a fixed time along a given field line between points P_1 and P_2 both lying in the ideal region on the same field line but on opposite sides of the nonideal region D_R (Fig. 11.24) and define the quantity

$$U(\alpha, \beta) = -\int_{P_1}^{P_2} E_\parallel(\alpha, \beta, s)\,\mathrm{d}s = -\int_{P_1'}^{P_2'} E_\parallel(\alpha, \beta, s)\,\mathrm{d}s\,, \qquad (11.84)$$

where the time-dependence is suppressed. Since the ideal sections on the field line do not contribute to the integral, the integration interval can be replaced by the interval between the points P_1', P_2' located on the boundary of D_R (Fig. 11.24), so that U is a function of α and β alone. Then, from (11.83) it follows that

$$U(\alpha, \beta) = [\psi]_{\alpha\beta}\,, \qquad (11.85)$$

where the bracket $[q]_{\alpha\beta}$ of a quantity q denotes the difference $q(P_2') - q(P_1')$ on the field line α, β. Further, one finds from (11.81) and (11.82)

$$\left[\frac{\mathrm{d}\beta}{\mathrm{d}t}\right]_{\alpha,\beta} = \frac{\partial U}{\partial \alpha} \qquad (11.86)$$

$$\left[\frac{\mathrm{d}\alpha}{\mathrm{d}t}\right]_{\alpha,\beta} = -\frac{\partial U}{\partial \beta} \qquad (11.87)$$

implying

$$\frac{\partial}{\partial \alpha}\left[\frac{\mathrm{d}\alpha}{\mathrm{d}t}\right]_{\alpha,\beta} + \frac{\partial}{\partial \beta}\left[\frac{\mathrm{d}\beta}{\mathrm{d}t}\right]_{\alpha,\beta} = 0\,. \qquad (11.88)$$

These results allow us to investigate breaking of line conservation for 3D time-dependent fields. Consider the case where U does not vanish identically. As U vanishes for field lines that are tangential to D_R, there must exist field lines for which $\partial U/\partial \alpha$ or $\partial U/\partial \beta$ or both are different from zero. By (11.86) and (11.87), the same holds for $[\mathrm{d}\alpha/\mathrm{d}t]_{\alpha,\beta}$ and $[\mathrm{d}\beta/\mathrm{d}t]_{\alpha,\beta}$. So, two plasma elements on opposite sites of D_R that share a field line with this property at some time t (Fig. 11.24) will be found on different field lines after an arbitrarily small time interval. Thus, the presence of the local nonideal region with U not identically vanishing has the *global effect* that line conservation is broken.

The breakdown of line conservation involves non-vanishing values of $\mathrm{d}\alpha/\mathrm{d}t = \partial \alpha/\partial t + \boldsymbol{v}\cdot\nabla\alpha$ or of $\mathrm{d}\beta/\mathrm{d}t = \partial \beta/\partial t + \boldsymbol{v}\cdot\nabla\beta$ or both. So, there are contributions from time-dependence and from the plasma motion. Fig. 11.25 illustrates the two limiting cases of plasma motion in a time-independent field (a) and plasma elements at rest in a localized time-dependent magnetic field (b). In both cases the plasma elements that originally shared

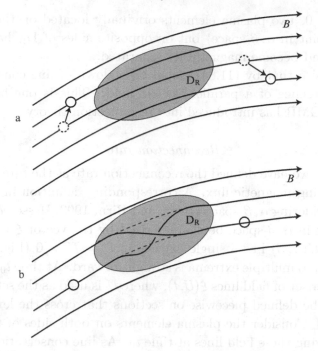

Fig. 11.25 Breakdown of magnetic line conservation due to a localized nonideal region (shaded) with $U \neq 0$. Two limiting cases are those of a time-independent electromagnetic field with plasma elements moving to different field lines (a) and localized time-dependence of the field with non-moving elements. In both cases the elements sharing a field line initially end up on different lines. Broken lines refer to initial states. See also Hesse *et al.* (2005a).

a field line end up on different field lines and by (11.86) and (11.87) both cases require non-vanishing U on the field lines involved. The general case combines both effects.

Another way of looking at the breakdown of line conservation uses a representation in α, β-space, a plane with Cartesian coordinates α, β (Hesse and Birn, 1993). There the dynamics of plasma elements in the ideal region is governed by Hamiltonian equations of motion obtained from (11.81) and (11.82) for $\boldsymbol{R} = 0$

$$\frac{\mathrm{d}\alpha}{\mathrm{d}t} = -\frac{\partial \psi}{\partial \beta}, \quad \frac{\mathrm{d}\beta}{\mathrm{d}t} = \frac{\partial \psi}{\partial \alpha} \tag{11.89}$$

where $\psi(\alpha, \beta)$ is the Hamiltonian. That Hamiltonian, however, is different on both sides of the nonideal region, because (11.83) or (11.85) give

$$\psi_2(\alpha, \beta, t) - \psi_1(\alpha, \beta, t) = U(\alpha, \beta, t) . \tag{11.90}$$

Thus, if $U \neq 0$, two plasma elements originally located on the same field line (same point in α, β-space) but on opposite sides of D_R have different trajectories and become magnetically separated.

We emphasize that by (11.84) global breakdown of line conservation requires the presence of a parallel electric field. This is one of the major aspects that GMR has introduced into reconnection theory.

Reconnection rate

In the 2D case we have defined the reconnection rate as the rate of change of the reconnecting magnetic flux. A corresponding definition in 3D is found by again considering α, β-space (Hesse and Birn, 1993; Hesse *et al.*, 2005a).

Let a point in α, β-space be characterized by the vector $\boldsymbol{\xi} = (\alpha, \beta)$, and assume that $U(\alpha, \beta)$ has a single extremum $U = \hat{U} \neq 0$ (Fig. 11.26); the generalization to multiple extrema is straightforward. At time t_0 we choose a one-parameter set of field lines $\boldsymbol{\xi}(U, t)$, where U is used as the set parameter. ($\boldsymbol{\xi}(U, t)$ is to be defined piecewise on sections that cross the level curves of U only once.) Consider the plasma elements on both sides of the nonideal region occupying these field lines at time t_0. As line conservation is broken, the function $\boldsymbol{\xi}(U, t)$ will develop differently on the different sides of D_R, so that we have to distinguish between $\boldsymbol{\xi}_1(U, t)$ and $\boldsymbol{\xi}_2(U, t)$, where the subscript refers to the side of D_R. By assumption

$$\boldsymbol{\xi}_1(U, t_0) = \boldsymbol{\xi}_2(U, t_0) \tag{11.91}$$

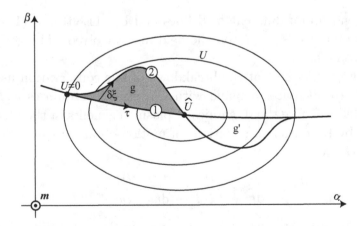

Fig. 11.26 Reconnection in α, β-space. The thin lines are level curves of U, the lines marked as 1 and 2 represent a set of plasma elements on opposite sides of D_R at time $t_0 + \delta t$ that started on the same field lines at time t_0. The reconnection rate is obtained from the rate of change of the shaded area g.

holds. The quantities $\boldsymbol{\xi}_1$ and $\boldsymbol{\xi}_2$ define lines in α, β-space, which coincide at $t = t_0$. At time $t_0 + \delta t$ these lines have moved to positions labelled as 1 and 2 in Fig. 11.26. The original line (corresponding to $t = t_0$, not shown in the figure) has been arranged in such a way that if line 1 (i.e., $\boldsymbol{\xi}_1(U, t_0 + \delta t)$) passes through the point where $U = \hat{U}$, then line 2 must also pass through that point. Before establishing this property, let us define the reconnection rate R_{rec} as $|\mathrm{d}g/\mathrm{d}t|$, where g is the shaded area in Fig. 11.26. Since areas in the α, β-plane correspond to magnetic fluxes in configuration space (Section 5.1.2), that area is an appropriate measure of the reconnected magnetic flux. As we are interested in the reconnection rate at time t_0, and therefore in the limit $\delta t \to 0$, it suffices to consider only terms in lowest significant order in δt. Taking (11.91) into account, one then finds

$$\frac{\mathrm{d}g}{\mathrm{d}t} = \frac{1}{\delta t} \int \boldsymbol{m} \cdot \boldsymbol{\tau} \times \delta\boldsymbol{\xi}\, \mathrm{d}\sigma, \qquad (11.92)$$

where $\boldsymbol{m}$ is a unit vector perpendicular to the α, β-plane, which allows us to use the vector product to define a (differential) area, and $\mathrm{d}\sigma$ is the arclength element. Note that $\boldsymbol{\tau}$ is the unit tangential vector on line 1. The integration is extended from the point where $U = 0$ to the point where $U = \hat{U}$ (Fig. 11.26). The vector $\delta\boldsymbol{\xi}$ is given by

$$\delta\boldsymbol{\xi} = \boldsymbol{\xi}_2(U, t_0 + \delta t) - \boldsymbol{\xi}_1(U, t_0 + \delta t)$$

$$= \left[\frac{\mathrm{d}\boldsymbol{\xi}}{\mathrm{d}t}\right]_{t_0} \delta t$$

$$= -\nabla_{\alpha\beta} U \times \boldsymbol{m}\delta t, \qquad (11.93)$$

where equations (11.86) and (11.87) were used in the last step. Using this expression in (11.92), we obtain (Hesse and Birn, 1993)

$$\frac{\mathrm{d}g}{\mathrm{d}t} = \int \nabla_{\alpha\beta} U \cdot \boldsymbol{\tau}\, \mathrm{d}\sigma \qquad (11.94)$$

$$= \int_0^{\hat{U}} \mathrm{d}U = \hat{U}, \qquad (11.95)$$

so that

$$R_{\text{rec}} = |\hat{U}| = \max \left| \int E_\parallel\, \mathrm{d}s \right|. \qquad (11.96)$$

This expression is consistent with the rate $|E_y|$ of stationary 2D states; note that the 2D reconnecting flux is the corresponding 3D flux per unit length in the y-direction.

It remains to justify the crossing of the curves 1 and 2 at $\hat{U}$; this immediately follows from (11.93).

The reconnection rate could also have been based on the area g' in Fig. 11.26 with the same result, because $g + g' = 0$. (Note that the area in α, β-space, as defined here, carries a sign.) Although Euler potentials are not gauge invariant themselves, the reconnection rate is an invariant quantity.

The derivation given above can be carried out in configuration space by using the original kinematic equations. However, this is a three-dimensional problem and considerably more complicated than the present approach which profits from the reduction to a planar problem. For steady state reconnection the derivation in configuration space simplifies somewhat. A detailed analysis of an example was presented by Hornig and Priest (2003).

Note that the relative velocity $[d\boldsymbol{\xi}/dt]$ in α, β-space is directed parallel to the level curves of U, as follows from (11.93). The resulting rotational transport typically translates to configuration space. For instance, in Fig. 11.25a the two plasma elements perform a relative rotation and in the time-dependent case (b) the internal magnetic field typically develops a twist. Among others, this effect was analysed by Hornig and Priest (2003) in their model. The figure 11.27 shows the rotation of fluid elements on opposite sides of the nonideal region represented by the field lines on which they are instantaneously located. Employing a picture where the ideal field line transport is extrapolated into D_R from both sides, the authors visualize that process as a *splitting* of flux tubes.

Hesse *et al.* (2005a) constructed models for both limiting cases. Fig. 11.28 illustrates the mechanism of their time-dependent model, which addresses reconnection in the solar corona. The process is similar to the case of a 3D plasmoid in the Earth's magnetotail, where transitions between the field topologies shown in Fig. 11.19 occur. A corresponding change of topology has been suggested for coronal mass ejections (Gosling *et al.*, 1995).

A further important aspect that the present approach offers is that the integral $\int E_\parallel \, ds$ can be estimated from a known reconnection rate. For a sufficiently conducting ionosphere the ionospheric plasma is frozen into the geomagnetic field rotating with the Earth. Yet, the plasma in the tail does not participate in that rotation. An estimate of the corresponding reconnection rate gives $\int E_\parallel \, ds$ in the kV range. Similar conditions can be expected to apply to other astrophysical objects (Schindler *et al.*, 1991). For example, this concept provides values of U that could be relevant for solar particle acceleration.

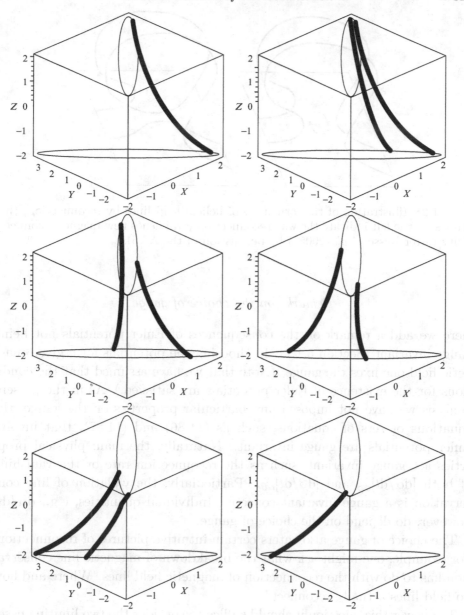

Fig. 11.27 Relative rotation of plasma elements located on opposite sides of the nonideal region, represented by the field lines on which they are instantaneously located (Hornig and Priest, 2003). (Reproduced with permission from G. Hornig. Copyright 2003, AIP.)

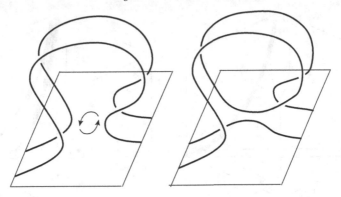

Fig. 11.28 Illustration of the formation of helical field lines by reconnection; the arrows on the left indicate the way reconnection establishes new magnetic connectivity (after Hesse *et al.* (2005a) by permission of the AAS).

Remarks on the choice of gauge

Here we add a remark on the consequences of Euler potentials not being gauge invariant (Section 5.1.2). In choosing the potentials for a given magnetic field one fixes the gauge. (Note that we have assumed that the conditions for the existence of Euler potentials are satisfied.) As in the present analysis we have not imposed any particular properties on the gauge, the equations or sets of equations, such as (11.86) and (11.87), that involve Euler potentials are gauge invariant. Naturally, the main physical properties are gauge invariant, such as the reconnection rate or the vanishing of both $[d\alpha/dt]_{\alpha\beta}$ and $[d\beta/dt]_{\alpha\beta}$. Particularly, the violation of line conservation is a gauge invariant concept. Individual quantities, e.g., $d\alpha/dt$, however, do depend on the choice of gauge.

The choice of gauge also enters certain intuitive pictures of reconnection. For example, one might ask what the breakdown of magnetic line conservation has to do with the reconnection of magnetic field lines. Where and how do field lines actually reconnect?

To answer this question it should suffice to consider the two limiting cases of Fig. 11.25, which we related to the two contributions from $d/dt = \partial/\partial t + \boldsymbol{v}\cdot\nabla$, applied to α and β. Already here we encounter the first gauge problem: this distinction is not gauge invariant. Even a time-independent magnetic field can be represented by time-dependent Euler potentials, for example, by just adding functions of time. Here the essential property is that it is possible to find time-independent potentials if and only if the magnetic field is time-independent. In case (a) we have tacitly made use of that possibility, choosing potentials with $\partial\alpha/\partial t = 0$, $\partial\beta/\partial t = 0$. In case (b) the plasma velocity was set to zero, which is a gauge invariant condition.

Conspicuous reconnection of field lines occurs in case (b); due to the internal time-dependent evolution the connectivity of the external field line sections changes.

In case (a) there is no such change. Instead the plasma elements move to different field lines. However, this is the result of having chosen a time-independent gauge. If, instead, on each side separately one chooses a gauge for which the field lines are frozen into the plasma motion, then the assignment of field line labels α and β to the field line on each side of D_R is different and changes with time, so that the connectivity changes, too. In this way the reconnection of field lines becomes conspicuous also in case (a). It should be noted, however, that this notion does not necessarily reflect a real physical process, because it exists only for a particular choice of gauge. What counts physically is the reconnection rate, which is the same for all gauges.

Although the frozen-in picture outlined above is quite popular in reconnection physics, it has the complication that the field line identification inside D_R is no longer unique. So the internal field line identification depends on the potential assignment that is used, that of side 1 or that of side 2. This complication led some authors to choose an artificial surface separating D_R into two parts. On each side the frozen-in gauge is used up to that surface; at the surface a jump of the field identification occurs, which visualizes the reconnection.

Another possibility is to accept the double identification throughout D_R, which implies that an originally chosen field line with matching identification splits into two field lines, which perform a relative rotation. Figure 11.27 gives an example of field line splitting and rotation.

Although these pictures seem attractive from an intuitive point of view, they are less useful for formal developments. At least one should be aware that one describes an electromagnetic field in two conflicting gauges. For any choice of a uniformly valid gauge, the appropriate (formal and intuitive) signature of reconnection is the violation of line conservation in the way we have described it in the previous section.

11.5.2 Reconnection based on magnetic nulls

Here we turn to the second class of candidates for 3D reconnection, characterized by (11.78). For three-dimensional magnetic fields, isolated points where B vanishes play a role analogous to the neutral points in two-dimensional fields with vanishing component in the invariant direction. It is common to refer to such a point as a *null point* or simply as *null*. The structure of magnetic fields in the vicinity of nulls has been studied thoroughly (Cowley, 1973; Greene, 1988; Lau and Finn, 1990; Parnell *et al.*, 1996; Priest and Titov, 1996). In the following we give a brief overview and add a few remarks on kinematic properties.

The magnetic field near a null point

Assuming that the magnetic field has a Taylor expansion with respect to the radius vector $\boldsymbol{r}$ originating at the null point, to lowest order the field has the form

$$\boldsymbol{B} = \mathcal{M} \cdot \boldsymbol{r} \,, \tag{11.97}$$

where the tensor $\mathcal{M}$ is given by

$$\mathcal{M} = \nabla \boldsymbol{B}|_{\boldsymbol{r}=0} \,. \tag{11.98}$$

The condition $\nabla \cdot \boldsymbol{B} = 0$ requires that the trace of $\mathcal{M}$ vanishes,

$$\mathrm{Tr}(\mathcal{M}) = 0 \,. \tag{11.99}$$

Of particular interest are nulls that are structurally stable. This property implies that the null cannot be removed by the superposition of an arbitrary perturbation (e.g., Lau and Finn, 1990). For a small perturbation $\boldsymbol{b}$, which can be considered as constant in the vicinity of the null, this means that the null is shifted to a new location $\boldsymbol{r}'$ determined by

$$\mathcal{M} \cdot \boldsymbol{r}' + \boldsymbol{b} = 0 \,. \tag{11.100}$$

So, structural stability requires that (11.100) has a solution $\boldsymbol{r}'$ which means that

$$\det(\mathcal{M}) \neq 0 \,. \tag{11.101}$$

Invariant properties of $\mathcal{M}$ are appropriately expressed in terms of its eigenvalues γ_j and eigenvectors $\boldsymbol{\rho}_j$, i.e., the solutions of the eigenvalue problem

$$\mathcal{M} \cdot \boldsymbol{\rho} = \gamma \boldsymbol{\rho} \,. \tag{11.102}$$

In view of (11.99) and (11.101) the tensor $\mathcal{M}$ has three non-vanishing eigenvalues with their sum vanishing. Since $\mathcal{M}$ is real, the eigenvalues are either real or conjugate complex pairs. So there must be two eigenvalues, γ_1, γ_2, with the same sign of their real parts and a third eigenvalue, γ_3, must be real and must have the opposite sign. For concreteness let us assume that $\gamma_3 < 0$; the cases with $\gamma_3 > 0$ are generated by reversal of the field direction.

There are field lines and flux surfaces that separate field regions of different topological structure (analogous to the separatrices in 2D fields). These are the axis defined by $\boldsymbol{\rho}_3$, called *spine*, and the plane defined by $\boldsymbol{\rho}_1$ and $\boldsymbol{\rho}_2$, called *fan*. (Note that in the case of complex γ_1 and γ_2, the complex eigenvectors still define two preferred real directions, again a consequence of $\mathcal{M}$ being real.)

In a frame of reference that has the spine oriented along the z-axis, after suitable normalization and rotation in the x, y-plane, the component matrix of $\mathcal{M}$ assumes the general form (Parnell *et al.*, 1996)

$$M = \begin{pmatrix} 1 & (q - j_\mathrm{s})/2 & 0 \\ (q + j_\mathrm{s})/2 & b & 0 \\ 0 & j_\mathrm{n} & -1 - b \end{pmatrix} , \tag{11.103}$$

where j_s is the current density component in the spine direction and j_n the component normal to the spine. The conditions $4b - q^2 + j_\mathrm{s}^2 > 0$ and $b > -1$ must be satisfied to ensure that the spine lies on the z-axis and that the inflow occurs along the spine ($\gamma_3 < 0$). For vanishing perpendicular current the spine is perpendicular to the fan. If $j_\mathrm{s}^2 < q^2 + (b - 1)^2$, all eigenvalues are real. For $\gamma_1 \neq \gamma_2$ the fan lines emanate from the null tangent to one of the eigenvector directions, and in the degenerate case $\gamma_1 = \gamma_2$ all fan lines are straight lines emanating from the null. If $j_\mathrm{s}^2 > q^2 + (b - 1)^2$, γ_1 and γ_2 are conjugate complex and the fan lines spiral out of the null.

Away from the null point, where the lowest approximation (11.97) becomes invalid, the spine generally continues as a curved field line and the fan as a curved flux surface.

The case that in several respects has the simplest structure is characterized by the choice $\boldsymbol{j}_\mathrm{s} = 0$, $q = 0$, $b = 1$, so that

$$\boldsymbol{B} = x\boldsymbol{e}_x + y\boldsymbol{e}_y - 2z\boldsymbol{e}_z , \tag{11.104}$$

or in cylindrical coordinates r, θ, z

$$\boldsymbol{B} = r\boldsymbol{e}_r - 2z\boldsymbol{e}_z . \tag{11.105}$$

This field can be generated by Euler potentials

$$\alpha = -r^2 z, \quad \beta = \theta, \tag{11.106}$$

it has rotational symmetry about the spine axis and the fan lines are straight. Note that in spite of the discontinuity of β the magnetic field $\boldsymbol{B} = \nabla\alpha \times \nabla\beta$ remains smooth. Fig. 11.29 shows the flux surface $\alpha = 0.1$. The spine is the

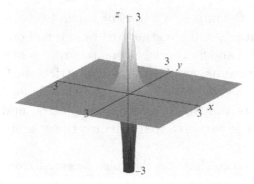

Fig. 11.29 Flux surfaces $\alpha = 0.1$ of a magnetic null (located at the origin), represented by the potentials (11.106).

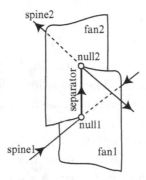

Fig. 11.30 A configuration with two nulls; they are connected by the separator, the intersection line of the fan surfaces.

z-axis. The surfaces rapidly approach the fan (x, y-plane) from above and below as $r = \sqrt{x^2 + y^2}$ increases.

Sufficiently structured magnetic fields may have more than one null point. In the case of a pair of nulls a generic configuration consists of a null with spine inflow and one with spine outflow (Lau and Finn, 1990). There is a topologically preferred field line, the *separator*, that connects the nulls, being a fan line of both nulls (Fig. 11.30). A simple example has the magnetic field (Priest and Forbes, 2000)

$$\boldsymbol{B} = x(z - 3)\boldsymbol{e}_x + y(z + 3)\boldsymbol{e}_y + (1 - z^2)\boldsymbol{e}_z \, , \tag{11.107}$$

the nulls are located on the z-axis at $z = \pm 1$.

A pair of nulls can be generated by superimposing a homogeneous magnetic field on a dipole field, when the dipole axis is oblique to the homogeneous field (Cowley, 1973). This case is a simple model of a planetary dipole field located in the large scale interplanetary field.

Reconnection in fields with null points

A smooth ideal plasma flow cannot cross the fan nor the spine of a null. In other words, if one forces a flow to cross the fan or the spine, it cannot remain smooth. We illustrate this property for the example of (11.104).

Let us assume ideal steady state flow. Under these conditions the electric field has a potential ϕ that is constant on field lines, so that we can understand ϕ as a function of α and β. Then we obtain for the perpendicular plasma velocity $\boldsymbol{v}_\perp = \boldsymbol{E} \times \boldsymbol{B}/B^2$

$$v_\perp = \frac{\partial \phi}{\partial \alpha} r e_\theta + \frac{\partial \phi}{\partial \beta} \frac{1}{r^2 + 4z^2} \left(\frac{2z}{r} e_r + e_z \right) . \tag{11.108}$$

A simple example of a flow that crosses the fan plane $z = 0$ has the potential $\phi = \phi_0(\theta)$, where $\phi_0(\theta)$ is 2π-periodic but arbitrary otherwise. This gives (Priest and Forbes, 2000)

$$v_\perp = \frac{d\phi_0}{d\theta} \frac{1}{r^2 + 4z^2} \left(\frac{2z}{r} e_r + e_z \right) . \tag{11.109}$$

Except at the null the plasma crosses the fan with $v_\perp = d\phi_0/d\theta (1/r^2) e_z$. The velocity becomes singular along the spine ($r = 0$), so that here the ideal dynamics breaks down.

Interpreting this breakdown as evidence for reconnection, this case has been termed *spine reconnection*. Note, however, that the present simple model has no current so that a finite resistivity would not cause resistive diffusion. Craig and Fabling (1996) have solved the resistive MHD equations for a model that includes currents and found that resistive diffusion can indeed support spine reconnection.

The second example has flow crossing the spine, with the choice $\phi = \sqrt{-\alpha} \sin \beta$, valid for $z > 0$. In Cartesian coordinates this gives

$$v_\perp = \frac{1}{x^2 + y^2 + 4z^2} \left(\frac{y^2 + 4z^2}{2\sqrt{z}} e_x - \frac{xy}{2\sqrt{z}} e_y + x\sqrt{z} e_z \right) . \tag{11.110}$$

On the spine the perpendicular velocity is $1/(2\sqrt{z}) e_x$ and it becomes singular at the fan. This case has been termed *fan reconnection*. Craig and Fabling (1996) have verified that resistive diffusion can regularize the flow field of fan reconnection as well.

The singularities of the ideal flow have a rather simple explanation. Let us consider spine reconnection. In the absence of a B_θ-component the velocity component v_z crossing the fan is given by $v_z = (\partial \phi/\partial \theta)(B_r/B^2)$, so that ϕ must vary in the azimuthal direction. Fig. 11.31a illustrates the level curves of ϕ in a plane cutting through the spine for the choice $\phi = \sin \theta$. Lines

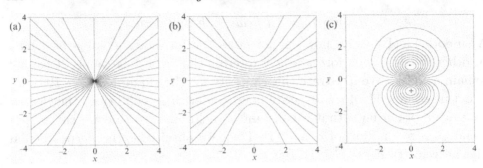

Fig. 11.31 The electric potential ϕ and $\boldsymbol{E} \cdot \boldsymbol{B}$ in the x, y-plane for a kinematic model of spine reconnection: (a) level curves of the singular ideal flow potential, (b) the expected structure of the potential smoothed by nonideal processes, and (c) the resulting parallel electric field, represented by level curves of $\boldsymbol{E} \cdot \boldsymbol{B}$.

carrying different values of ϕ join at the spine, which causes the singularity. So, $\nabla \phi$ and with it the electric field and the plasma velocity diverges on the spine.

For reconnection to occur, this divergence has to be removed by embedding the spine in a nonideal region. In that region the electric potential must be smoothed; the expected structure is shown in Fig. 11.31b, where lines of equal potential are connected in the most direct way, so that they do not continue into the spine, thereby avoiding the singularity. This implies an electric field component parallel to the magnetic field. Fig. 11.31c illustrates the distribution of $\boldsymbol{E} \cdot \boldsymbol{B}$.

In the model of Fig. 11.31 the smoothing is based on the assumption that near the spine the potential lines must connect along the x-axis, so that there $\phi = r \sin \theta$ holds. This expression is smoothly matched with the external ideal potential $\phi = \sin \theta$ by the choice

$$\phi = \tanh r \sin \theta \,, \tag{11.111}$$

so that

$$\boldsymbol{E} = -\frac{\sin \theta}{\cosh^2 r} \boldsymbol{e}_r - \frac{\tanh r}{r} \cos \theta \, \boldsymbol{e}_\theta, \quad \boldsymbol{E} \cdot \boldsymbol{B} = -\frac{r \sin \theta}{\cosh^2 r} \,. \tag{11.112}$$

It is a general property of kinematic models that the question of how the nonidealness can be sustained self-consistently is set aside. Nevertheless, it is interesting that the resistive MHD model of Craig and Fabling (1996) yields a double structure of the electric current density that is qualitatively consistent with that of Fig. 11.31c.

Resistive MHD models require currents to create the appropriate nonidealness, which excludes simple potential fields as we discussed here. Kinematic models are not subject to that constraint and the reconnection process might become more transparent. This has motivated the following global kinematic model.

A global kinematic model of null point reconnection

For a global kinematic example we use (11.105) with potentials (11.106) and extend the electric potential (11.111), which has only local significance, by the choice

$$\phi = \tanh r \sin\theta \, \frac{e^{-1/B^2}}{\cosh^2 \alpha} \,, \tag{11.113}$$

where the extra factor avoids the singularity at the null and concentrates the effect to the vicinity of $\alpha = 0$. Further, we assume that the nonidealness $\mathbf{R}$ is parallel to $\mathbf{B}$.

This model generates an electric field and a velocity that are smooth everywhere. The plasma crosses the fan ($z = 0$) with the velocity

$$\mathbf{v}_{\text{fan}} = \frac{x \tanh r}{r} e^{-\frac{1}{r^2}} \mathbf{e}_z \tag{11.114}$$

and the spine ($r = 0$) is crossed with

$$\mathbf{v}_{\text{spine}} = \frac{1}{2z} e^{-\frac{1}{4z^2}} \mathbf{e}_x \,. \tag{11.115}$$

So, both spine and fan reconnection occur simultaneously. The velocity signature in the x, z-plane is qualitatively equivalent to that of 2D x-point reconnection (Fig. 11.32a), while in the y, z-plane the spine is not crossed (Fig. 11.32b). At crossing the plasma elements swap partners over large

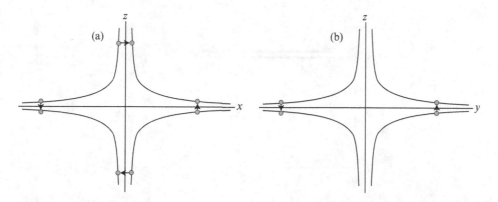

Fig. 11.32 Null point reconnection based on (11.105) and (11.113): motion of plasma elements, (a) crossing the spine and the fan in the x, z-plane, (b) crossing only the fan in the y, z-plane.

distances. In the picture of frozen-in field line motion the field lines reconnect at the null in plane (a) and *flip* (Priest and Forbes, 1992b) across the fan in plane (b).

The quantity $\boldsymbol{E} \cdot \boldsymbol{B}$, which follows as

$$\boldsymbol{E} \cdot \boldsymbol{B} = -e^{-1/B^2} \frac{\sin \theta}{\cosh^2(r^2 z)} \left(2 \frac{\tanh r(r^2 - 8 z^2)}{B^4} + \frac{r}{\cosh^2 r} \right), \quad (11.116)$$

emphasizes the spine and fan regions (Fig. 11.33).

Although the nonideal region extends infinitely along spine and fan, some of the tools that were applied to $B \neq 0$ reconnection are still applicable because the potential $U(\alpha, \beta)$, given by (11.84), is well defined. Along every field line the nonidealness is concentrated, so that the integral converges and U becomes

$$U(\alpha, \beta) = \frac{\sin \beta}{\cosh^2 \alpha} . \quad (11.117)$$

The level curves of U are shown in Fig. 11.34, indicating a double structure with a positive peak ($U_{\max} = 1$) and a negative peak ($U_{\min} = -1$). In this case the maximum reconnection rate occurs on a flux surface that corresponds to a line connecting the two peaks, in the figure represented by the straight vertical line. The reconnection velocity displaces this line

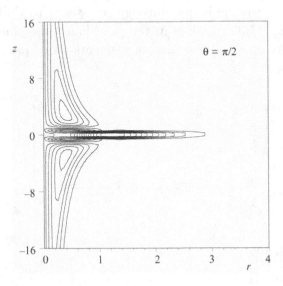

Fig. 11.33 Level curves of $\boldsymbol{E} \cdot \boldsymbol{B}$, given by (11.116), of the model (11.105) and (11.113).

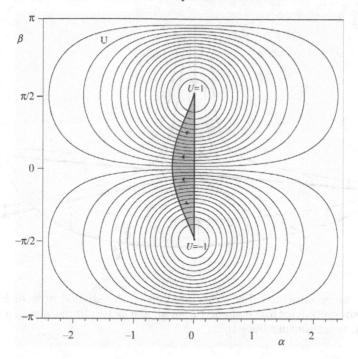

Fig. 11.34 The potential U in α, β-space. The lines are level curves of U and show a double structure with a positive and a negative peak. The shaded area indicates the reconnected flux.

as shown and the shaded area is the reconnected magnetic flux in a time interval dt. An argument analogous to that applied to the case of Fig. 11.26 shows that the reconnection rate is given by $|U_{\max} - U_{\min}| = 2$.

Let us confirm this rate by leaving α, β-space and computing the reconnected flux explicitly. Although the spine is crossed by plasma flow, that flow generates a line rather than a surface, so that no flux is involved. This is contrasted by the flow across the fan which generates a finite magnetic flux connected with the shaded surface of Fig. 11.35. This flux is readily evaluated as

$$\mathrm{d}F = \int_{-\pi/2}^{\pi/2} v_z B_r r \, \mathrm{d}\theta \, \mathrm{d}t = 2\mathrm{e}^{-\frac{1}{r^2}} \tanh(r) \, \mathrm{d}t \, , \tag{11.118}$$

so that the reconnection rate ($|\mathrm{d}F/\mathrm{d}t|$), which is to be evaluated in the ideal region, i.e., for $r \to \infty$, assumes the value 2, in agreement with the result obtained from $U(\alpha, \beta)$. This agreement had to be expected, because we considered the same problem in two different representations.

We emphasize that these representations show a distinct geometrical dif-

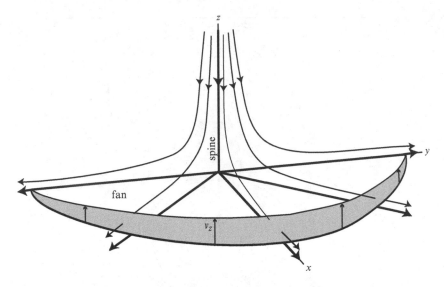

Fig. 11.35 The reconnected flux corresponding to the shaded area in Fig. 11.34 is the flux connected with the shaded area shown here. Spine and fan lines are drawn thick, neighbouring lines thin.

ference. In coordinate space the reconnection has dramatic consequences: plasma elements swap partners over large distances or, equivalently, branches of field lines of grossly different shape reconnect or field lines undergo flipping with a drastic change of their shape. None of these effects appears in α, β-space, where the reconnection process is represented by a smooth distortion of a curve. We return to this aspect in the following section.

Fan and spine reconnection also have been studied in a resistive kinematic approach in null configurations with currents by Pontin *et al.* (2004, 2005). Their modelling included resolving the singularity at the null; they arrived at similar conclusions.

From general magnetic reconnection to magnetic reconnection

We saw that the framework of the GMR approach with its two classes (11.77) and (11.78) covers a wide range of processes that have important aspects of the intuitive reconnection picture described in the introduction to this chapter. Yet, there are striking differences. Fig. 11.36 illustrates a fundamental difference between two steady state examples. In the case shown on the left the magnetic field is homogeneous and the plasma exhibits the characteristic plasma rotation in the magnetic flux tube passing through the nonideal region. This rotation directly reflects the relative rotation occurring

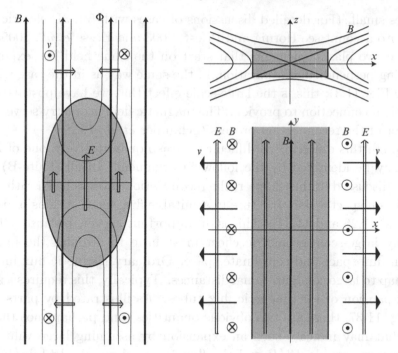

Fig. 11.36 Examples of steady state $B \neq 0$ configurations. The case shown on the left has a homogeneous magnetic field. The localization of the parallel electric field leads to the characteristic plasma rotation above and below the nonideal region. The thin lines are lines of constant electrostatic potential ϕ. The configuration on the right is quasi-two-dimensional, locally approximated by a 2D standard x-line configuration with a constant guide field. The particular field structure leads to plasma flow away from the nonideal region. The nonideal region is shaded in both cases.

in α, β-space. It extends along the magnetic field in the ideal region on both sides of D_R, associated with perpendicular electric fields that become strong for nonideal regions that are elongated along the magnetic field lines. The relative perpendicular displacement between plasma elements on different sides of D_R remains of the order of the perpendicular dimension of D_R, which (by definition) is much smaller than the overall scale length.

The example shown on the right has the familiar magnetic field structure of the standard 2D x-type neutral line configuration (Fig. 11.1) with a constant guide field superimposed. The electric field is assumed to have a parallel component in a region which extends sufficiently in the y-direction, so that the electric field near $y = 0$ can be approximated by the 2D field, i.e., constant E_y. The nonidealness is localized in the x, z-plane, because away from the x-point the magnetic field strength increases, so that $E_\parallel = E_y B_y / B$

becomes small. (For detailed discussions of cases with full 3D electric fields taken into account see Hornig and Priest, 2003 and Hesse *et al.*, 2005a.) In this case two plasma elements that start on the same field line experience increasing perpendicular separation, in the same way as in the case without B_y (see Fig. 11.3); this is the large-scale effect that we have postulated for magnetic reconnection to provide. The magnetic field geometry serves as an amplifier of the reconnection process (Schindler *et al.*, 1988).

Although this difference is difficult to grasp quantitatively, one of its aspects may be identified by the following argument. In all (finite-B) cases that we discussed in this chapter, the reconnection process has a rather simple and non-spectacular structure in a suitable flux space such as α, β-space (see Figs. 11.26 and 11.34). This is an important aspect, because it means that any large-scale geometric effect must have to do with the mapping between flux space and coordinate space. Ordinary magnetic flux intervals must map to large configurational distances. Typically, this requires a strong local expansion of the magnetic flux tubes. As illustrated by parts a and b of Fig. 11.37, there are two obvious quantities (and perhaps not the only ones) that may indicate such an expansion by assuming large values, the cross-section area $d\alpha\,d\beta/B$ and the flux tube volume $d\alpha\,d\beta\int ds/B$. The reconnection site and the region of flux tube divergence can, but must not spatially coincide. They may well be separated by a macroscopic distance (part c).

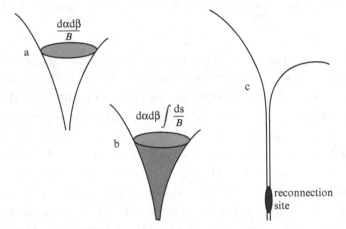

Fig. 11.37 Schematic sketches illustrating large scale effects of reconnection resulting from diverging field lines. The divergence may be due to large cross-section area (a) or large flux tube volume (b). The region of flux tube divergence may be well separated from the reconnection site (c).

Is there a causal relationship between the reconnection process of part c of Fig. 11.37 and the presence of the amplifying field divergence? This question cannot be answered on the present kinematic level. But using additional dynamic information, discussed earlier, the following picture arises. The two field lines shown in part c can be visualized to be located on opposite sides of a separatrix. If that separatrix carries a current sheet (Section 8.6.2) it would favour reconnection to occur somewhere on the separatrix. The location and the details of the reconnection physics, however, would be determined entirely by the local conditions. For instance, the reconnection might be initiated by some small-scale fluctuation, enhancing the current density locally. The resulting reconnection pattern would evolve independent of the gross magnetic field structure (see Section 11.2.7) and its dynamics would not require any field singularity. In that sense there is no causal relationship between the reconnection process and a distant amplifying field singularity such as a magnetic null. This picture is applicable to the magnetopause of the Earth's magnetosphere (Chapter 13).

For most reconnection cases it is not necessary to distinguish between possibilities a and b. For instance, in the presence of a 2D x-type neutral point without a guide field or of a null point in a 3D field, both quantities diverge. However, in 2D reconnection with a poloidal x-point and a guide field superimposed, which in many ways behaves similarly to the case without the guide field, it is the flux tube volume that diverges, while the cross-section area remains bounded. Fig. 11.38 illustrates the reason why the flux tube volume becomes unbounded although B is bounded away from zero. The distance between the entrance and exit points on the surface of a test cylinder diverges on the separatrices. Correspondingly, on the separatrices the integration interval of the flux tube volume becomes infinite, so that the flux tube volume diverges.

One can understand this divergence as a singularity of the Jacobian of the map between entrance and exit (foot) point locations of the field lines. In cases where the Jacobian of such a map remains bounded but becomes locally large, often a layer appears that qualitatively resembles a separatrix and has been termed *quasi-separatrix layer* (Priest and Démoulin, 1995). Corresponding reconnection processes have been shown to involve field line flipping through such layers.

Diverging flux tubes is only one aspect of the need for a large-scale effect. A suitable plasma flow can also lead to a large-scale separation of fluid elements that are close to each other originally (Hornig, 2006). A standard example is the stagnation flow of x-line reconnection, with or without a guide field.

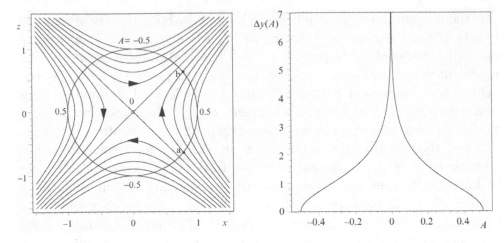

Fig. 11.38 Singularity of the x-line configuration with a guide field. The magnetic field has the components $(z, 1, x)$. Left part: poloidal projection of field lines; the field lines enter a cylinder (circle) at point a and leave the cylinder at b. Right part: distance Δy between the entrance and exit points in the y-direction versus the poloidal flux function A. At the separatrices ($A = 0$) the displacement and hence the flux tube volume becomes unbounded.

So it seems that the additional requirement, necessary for a general magnetic reconnection process to become a genuine reconnection process (i.e., reconnection with large-scale consequences), is that the separation of fluid elements, provided by global breakdown of line conservation, must be amplified. As we have seen, this can be achieved by strongly diverging magnetic field lines or by a suitable plasma flow geometry.

The attempt to confine the notion of reconnection to the presence of a null as a necessary feature (Greene, 1988; Lau and Finn, 1990) seems too restrictive, considering that examples such as 2D x-point configurations with a guide field or fields with quasi-separatrix layers would not be covered. (See also the discussion of magnetopause reconnection in Section 13.1.2.)

A final remark concerns a degenerate case. Using Euler potentials, it is necessary for reconnection to occur that there must be field lines with $[\mathrm{d}\alpha/\mathrm{d}t]_{\alpha,\beta}$, $[\mathrm{d}\beta/\mathrm{d}t]_{\alpha,\beta}$ or both non-vanishing. Any relative motion of pairs of plasma elements with one partner inside $\mathrm{D_R}$ is not relevant for reconnection. So, the one-dimensional steady state MHD shock waves would not count as a reconnection site. This is directly evident by viewing the shock wave in the de Hoffmann–Teller frame (see Section 3.9), where in the upstream and downstream regions the velocity is parallel to the magnetic field, so that $\mathrm{d}\alpha/\mathrm{d}t$ and $\mathrm{d}\beta/\mathrm{d}t$ both vanish. If one follows a pair of plasma elements starting upstream on the same field line, one finds that they get separated when the first element enters the shock region, but they are on the same field line again after both elements have entered the downstream region. There is no effect on scales larger than the shock thickness. (Such effects have been termed *GMR local effects* (Schindler *et al.*, 1988).)

Summarizing, we conclude that magnetic reconnection can be characterized by the following kinematic properties:

- presence of a localized nonideal region D_R,
- global breakdown of magnetic line conservation,
- spatial amplification of the effect of breakdown of line conservation, provided by strong divergence of magnetic field lines connected to D_R and/or a suitable plasma flow geometry, such as stagnation point flow.

The significance of diverging magnetic fields (third property) makes separatrices and separatrix layers preferred locations of magnetic reconnection. This is consistent with the (dynamic) property that separatrices often are associated with current sheets (Section 8.6.2).

11.5.3 Approximate conservation of helicity

Magnetic helicity

$$K = \int_D \boldsymbol{A} \cdot \boldsymbol{B} \, \mathrm{d}^3 r \,, \tag{11.119}$$

as already defined in (3.92), was introduced into magnetohydrodynamics in the context of dynamo theory (Elsasser, 1956) and force-free fields (Woltjer, 1958). In Section 3.8.3 we have seen that in ideal MHD magnetic helicity is a conserved quantity. So the question arises what role helicity plays in reconnection.

The helicity definition (11.119) in general is not gauge invariant. There are several modifications by which gauge invariance of K and its time derivative (3.93) can be achieved (Berger and Field, 1984; Finn and Antonsen, 1985); here we mention two possibilities. The first is to restrict the applications to configurations with special boundary conditions. This, for instance, applies to plasmas surrounded by an infinitely conducting static boundary with $\boldsymbol{n} \cdot \boldsymbol{B} = 0$, which gives

$$\frac{\mathrm{d}K}{\mathrm{d}t} = \Theta, \quad \Theta = -2 \int_D \boldsymbol{E} \cdot \boldsymbol{B} \, \mathrm{d}^3 r \,. \tag{11.120}$$

The second method is to use modified versions of helicity, often called *relative helicity*, which lead to gauge invariance for other boundary conditions, too

(Berger and Field, 1984; Finn and Antonsen, 1985). A similar possibility applies to instability processes in systems embedded in a static magnetic field $\boldsymbol{B}_0$ with a vanishing electric field. Then, defining (Schindler *et al.* (1988))

$$\bar{K} = \int_D (\boldsymbol{A} + \boldsymbol{A}_0) \cdot (\boldsymbol{B} - \boldsymbol{B}_0)\, \mathrm{d}^3 r\,, \qquad (11.121)$$

one finds $\mathrm{d}\bar{K}/\mathrm{d}t = \Theta$ as in (11.120), so that, analogous to K, under the assumed conditions $\bar{K}$ is gauge invariant and conserved in ideal MHD.

Regarding magnetic reconnection processes, we first note that, strictly speaking, typical reconnection processes would change magnetic helicity because of the presence of a parallel electric field entering the integral on the right side of (11.120). However, the integrand is different from zero only in the nonideal region. So, if the nonideal region is small compared with the overall extent of the configuration, the integral might be small such that helicity is approximately conserved. Quantitatively, this effect will depend on how the integrand varies as the nonideal region shrinks. For an MHD model exact limits on the helicity dissipation were derived by Berger (1984). We outline the essence of the argument.

With Ohm's law given by (3.67), one obtains

$$\Theta = -2 \int_D \eta(\boldsymbol{r}) j_\parallel(\boldsymbol{r}) B(\boldsymbol{r})\, \mathrm{d}^3 r\,. \qquad (11.122)$$

The integral can be estimated using Schwarz's inequality,

$$\left(\int_D \eta j_\parallel B\, \mathrm{d}^3 r \right)^2 \leq \left(\int_D \eta B^2\, \mathrm{d}^3 r \right) \left(\int_D \eta j_\parallel{}^2\, \mathrm{d}^3 r \right)\,, \qquad (11.123)$$

so that one obtains from (11.122)

$$\Theta^2 \leq 4\bar{\eta} \int_D \eta j_\parallel{}^2\, \mathrm{d}^3 r \int_D B^2\, \mathrm{d}^3 r\,, \qquad (11.124)$$

where

$$\bar{\eta} = \frac{\int_D \eta B^2\, \mathrm{d}^3 r}{\int_D B^2\, \mathrm{d}^3 r} \qquad (11.125)$$

is an average value of the resistivity η. Assuming that in the nonideal region the rate of change $\dot{W}_\mathrm{m}$ of magnetic energy W_m approximately equals $\int \eta j^2\, \mathrm{d}^3 r$ (the significance of that assumption is discussed below), (11.124) can be put in the form

$$\Theta^2 \leq 8\,\mu_0 \bar{\eta} W_\mathrm{m} |\dot{W}_\mathrm{m}|\,. \qquad (11.126)$$

An expression for the magnitude of the change of helicity, $|\Delta K|$, during the time interval $(0, \Delta t)$ is obtained by integrating (11.126) with respect to time,

$$|\Delta K| \leq 2\sqrt{\mu_0 \bar{\eta}} \int_0^{\Delta t} \sqrt{2 W_{\mathrm{m}} |\dot{W}_{\mathrm{m}}|} \, dt . \qquad (11.127)$$

By a straightforward variational procedure the integral in (11.127) can be seen to be maximized for

$$W_{\mathrm{m}} = W_{\mathrm{m,i}} \sqrt{1 - \frac{t}{\Delta t} \left(1 - \left(\frac{W_{\mathrm{m,f}}}{W_{\mathrm{m,i}}} \right)^2 \right)} , \qquad (11.128)$$

such that

$$|\Delta K| \leq 2\sqrt{\mu_0 \bar{\eta} \Delta t \left(W_{\mathrm{m,i}}^2 - W_{\mathrm{m,f}}^2 \right)} . \qquad (11.129)$$

Here, $W_{\mathrm{m,i}}$ and $W_{\mathrm{m,f}}$ denote the initial and the final value of W_{m}, respectively. Solving (11.129) for Δt gives an estimate of the time required for a given change of helicity,

$$\Delta t \geq \frac{|\Delta K|^2}{4 \mu_0 \bar{\eta} \left(W_{\mathrm{m,i}}^2 - W_{\mathrm{m,f}}^2 \right)} . \qquad (11.130)$$

For a small nonideal region with volume V_{R} and resistivity η_0 embedded in an external ideal region of volume V_0, one finally obtains

$$\Delta t \geq \frac{|\Delta K|^2 V_0}{4 \mu_0 \eta_0 V_{\mathrm{R}} \left(W_{\mathrm{m,i}}^2 - W_{\mathrm{m,f}}^2 \right)} , \qquad (11.131)$$

where it was assumed that B^2 averaged over $\mathrm{D_R}$ is not larger than averaged over the macroscopic domain D. Clearly, for a given change of helicity and initial and final magnetic energy the required time becomes arbitrarily large as the volume V_{R} of the nonideal region approaches zero.

The main assumption that was made in deriving (11.131) was to identify $\dot{W}_{\mathrm{m}}$ with $\int \eta j^2 \, d^3 r$. In resistive MHD the Poynting theorem for region D with boundary ∂D reads (see (3.43))

$$\dot{W}_{\mathrm{m}} = -\frac{1}{\mu_0} \oint_{\partial \mathrm{D}} \boldsymbol{E} \times \boldsymbol{B} \cdot \boldsymbol{n} \, dS - \int_{\mathrm{D}} \boldsymbol{v} \cdot \boldsymbol{j} \times \boldsymbol{B} \, d^3 r - \int_{\mathrm{D}} \eta j_{\parallel}^2 \, d^3 r - \int_{\mathrm{D}} \eta j_{\perp}^2 \, d^3 r .$$
$$(11.132)$$

Thus, the Poynting flux, the work associated with the Lorentz force and the dissipation resulting from $j_{\perp}$ have been neglected. This can be justified for energetically isolated force free fields, for which the Poynting flux through the boundary and the terms involving $j_\perp$ are negligible. (In fact, the estimate (11.131) was derived in view of coronal magnetic fields (Berger, 1984), which largely can be idealized as being force-free.) If these terms are not negligible, (11.131) can still be regarded as an order of magnitude estimate as long as the dissipation term due to $j_\parallel$ is not small compared to the other terms in (11.132).

Note that for $V_R/V_0 \ll 1$ approximate helicity conservation is consistent with finite changes of magnetic energy, but it puts a constraint on the final configuration. For instance, magnetic energy accumulated in field line linkage can relax by reconnection into a twisted field configuration. Such relaxations were found in 3D nonlinear current sheet reconnection (Fig. 11.21). For a covariant kinematic model see Fig. 11.39.

The general process of magnetic field relaxation under conservation of helicity is known as *Taylor relaxation* (Taylor, 1974). Under suitable boundary conditions (e.g., vanishing normal magnetic field component) the relaxed state is a force-free state with constant κ (see Section 5.1.5).

11.6 Remark on relativistic covariance

Here we briefly address relativistic reconnection.

In general, the laws of physics must be covariant, i.e., independent of the observer's frame of reference. (We confine the discussion to special relativity where the relevant frames are the inertial frames, generated by the Lorentz transformation.)

One could object that in most applications reconnection occurs in non-relativistic plasmas, and that therefore relativistic covariance is not an important requirement. However, some problems persist even for non-relativistic frame velocities. In any case, it is of interest to see whether or not a convincing relativistic notion of magnetic reconnection does exist (Hornig, 1997a,b).

The problem originates from the fact that the notion of transport of magnetic field lines concerns only the magnetic field, which has no covariant meaning. This has led Hornig (1997a) to investigate the transport of the electromagnetic field, for instance, represented by the field tensor $\mathcal{F}$ with components $F^{\mu\nu}$ in Minkowski space. Here we briefly discuss that concept (without derivation) and its consequences for magnetic reconnection.

Mathematically, the condition of existence of a line conserving transport velocity of the magnetic field (Section 3.8.2) can be expressed in terms of the Lie derivative of $\mathbf{B}$. This formulation has a straightforward generalization to the transport of the field tensor. It turns out that a parameter that corresponds to Λ in (3.90) would be zero in many cases. Ignoring that parameter altogether, one obtains the following condition for smooth transport of the electromagnetic field

$$V^0 \frac{\partial \Phi}{\partial x^0} + \boldsymbol{V} \cdot \nabla \Phi = 0 \tag{11.133}$$

$$V^0 \boldsymbol{E} + \boldsymbol{V} \times \boldsymbol{B} = \nabla \Phi \,, \tag{11.134}$$

where $\mathcal{F}$ is transported by the 4-vector $V = (V^0, \boldsymbol{V})$. Here Φ is an arbitrary function of $\boldsymbol{r}, t$, combined to the 4-vector x_λ, where x_1, x_2, x_3 are the components of $\boldsymbol{r}$ and $x_0 = ct$. By construction, this condition is covariant, implying that under (11.133) and (11.134) the Lorentz transformation conserves the property of smooth transport of the electromagnetic field.

In the non-relativistic limit with $V = (1, \boldsymbol{v})$ the conditions (11.133) and (11.134) reduce to

$$\frac{\partial \Phi}{\partial t} + \boldsymbol{v} \cdot \nabla \Phi = 0 \tag{11.135}$$

$$\boldsymbol{E} + \boldsymbol{v} \times \boldsymbol{B} = \nabla \Phi \,. \tag{11.136}$$

Clearly, ideal MHD is contained as a special case. More generally, (11.131) states conservation of magnetic flux rather than magnetic line conservation.

In fact, under the conditions (11.133) and (11.134), magnetic flux conservation generalizes to conservation of electromagnetic flux

$$\int_C \mathcal{F} \cdot \mathrm{d}\mathcal{A} = \int \boldsymbol{B} \cdot \mathrm{d}\boldsymbol{a} + \frac{1}{c} \int \boldsymbol{E} \cdot \mathrm{d}\boldsymbol{a}_0 = const \,. \tag{11.137}$$

Here, the integrals are taken over projections of a selected surface in Minkowski space (for details see Hornig and Schindler (1996b)).

Flows conserving electromagnetic flux may transform magnetic flux (first term in (11.137)) into *electric flux* (second term in (11.137)). In that sense conservation of electromagnetic flux can break the frozen-in constraint implied by magnetic flux conservation. Obviously, this is of interest for reconnection processes.

Another important feature of electromagnetic flux conservation is conservation of helicity for flows with $V^0 > 0$. (For a more precise statement see Hornig, 1997a.)

The physical properties of the transporting flow depend on the sign of $V^{0^2} - \boldsymbol{V}^2$. For magnetic reconnection it turns out that the most relevant cases are those with coexisting time-like (positive sign) and space-like (negative sign) subdomains and with lines on which both V^{0^2} and $\boldsymbol{V}^2$ vanish.

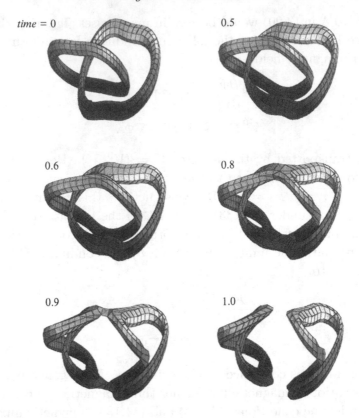

Fig. 11.39 Kinematic example of magnetic reconnection under helicity conservation (Hornig, 1997b): Linked flux tubes without twist turn into unlinked twisted
tubes (by permission from G. Hornig).

Fig. 11.39 shows an example illustrating a reconnection process described
by a covariant transporting flow, where the magnetic field does not involve
nulls. Helicity is conserved and flux linkage turns into twist.

The non-relativistic limit of this approach corresponds to violation of magnetic flux conservation, as is clear from (11.136). To cover violation of line
conservation it seems that the function Λ, mentioned above, has to be incorporated in a nontrivial way.

12

Aspects of bifurcation and nonlinear dynamics

Important aspects of activity of space plasmas can be described in terms of transitions from stable to unstable states. Therefore, it was necessary to deal with the stability properties of selected equilibria, which has been a major aspect in this part of the book.

However, to obtain a deeper physical understanding of the dynamic properties of a given system it is desirable to have available a complete overview of all equilibrium states and their stability properties for every choice of a suitable (*control*) parameter. Such information is provided by *bifurcation diagrams*. They are particularly useful to assess the qualitative behaviour of nonlinear systems. Here we can give only an elementary introduction aimed at clarifying notions that will be used later. For rigorous treatments the reader should consult the literature (e.g., Berge *et al.*, 1986; Guckenheimer and Holmes, 1983).

Statistical mechanics and nonlinear dynamics provide additional techniques that have been applied to space plasma activity.

12.1 Bifurcation

For illustration of bifurcations let us begin with a set of simple examples shown in Fig. 12.1. A point mass moves in a potential $U(x, \lambda)$ subject to the force $-\partial U/\partial x$, λ being the control parameter. On the right, the figure also shows the bifurcation diagrams, which are the plots of the equilibrium positions versus λ. At bifurcation points the solution structure (here manifested by the number of solutions for a given value of λ) as well as stability undergoes qualitative changes. In the present model bifurcation points are characterized by the simultaneous vanishing of the first and second derivatives of U with respect to x. The cases a and b are bifurcations of the pitchfork type (*pitchfork bifurcation*) while c is the simplest version of a

343

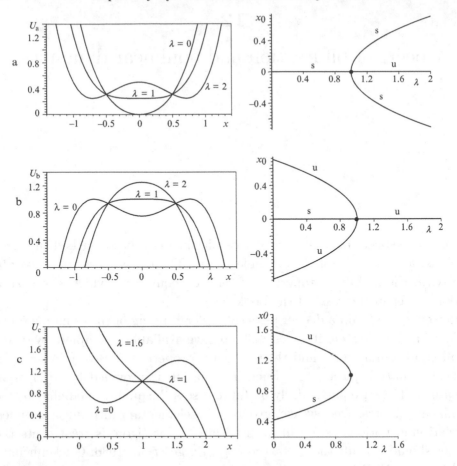

Fig. 12.1 Examples of bifurcation diagrams (on the right) for one-dimensional mechanical systems with several choices of the potential $U(x, \lambda)$ (on the left); x_0 denotes the equilibrium positions and λ is the control parameter. Stable branches are marked as s and unstable branches as u. The black dots indicate the bifurcation points. For the cases a, b, c, the potentials are chosen as $U_a = x^4 - (\lambda - 1)x^2 + \lambda/4$, $U_b = -x^4 - (\lambda - 1)x^2 + (\lambda + 3)/4$, $U_c = -(x - 1)^3 - (\lambda - 1)(x - 1) + 1$, respectively.

catastrophe, characterized by loss of equilibrium on one side of the bifurcation point.

A stable equilibrium is *globally stable* if it corresponds to the lowest possible value of the potential (*minimum energy state*). For $\lambda < 1$ the stable states of Fig. 12.1a are globally stable while there are no globally stable states in Figs. 12.1b and 12.1c.

The dynamical solutions $x(t)$ evolve as expected from the bifurcation diagrams. Fig. 12.2 shows an example, where λ is slowly increased with time,

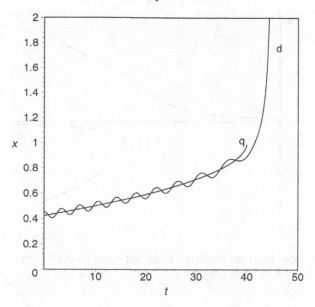

Fig. 12.2 A dynamic solution for the catastrophe scenario of case c of Fig. 12.1, with λ slowly increasing with time so that the bifurcation point is reached at $t = 40$. Initially the dynamic solution (d) oscillates around the quasi-static solution (q); when the bifurcation point is reached, the orbit 'catastrophically' changes its structure and the point mass rapidly moves to large distances, away from the bifurcation position.

so that the system moves toward a catastrophe point. The sudden change of the orbit near the bifurcation point is typical for a *catastrophe*.

In the cases shown in Fig. 12.1 the potential is symmetric with respect to the sign of the coordinate x. Each one of the branches bifurcating away from the equilibria $x_0 = 0$ breaks that symmetry. A different type of symmetry breaking may apply to the potential itself. A symmetry breaking pertur- bation of the potential may cause a structural instability of the bifurcation diagram obtained with the original symmetric potential. For example, a small perturbation turns the bifurcation diagram of case a of Fig. 12.1 into that shown in Fig. 12.3. However, for sufficiently small perturbations typical solutions $x(t)$ remain qualitatively unaffected.

It is essential that the parameter λ is an appropriate control parameter in the sense that it must be freely adjustable. For the present one-dimensional example it is not difficult to imagine that the potential can be adjusted by external measures. In more complex situations discussed later, this require- ment plays a crucial role for the interpretation of the diagrams.

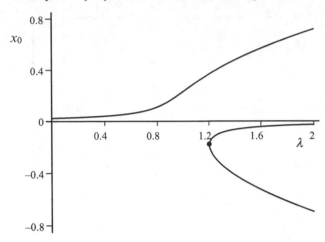

Fig. 12.3 Bifurcation diagram resulting from applying a small perturbation to case a of Fig. 12.1; the potential is chosen as $U = U_a - 0.05x$.

A final remark concerns dynamical constraints. We will encounter a case where two models, in a simplified description, differ only by the absence or presence of a constraint imposed on the perturbations. The second model is unstable, in the first model the unstable perturbations are suppressed by the constraint. Both models have an identical equilibrium branch but with different stability properties. The sudden breakdown of the constraint will have dramatic consequences. Fig. 12.4 illustrates this possibility for the example of a breaking dam of a water reservoir.

12.1.1 Bifurcation properties of Grad–Shafranov theory

Let us consider the Grad–Shafranov equation for one- or two-dimensional Cartesian coordinate spaces. For concreteness, let us assume that the domain of interest Ω is bounded with a smooth boundary $\partial\Omega$. Equilibrium sequences with a sequence parameter λ can be generated by choosing the pressure of the form $p(x, \lambda)$, where x stands for the coordinate(s). A simple prototype case corresponds to $p = \lambda \exp(2A)/2$ so that the Grad–Shafranov equation becomes

$$-\Delta A = \lambda e^{2A} . \tag{12.1}$$

This case is equivalent to the choice (5.88), the two cases differ only in that the magnetic fields have opposite directions. The following section addresses the bifurcation properties of a more general class of boundary value problems.

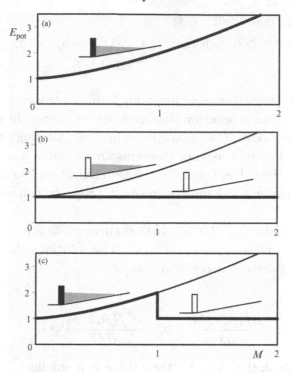

Fig. 12.4 Symbolic illustration of the role of a constraint. A dam (rectangle) of a water reservoir may be stable (a) or unstable (b) against perturbations that cause a leakage. In case (a) there is a single (stable) equilibrium branch represented by the plot of the potential energy E_{pot} of the supplied water vs. its mass M. If in the absence of the constraint the leakage instability is active (b), the solution where the reservoir still contains water is unstable (thin line) and the stable solution corresponds to complete unloading (thick line, the entire mass is lost to the ground level, so that its potential energy is 1). If for some reason the constraint breaks down at some point in the evolution ($M = 1$ in part (c)), the breakdown of the constraint leads to a sudden dramatic unloading of the reservoir, releasing the previously stored potential energy.

Convex currents

Consider Grad–Shafranov equations of the following type:

$$-\Delta A = f(A, x, \lambda) \quad x \in \Omega \tag{12.2}$$
$$A = 0 \quad x \in \partial\Omega . \tag{12.3}$$

Here the current density is represented by $f(A, x, \lambda)$, a smooth function (continuous first derivatives) with the following properties:

(*i*) $f(0, x, 0) = 0$, all $x \in \bar{\Omega}$,   (12.4)

(*ii*) $\exists\, \mu > 0, \nu \geq 0$ such that $f(A, x, \lambda) \geq \lambda(\mu A + \nu)$,

all $A \geq 0, x \in \bar{\Omega}, \lambda \geq 0$,   (12.5)

where (*ii*) can be regarded as a particular form of convexity.

As there is no smooth solution that assumes negative values (absence of a minimum), one concentrates on positive solutions. Although this problem is nonlinear, several useful existence theorems are available (e.g., Fujita, 1969). (For a review of the subject, details and generalizations see Amann, 1976.) In the present context the most important property can be expressed as follows:

The conditions (12.4), (12.5) imply that there exists a parameter value λ^* such that there is no solution for $\lambda > \lambda^*$. The solutions that exist in some λ-range below λ^* are not necessarily unique.

If, in addition,

$$\frac{\partial f(A, x, \lambda)}{\partial \lambda} \geq 0, \quad \frac{\partial^2 f(A, x, \lambda)}{\partial A^2} \geq 0 \qquad (12.6)$$

holds for all $A \geq 0$, $0 \leq \lambda \leq \lambda^*$, then there is a smallest λ^* such that for values of λ satisfying $0 \leq \lambda \leq \lambda^*$ there is at least one positive solution. The smallest (with respect to maximum norm) positive solution is stable in the sense that the functional

$$\int \left((\nabla \xi)^2 - \frac{\partial f}{\partial A} \xi^2 \right) \mathrm{d}^2 r \qquad (12.7)$$

is positive definite. As this functional, except for an irrelevant factor, is a generalized version of (10.58) we refer to this stability definition as *F-stability*.

Fig. 12.5 illustrates these properties in the *bifurcation diagram* $\|A\|_{\max}$ vs. λ. If λ is a control parameter (i.e., a freely adjustable parameter), λ^* is a catastrophe point. Crossing that point the system loses equilibrium and necessarily becomes dynamic, at least within the present 2D picture. Also, note that the 2D variational expression (10.57) for MHD and the corresponding functional of the Vlasov theory (10.214) consist of F_2 plus positive terms. Therefore, the smallest solution is stable within 2D MHD and Vlasov theories.

So the full set of equilibrium solutions provides interesting stability properties, a fact that we encountered already in the elementary examples discussed above.

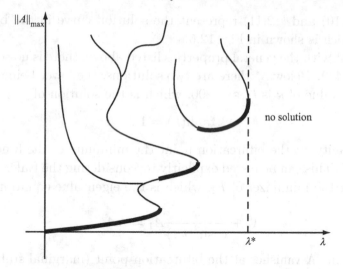

Fig. 12.5 Qualitative sketch illustrating the bifurcation properties of the solutions of (12.2), (12.3) with (12.4)–(12.6). The solutions on the thick line sections are F-stable.

Harris sheet: Loss of equilibrium

As the conditions (12.4)–(12.6) are satisfied for the example of (12.1), we can conclude that the Harris sheet problem of Section 5.3.2, understood as a boundary value problem with A vanishing on the boundary, must show the properties listed above. Here A is a function of z alone so that (12.1) reduces to

$$-\frac{\partial^2 A}{\partial z^2} = \lambda\, e^{2A}, \quad A(\pm 1) = 0 . \tag{12.8}$$

Concentrating on the region $z \geq 0$, the boundary condition at $z = -1$ is replaced by vanishing derivative with respect to z at $z = 0$.

Equation (12.8) has the general symmetric solution

$$A = -\ln\frac{\cosh(kz)}{\cosh(k)} , \tag{12.9}$$

where k is related to λ by

$$\lambda = \frac{k^2}{\cosh^2(k)} . \tag{12.10}$$

The maximum norm of A is

$$\|A\|_{\max} = \ln\cosh(k), \tag{12.11}$$

so that (12.10) and (12.11) represent the solution curve in the bifurcation diagram which is shown in Fig. 12.6.

Consistent with the general properties listed above, there is no solution for $\lambda > \lambda^* \approx 0.439$. Below λ^* there are two solutions, the lower being F-stable. The critical value of k is $k^* \approx 1.200$, which is the solution of

$$k \tanh(k) = 1 . \tag{12.12}$$

By continuity, at the bifurcation point the minimum of the functional F_2 goes to zero. This can be tested explicitly by considering the Euler–Lagrange equation for the minimizer of F_2, which is the eigenvalue equation

$$A_1'' + \frac{2k^2}{\cosh^2(kz)} A_1 = \Lambda A_1 . \tag{12.13}$$

The eigenvalue Λ vanishes at the bifurcation point (marginal stability) and the symmetric solution is

$$A_1 = k^* z \tanh(k^* z) - 1 , \tag{12.14}$$

where (12.12) ensures that A_1 satisfies the boundary condition $A_1(1) = 0$.

The catastrophe of the Harris solution has attracted much attention in the context of space plasma activity as a simple model of a current sheet that from quasi-static stable evolution (lowest solution, λ increasing) suddenly

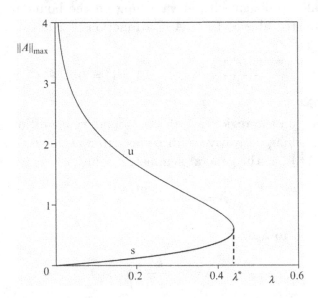

Fig. 12.6 Bifurcation diagram of the problem (12.8). The critical parameter is $\lambda^* = 0.439$. The lower solution (thick line) is F-stable, the upper one (thin line) F-unstable.

enters a dramatic dynamic phase. (For details see Low, e.g., 1977; Birn and Schindler, e.g., 1981.) However, it also has become clear that a region of nonexistence with respect to a particular parameter must not necessarily mean that a catastrophe actually occurs as a physical process.

The problem is that λ must be a control parameter which can be assigned values on both sides of λ^*. This is not obvious in the present case. At first sight, one might be misled to understand the parameter λ in (12.8) as controlling the current density in a monotonic relationship. However, the central current density j_c is given by k^2. Using j_c as control parameter (Fig. 12.7) the catastrophe has disappeared. In fact, its presence in the graph of $\|A\|_{\max}(\lambda)$ results from the non-monotonic relationship between λ and j_c, which is given by $\lambda = j_c / \cosh^2(\sqrt{j_c})$.

This case obviously does not fall under the general class described in the previous section. In fact, the convexity condition (ii) is violated.

A monotonic bifurcation diagram also arises if the integrated current of the sheet is used as the control parameter.

Note that in both cases a slow temporal increase of the control parameter will lead to a transition from a quiescent to a dynamic phase. The reason for the stability transition in the diagram of Fig. 12.7 is that with increasing j_c the current sheet becomes more and more concentrated between the (fixed)

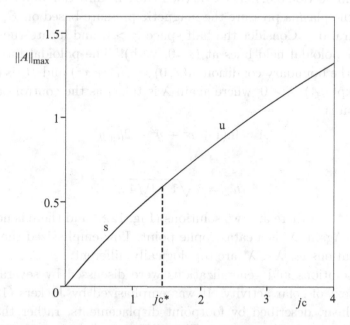

Fig. 12.7 Bifurcation diagram of the problem (12.8) with the central current density j_c as the control parameter; j_c^* corresponds to λ^*.

boundaries at $z = \pm 1$. This is obvious from (12.9), which shows that $1/k = 1/\sqrt{j_c}$ is the width of the Harris sheet profile. The stability transition occurs at the width $1/k^* = 0.834$.

If the sheet profile is expressed in terms of the coordinate kz rather than z, the evolution consists of the boundaries moving outward for a fixed Harris profile, and the stability transition occurs at the critical location 1.200.

It might surprise that the graph in Fig. 12.7 shows a stability transition without a bifurcating solution branch. In fact, there is a bifurcation, but the corresponding solutions are two-dimensional, they do not appear in the present one-dimensional treatment. We will return to this question in Section 12.2.2.

In spite of the strong simplifying assumptions that were imposed on the sequences described above, some of their features will persist when more realistic models are described in the course of this chapter. In particular, the nonequilibrium scenario of Fig. 12.6 and the stability transition of Fig. 12.7 are prototypes of two basic bifurcation processes.

Low's solution

Low (1977) considered a sequence of cylindrical states based on the Bennett pinch solution (5.129). A selected equilibrium of that sequence was already considered in Section 5.3.4. Let z be the invariant direction and assume that instead of the plasma pressure the magnetic pressure based on B_z supports the configuration. Consider the half space $x > 0$ and locate the centre of the circular poloidal field lines at (x=0, y=h). The poloidal flux function A satisfies the boundary condition $A(x,0) = ln(1 + x^2)$ and B_z is chosen as $B_z = \sqrt{\lambda}\exp(-A)$, $\lambda \geq 0$, where again λ is taken as the control parameter. The solution is

$$A = \ln(1 + x^2 + y^2 - 2h_\pm y) \tag{12.15}$$

with

$$h_\pm = \pm\sqrt{1 - \lambda/4}. \tag{12.16}$$

For $\lambda < \lambda^* = 4$ there are two solutions (Fig. 12.8) and there is no solution for $\lambda > \lambda^*$. Again, λ^* is a catastrophe point. Low emphasized the fact that the two solutions for $\lambda < \lambda^*$ are topologically different.

Similar solutions and generalizations were discussed by several authors in the context of solar activity. It was emphasized by Jockers (1978) that magnetic shear, described by footpoint displacements, rather than B_z, is an appropriate control quantity. This difference is a serious problem for $\lambda \geq \lambda^*$; for further remarks on this problem see Section 14.2. Priest and

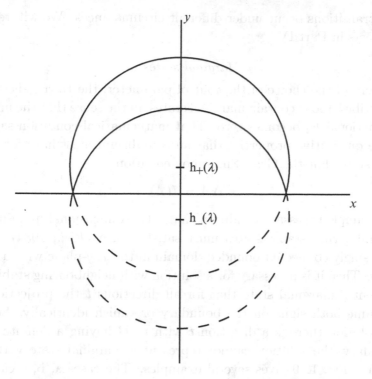

Fig. 12.8 Two field lines of Low's solution (12.15) belonging to the same value of λ. The field lines are helices with circular poloidal projections. There is no solution for $\lambda > 4$.

Milne (1980) found non-unique solutions for fixed shear. A rather general study by Heyvaerts *et al.* (1982) addressed the existence of closed and open MHS configurations. For small current or pressure there always exist solutions with closed topology, but they disappear at a bifurcation point. Open solutions exist for arbitrary currents or pressures. Aly (1995) considered sheared axisymmetric force-free fields. After an initial period of quiet energy gain the field starts expanding at an increasing rate, so that at some stage the quasi-static approximation breaks down. This is interpreted as the *global singular nonequilibrium*. Notably, loss of equilibrium occurs in close connection with the formation of a current sheet. For two-dimensional configurations supported by pressure, Birn and Schindler (2002) constructed a quasi-static sequence that also led to loss of equilibrium by the formation of a singular current sheet (see Section 8.4.2).

Bifurcation-relevant results for three-dimensional configurations largely come from numerical simulations. It seems that both loss of equilibrium and stability transitions occur under different circumstances. We will return to these aspects in Part IV.

Marginal states

Independent of the choice of the control parameter, the Harris sheet evolutions described above contain marginal states, in the sense that the minimum of the functional F_2 becomes zero. That mathematical condition says little about the qualitative properties that an equilibrium must have for being a marginal state. For the Grad–Shafranov equation

$$-\Delta A = J(A) \tag{12.17}$$

there is a simple necessary condition that a two-dimensional marginal state with neighbouring stable states must satisfy (Birn *et al.*, 1984). Let us assume a singly connected bounded domain in the x, y-plane with a smooth boundary. Then it is necessary for a solution with neighbouring stable states to represent a marginal state that for all directions t the projection $t \cdot B$ must assume both signs on the boundary or vanish identically. So, if for a given solution there is a direction t with $t \cdot B$ having a definite sign on the boundary, the solution cannot represent a marginal state with stable neighbours. Fig. 12.9 gives several examples. The cases a, b, c cannot be marginal states because there is a direction t with no change of sign of $t \cdot B$ on the boundary; in the cases d, e, f the criterion is satisfied.

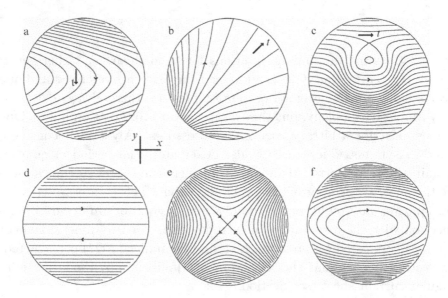

Fig. 12.9 Field lines of 2D magnetic fields that do not satisfy the marginal state condition (a, b, c) and of structures that do satisfy that condition (d, e, f).

This criterion is easily obtained in the following way. In the marginal state the Euler–Lagrange equation for the minimizer A_1 of F_2 is

$$-\Delta A_1 = J'(A)A_1 \,, \tag{12.18}$$

A_1 vanishing on the boundary. If the state A has a stable neighbourhood, A_1 has no zeros. (A_1 is the eigenfunction associated with the lowest eigenvalue of a Sturm–Liouville eigenvalue problem, here the eigenvalue vanishes.)

Let us first choose the Cartesian coordinate direction e_x as the direction t. Differentiating (12.17) with respect to y we find

$$-\Delta B_x = J'(A)B_x \,. \tag{12.19}$$

Eliminating J' from (12.18) and (12.19) one can write the result as

$$\nabla \cdot (B_x \nabla A_1 - A_1 \nabla B_x) = 0 \,. \tag{12.20}$$

With the help of Gauss's theorem and the boundary condition of A_1 one finally obtains

$$\oint B_x \boldsymbol{n} \cdot \nabla A_1 \, dl = 0 \,, \tag{12.21}$$

where the integral is extended over the boundary with normal $\boldsymbol{n}$. As A_1 has no zeros inside the boundary, $\boldsymbol{n} \cdot \nabla A_1$ does not change sign. Therefore, B_x must change its sign or be zero for the integral to vanish. Considering that we could have chosen any direction as the x-direction completes the proof.

Note that for this derivation it is essential that Δ commutes with ∇, which is the case for the Cartesian coordinates employed here. We add that the examples d, e, f in Fig. 12.9 contain neutral points, but this is not a strict requirement for the criterion to be satisfied. A neutral point is present in case c, although the criterion is violated. It seems that the magnetic field of a marginal equilibrium has to be sufficiently structured on the boundary.

12.2 A statistical mechanics approach to 2D current sheet bifurcation

A closed ensemble of particles will eventually relax into thermodynamic equilibrium, in which the current density vanishes. So particular circumstances must prevail to allow a current sheet to persist. Kiessling (1995) has obtained a class of kinetic equations with effective collision operators where the electric field term and the collision term cancel each other, so that the model formally reduces to a two-dimensional Vlasov description with no electric field in the current direction. For such systems the minimum formal requirement for avoiding relaxational decay of the current is the conservation of the total canonical momenta of electrons and ions. The statistical mechanics of such systems was developed and applied to current sheets by Hesse and Schindler (1986); Kiessling *et al.* (1986); Brinkmann (1987); Kiessling and Schindler (1987); Kiessling (1993). The work by Kiessling *et al.* (1986) includes a discussion of the requirements for fluctuation spectra of weak turbulence. Here we use that approach for a two-dimensional bifurcation analysis of current sheets.

12.2.1 Brief outline of the model

Formally, the plasma is treated as a two-dimensional thermodynamic equilibrium in a finite poloidal domain Ω with boundary $\partial\Omega$, coupled to a heat reservoir of temperature T. The system is invariant in the y-direction and it is assumed that the y-component of the canonical momentum of each particle species s remains constant. This is the appropriate statistical mechanics constraint for making sure that the system relaxes into an equilibrium that carries a non-vanishing current. By applying the standard canonical ensemble methods one obtains the following expression for the free energy

$$\mathcal{F} = \int_\Omega \left(\frac{1}{2\mu_0}(\nabla A)^2 - \frac{\epsilon_0}{2}(\nabla\phi)^2 \right) \mathrm{d}^2 r - \frac{1}{2}\sum_s N_s m_s u_s^2$$

$$- \sum_s N_s k_\mathrm{B} T \left(\ln \frac{U_s}{N_s l_s^2} + 1 \right) \quad (12.22)$$

where

$$U_s = \int_\Omega \exp\left(-\frac{q_s}{k_\mathrm{B}T}(\phi - u_s A) \right) \mathrm{d}^2 r \quad (12.23)$$

$$l_s = \sqrt{\frac{h^2}{2\pi m_s k_\mathrm{B}T}} \, . \quad (12.24)$$

Here N_s is the number of particles per unit length in the invariant direction and u_s the y-component of the bulk velocity for species s; the magnetic and electric fields are poloidal with flux function A and potential ϕ, respectively; l_s is the thermal de Broglie wavelength of species s.

Introducing quasi-neutrality with $N_\mathrm{i} = N_\mathrm{e} = N$ and dimensionless notation, assuming an electron–proton plasma and omitting an additive constant, one finds that (12.22) reduces to

$$\mathcal{F} = \frac{1}{\lambda} \int_\Omega (\nabla A)^2 \, \mathrm{d}^2 r - \ln \int_\Omega \mathrm{e}^{2A} \, \mathrm{d}^2 r \quad (12.25)$$

with

$$\lambda = \frac{\mu_0 I^2}{4 N_s k_B T} . \tag{12.26}$$

Here $I = Ne(u_i - u_e)$ is the electric current; A is normalized by $4Nk_B T/I$ and coordinates by an arbitrary length.

Importantly, a single parameter λ appears, which is interpreted as the natural control parameter.

Equilibria are characterized by extrema of the free energy, obtained from setting the first variation of (12.25) to zero. This gives the equilibrium equation

$$-\Delta A = \lambda \frac{e^{2A}}{\int_\Omega e^{2A} \, d^2 r} , \tag{12.27}$$

which is an integro-differential equation for A. For any given solution the integral in (12.27) is constant and (12.27) has the form of a Grad–Shafranov equation, but the integral plays an important role in the stability and bifurcation properties.

In the sense of statistical mechanics, an equilibrium described by a solution of (12.27) is locally stable if at the equilibrium the free energy (12.25) assumes a local minimum, it is globally stable if the minimum is a global one. (Note that this is consistent with the stability concept of Section 10.1 and the elementary examples of Fig. 12.1.)

So, local stability is determined by the second variation of $\mathcal{F}$

$$\mathcal{F}_2 = \frac{1}{\lambda} \int_\Omega \left((\nabla A_1)^2 - \frac{2\lambda \exp(2A)}{\int_\Omega \exp(2A) \, d^2 r} (A_1^2 - \widetilde{A_1}^2) \right) d^2 r , \tag{12.28}$$

where the tilde symbol denotes the average

$$\widetilde{\cdots} = \frac{\int_\Omega \cdots \exp(2A) d^2 r}{\int_\Omega \exp(2A) d^2 r} . \tag{12.29}$$

An equilibrium is locally stable if and only if $\mathcal{F}_2$ is positive definite with respect to test functions A_1 vanishing on the boundary. As it is the case for the variational expressions considered in Sections 10.2 and 10.5 this criterion can also be expressed in terms of the Euler–Lagrange equation for the minimizer, which becomes the eigenvalue problem

$$-\Delta A_1 - \frac{2\lambda \exp(2A)}{\int_\Omega \exp(2A) d^2 r} (A_1 - \widetilde{A_1}) = \eta A_1 , \tag{12.30}$$

where η is the eigenvalue. The sign of the lowest eigenvalue determines local stability.

Notably, this model can also be based on a statistical mechanics treatment of interacting line currents (Kiessling, 1995). This approach most directly exhibits the close analogy with 2D gravitational interaction and with systems of line vortices. The common aspect is 2D attractive interaction with the force varying as $1/r$ with separation r.

12.2.2 Bifurcations of the Harris sheet

Let us apply this approach to the Harris sheet equilibrium. For a rectangular domain $0 \leq x \leq a$, $-b/2 \leq z \leq b/2$ the corresponding solution of (12.27) is

$$A_{\mathrm{H}} = - \ln \cosh(\nu z) \tag{12.31}$$

$$\lambda = 2a\nu \tanh(\nu b/2) \,, \tag{12.32}$$

where $\nu > 0$ (the inverse thickness of the sheet) acts as a parameter so that A can be understood as a function of x, z, λ and A_{H} defines the (Dirichlet) boundary condition.

As in the case of fixed current density of Fig. 12.7, but unlike the case of the catastrophe of Fig. 12.6, the solution exists for all positive λ. So, the next problem is to find the stability transitions.

The eigenfunction corresponding to the lowest eigenvalue was found to be (Zwingmann, 1983; Hesse and Schindler, 1986)

$$A_{1,0} = \bigl(\sinh(\mu\nu z) \tanh(\nu z) - \mu \cosh(\mu\nu z) \bigr) \sin(2\pi z/a) \tag{12.33}$$

where μ is determined by the boundary condition, such that

$$\tanh(\mu\nu b/2) \tanh(\nu b/2) - \mu = 0 \,. \tag{12.34}$$

The lowest eigenvalue is

$$\eta_0 = \frac{4\pi^2}{a^2} - \mu^2 \nu^2 \,. \tag{12.35}$$

$A_{1,0}$ was found to be the eigenfunction corresponding to the lowest eigenvalue (Hesse and Schindler, 1986; Kiessling and Schindler, 1987) although it has zeros inside the domain, which would not occur for Schrödinger-type problems.

Combining (12.32), (12.34) and (12.35) one finds

$$\lambda = \frac{2\sqrt{4\pi^2 - \eta_0 a^2}}{\tanh(\sqrt{4\pi^2 - \eta_0 a^2}\,\frac{b}{2a})} \tag{12.36}$$

which shows that η_0 decreases monotonically from its value $4\pi^2/a^2$ and changes its sign at

$$\lambda^* = \frac{4\pi}{\tanh(\frac{\pi b}{a})} . \tag{12.37}$$

This is a bifurcation point, where a new branch bifurcates from the Harris solution (Hesse and Schindler, 1986). The branch is back-bending and therefore unstable. The full bifurcation diagram is shown in Fig. 12.10 for three different values of the parameter b/a.

For $b/a = \infty$ the bifurcating solutions are the cat's eye solutions (5.131). Here they take the form

$$A = -\ln\left(\frac{\cosh(2\pi z/a) + p\sin(2\pi x/a)}{\sqrt{1-p^2}}\right). \tag{12.38}$$

The (maximum) norm becomes $\|A - A_H\| = 0.5\ln[(1+p)/(1-p)]$. As this norm is independent of λ the branch is vertical. On this branch the parameter p increases from $p = 0$ (Harris sheet) to $p = 1$, where the current becomes concentrated as a line current located at $(x = 0.75a, z = 0)$ so that the norm becomes infinite. Including this solution, we have left the realm of continuous functions and turned to generalized functions (distributions). Remarkably, here we are dealing with one of the rather rare occasions where

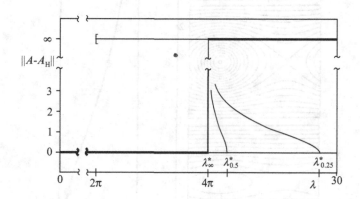

Fig. 12.10 Bifurcation diagram (maximum norm of $(A - A_H)$ vs. λ) of the problem (12.27). On the boundary of a rectangular domain with sides a, b the solutions assume the values of the Harris sheet solution A_H given by (12.31). The lower horizontal line (vanishing norm) corresponds to the Harris sheet solution, the upper horizontal line refers to line current solutions with infinite norm. Branches bifurcating off the Harris sheet solution are shown for $b/a = 0.25$, $b/a = 0.5$, $b/a = \infty$, with bifurcation points at $\lambda^*_{0.25} = 19.162$, $\lambda^*_{0.5} = 13.702$, $\lambda^*_{\infty} = 4\pi$. Thick lines indicate global stability. (Reused with permission from A. Schröer *et al.*, copyright 1994, AIP.)

such mathematical subtlety has important physical implications (Hesse and Schindler, 1986; Kiessling, 1989, 1993; Schröer *et al.*, 1994).

The singular cat's eye solution turns out to be a member of a whole branch of infinite norm solutions covering the range of $\lambda \geq 2\pi$. For these states free energy has the following signature: $\mathcal{F} = +\infty$ for $2\pi \leq \lambda \leq 4\pi$ and $\mathcal{F} = -\infty$ for $\lambda > 4\pi$. Assuming that the bifurcation diagram of Fig. 12.10 is complete, we find that for $\lambda < 4\pi$ the Harris sheet has lowest free energy and therefore is globally stable, while for $\lambda > 4\pi$ the singular branch has lowest free energy and is the globally stable solution. Local stability changes at $\lambda = \lambda^*$, as discussed above. In the range $4\pi < \lambda < \lambda^*$ the Harris sheet is metastable, i.e., locally but not globally stable.

The qualitative structure of the solutions lying on the bifurcating branches corresponds to the structure of the cat's eye solutions at $\lambda = 4\pi$. This is illustrated in Fig. 12.11 for the choice $b/a = 0.85$. In fact the bifurcating

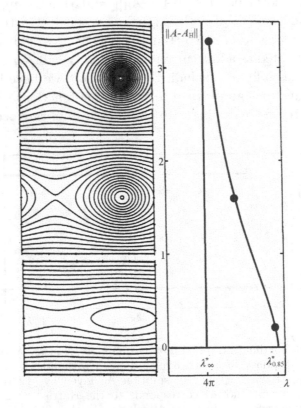

Fig. 12.11 Details of solutions on the bifurcating branch for $b/a = 0.85$ (Fig. 12.10), shown on the right; the dots mark the solutions (level curves of A) that are plotted on the left in the same vertical sequence. (Reused with permission from A. Schröer *et al.*, copyright 1994, AIP.)

branch seems to approach the cat's eye branch asymptotically as the norm becomes large.

12.2.3 Breakdown of a constraint

In applications one sometimes finds that the actual control parameter values lie far on the unstable side of a bifurcation point. Let us assume, for example, that in the model of the previous section one chooses $b/2 = 10$ and $a = 100$, with the width of the current sheet set to 1, so that $\nu = 1$. This gives $\lambda^* \approx 22.6$ and $\lambda \approx 200$. This would mean that the sheet is in the unstable regime. If such states exist it must mean that the relaxation process is inhibited; in other words, a realistic description is obtained by imposing a suitable constraint. In the present case such a constraint could be the conservation of magnetic topology. This would exclude the appearance of closed field lines, which seem to be typical for the bifurcating states (see Fig. 12.11). If for some reason (not included in the model) at some value of the control parameter this constraint breaks, the Harris sheet will collapse into a configuration with strong localization of the current. Note that this is qualitatively the same conclusion that was reached in the plasmoid formation scenario of Section 11.2.10, although here we are not referring to a particular instability such as the tearing mode. So, the breakdown of a constraint appears as a powerful notion in space plasma activity.

12.3 Role of perturbations

Here we address external perturbations superimposed on systems in quasi-static evolution. As we will see, such perturbations will have drastic effects near bifurcation points.

For orientation let us again consider the Grad–Shafranov problem (12.1) with Dirichlet conditions on the boundary $\partial\Omega$ of a finite 2D domain Ω. The flux function A_0 of the unperturbed problem satisfies

$$\Delta A_0 + \lambda e^{2A_0} = 0, \tag{12.39}$$

with

$$A_0 = g_0(r), \quad r \in \partial\Omega. \tag{12.40}$$

A solution $A_0(r, \lambda)$ is assumed to be located on the lower branch of the bifurcation diagram of Fig. 12.6.

The perturbation consists in a change of the boundary condition with the differential operator unchanged, so that

$$\Delta A + \lambda e^{2A} = 0, \qquad A = g_0(\boldsymbol{r}) + h(\boldsymbol{r}), \quad \boldsymbol{r} \in \partial\Omega, \tag{12.41}$$

where h is the perturbation.

It is convenient to make a transformation to a problem with unperturbed boundary values and perturbed differential equation. This is easily done by introducing the potential field $u(\boldsymbol{r})$ with $u(\boldsymbol{r}) = h(\boldsymbol{r})$, $\boldsymbol{r} \in \partial\Omega$. Then, let us write

$$A = A_0 + u + a \tag{12.42}$$

where a represents the perturbation of the flux function with $a = 0$ on the boundary. Substituting (12.42) in (12.41) and linearizing in the perturbation quantities u and a, we obtain

$$Ga = I, \qquad a(\boldsymbol{r}) = 0, \quad \boldsymbol{r} \in \partial\Omega, \tag{12.43}$$

where G is the operator $G = \Delta + 2\lambda \exp(2A_0)$ and I the inhomogeneity $I = -2\lambda \exp(2A_0)u$, which is treated as known.

G generates the eigenvalue problem

$$Gv_j = \eta_j v_j \tag{12.44}$$

with eigenvalues η_j and eigenfunctions v_j.

The problem (12.43) is integrable because on the lowest branch for $\lambda < \lambda^*$ all eigenvalues are positive (a direct consequence of F-stability). So there is no vanishing eigenvalue (empty kernel) and the integrability condition is satisfied trivially.

However, as λ approaches λ^* the perturbation a becomes larger and larger. This divergence can only be avoided for the case where I and v_0 are orthogonal at $\lambda = \lambda^*$; then a solution exists for $\lambda = \lambda^*$ also. This case, however, is highly exceptional and can be disregarded in view of the arbitrariness of the perturbation.

These properties are obtained explicitly by expanding a and I with respect to v_j (assuming absence of degeneracy for simplicity)

$$a = \sum_j a_j v_j, \quad I = \sum_j I_j v_j, \tag{12.45}$$

which gives $a_j = I_j/\eta_j$ so that

$$a = \sum_j \frac{I_j}{\eta_j} v_j. \tag{12.46}$$

The term $j = 0$ diverges as $\eta_0 \to 0$ as $\lambda \to \lambda^*$. (The exception occurring for $I_0 = 0$ is disregarded.) Near the bifurcation point a becomes approximately $I_0 v_0 / \eta_0$.

For an illustration let us consider a sheet $A_0 = -\ln \cosh z$ with varying extent in the x-direction. The domain is given by $-x_{\mathrm{m}} \leq x \leq x_{\mathrm{m}}$, $-z_{\mathrm{m}} \leq z \leq z_{\mathrm{m}}$, where z_{m} is fixed and x_{m} takes the role of the control parameter (Fig. 12.12). From the examples discussed above we expect destabilization for increasing domain size. The perturbation is a triangular indentation, vanishing at $x = \pm x_{\mathrm{m}}$, which is represented by an infinite Fourier series. One finds

$$A = A_0 + \hat{a}\frac{N}{2N_{\mathrm{m}}} + \frac{4\hat{a}}{\pi^2} \sum_{j=1}^{\infty} \frac{1}{(2j-1)^2} \frac{w_j}{w_{j\mathrm{m}}} \cos(k_j x), \qquad (12.47)$$

where $k_j = \pi(2j-1)x_{\mathrm{m}}$, $N = z\tanh z - 1$, $N_{\mathrm{m}} = z_{\mathrm{m}}\tanh z_{\mathrm{m}} - 1$, $w_j = k_j \cosh(k_j z) - \tanh(z)\sinh(k_j z)$, $w_{j\mathrm{m}} = k_j \cosh(k_j z_{\mathrm{m}}) - \tanh(z_{\mathrm{m}})\sinh(k_j z_{\mathrm{m}})$ and $\hat{a}$ is a constant amplitude factor.

A singularity of A would occur if one of the k_j coincided with k^*, the root of

$$k^* \cosh(k^* z_{\mathrm{m}}) - \tanh(z_{\mathrm{m}})\sinh(k^* z_{\mathrm{m}}) = 0, \qquad (12.48)$$

so that one of the denominators $w_{j\mathrm{m}}$ vanishes. At the bifurcation point this takes place for the smallest value of k so that the critical value of x_{m} is given by $x_{\mathrm{m}}^* = \pi/k^*$. In Fig. 12.12 z_{m} was chosen as 2.5, which corresponds to $k^* = 0.971$ and $x_{\mathrm{m}}^* = 3.23$. The increase of the effect of the perturbation is seen clearly. Note that the maximum boundary indentation is kept fixed.

It is a rather general property that near bifurcation points a system becomes extremely sensitive to external perturbations. This is analogous to

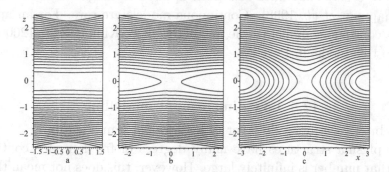

Fig. 12.12 Field lines of a perturbed Harris sheet for $x_{\mathrm{m}} = 0.5x_{\mathrm{m}}^*$ (a), $x_{\mathrm{m}} = 0.75x_{\mathrm{m}}^*$ (b), $x_{\mathrm{m}} = 0.95x_{\mathrm{m}}^*$ (c); x_{m} is the halfwidth in the x-direction and x_{m}^* the bifurcation point value. The z-halfwidth is 2.5, so that x_{m}^* becomes 3.23.

the resonances of forced oscillators. The linear perturbation analysis breaks down, so that the dynamics becomes nonlinear. Therefore, realistic theoretical descriptions of the bifurcation dynamics must take nonlinear processes into account (Kropotkin *et al.*, 2002a,b).

12.4 Self-organized criticality

As discussed above, a system that is in a quasi-statical development along a stable branch by external driving, typically turns to a fast dynamic evolution at a bifurcation point. The state corresponding to the bifurcation point is called *critical state*. In most cases the subsequent dynamics carries the system away from the critical state.

However, there is the possibility of an interesting exception. The nonlinear evolution might have the effect that it restores criticality in some average sense, so that the system can be described as a critical state with a superimposed noise. Such systems were suggested by Bak *et al.* (1988). They have critical states that are scale invariant, corresponding to power law spectra of the noise and remain close to the critical state. This property has been termed *self-organized criticality*.

A widely discussed (however possibly over-idealized) prototype model of self-organized criticality is a sand pile, with sand grains added continuously, so that the accretion leads to a steepening of the slope. If a critical slope is surpassed, sand avalanches are visualized to restore the critical slope as the background state, the noise being represented by the avalanches. In a large system (formally infinite) there will be no preferred scale of the avalanches.

Typical statistical realizations of self-organized criticality use cellular automata models (Dendy and Helander, 1998), where the system is represented by discrete cells. They have also been applied to magnetospheric activity (e.g., Chapman *et al.*, 1998). Continuous models are rare. For an application of a continuous model to the magnetotail see Klimas *et al.* (2000). We return to this model in Part IV.

12.5 Low-dimensional modelling

Here we briefly discuss a different type of model simplification.

Space plasma structures have many degrees of freedom; for continuum models that number is infinitely large. However, this does not mean that all phenomena involve the entire set. In fact, it is possible that the time-series characteristics of a given observable can approximately be reproduced from a low-dimensional description. In the simplest case the model is assumed to be autonomous, i.e., the dynamics is determined by internal properties alone. Another possibility is to take external influences into account explicitly. In the space plasma context, such models are usually referred to as *input–output models*.

A famous example of an autonomous low-dimensional representation of a continuous system is the Lorentz attractor (e.g., Lichtenberg and Liebermann, 1983), a model of atmospheric convection described by an autonomous system of three ordinary differential equations. Its chaotic solutions lead to an essentially unpredictable long-term behaviour.

In view of the obvious advantages of a low-dimensional autonomous description considerable effort was devoted to apply this method to space plasmas. For magnetospheric activity the AE index, a measure of the auroral electrojet, was used to determine whether a low-dimensional representation exists and, if so, what dimension could be assigned to it. However, despite a considerable number of statistical studies, a clear picture has not emerged. A review is contained in Klimas *et al.* (1996). Intuitively, it seems plausible to expect that the solar wind conditions have to be taken into account as an external agent to explain AE statistics. The solar wind could keep the magnetosphere from converging to a low-dimensional attractor even if it existed (Takalo *et al.*, 1994).

Low-dimensional input–output models have been more successful than the autonomous descriptions. There is no general recipe for the construction of such models. It is based on intuition and trial and error. For magnetospheric activity, there are models using analogous systems from different contexts for orientation (e.g., the *dripping faucet model* (Hones, 1979; Baker *et al.*, 1990)) and models selecting particular physical aspects of the actual system (e.g., the *Faraday loop model* (Klimas *et al.*, 1992) or the energy conserving model by Horton and Doxas (1996)).

Input–output models can play a valuable role in improving the understanding of the process in question on an intuitive level. However, it cannot fully replace a description that is justified by physical principles entirely.

Part IV
Applications

The main challenge for explaining space plasma activity is to understand the transition from a quiescent phase, during which energy is slowly accumulated, to a dynamic phase, during which energy is released in a fast evolution of the system. Examples have been described phenomenologically in Part I. Here we will use the theoretical tools provided in Parts II and III to address this problem in a more fundamental way.

13

Magnetospheric activity

In this chapter we will address the activity of the Earth's magnetosphere, with emphasis placed on the substorm, which is regarded as the dominant dynamical process of the magnetosphere. What can be expected from our theory-oriented approach? Certainly not deterministic predictions, which are excluded not only by the limitations of the present state of the theory, but also by more fundamental properties such as the chaotic nature of large particle ensembles. Rather, the realistic question is this: Equipped with the tools of Parts II and III, how far do we get? Can the tools help us to make the step from a mere phenomenological picture to a description which allows insights and interpretations in terms of physical processes? Where this goal is not reached, can we at least identify realistic possibilities?

To pursue this line, substantial observational input is required, which means that we abandon the strict theoretical point of view, which was appropriate in Parts II and III to generate a set of tools. (Even there, the selection of problem areas, the choice of parameter regimes or of simplifications were influenced by observations.)

A full discussion of all processes observed to be related to magnetospheric activity is far beyond the present scope. In particular, this applies to the wealth of ionospheric phenomena. The aim is to understand the physics of the large-scale magnetospheric phenomena.

13.1 Interaction between the solar wind and the magnetosphere

We begin with some fundamental aspects of the interaction between the solar wind and the magnetosphere. We largely use geocentric solar-magnetospheric coordinates, the x-axis pointing sunward, the z-axis arranged so that the Earth's dipole moment lies in the x, z-plane and the y-direction completing a right-handed Cartesian coordinate system. For most purposes we consider the simplified configuration with the Earth's dipole moment pointing in the z-direction. (Note that in some of the theoretical studies of the magnetotail the x-axis points tailward.)

13.1.1 *Open versus closed magnetosphere*

Let us, for a moment, consider the consequences of a hypothetical model, which describes the solar wind/magnetosphere interaction by ideal MHD, including MHD discontinuities (Section 3.9). As the Earth's dipole field is an obstacle in the supersonic solar wind, a fast MHD shock wave, the bow shock, stands in front of the magnetosphere. The magnetopause is a tangential discontinuity, confining the geomagnetic field. It is a necessary element of the initial value problem (an analogous case was discussed in Section 11.2.6); the magnetized solar and magnetospheric plasmas cannot mix, because magnetic line conservation (Section 3.8.2), which holds in our hypothetical model, would be violated. The solar wind, when switched on, would sweep the magnetospheric medium toward the Earth until pressure equilibrium is reached. The magnetosphere would be completely closed and, as the magnetopause would separate media of different origin, we can expect the magnetopause to carry a current (Section 8.6.2).

In spite of the fundamental deficiencies of such a model, which are described below, it is not completely off the mark; it can explain global features such as the positions of the bow shock and of the frontside magnetopause reasonably well (Spreiter *et al.*, 1966). It drastically fails, however, in other respects. For example, for a closed magnetosphere the interaction with the solar wind would be independent of a reversal of the directions of the interplanetary magnetic and electric fields. It follows from the structure of the ideal MHD equations (Section 3.3.2) that this reversal would not affect the hydrodynamic variables and the magnitude of the magnetic field B. In particular the total pressure $(p + B^2/2\mu_0)$ at the magnetopause remains unchanged. Since in MHD, besides vanishing normal velocity and magnetic field components, total pressure balance is the only condition for a tangential discontinuity, the field reversal has no consequences for the interaction.

The closed magnetosphere is in conflict with both observational facts and theoretical results. Observationally, one finds that magnetospheric dynamics, and in particular substorm dynamics, is strongly correlated with the sign of the z-component of the interplanetary magnetic field (IMF). This is based on an impressive history of studies investigating the correlation between interplanetary signatures with geomagnetic disturbances in the auroral zones (e.g., Bargatze *et al.*, 1985, 1999). In one of the earlier studies

Arnoldy (1971) correlated the *AE index*, a quantitative measure of auroral zone magnetic activity (Davis and Sugiura, 1966), with B_s, which is set to zero for IMF $B_z \geq 0$ and to $|B_z|$ for southward orientation of B_z. On the basis of hourly averages a correlation coefficient of 0.8 was obtained, while other solar wind variables produced coefficients < 0.4. Further studies introduced more refined coupling expressions (e.g., Perreault and Akasofu, 1978; Bargatze *et al.*, 1985); for instance, B_s was replaced by the dawn–dusk electric field component $V_{sw}B_s$ (Burton *et al.*, 1975), where V_{sw} is the solar wind velocity. All studies, although varying in details, confirmed that the southward directed IMF component plays a central role in determining the strength of energy coupling between the solar wind and the terrestrial magnetosphere (Bargatze *et al.*, 1999), so that these (as many other) observations are in conflict with the closed magnetosphere model.

A conceptual model of an open magnetosphere by Dungey (1961) removes this difficulty, at least qualitatively. The magnetic topology and the resulting flow pattern are illustrated in Fig. 13.1 for a southward-pointing IMF. If the IMF was northward the structure would be quite different, as discussed later. So, in this model the magnetospheric structure is strongly influenced by the direction of the IMF z-component.

Clearly, Dungey's model involves magnetic reconnection (Section 11.2.1). (In fact, the term *magnetic reconnection* was suggested by Dungey in this

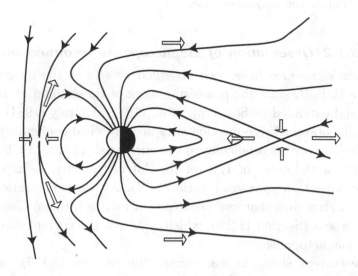

Fig. 13.1 Sketch of magnetic field lines of Dungey's open magnetosphere in the non-midnight meridian plane; the plasma flow is indicated by the open arrows.

context.) In the midnight meridian plane shown in Fig. 13.1 there is a recon-
nection site at the front side and one at the night side of the magnetosphere.
As reconnection involves nonideal processes, the open topology violates ideal
MHD, consistent with the above reasoning.

What do the theoretical tools teach us about the issue of closed versus
open magnetosphere? An important fact is that a collisionless current sheet
like the magnetopause can become the site of magnetic reconnection. The
collisionless tearing instability (Section 10.4) occurs under a variety of circumstances. Although single modes might saturate at small amplitudes,
mode coupling can lead to further growth; more details are given below.
Also, external driving by local perturbations can speed up the instability,
so that a nonlinear reconnection pattern would arise (Section 11.3.3). This
scenario would have the consequence that an initially closed magnetopause
would open by magnetic reconnection. The details would depend on the ef-
fectiveness of the opening and on the orientation of the interplanetary field.
Intuitively, one expects most efficient reconnection for southward interplanetary field orientation, when the magnetospheric and interplanetary fields
are antiparallel at the subsolar magnetopause region. These arguments give
strong support to Dungey's field topology. For an MHD model of a complete
open magnetotail boundary see Siscoe and Sanchez (1987).

On purely theoretical reasoning, the opening of the magnetosphere, although fairly plausible, cannot be considered as rigorously established.
Therefore, it seems advisable to test it by looking at *in situ* observations
of reconnection at the magnetopause.

13.1.2 *Observation of magnetopause reconnection*

Reconnection signatures have been observed *in situ* at the magnetopause
for negative IMF B_z since the pioneering work of Paschmann *et al.* (1979),
confirmed and extended by Sonnerup *et al.* (1981), Cowley (1984) and others. Typically, observed reconnection sites are stretched configurations with
strong variation only perpendicular to the current sheet (Fig. 13.2). Plasma
velocities are of the order of 1/10 of the Alfvén velocity, as expected for
standard reconnection processes (Section 11.2). As the configuration is non-
symmetric, a structure that can be idealized as a rotational discontinuity
should be present (Section 11.2.6), which locally is the current sheet identified as the magnetopause.

Near the current sheet, in many cases the relation (3.113), called the
Walén relation, is approximately satisfied, which is an important consistency
check.

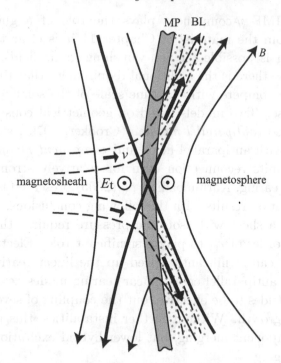

Fig. 13.2 Magnetopause reconnection layer; MP refers to the magnetopause, BL to the boundary layer, full lines are magnetic field lines, the plasma flow is indicated by velocity vectors (v) and stream lines (dashed lines), E_t is the tangential electric field component (after Sonnerup *et al.* (1981) by permission of the American Geophysical Union).

Many more observations have confirmed the occurrence of reconnection at the magnetopause, but they have also shown that more details have to be added to the simple picture of Fig. 13.2. Let us look at a few examples.

Evidence for slow shocks (Section 3.9), which are to be expected to form in Petschek-type reconnection (Section 11.2.4), has been reported in a few cases. In one of the best documented cases pressure anisotropy was found to play an important role (Walthour *et al.*, 1994). Further, a slow shock may form an intermediate shock (Section 3.9) by combining with the rotational discontinuity (Wu, 1987).

Reconnection does take place not only at the subsolar magnetopause with the interplanetary magnetic field pointing southward, but under much more general conditions. If the IMF z-component is positive, there are still positions at the magnetopause favourable for reconnection. Particularly, this applies to the high latitude magnetosphere tailward of the cusp region (see Fig. 2.1); for an example of a corresponding observation see Retinò *et al.* (2005).

A significant IMF y-component plays the role of a guide field (Section 11.2.7). From the material of Chapter 11 it is clear that, in principle, reconnection is possible for non-vanishing guide fields, also. But at the magnetopause there is the additional requirement that the reconnection time scale has to compete with the time scale of the solar wind flow along the magnetopause. Two models based on geometrical considerations have been distinguished, *antiparallel merging* (Crooker, 1979), strongly emphasizing locations with antiparallel fields and *component merging* (Sonnerup, 1974), which admits reconnection even for relatively strong guide fields. For collisionless tearing, Karimabadi *et al.* (2005) presented theoretical estimates and simulation results with the following conclusions. A single linear tearing mode in a sheet with isotropic pressure requires thin sheets with thickness of the order of r_{gi} to play a significant role. Electron anisotropy with $T_\perp/T_\parallel > 1$ can significantly speed up the linear tearing mode (Section 10.4) in the antiparallel case. Linear tearing modes generally saturate at too low amplitudes to be effective, but the coupling of several modes can lead to realistic growth. Without further instabilities, this picture is compatible with component merging, but lower-hybrid excitation could favour anti-parallel merging.

Significant parallel electric field components (Section 11.5.1) were found in global simulations (Siscoe *et al.*, 2001) as well as in local magnetopause observation (Scudder *et al.*, 2002).

Multiple satellite observations have discovered many more details than single spacecraft can provide. For instance, observations by the CLUSTER satellites have revealed a case with a complex magnetic structure (Louarn *et al.*, 2004), which is similar to the flux linkage shown in Fig. 11.21. Also the interpretation, in terms of the interaction between two neighbouring reconnection events, is the same.

13.1.3 Alternative processes opening the magnetopause

The direct observation of reconnection at the magnetopause does not rule out other modes of interaction between the solar wind and the magnetosphere, taking place at other times and locations. Observations pertaining to that question relate the polar cap potential ϕ_{PC} to the interplanetary electric field (IEF). This is relevant for the following reason. In the closed ideal MHD magnetosphere there is no tangential electric field at the magnetopause, which follows directly from ideal Ohm's law (3.3.2). (If the magnetopause is locally in motion, this statement applies to a local co-moving frame.) In other words, in a quasi-steady state the interplanetary electric field cannot penetrate the magnetopause, it is completely shielded. Magnetic reconnection changes this picture drastically, because it leads to a tangential electric field at the magnetopause (Fig. 13.2). If one makes the (questionable) assumption that the inner magnetospheric flow is ideal and stationary, the electric potential ϕ would be constant on magnetic field lines. (Ideal Ohm's law (Section 3.3.2) gives $\boldsymbol{B} \cdot \nabla\phi = 0$.) Therefore, the tangential magnetic field inside the magnetopause would lead to a potential difference across the polar cap. If the observed polar cap potential ϕ_{PC} was entirely due to reconnection, it would be (approximately) proportional to the IEF (for IMF $B_z < 0$). Observations confirm that property, but only above some background value ϕ_0. The latter is interpreted as resulting from entry processes that do not involve reconnection as the primary process. For pronounced negative values of the IMF z-component, Burke *et al.* (1999) found values of ϕ_{PC} above 200 kV, with ϕ_0 in the range of 30 to 40 kV; this is just one example of a substantial number of similar investigations. The results indicate the role of magnetopause reconnection as the dominant solar wind/magnetosphere coupling process.

The processes that generate ϕ_0 are not yet clearly identified. Viscous drag (Axford and Hines, 1961), possibly in combination with Kelvin–Helmholtz instabilities, might play a role.

If Kelvin–Helmholtz modes are important one would expect to observe them when the magnetopause reconnection is weak, i.e. for strongly northward IMF. This has indeed been observed (Fairfield *et al.*, 2000). It was also seen in simulations (Otto and Fairfield, 2000), suggesting that some plasma entry is made possible by reconnection occurring inside the Kelvin–Helmholtz vortices.

13.1.4 Flux transfer

As a result of magnetopause reconnection the IMF connects to the geomagnetic field. If the medium away from the reconnection site has scales much larger than the intrinsic plasma scales (Section 2.1) one can expect that there the flow can be modelled approximately as an ideal fluid. Then, due to the frozen-in property (Section 3.3.2), the solar wind would drag the reconnected flux along the magnetopause to the night side, so transferring magnetic flux to the magnetotail. If the original reconnection takes place as an individual burst, the result is a *flux transfer event* (FTE) (Russell and Elphic, 1978). For *component merging* a rope-like structure with a spiralling internal field would form, called *magnetic flux rope* (Fig. 13.3).

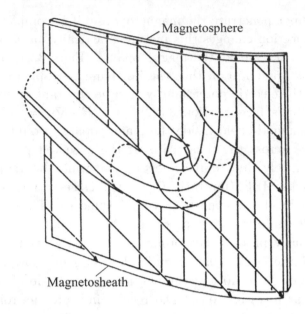

Fig. 13.3 Sketch of the formation of a magnetic flux rope during a flux transfer event (from Russell and Elphic (1979) by permission of the American Geophysical Union).

Multiple spacecraft observations have largely confirmed that picture (e.g., Sonnerup *et al.*, 2004). Wild *et al.* (2005) analysed FTE signatures that were observed by both CLUSTER and GEOTAIL missions.

13.2 Convection and pressure crisis

Here we enter the central part of our discussion of geomagnetic activity, dealing with the further fate of the magnetic flux and energy supplied to the magnetotail by flux transfer from the front side magnetosphere.

13.2.1 Convection

Dungey's open topology implies that the magnetic flux which is transferred to the magnetotail again reconnects at the distant neutral line and the earthward part (termed *closed field region*) returns to the inner magnetosphere. Let us first concentrate on the midnight meridian plane. There, a simple flow field that provides this return is indicated in Fig. 13.1. The plasma velocity component perpendicular to the magnetic field is referred to as *convection velocity*.

In this picture the flux return in the closed field region is of particular interest. For getting insight into the underlying physics it is convenient to proceed in the following, admittedly somewhat unusual, fashion. We will formulate a set of test assumptions and prove that under these assumptions the flux return is not possible. This will teach us that at least one of the assumed properties must be violated, which will provide valuable input to the further reasoning.

Test assumptions

The test assumptions concern a smooth tail configuration with a quasi-neutral collisionless electron/proton plasma that evolves in response to the flux transfer from the front side magnetosphere. The main assumptions are

  (i) large scale quasi-static flux return,
  (ii) isotropic pressure tensors, adiabatic pressure laws with $\gamma = 5/3$,
  (iii) $U \gg k_B T_i/e$,
  (iv) temperatures constant on field lines and negligible parallel losses from flux tubes.

Here, U is the cross-tail electric potential, *parallel* refers to the direction parallel to the magnetic field and *large scale* means that spatial and temporal scales are large compared with the intrinsic plasma scales (Section 2.1). Gravity is ignored which is an excellent approximation for most magnetospheric phenomena. (For plasma sheet conditions (Section 2.1) the gravity scale height (Section 5.1.1) exceeds 1000 R_E.)

The assumptions (i)–(ii) allow us to specialize the fluid equations of Section 3.3 as ($s = i, e$)

$$\frac{\partial n}{\partial t} + \nabla \cdot (n \boldsymbol{v}_s) = 0 \tag{13.1}$$

$$-\nabla p_s + q_s n(\boldsymbol{E} + \boldsymbol{v}_s \times \boldsymbol{B}) = 0 \tag{13.2}$$

$$\frac{\partial}{\partial t}\left(\frac{p_s}{n^\gamma}\right) + \boldsymbol{v}_s \cdot \nabla\left(\frac{p_s}{n^\gamma}\right) = 0 . \tag{13.3}$$

Adding the equations (13.2) for electrons and ions to form the total momentum equation and using the assumptions (iii) and (iv) in (13.3) one finds

$$0 = -\nabla p + \boldsymbol{j} \times \boldsymbol{B} \tag{13.4}$$

$$pV^\gamma \text{ conserved for each flux tube,} \tag{13.5}$$

where V is the flux tube volume (8.26).

The significance of property (iii) can be illustrated by combining (13.2) and (13.3) to obtain

$$\frac{\partial}{\partial t}\left(\frac{p}{n^\gamma}\right) + \boldsymbol{u}_E \cdot \nabla\left(\frac{p}{n^\gamma}\right) + \frac{\gamma}{2B^2 en^{\gamma+2}}\boldsymbol{B} \times \nabla n \cdot \nabla(p_i^2 - p_e^2) = 0 \,, \qquad (13.6)$$

where $\boldsymbol{u}_E$ is the $\boldsymbol{E} \times \boldsymbol{B}$ drift velocity (Section 3.2). Under assumption (iii) the last term is negligible (for the scaling (5.171)). Also, Ohm's law reduces to its ideal form.

The pressure crisis and its consequences

As shown in Section 8.3.3 in detail, a 2D model satisfying the test assumptions predicts a serious problem for the assumed steady state convection, manifested by the *pressure crisis* or *pressure balance inconsistency*. This crisis has a simple reason (Erickson and Wolf, 1980; Schindler and Birn, 1982): During flux return, the plasma that originally was contained in a large flux tube extending far into the tail would be strongly compressed during the earthward convection, when the flux tube volume decreases drastically.

Fig. 8.3 illustrates the pressure crisis by a plot of the pressure in the centre of the plasma sheet versus distance from the Earth for the 2D steady state polytropic model. In the near-Earth plasma sheet the adiabatic curve ($\gamma = 5/3$) shows an extremely large deviation from a pressure profile in the observed regime. The model does not provide a realistic description. If one formally considers γ as an open parameter, one finds values well below 1, indicating strong losses.

This interpretation together with the extremely large effect suggest that the crisis applies not only to exact steady states but also to the case of slowly time-dependent flux return, where the same large change of volume would occur. For three-dimensional systems the pressure crisis was confirmed by Birn and Schindler (1983), so that one arrives at the same conclusion.

The real process in the magnetosphere must find a way to avoid the pressure crisis; corresponding observational evidence was presented by Borovsky *et al.* (1998), Garner *et al.* (2003) and others.

For our theoretical reasoning the crisis argument means that at least one of the test properties must be violated. We will argue that, if (i) is satisfied, significant violation of (ii), (iii) and (iv) will not occur. This will lead us to conclude that (i) must be violated in magnetic flux return in the magnetotail.

Let us begin with assumption (ii). A collisionless plasma, in general, cannot be expected to be isotropic. Typically, isotropizing instabilities become efficient only beyond a critical anisotropy. For an important example, let us look at the fire hose instability (see, e.g., www.tp4.rub.de/~ks/ta.pdf). It requires $p_\parallel - p_\perp - B^2/\mu_0 > 0$ to become excited. One can expect that the turbulence arising from the instability keeps the anisotropy near a finite threshold (Gary *et al.*, 1994). In the centre of the plasma sheet the value of B is small, of order ϵ in the tail asymptotic expansion (Section 5.4.1), so that the anisotropy $2(p_\parallel - p_\perp)/(p_\parallel + p_\perp)$ should be small, formally of order ϵ^2. Away from the centre, and in the absence of parallel electric fields, equilibrium momentum balance requires a characteristic variation of the anisotropy along field lines, described by (7.11). This holds both in fluid and kinetic descriptions. On that basis Nötzel *et al.* (1985) investigated a plasma sheet model with local Maxwellian distribution functions, including anisotropy resulting from different temperatures along and perpendicular to the magnetic field direction. They found that the anisotropy decreases away from the plasma sheet centre. 3D anisotropic MHD simulations by Birn *et al.* (1995), taking isotropizing turbulence into account by effective forces, confirmed that the centre region of the plasma sheet remains close to isotropic, with a small $p_\parallel > p_\perp$ anisotropy left. Instabilities with $p_\parallel < p_\perp$ remained limited to the boundary of the plasma sheet and the lobe regions.

Observations also indicate small pressure anisotropies in the central plasma sheet (Stiles *et al.*, 1978), particularly during quiet times (Mitchell *et al.*, 1990), i.e., intervals with a low level of geomagnetic perturbations in the auroral zones.

Here, it is an interesting aspect that in the presence of isotropizing turbulence, fluid pictures which take care of the isotropization by postulating an isotropic pressure or suitable threshold states, tend to be more realistic than collisionless plasma models that ignore the turbulence.

Regarding losses from flux tubes, the situation is less clear. Let us first discuss perpendicular losses (losses by motion perpendicular to the magnetic field). Kivelson and Spence (1988) obtained results suggesting that losses by gradient/curvature drift in a tail of finite width can avoid the pressure crisis (see also Wang *et al.* (2001)). Doubts regarding the significance of losses in this context were expressed by Erickson *et al.* (1991) and Garner *et al.* (2003). The arguments can be understood considering our test assumptions, which exclude perpendicular losses by property (iii). This means that perpendicular losses can play a significant role only for rather small cross-tail electric potentials, not much exceeding $k_B T_i/e \approx 5$ kV. This range is considerably below typical potentials, which can reach the order of 100 kV. So, perpendicular losses are unlikely to play the dominant role in avoiding the pressure crisis.

Substantial parallel losses of particles precipitating along field lines into the ionosphere are considered to apply mainly to electrons (e.g., Borovsky *et al.*, 1998), so that the mass content of the plasma sheet would not be affected significantly, the same applies to entropy because of the electron temperature (on plasma sheet field lines) being considerably smaller than that of the ions (Section 2.1). The assumption of temperature being constant on field lines was made to arrive at the MHD equations used in Section 8.3.3 to demonstrate the pressure crisis. It can be expected that this assumption is not crucial for the crisis to occur.

As at least one of the test assumptions must be violated to avoid the pressure crisis, we conclude that our discussion, although not rigorous, strongly suggests property (i) as the most likely candidate for being broken. It seems that the flux return cannot occur as a large scale quasi-static process. It will necessarily involve dynamical phenomena on length and/or time scales shorter than those of the external flux supply, which we have assumed to be a large scale phenomenon. Note that in ruling out significant violation of properties (ii), (iii) and (iv) we had to assume that (i) holds, so that dynamic processes violating (i) can violate the other properties, too.

Two major possibilities have been suggested for dynamical processes relevant for large-scale flux return; these are non-quasi-static flow events, widely referred to as *bursty bulk flows*, and near-Earth magnetic reconnection.

13.2.2 *Bursty bulk flows*

A bursty bulk flows event (BBF) (Baumjohann *et al.*, 1990; Angelopoulos *et al.*, 1992) shows large bulk plasma flows during periods of the order of 10 min, with a peak bulk velocity above 300 or 400 km/s, details depending on the definition that is applied. In principle, the presence of such flows can drastically invalidate quasi-static modelling. A rough quantitative criterion is found from the scaling (8.3) and (8.4), where δ may be understood as v/v_0, where v_0 is a typical MHD phase velocity. Choosing the sound speed in the plasma sheet and using pressure balance (Section 8.3.1), one conveniently identifies v_0 with the Alfvén velocity v_A based on central plasma sheet density and lobe magnetic field strength. The inertial term in the momentum equation (3.59) is of the order of δ^2; so it can be ignored if $(v/v_A)^2 \ll 1$, assuming that the flow has the same length and time scales as the ambient medium. Typical Alfvén velocities are near 1000 km/s (see data in Table 2.1). Admitting an error of 10%, this means that flow should not exceed about 320 km/s to keep quasi-static conditions, which is surpassed by the peak values given above. But it must be realized that spatial or temporal localization of a BBF can modify that threshold considerably. Even if the flow is below the threshold instantaneously, turbulent flows with a preferred direction, or flow fluctuations that are correlated with magnetic and/or density fluctuations, might still lead to significant secular transport of mass, energy and magnetic flux (Baumjohann, 2002).

During quiet times the secular consequences of transient flow events in the plasma sheet may well be small. Indeed, during quiet times, BBFs with a large flow component perpendicular to the magnetic field (necessary for efficient flux transport) appear to have a rather low occurrence rate in the high-beta plasma sheet. For $\beta > 0.5$ and in time bins of duration of 1 min or a fraction of a minute (depending on the data set), peak velocities above 300 km/s were observed only in $\approx 0.1\%$ of the bins, which caused doubts about their relevance for large scale transport (Paterson *et al.*, 1998, 1999). (See Angelopoulos *et al.* (1999) for an alternative view.)

Other studies also find that the BBF occurrence correlates positively with terrestrial geomagnetic disturbances (Schödel *et al.*, 2001; Baumjohann, 2002). For an example, Figure 13.4, taken from Tanskanen *et al.* (2005), includes a period that was identified as an extended loading phase with negative IMF z-component. The velocity component along the Earth–Sun line

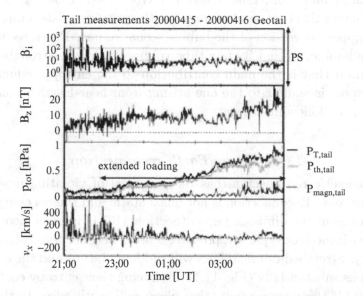

Fig. 13.4 An extended loading phase from GEOTAIL data (Tanskanen *et al.*, 2005). The loading process is manifested by a significant increase of $p_{T,\text{tail}}$, the sum of magnetic and ion pressures. (From Tanskanen *et al.* (2005) by permission of the American Geophysical Union.)

(v_x) assumes rather small magnitudes in the plasma sheet, particularly so during the last two hours of the interval.

To avoid misinterpretations it is necessary to distinguish between strong parallel flows in the plasma sheet boundary layer and BBFs. The boundary layer flows are mainly parallel to the magnetic field, possibly originating from the reconnection site in the distant tail (Fig. 13.1). A possible mechanism is discussed in Section 11.2.9.

At present, the origin of BBFs is not well understood. A possible cause is localized reconnection (Borovsky *et al.*, 1998; Chen and Wolf, 1999). Indeed, reconnection is one of the most efficient generators of localized large bulk flows (Chapter 11).

Apart from reconnection, can BBFs be caused by some plasma instability? A smooth tail configuration, which is stretched but has a sufficiently broad plasma sheet for ideal MHD to be applicable, is stable. For two-dimensional equilibria with 3D perturbations this is shown explicitly in Section 10.2.5. Also, 3D MHD large-scale simulations did not produce evidence for an instability.

However, in the case of strong inhomogeneous flux transfer an interchange instability (Section 10.2.5) seems to be possible. It has been suggested that in this case *bubbles* of low entropy content move toward the Earth with high speed (Pontius and Wolf, 1990; Chen and Wolf, 1993, 1999). It is unclear, though, whether this process alone removes the pressure crisis completely. In a widely supported version of the bubble scenario, near-Earth reconnection creates the bubble, so that its flux tube volume is already strongly reduced at its birth. If that is the main contribution to the entropy reduction, the basic transport is similar to the one arising from near-Earth reconnection, the topic of the following section.

13.2.3 Near-Earth reconnection

Clearly, magnetic reconnection has the potential of violating most of our test assumptions. Reconnection is not quasi-static and the necessity of non-ideal processes in the diffusion region (Section 11.2.1) implies inconsistency with adiabatic models; typically, pressure tensors become non-isotropic (Section 11.3.2). Stretched-out field lines will suddenly be cut and the earthward part shortens substantially (Fig. 11.11), reducing their entropy content by a large factor. If the reconnection takes place sufficiently close to the Earth, the return of the shortened flux tubes will not suffer from a pressure crisis.

Motivated by observations, it seems that one has to distinguish two limiting cases, global reconnection at a newly forming near-Earth neutral line affecting the entire plasma sheet and localized reconnection events with consequences only for a limited portion of the plasma sheet. The former is widely identified with the tail signature of a magnetospheric substorm, the latter has more local significance, and is a possible source of BBFs. Although there are indications of intermediate cases, it proves useful to concentrate on the two limiting cases. As we will see, basic observational facts can be discussed in such a framework.

13.3 Magnetospheric substorms

Here we discuss the presence of global reconnection in the magnetotail at a newly formed reconnection site in the region of closed flux. Our aim is to understand the underlying physics of the substorm phases (Section 2.1).

13.3.1 Growth phase

As the initial state consider a simplified tail configuration that possesses our test properties. In particular, the length scales are large compared with the intrinsic plasma scales, which is satisfied for tail conditions if $L \gg c/\omega_{\mathrm{pi}}$ or equivalently (because of pressure balance) $L \gg r_{\mathrm{gi}}$. Then, the tail is in a state that can approximately be considered as an ideal MHD state. If sufficiently smooth and stretched, that state is stable (Section 10.2.5). Models of this type have proven successful in describing the magnetotail under quiet time conditions, at least from a global point of view (e.g., Birn and Schindler, 1983)

Under those conditions, the magnetic flux that transferred to the tail cannot return to the near-Earth region because of the pressure crisis. The flux will get piled up in the tail, which becomes loaded with magnetic flux and energy. It is natural to identify this phase with the growth phase of a magnetospheric substorm (McPherron, 1970) (Section 2.1).

The temporal development can be expected to be reasonably approximated by the model of Section 8.3, building on a first, extremely simplified, theoretical model (Schindler, 1974). The selfsimilar solution, which can give a rough indication of the convection properties under the assumed conditions, indicates that the lobe magnetic field increases and the plasma sheet thins. Due to induction, the cross-tail electric field becomes small in the centre plasma sheet so that there the bulk flow velocities are small also. For orientation, let us choose a magnetopause electric field of 0.2 mV/m, a lobe field of 20 nT, the magnetopause location at $z = 20R_{\mathrm{E}}$. Then with the help of Table 8.1 and equations (8.40), (8.42), v_x becomes -4 km/s at

$x = 40R_{\mathrm{E}}$. Here x is positive in the tailward direction, so that the flow is earthward. This direction follows from the fact that for $\gamma < 2$ the plasma pressure increase is less than that of the lobe magnetic pressure, so that an additional earthward compression is required. In the similarity solution $|v_x|$ increases proportional to x. There is no flux return to the near-Earth tail boundary $x = 0$. However, there may be flow parallel to $\boldsymbol{B}$ in the plasma sheet boundary layer, as discussed above.

The growth phase cannot last forever, at least not in the framework of Dungey's topology, where (for periods of negative IMF B_z) there is continuous earthward flux transport issuing from the reconnection site in the distant tail (Fig. 13.1). How can the flux return? In stretched field configurations the reconnection at Dungey's distant neutral line cannot be reversed (Section 11.2.10). (This argument assumes that reconnection is generally irreversible, as it is for resistive fluids.) Further, a sufficiently smooth and stretched tail configuration is stable from the ideal MHD point of view (Section 10.2.5).

Also, reconnection cannot set in under the ideal conditions that we found as characteristic for the growth phase. Reconnection requires nonidealness, which in the collisionless tail plasma needs scales of the order of c/ω_{pi} or smaller. Interestingly, such thin sheets will actually form during the later stages of the growth phase. This was shown, under various different assumptions, in Chapter 8. In fact, the formation of thin current sheets seems to be a natural part of the ideal dynamics of the tail configuration. (The selfsimilar solutions are untypical in this respect, because current sheet formation is excluded by the imposed selfsimilarity.) As in the pressure crisis argument, the radial variation of the flux tube volume plays an important role in the process of current sheet thinning (Section 8.3.4). Loosely speaking, the same property that is responsible for the development of the pressure crisis also offers the means of its resolution!

In the model of Birn *et al.* (2004) the current can even become singular at a finite time, when the plasma sheet experiences loss of equilibrium (Section 8.4). Even in more regular MHD cases, the build-up of the current sheet is a rapid process (see (8.54)).

It has been verified that thin current sheets also exist in kinetic equilibrium theory. They are characterized by strong electric fields perpendicular to the magnetic field (Section 8.5.3). The corresponding potentials would extend along the magnetic field lines into the inner magnetosphere. It has been speculated that these potentials might close through parallel electric fields in the auroral acceleration region, as illustrated in Fig. 13.5. Further, magnetic field lines, frozen into the electron motion in a time-dependent situation,

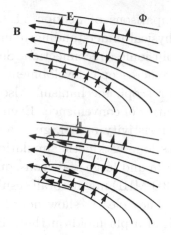

Fig. 13.5 Possible generation of parallel electric fields and field-aligned currents by kinetic current sheet formation (after Schindler and Birn (2002) by permission of the American Geophysical Union).

would become bent out of the x, z-plane causing a pattern of field-aligned currents (dashed arrows). This would suggest a direct link between thin current sheets in the tail and auroral arcs. A thin current sheet of a width of 500 km would map to an arc width of about 10 km.

13.3.2 Expansion phase

In the real plasma sheet, reconnection will set in when the current sheet has become sufficiently thin. The actual onset process is still unclear. But the collisionless tearing instability will start when the normal magnetic field component (B_z) has become sufficiently small for the electrons to become significantly non-adiabatic (Section 10.5.5). Sufficiently thin sheets are also subject to microinstabilities (Section 9.3.1), which also may lead to nonideal effects (Section 9.3.2).

For a sufficiently thin sheet the tearing instability is stabilized only for sheets enclosed in a narrow region by boundaries. Close boundaries would lead to F-stability in the sense of Section 12.1.1, which also applies to the collisionless tearing instability, as seen from the structure of the variational principles of Section 10.5. The presence of pronounced lobes and a tailward scale large compared to the sheet width avoid boundary stabilization.

The onset of a linear instability is likely to be obscured by external perturbations, which will have an increasingly large effect as marginality is approached (Section 12.3).

As the instability has developed into its nonlinear regime, reconnection will proceed at a rate that depends on the details of the diffusion mechanism only weakly (see the comparisons of Section 11.3.3). This means that, although a satisfactory physical understanding requires detailed knowledge of the plasma physics of collisionless reconnection (Section 11.3), we can continue the discussion of the macroscopic development in an approximate sense without specifying the microscopic mechanism. Also, choosing a mechanism for a specific model is a matter of convenience. Even resistive MHD, with an appropriate choice of the resistivity model, gives acceptable answers. This is essential, particularly as long as kinetic simulations are too large or too costly to cover the long term and large scale consequences of reconnection.

Corresponding 2D and 3D MHD magnetotail results are presented in Sections 11.2.10 and 11.4.2. They clearly show near-Earth reconnection with the formation and growth of a plasmoid on the tailward side and dipolarization earthward of the reconnection site. Fig. 11.11 shows a qualitative sketch of snapshots of this process.

The dipolarization is associated with earthward plasma flows, generated by the reconnection process (Chapter 11). The flow runs against the dipolar field region closer to the Earth and (approximately treated as ideal) compresses the B_z-component in the midnight meridian plane, which is a major signature of dipolarization. Away from the $y = 0$-plane, the flow is deflected by the dipolar field region and deforms the field lines in the near-Earth plasma sheet accordingly. This deformation is associated with field-aligned currents (Fig. 13.6). Mapping these currents along field lines and assuming that they close in the ionosphere generates the electrojet (Section 2.1), the

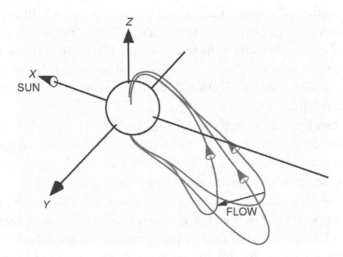

Fig. 13.6 Sketch illustrating the origin of field-aligned currents during dipolarization (from Birn and Hesse (1996) by permission of the American Geophysical Union).

main ionospheric current signature during the expansion phase, causing enhanced auroral light emission. This picture also suggests gross conjugacy of expansion phase auroral forms in the two hemispheres, which is an observed fact (e.g., Pulkkinen *et al.*, 1995).

This current system deviates a near-Earth fraction of the tail current through the ionosphere and generates the observed *current wedge* (Section 2.1). This feature also has been termed *current disruption* (Lui, 1996). Resistive MHD simulations (Hesse and Birn, 1991a) have shown that the dipolarization and the current wedge can be understood as a feature of the large-scale reconnection dynamics. The results are consistent with corresponding observations (McPherron *et al.*, 1973; Lopez and Lui, 1990). Current disruption has also been viewed as a process caused by microinstabilities (Section 11.4.1), independent of reconnection (Lui, 1996). However, this view has not been supported by simulations. If a microinstability causes a significant collective resistivity, corresponding resistive MHD models have always shown reconnection (e.g., Birn and Hesse, 1996).

What causes the electric fields that accelerate the auroral electrons during substorms? A possible cause could be plasma slippage with respect to the magnetic field in the dipolarization. In fact, it is difficult to visualize an efficient dipolarization with the frozen-in condition satisfied. Such slippage would pose a typical GMR problem (Section 11.5.1). The corresponding parallel potential has been estimated by Schindler *et al.* (1991) to be of the order of a few kV. (A similar problem arises from the coexistence of ionospheric plasma rotating with the Earth and non-rotating magnetospheric plasma on the same field lines.)

Also, large perpendicular electric fields in combination with demagnetized electrons, as observed by Scudder and Mozer (2005) near the magnetopause, need to be explained. The authors point out that strong perpendicular electric fields would be required to cause nongyrotropic electrons in reconnection with a strong guide field (Section 11.3.2). Alternatively, it could be relevant that, under quasi-steady conditions, reconnection with a bounded nonideal region elongated along magnetic field lines leads to strong perpendicular electric fields in the ideal region, even at considerable distances from the reconnection site (Fig. 11.36).

13.3.3 Recovery phase

During the recovery phase the magnetosphere relaxes from the state after the expansion phase to a state more typical for quiet times. It can best be observed during isolated substorms.

The pronounced dipolar magnetic field at, say, the geostationary orbit, relaxes toward the quiet time configuration. Also, the near-Earth neutral line seems to retreat tailward, as originally suggested by Hones *et al.* (1984). Interestingly, during recovery, a double oval configuration has been observed in the ionosphere, where the higher latitude structure maps to the plasma sheet boundary layer (Mende *et al.*, 2002), which should contain the magnetic separatrix.

It has been suggested that the recovery phase is not just a simple relaxation phase. There are indications that the northward-turning of the IMF might have an influence on recovery (Rostoker, 1983). But from their magnetic field measurements at the geostationary orbit during 11 recovery phases Pulkkinen *et al.* (1994) concluded that the start time of the increase of B_z as well as the duration of the relaxation did not depend on the direction of the IMF z-component.

From ionospheric signatures Opgenoorth *et al.* (1994) concluded that the recovery phase even involves its own form of activity, distinct from the expansion phase.

13.3.4 *Extended simulations*

There are MHD simulations that cover both growth and expansion phases. Fig. 13.7 illustrates corresponding results from a 3D resistive MHD simulation by Birn and Hesse (1996) with $S = 500$ (Section 10.3) of the magnetotail including the transition to a dipolar field. Slow external driving leads to a growth phase which lasts until about $t = 80$ (in non-dimensional units (Section 10.3.1)). The accumulation of magnetic flux and the formation of a thin current sheet is clearly seen. At $t = 80$ the current sheet becomes unstable, which initiates the expansion phase. A resistive tearing mode (Section 10.3) grows into a nonlinear reconnection regime, resulting in plasmoid formation and tailward ejection, and dipolarization in the near-Earth region. The current sheet bifurcates in the later stages. The time-dependence of the reconnected flux and of the maximum values of the cross-tail electric field E_y, current density j_y and velocity v_x are given in Fig. 13.8. Due to the thin current sheet the reconnection pattern grows very fast, on the time scale of a few minutes.

The reconnection has a finite extent in the y-direction. It starts in the midnight meridian plane, where the structure qualitatively agrees with plasmoid formation in systems with translational invariance (y-independence). However, instead of x- and o-lines extending to infinity, they join at an increasing distance from the centre (Fig. 13.9) and, in the present symmetric geometry, form a closed loop (Vasyliunas, 1976).

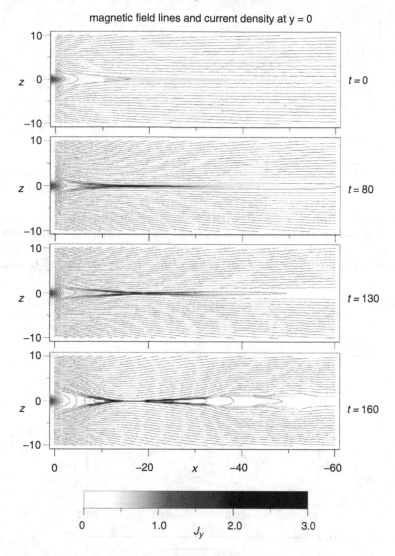

Fig. 13.7 Results from a resistive MHD simulation by Birn *et al.* (1996). Shown are the magnetic field lines and the cross-tail current density $|j_y|$ (shading); note that the x-axis points sunward. (From Birn *et al.* (1996) by permission of the American Geophysical Union.)

In the presence of a unidirectional B_y-component the plasmoid assumes a flux rope structure with complicated magnetic connectivity properties at its ends (Section 11.4.2). Fig. 13.10 gives a perspective view of such a plasmoid. The absence of neutral points (Hughes and Sibeck, 1987) has been the main motivation for studying finite-B reconnection (Section 11.5.1).

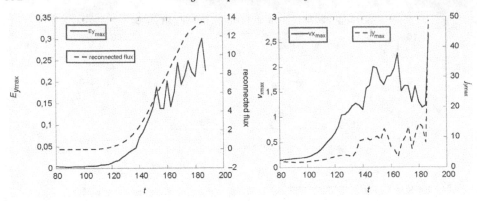

Fig. 13.8 Results from a resistive MHD simulation; shown are the reconnected magnetic flux and maximum values of E_y (left) and of j_y and v_x (right), measured in units of $3.2 \cdot 10^6$ Wb, 80 mV/m, 10^{-8} A/m^2 and 1000 km/s, respectively, the time unit is 6.3 s; note that the x-axis points sunward. (Courtesy of J. Birn.)

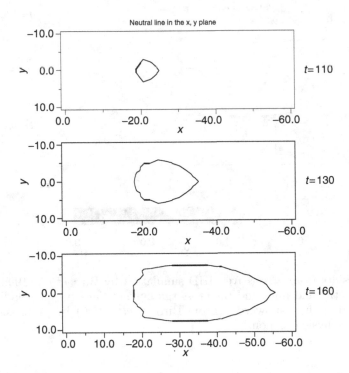

Fig. 13.9 Results from a 3D resistive MHD simulation; shown is the temporal evolution of the neutral line ($B_z = 0$) in the equatorial plane. (From Birn *et al.* (1996) by permission of the American Geophysical Union.)

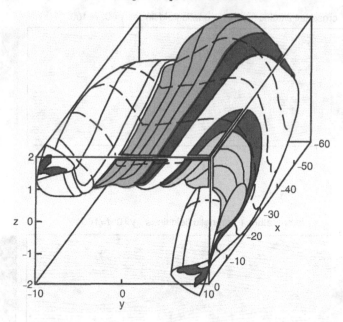

Fig. 13.10 Perspective view of a partially separated plasmoid as seen from the Earth (Birn *et al.*, 1996). Indicated are regions of field lines with both feet connected to the Earth (white), with one foot connected to the Earth (black) and of field lines that are completely disconnected (gray). (From Birn *et al.* (1996) by permission of the American Geophysical Union.)

As already mentioned, the dipolarization is associated with a field-aligned current system (Fig. 13.6) and with particle acceleration. The latter largely results from the cross-tail electric field. While in quasi-steady reconnection the electric field E_y is rather smoothly distributed over a wide region (in 2D steady states E_y is a constant), the pronounced time-dependence leads to a strong localization. This is illustrated in the upper panel of Fig. 13.11.

Surprisingly, the strongest electric fields do not appear at the reconnection site, where one might be tempted to expect them, but closer to the Earth. These fields are generated by the strong time-dependence of the magnetic field associated with the dipolarization. These strong fields lead to intense acceleration of both ions and electrons, and are largely consistent with the observed injection events (Birn *et al.*, 1997, 1998b) (Section 2.1).

The simulations discussed so far address intermediate magnetospheric scales. They cover a relevant fraction of the magnetotail and, in some cases, include a part of the inner magnetosphere. They start out from realistic self-consistent configurations and have good resolution of the reconnection site. Alternatively, global simulations include the solar wind and the ionosphere.

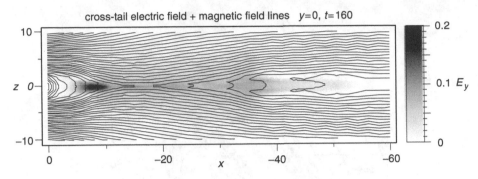

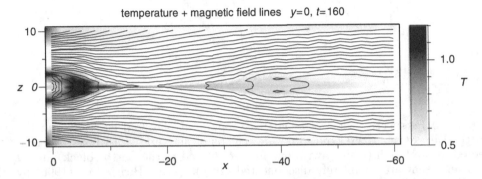

Fig. 13.11 A snapshot from a 3D MHD simulation; shown are the magnetic field lines in the midnight meridian plane and the cross-tail electric field (upper panel) and the temperature (lower panel). (From Birn *et al.* (1996) by permission of the American Geophysical Union.)

Naturally their resolution tends to be smaller than that of the intermediate simulations, but they already have reproduced large-scale substorm signatures in the magnetosphere. Fig. 13.12 shows an example. The substorm was initiated by a southward turning of the IMF after a long northward period. It shows complete growth and expansion phases, qualitatively consistent with the picture outlined above and with the intermediate simulations (e.g., Fig. 13.7).

In a global simulation (Ogino *et al.*, 1990) the formation of a flux rope for a large y-component of the IMF was confirmed. Ashour-Abdalla *et al.* (2002) reported a simulation, where the earthward flow from the reconnection line initiated a vortical pattern in the inner magnetosphere.

Global MHD simulations have reached a stage that allows quantitative comparison with satellite observations (Ohtani and Raeder, 2004), defining areas of agreement, but also shedding light on the reasons for the deviations.

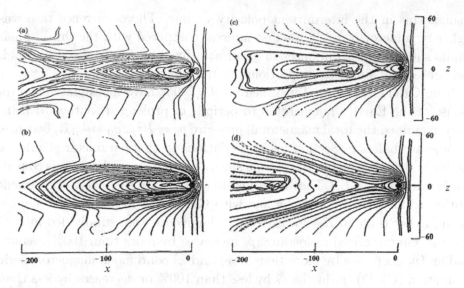

Fig. 13.12 A substorm in a global simulation. (From Slinker *et al.* (1995) by permission of the American Geophysical Union.)

The most difficult part seems to be the coupling with the ionosphere (Raeder *et al.*, 2001). For northward-pointing IMF a global MHD simulation by Song *et al.* (1999) demonstrated that large parts of the magnetopause are closed and that reconnection is limited to regions near the cusps.

Obviously, the described theoretical and simulation results are grossly consistent with the phenomenological near-Earth neutral line model (Section 2.1), but have reached a much more developed stage offering many more details.

13.3.5 Observations

The last step is to compare the substorm picture, as it emerged from theoretical reasoning and simulations (enriched by some key observational facts), with further observations. This is a vast field by itself, so that here we must content ourselves with a few typical examples.

Statistical results demonstrating that a negative IMF z-component favours geomagnetic activity have already been mentioned. This has also been confirmed by looking at individual cases. Freeman and Farrugia (1999) investigated a period consisting of a 14 h interval of continuous and strong IMF $B_z < 0$, a 16 h interval of continuous and strong IMF $B_z > 0$ and a 22 h interval of intermittent IMF polarity. They found 5 substorms in the interval of strong negative B_z, no substorm in the interval of strong positive B_z and

6 substorms in the intermittent polarity regime. The occurrence rate was higher by a factor of 1.4 in the first interval compared with the third. Their results suggest substorm onset to occur at a fixed energy or flux threshold, and subsequent flux loss proportional to the input rate at onset.

Using GEOTAIL data, Tanskanen *et al.* (2005) have investigated the response of the Earth's magnetotail to periods of prolonged southward IMF. They examined the total magnetotail pressure $p_t = B^2/2\mu_0 + nk_B T_i$, because, by pressure balance (Section 5.4.1), variations in this parameter should be similar in the lobes and in the plasma sheet. (T_e is disregarded as it is considerably smaller than T_i.) They identified 13 events with the IMF z-component southward for 8 hours or longer and with GEOTAIL located in the magnetotail farther than 10 R_E downstream. The events were subdivided into 37 intervals characterized as loading (p_t increases by more than 100%), as unloading (p_t decreases by more than 50%) and as continuous magnetospheric dissipation (CMD) (p_t increases by less than 100% or decreases by less than 50%). With these definitions, 37 loading, 37 unloading and 28 CMD events were found. So, the loading/release scenario was verified, but the results also indicate the existence of a different type of magnetospheric response to flux transfer from the solar wind. We return to this aspect farther below.

Another important question is whether there is *in situ* evidence of a coherent reconnection process in the magnetotail. Part of the GEOTAIL mission was especially designed for this task by exploring the plasma sheet in the region $-20R_E > x > -30R_E$, where the near-Earth reconnection site is to be expected. The satellite has encountered the near x-line region frequently. It has confirmed rapid flow reversals, interpreted as the reconnection site crossing the satellite. In a statistical study by Asano *et al.* (2004), substorm-associated x-line crossings were analysed in detail using the velocity moments of both ions and electrons simultaneously. Ampère's law was used to determine the current sheet width near the x-line. In some cases it was near 500 km, that is, below $c/\omega_{pi} \approx 750$ km. Accordingly, a Hall zone (Section 11.3.1) with its magnetic quadrupolar signature (see Fig. 11.13 and Fig. 11.14) were identified. Near the x-line the electric current density was found to be dominated by the electron Hall current, generated by strong electric fields in the inflow direction. Away from the x-line the current density bifurcates (see also Lottermoser and Scholer, 1997). Similar conclusions have been reached from 4-satellite CLUSTER data by Runov *et al.* (2003).

Other studies focussed on plasmoids in the magnetotail. A statistical study by Ieda *et al.* (1998) covered 824 plasmoid events in the range $-210\ R_E < x < -16\ R_E$, a plasmoid being defined by a rotating magnetic field structure and enhanced total pressure (the enhancement is due to the curvature term in (5.9)). In the near tail the plasmoids were observed around the tail axis and found to expand in the y-direction. The authors interpret their findings as strongly supporting the view that the plasmoids are initially formed at the near-Earth neutral line, which has a limited extent in the y-direction. The plasmoids move tailward with velocities up to 700 km/s, but with a considerable spread and y-dependence. The findings are summarized in Fig. 13.13. Attributing on average 1.8 plasmoids to a substorm, it was concluded that the average energy carried away by plasmoids in a typical substorm is $\sim 3 \times 10^{14}$ J. However, fast flows accompanying the plasmoids are estimated to carry even more energy so that the total energy released tailward in the course of a substorm is estimated as roughly $\sim 10^{15}$ J. This is about the same energy as has been estimated for the substorm energy released into the inner magnetosphere and the ionosphere.

The results from *in situ* satellite observations are generally consistent with the phenomenological near-Earth neutral line model (Section 2.1), but they have detected many more details and new aspects. Note that we arrived at largely the same conclusion from theory and simulation points of view. This means that the near-Earth neutral line model of substorms has found strong support from all three sources of information.

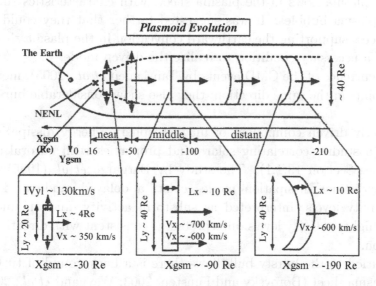

Fig. 13.13 Schematic illustration of plasmoid properties (from Ieda *et al.* (1998) by permission of the American Geophysical Union).

13.4 Further large-scale processes

Here we briefly describe two other large-scale phenomena of the magnetosphere that have to be addressed for the sake of a more complete picture of magnetospheric dynamics.

13.4.1 Quasi-steady convection

The pressure crisis eliminates steady state convection under our test assumptions. As discussed, substorms represent a major resolution of this problem. They drastically shorten the flux tubes and thereby strongly reduce their entropy content. Also we have seen that the remaining possibility is likely to be bursty flows. As discussed, and following Garner *et al.* (2003), one would discard losses as the main entropy reducing process, because losses may play a significant role only for rather weak convection states, although this view might still require final confirmation (Wang *et al.*, 2004).

There is considerable evidence regarding non-laminar convection states, not directly related to substorm activity (Sergeev *et al.*, 1996). Corresponding events carry different names (*convection bay, steady magnetospheric convection, SMC, continuous magnetospheric dissipation, CMD*), reflecting differences in the operational definitions. In Section 2.1 Fig. 2.3 shows a convection bay. The sporadic appearance of intense auroral streamers point at bursty plasma flows in the plasma sheet, with characteristics similar to those of plasma bubbles. It was suggested further that they could be the main process supporting the earthward convection in the plasma sheet, and represent a necessary condition to realize the driven mode.

The occurrence of the CMD events by Tanskanen *et al.* (2005), mentioned above, points in the same direction; they also show considerable bursty flow activity.

A directly driven component of magnetospheric energy dissipation also appears in studies correlating solar wind properties with auroral activity indices (see Section 13.1.1). The study by Bargatze *et al.* (1985) showed that two distinct dissipation modes exist, a delayed mode at intermediate activity levels, interpreted as substorm activity, and an undelayed mode at higher activity levels, interpreted as consistent with directly driven dissipation.

In addition to the bursty bulk flows there is a background of turbulence in the plasma sheet (Borovsky and Funsten, 2003; Weygand *et al.*, 2005). A possible origin is shear flow, but it is not clear whether the turbulence plays a significant role for the large-scale dynamics.

13.4.2 Magnetic storms

Although the basic eruption phenomenon in the magnetosphere, and therefore our main object of interest, is the substorm, it is appropriate to add at least a brief description of the geomagnetic storm and its relation to substorms.

In a traditional picture there is a direct causal relationship between substorms and ring current intensifications (Akasofu *et al.*, 1974). Here, the injection of energized particle fluxes generated by substorms feed the ring current in an accumulative way. This concept would imply a corresponding correlation between $V_{\text{sw}}B_{\text{sw}}$ Section 3.1.1 and Dst, which in fact is observed (e.g., Kamide *et al.*, 1998).

However, as we have seen, periods of southward IMF may cause not only substorms but also quasi-steady convection. Does that phenomenon contribute to storms also? A case where quasi-steady convection led to a (weak) magnetic storm has been identified and investigated in a multisatellite study (Zhu *et al.*, 2003).

Considerable efforts have been devoted to the simulation of the inner magnetospheric dynamics with emphasis on magnetic storms. Rather complex models have been devised, such as the renowned *Rice convection model* (Wolf, 1983). Until recently, the models concentrated on the electric field, prescribing the magnetic field. Such modelling (e.g., Wolf *et al.*, 1997) suggested that the IMF-driven convection would play a more important role than substorms. This was surprising, as the convection should be inhibited by the need to avoid the pressure crisis, as discussed above. However, to cover that effect requires taking into account field lines that are stretched out into the tail region (Section 8.3.3). Indeed, a more recent study (Lemon *et al.*, 2004), which included the near-Earth tail and computed the magnetic field self-consistently, confirmed that adiabatic convection did not cause an increase of the ring current particle flux. To achieve a significant intensification of the ring current, the pressure crisis was shown to be avoided by an artificial lowering of the flux tube entropy. So, this leads us back to the (as yet unanswered) question of the dominant mode of plasma sheet transport. We can expect that, once this problem is solved, it will also become clear whether the substorm plays a leading or just a supporting role in determining the intensity of magnetic storms (Kamide *et al.*, 1998; Moon *et al.*, 2004).

13.4.3 Spontaneous versus directly driven processes

The presence of directly driven processes and of processes that are not directly driven raises questions about classification. Spontaneous processes occur without requiring a particular external trigger signal in addition to a broad noise spectrum. As a typical example consider an instability that occurs when a system passes a bifurcation point (Chapter 12) and becomes unstable. Now suppose that one observes a perturbation pulse at the time of the onset of the instability. Is this pulse evidence against the instability scenario? The answer is *no*, because otherwise a system, which has almost reached the bifurcation point and is helped over the last threshold by the external perturbation pulse, would be considered as directly driven, although after a short time interval the instability would have started without the trigger process. Here, a meaningful distinction requires a deeper consideration. A possible criterion is based on the energetics of the system. A system that releases more energy than the external pulse supplies is unstable; if there is no release of system energy, the system is stable and the dynamic response can be regarded as directly driven. This is the point of view that motivated the stability concept described in Section 10.1.

This point of view implies for the magnetosphere that the appropriate distinction is to be made between directly driven energy supply and the loading/release scenario (Bargatze *et al.*, 1985; Tanskanen *et al.*, 2005). The search for triggers does not carry far enough to allow a fundamental distinction. Nevertheless, trigger studies provide information on the sensitivity of the magnetosphere with respect to the different perturbation forms.

13.5 Statistical and nonlinear dynamics aspects

Magnetospheric activity is based on the nonlinear mean field interaction of many particles. In principle, for the collisionless magnetospheric plasma the Vlasov theory is the most appropriate description, but only if fluctuations are properly taken into account. As we have seen, in some cases fluctuations may cause a fluid behaviour of the collisionless plasma. There are other concepts developed in statistical mechanics and nonlinear dynamics that might also be useful for magnetospheric physics. Some of these approaches have been summarized in Chapter 12. Here we will give a brief account of their applications to magnetospheric activity.

As we will see, they provide valuable conceptual input, but they are not particularly strong in the area of quantitative morphology.

13.5.1 Statistical mechanics with constraint

As already discussed (Section 12.2), several stability properties of a current sheet can be understood, at least qualitatively, by treating it as a canonical ensemble of particles. The control parameter is $\lambda = \mu_0 I^2/(4Nk_\mathrm{B}T)$. A non-vanishing current per unit length I is sustained by the different particle species having different total canonical momentum in the invariant direction. Without further restrictions the sheet is stable only in the regime of wall-stabilization, which is unrealistic for the plasma sheet. However, the instability is removed by introducing a constraint that is equivalent to topology conservation. That constraint would be relevant during the growth phase. Then the plasma is considered as an ideal medium, so that the sheet stably evolves as the current I increases due to flux transfer to the tail. The instability will set in, as soon as the topology constraint breaks down. This occurs by the formation of a thin current sheet (Chapter 8), which sets an end to the ideal evolution. (Within Vlasov theory this might correspond to the onset of a collisionless tearing mode for sufficiently thin sheets. (Section 10.4)). Formally, at this point the sheet would develop toward its state of lowest free energy (Fig. 12.10). That state is characterized by the current being concentrated into a single filament. This concentration can be understood as the result of the attractive force between parallel current filaments. So the final state would be an extreme form of a plasmoid (Fig. 12.11).

It is interesting that the growth phase, the instability onset and the plasmoid formation are contained in this model in a qualitative sense. The role of current filament attraction is brought out without the need for considering a particular dynamical model.

Apparently, in the real magnetosphere the current collapse cannot proceed to a singular state. This is due to the influence of the environment of the sheet, which is not well represented in the model, in which the current sheet is simply enclosed in a box, in thermal contact with the environment.

13.5.2 Low-dimensional modelling

Attempts to identify magnetospheric activity as a low-dimensional autonomous system (Section 12.5) have not yielded a clear result (Klimas *et al.*, 1996). It seems that the interaction with the solar wind counteract the relaxation into an attractor.

Low-dimensional input–output models have been more successful (Section 12.5). In particular, this applies to the model by Horton and Doxas (1996). This model is completely based on physics, although it builds on earlier models containing analogies such as the dripping faucet; for a review see Klimas *et al.* (1996).

In its improved version (Horton and Doxas, 1998) the Horton–Doxas model has a 6-dimensional state space and 13 parameters. The state space

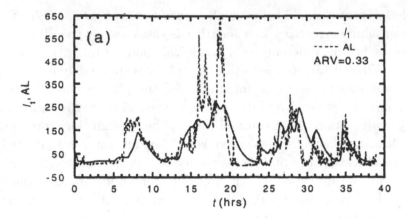

Fig. 13.14 Comparison of the electrojet current I_1 derived from the low-dimensional model by Horton and Doxas (1998) with corresponding AE results obtained from the database of Bargatze *et al.* (1985); ARV stands for average relative variance. (From Horton and Doxas (1998) by permission of the American Geophysical Union.)

variables are the total plasma sheet current, the auroral electrojet current, the cross-tail and ionospheric potentials, the kinetic energy in parallel plasma motion and the plasma sheet pressure. Integral versions of the basic momentum and energy conservation laws and of Faraday's law were used to derive a set of ordinary differential equations for the temporal evolution of the state variables. The model involves a critical plasma sheet current at which a bifurcation occurs into a fast unloading mode. Plasma sheet physics based on nongyrotropic particle orbits (Section 6.4) is also included.

Fig. 13.14 shows an example of a result produced by this model in comparison with corresponding data base results. The agreement seems fairly good. One might argue that with 13 parameters one should be able to fit many models to a given data set. But one should keep in mind that the parameters are physics-based and their values must lie in a realistic range. Again, the loading/unloading scenario is a basic aspect of the model, allowing for isolated substorms (see, for instance, the subinterval 5–12 h in Fig. 13.14).

13.5.3 Self-organized criticality

The concept of self-organized criticality (Section 12.4) also has been applied to the magnetospheric dynamics. Chapman *et al.* (1998) developed a cellular automaton model consisting of a scale-free part (corresponding to a sand pile model) and a non-selfsimilar systemwide discharge. The first part addresses non-substorm activity, the second describes substorms.

Klimas *et al.* (2000) abandoned the discrete cellular automata approach and adopted a continuous model by reinterpreting an earlier model developed by Lu (1995). The background is the assumption of localized short-duration reconnection events occurring on a broad range of scales in the plasma sheet (Baumjohann *et al.*, 1990). A simplified resistive MHD (Section 3.3.3) model is used and, by additional measures, adjusted to that situation. The resulting model is spatially one-dimensional. The basic process is magnetic annihilation at a neutral sheet, described by the diffusion/advection equation (11.1). The input velocity is prescribed and the resistivity, caused by a current-driven microinstability, jumps from a low value to a high value at a critical current density. Under suitable conditions this model shows typical aspects of self-organized criticality. The inflowing magnetic flux is dissipated in the form of annihilation avalanches, which, on average, restore the critical state. The model also shows a selfsimilar regime. The authors suggest that the actual plasma sheet can develop a similar state.

A model by Chang is based on overlapping resonances of plasma fluctuations (see the review by Chang (1999)). The corresponding fluctuations are considered as a model for BBF activity. The substorm is due to a *nonclassical* global instability and during the substorm evolution the magnetotail is in a state of forced and/or self-organized criticality.

Again, in view of the necessity of model assumptions, these approaches are unlikely to reproduce the morphology of magnetospheric dynamics in detail. But they do have the potential of contributing to the basic physical understanding of the processes involved. They concentrate on the statistical aspects, thereby complementing the deterministic studies, which represent the main stream of present theoretical and simulation studies. (This book is not an exception.)

13.6 Discussion

We have seen that the theoretical tools described in the main parts of this book can help to understand major phenomena of magnetospheric activity. The following picture becomes visible, at least in its gross contours.

The magnetosphere is necessarily open and the opening involves magnetic reconnection. A major consequence is magnetospheric convection with flux transfer to the magnetotail.

The flux return is a more complicated phenomenon. The straightforward return by quasi-steady convection is largely impeded, if not eliminated, by the need of magnetotail dynamics to avoid the pressure crisis. We have argued that (for not too small cross-tail potentials) the most promising candidates for avoiding the crisis and for allowing flux return are bursty bulk flow and near-Earth reconnection.

During periods where such processes are not available, the flux cannot return to the inner magnetosphere and piles up in the tail. This process is characterized by a small cross-tail electric field in the centre of the plasma sheet. As flux piles up, thin current sheets form. They bring a fundamentally new aspect into the picture. As current sheets assume a scale of the order of the ion gyro-scale or smaller, non-fluid properties of the plasma become important. This leads to a reconnection process, which shortens the earthward part of the field lines so that flux return becomes possible.

If there is a pronounced loading phase and if the thin current sheet favours a particular location in the plasma sheet, this picture provides the tail signature of a magnetospheric substorm. Consequences include tailward losses with a significant plasmoid contribution, earthward dipolarization, current wedge, field-aligned currents and particle acceleration.

In the presence of a sufficient level of turbulence, the flux return may occur in a quasi-steady way. Such states may represent the observed steady convection states or convection bays, with bursty bulk flows providing flux transport events localized in space and time.

Although there is substantial overlap between theory, simulation and observation, at least regarding the overall picture, substantial gaps still exist. One such gap concerns pseudo-breakups (Section 2.1). Qualitatively, one might describe them as local expansion phase signatures that do not evolve to ordinary substorm level, but a quantitative criterion is not available.

Also, it is unknown under what conditions magnetic flux transfer leads to substorms or to steady convection. There is no clear-cut hint available from our theoretical arsenal, but it suggests the following speculation. Assume that the quasi-steady convection states are characterized by the net effect of a set of individual reconnection events, distributed over the central plasma sheet, as discussed above. This would require several, if not many, thin current sheets to initiate reconnection. This is different from a (prototype) substorm, where the reconnection starts at a single current sheet. Further, there is no significant energy accumulation during steady state convection periods, while substorms have a growth phase. A possible property discriminating between the two phenomena is the magnetic configuration prevailing in the plasma sheet at the beginning of the southward turning of the IMF. If that configuration has a significant magnetic flux normal to the centre plane of the plasma sheet, thin current sheets are not available immediately. It takes a significant flux transfer and associated stretching, before current sheets form. (This is evident from Chapter 8.) In other words, a growth phase is required to set the conditions for reconnection to start.

The situation is different when initially the magnetic flux through the midplane of the plasma sheet is small enough so that thin current sheets can form more easily. This case can be understood best by looking at the extreme case of a locally one-dimensional sheet with no flux through the midplane. Then, thin current sheets form as the consequence of arbitrarily small external perturbations (Section 8.2). If the actual plasma sheet is sufficiently close to that limiting state, flux transfer will cause several, or even many, thin current sheets which then become local reconnection sites. How can such a process be kept in a stable steady state on average? If the reconnection transports too much flux toward the Earth, the plasma sheet field lines become a little more dipolar, which makes the current sheet formation less efficient. Flux transfer counteracts this development. So, it is conceivable that, once a thin sheet with multiple reconnection sites is established, it would stay in that state as long as the flux transfer continues. Clearly, sufficiently strong perturbing forces exerted by solar wind variations might be able to terminate a steady convection period. It should be noted that this scenario, although consistent with theoretical results, is still largely speculative.

Here we have concentrated on the activity mainly from the magnetospheric point of view. We have addressed the ionosphere essentially in two contexts. The first is determining the ionospheric electric fields by mapping the convection electric field along field lines (ignoring the parallel electric field component and the associated local plasma processes), the second is ionospheric closure of field-aligned currents in the electrojet. One could argue that this way of dealing with the ionosphere does not pay appropriate attention to the large wealth of research that has gone into the understanding of these processes. However, their omission is the price for giving a more detailed discussion of the overall magnetospheric eruption, which is our main topic.

As we have discussed, the variations associated with magnetospheric activity are related to the conditions of the solar wind, the direction of the interplanetary magnetic field playing an important role. The cause of these interplanetary variations can be traced back to phenomena occurring on the Sun. We will return to this aspect in the following chapter.

14

Models of solar activity

In the Solar System the most spectacular manifestations of space plasma activity are the large-scale solar eruptions, such as coronal mass ejections (CMEs), solar flares and prominence eruptions, as briefly described in Section 2.2. In this chapter we attempt to address the underlying physical processes. The approach leaves aside many details, although they would be exciting from a more morphological point of view. Instead, we are interested in the basic physical mechanisms and concentrate on the models and numerical simulations, which provide an excellent frame for our discussion. Naturally, as in the previous chapter on magnetospheric activity, the focus is on loading and release processes.

As we will see, the building blocks, such as ideal dynamics, magnetic reconnection, formation of thin current layers, plasmoid or flux rope formation are relevant elements also in current modelling of solar activity. However, in most solar activity models their role is different from their magnetospheric role. In other words, the building blocks are put together in a different way.

14.1 General aspects

Observations strongly suggest that solar eruption processes are of the loading/release type. The energy flux into the corona from below is considerably smaller than the energy flux that would be required if the eruptions were directly driven by the subphotospheric dynamics. In fact, it has been argued that models based on direct driving have been shown to be grossly inconsistent with observations (e.g., Forbes, 2000a).

As the eruptions originate from the lower corona, we can expect that the relevant energy reservoir is the energy of that region. In an active region, the magnetic energy density exceeds the thermal energy density by a factor above 1000; the kinetic energy density of the bulk plasma flow is

407

even smaller than the thermal energy density. So, the eruptions are believed to be fed by the free energy of the pre-eruption coronal magnetic field. Further, at sufficient temporal and spatial distance from an eruption the corona is thought to be in a state of relative quiescence. In that picture, as in the magnetospheric tail during the growth phase, one ignores superimposed fluctuations on smaller scales.

For further discussion we need quantitative information about a number of selected quantities. Again, it should be noted that this chapter is not meant to give a survey on solar atmospheric physics in general, but it concentrates on the basic physics of eruptions with emphasis on the most spectacular events. Where possible, the numbers are based on the values given in Section 2.2 and in Table 9.1.

The energies released by a CME can be as large as 10^{25} J and within a few solar radii CMEs assume velocities in the range of several tens to 1000 km/s (Hundhausen *et al.*, 1994). Photospheric plasma velocities are of the order of 1 km/s. Typically, the gravitational scale height (Section 5.1.1) is 60 000 km in the lower corona and 200 km in the chromosphere. So, in a broad range of length scales, gravity effects do not play an important role in the lower corona, while buoyancy of cooler plasma regions, magnetically separated from hotter parts, is an important transport mechanism below the coronal base (Parker, 1979). Higher up in the corona, where the temperature is lower, gravity can again become important, examples being the equilibrium of quiescent prominences, which are suspended in an equilibrium between gravity and magnetic forces. Also for the expanding solar wind gravity plays a crucial role (Parker, 1963a), as well as for the propagation of CMEs in interplanetary space (Gibson and Low, 1998).

Except on extremely short scales the coronal plasma is ideal (see Table 3.25). The Alfvén velocity in an active region with a magnetic field strength of 0.01 T and a density of $10^{15}/m^3$ is 7000 km/s, so that the changes resulting from footpoint motion induced by subphotospheric motions can be regarded as quasi-static.

Taking these facts together, one finds that in a wide range of scales quiescent coronal structures can be described as being in force-free magnetohydrostatic evolution without gravity. In thin current layers pressure can become important and on large scales (such as the solar radius scale) both pressure and gravity have to be taken into account.

There is a difficulty regarding terminology. In the solar context the term *current sheet* usually is reserved to the (idealized) case of an infinitely thin current structure, as represented by tangential discontinuities. In magnetospheric physics one uses the term *thin current sheet* for a structure with a large, but not necessarily singular, current density, an option we choose in other parts of this book (e.g., Chapter 8). To avoid confusion, in the present chapter we follow the solar physics terminology for *current sheet* and use the term *current layer*, when the current density is not necessarily singular.

14.1.1 Aspects of loading

It is widely accepted that the build-up of magnetic energy in the corona largely is the result of the motion of the field line footpoints, which are frozen into the photospheric medium with its fluctuating velocity field. This leads to complex deformations of the coronal magnetic field. The field gets braided, twisted and sheared (Parker, 1972). In addition, new magnetic flux emerges into the corona, where it is believed that buoyancy plays an important role in driving the upward motion (Parker, 1979). A considerable fraction of the rising structures may consist of largely contained fluxes so that they *bodily* enter the corona, possibly playing a significant role for the overall helicity budget (Low, 1996).

An often considered simple mode of energy transfer by photospheric motion is due to shear motion of footpoints of a magnetic structure lying above a reversal line of the normal magnetic field component, see Fig. 14.1; several examples are discussed below. The energy flux through the base of the corona is the Poynting flux $B^2 \boldsymbol{n} \cdot \boldsymbol{v}_\perp$, where $v_\perp$ is the velocity component perpendicular to the magnetic field. Note that in the figure the Poynting flux on the side of the larger velocity is upward, while downward for the smaller velocity, so that there is a net upward Poynting flux. A conspicuous effect of the energy transfer is an upward expansion of the field lines (e.g., Low, 1977; Inhester *et al.*, 1992) (Section 5.3.4).

As in the magnetospheric case, one may ask whether the loading can be reversed so that a steady state is possible, where on the global average the loading is balanced by its reverse process. For flux emergence based on buoyancy this is clearly not possible, due to the preferred direction of gravity.

Also a completely random photospheric velocity would lead to a secular input of energy (Parker, 1983). In the absence of dissipation, random twisting of flux tubes produces vanishing average twist (measured by the twist angle), but the mean square of the twist increases with time (Berger, 1991), as intuitively expected.

It should be noted that the convection velocity field is not completely random, it has a directed shear component resulting from solar rotation. Imagine the large velocity arrow in Fig. 14.1 to be parallel to the equator,

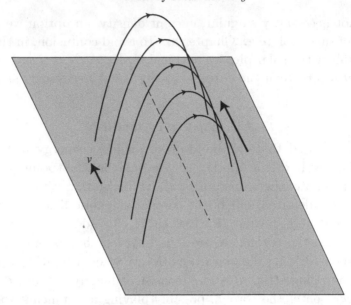

Fig. 14.1 The field lines of a magnetic arcade above a reversal line (broken line) are sheared by the photospheric velocity v.

which lies on the right side, the angular velocity vector pointing to the left. Then the velocity shear in the figure would be consistent with the shear produced by solar rotation. In the absence of stochastic motions this effect would lead to a monotonic increase of the shear of coronal arcades.

So, we can conclude that the stochastic convective shuffling of the footpoints of closed coronal flux tubes as well as directed effects based on gravity and rotation would secularly increase the magnetic energy in the corona. So, there must be one or more counteracting release processes at work.

Obviously, realistic coronal fields in which energy is accumulated can be much more complicated than the highly idealized arcade structure of Fig. 14.1. We will see examples for more realistic fields, for example, multiple arcades (Fig. 14.7) or twisted flux ropes. Many models assume translational or rotational invariance. Although this is a necessary step towards identification of physical processes, final answers require 3D modelling (e.g., Birn *et al.*, 2000).

14.1.2 Aspects of release

The literature on possible release processes conveys the following picture. The release processes cover a wide range of length, time and energy scales.

On the large end of these scales one finds the most spectacular phenomena, such as CMEs, large flares and eruptive prominences. The largest events often seem to involve all three phenomena. These processes will be addressed in Section 14.4. But there are also single events of each category and dual combinations. The detailed association, however, is not yet understood.

Release on small scales is intimately interwoven with the problem of coronal heating. The proposed heating scenarios fall into two main groups. One group is based on the release of energy in stressed magnetic fields, the other on heating by waves excited by photospheric motions. Mandrini *et al.* (2000) distinguish 14 stressing models and 8 wave models. These models are distinguished according to their scaling laws governing the dependence of the heating rate on the length of the flux tube considered. A comparison with observations seems to favour stressing models, although the authors emphasize the preliminary nature of their results in view of the imperfections of both the models and the observations.

Most stressing models of coronal heating invoke magnetic reconnection in one way or the other, which in essence goes back to suggestions by Gold (1964) and Parker (1979). In some of the models the reconnection process is incorporated only implicitly, for instance in the Taylor relaxation process (Section 11.5.3) (Heyvaerts and Priest, 1984) or in some of the current layer models, where current layers automatically develop into tangential discontinuities, decaying by reconnection (Parker, 1983). The reconnection process may occur in the form of small flare-like events (*nanoflares*), releasing energies smaller than a large flare by a factor of about 10^9 (Parker, 1988; Browning and Jain, 2004). Small-scale flare activity at network boundaries was invoked by Axford and McKenzie (2002). If flare mechanisms were confirmed as being relevant for coronal heating, this would bridge two historically different branches of solar physics (Vekstein and Katsukawa, 2000).

Let us close this brief discussion of coronal heating by looking at the simulation of Galsgaard and Nordlund (1996), which covers both loading and release. In full 3D MHD simulation with resistive and viscous dissipation a stochastic plasma flow was applied to opposing boundaries of the simulation box. One of the most pronounced effects was the rapid formation of thin current layers (Fig. 14.2), at which energy was dissipated by magnetic reconnection. In the initial phase (before reconnection becomes important) the root-mean-square of current density grows with time exponentially, consistent with results by Van Ballegooijen (1986).

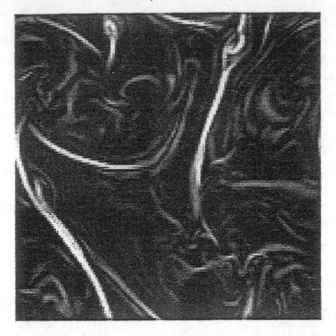

Fig. 14.2 Current layers obtained in a simulation by Galsgaard and Nordlund (1996). An initially homogeneous magnetic field in a cube is subject to a random motion at two opposing boundaries. The velocity amplitude was 0.2 v_A and the run time was about 15 Alfvén times. On average an approximate balance between stochastic energy input and dissipation by reconnection was reached. The current density was measured halfway between the driving boundaries. (From Galsgaard and Nordlund (1996) by permission of the American Geophysical Union.)

14.2 Model constraints and boundary conditions

Let us now turn to the solar eruptions. In view of the complex structure of the active corona it seems highly unlikely that one can construct a single theoretical model that explains all aspects of solar activity. So it is not surprising that there is a considerable variety of different models addressing particular features.

For models of solar activity, several constraints have been identified resulting from both theoretical arguments and observations.

14.2.1 Constraints

As already mentioned, the lower corona, from where the eruptions originate, is a low-β highly conductive medium. The plasma pressure is largely negligible, except in thin sheets, such as tangential discontinuities (Section 5.1.6). As discussed, gravity can be ignored on scales below, say, 60 000 km. In embedded cooler structures, such as prominences/filaments, gravity can become important on considerably smaller scales.

Missions such as SOHO or Yohkoh provide more and more constraining observational details. An important observational constraint concerns the CME-flare association. Although not all CMEs are accompanied by large flares, it seems that every large flare is associated with a CME (e.g., Forbes, 2000a). Here the fundamental question is that of causality. There is considerable evidence for the CME erupting from the lower corona before an associated large flare occurs (e.g., Gosling, 1993). In this picture models that predict the opposite sequence are not applicable.

In the earlier modelling the following question was raised (Aly, 1984). Consider a coronal structure such as a magnetic arcade which receives energy by footpoint motion. The structure typically grows in the radial direction. The question is, whether an eruption simply consists of a sudden opening so that eventually all field lines extend to infinity without change of topology. This possibility has been largely excluded by work of Aly (1991) and Sturrock (1991). Aly's result says that *any finite-energy force-free magnetic field occupying a half-space D (or the exterior of a 'star-shaped' region) and having all its lines unknotted and tied to the boundary ∂D of D must have an energy which is not larger than that of the 'open field' having the same flux distribution on ∂D.* Although the proof is not fully rigorous mathematically, the arguments are quite convincing. Also, the theorem has proven relevant in several numerical simulations.

In principle, the Aly–Sturrock constraint may be bypassed by violating its assumptions (Forbes, 2000a). Examples are partial opening of magnetic flux, change of magnetic topology so that a part of the energy is carried by flux not connected to the base, or energy supply during the eruption, analogous to the heating processes driving the solar wind expansion.

For given normal magnetic field component at the base the lowest energy configuration is the potential field. So, one might be tempted to conclude that the free energy of a stressed force-free configuration, available for release, is given by the difference between its energy and the corresponding potential field energy. This, however, is not the case, if the main relaxation process is magnetic reconnection. The reason is that the global magnetic helicity is approximately conserved (Section 11.5.3). Typically, the helicity will survive the relaxation in form of magnetic twist of large-scale magnetic flux ropes (Berger and Field, 1984; Berger, 1991).

14.2.2 Boundary conditions for quasi-static evolution

As discussed in Chapter 12, important dynamical information can be obtained from solution diagrams of quasi-static evolution. At bifurcation points, stability properties change in a systematic way and turning points suggest loss of equilibrium. So, in principle, the quantitative description of the loading phase can already give hints on possible release processes.

Despite these advantages, there is some danger of misinterpretation. The problem is that the instantaneous MHS equations must be complemented by conditions that single out the physical solution(s). Avoiding the immense complications of a self-consistent inclusion of the driving subphotospheric motions, one typically restricts the domain of interest to the lower corona (see, however, Zweibel (1985)). Then, the coupling to the driver is taken into account by boundary conditions at the base of the corona, which usually is identified with the photosphere. Unfortunately, the solution diagrams are sensitive to the choice of boundary conditions. Therefore, a detailed discussion is needed. The problem of choosing realistic boundary conditions is not restricted to eruptions, it also applies to the modelling of coronal loops and other coronal features.

A standard idealization is to treat the boundary as a *rigid wall* (e.g., Hood and Priest, 1979). Here the magnetic field is anchored (or *line-tied*) in the subphotospheric medium, which is treated as infinitely conducting and massive and there is no plasma motion across the boundary ($v_n = \boldsymbol{n} \cdot \boldsymbol{v} = 0$). A normal magnetic field component $B_n = \boldsymbol{n} \cdot \boldsymbol{B}$ must of course be allowed to be non-zero. Intuitively, one expects that the quantities that one can prescribe at the boundary in this case are B_n and the tangential velocity $\boldsymbol{v}_t$. In fact, these are the quantities that are continuous across the boundary, which immediately follows from the electromagnetic jump conditions stating that B_n and $\boldsymbol{E}_t$ must be continuous. One can also inspect the MHD jump conditions of Section 3.9, realizing that $m = 0$ and that the momentum balance does not apply, being replaced by the formal assumption of infinite mass of the wall. Other quantities, such as the tangential magnetic field at the coronal side of the boundary, are not necessarily continuous and are determined by the coronal dynamics alone. Thus, they cannot be used to present the subphotospheric driving (Jockers, 1978; Klimchuk and Sturrock, 1989; Finn and Chen, 1990). Therefore, earlier results based on prescribing $\boldsymbol{B}_t$ (Birn and Schindler, 1981; Low, 1982) were invalid (Jockers, 1978; Sturrock, 1989; Finn and Chen, 1990) and had to be reconsidered (Zwingmann, 1987; Platt and Neukirch, 1994), details will be given farther below. In the rigid-wall model the pressure would be determined by entropy conservation if the effects of heating and radiation can be ignored on the timescale considered (Finn and Chen, 1990).

Several authors have criticized the rigidity assumption $v_n = 0$ as being inappropriate. For instance, Einaudi and van Hoven (1983) refer to this boundary condition as 'overly restrictive'. In the context of stability An (1984) concludes that this constraint should be discarded and Karpen *et al.* (1990) arrived at a similar conclusion. Mok and van Hoven (1995) explained that the surface acts as a reservoir of mass and energy, determining density and temperature at the footpoints. In all those cases, although differing in details, the boundary is advocated to admit normal plasma flow ($v_n \neq 0$). The view taken by Mok and van Hoven (1995) would imply that the photospheric conditions determine the instantaneous pressure of the coronal flux tubes.

A different view was expressed by Finn and Chen (1990), who argue that there is no conservation law, based upon an energy equation for the plasma, by which the pressure in the corona can be specified. On that basis they consider results that were obtained with fixing the pressure on the surface (and thereby on coronal flux tubes if gravity can be ignored) as not corresponding to a physical process. This view favours the rigid-wall picture where the pressure is determined by entropy conservation or a corresponding law incorporating heating and radiation effects.

Mok and van Hoven (1995) have attempted to clarify the boundary condition issue by studying wave propagation in coronal loops from the corona into the chromosphere. They compared a continuous model with the rigid-wall model and a model in which the transition region is represented by a contact discontinuity, where density and temperature are discontinuous while the pressure is continuous. The former model is called the *discontinuous density model*. The conclusion is that none of the heuristic models is fully satisfactory. In particular, they found the rigid-wall model too restrictive, while the discontinuous density model at least reproduces the correct parameter scaling. Note that the discontinuous density model describes a physical situation where the pressure in a flux tube is continuous so that the pressure on the coronal side is determined by chromospheric conditions, the pressure being adjusted by normal flow. In other words, the equation for the internal energy is replaced by a boundary condition; in the absence of gravity the pressure becomes a conserved quantity. In terms of Euler potentials α, β (Section 5.1.2), defined as comoving with the subphotospheric medium ($d\alpha/dt = d\beta/dt = 0$), one has $p = p(\alpha, \beta)$ so that $dp/dt = 0$. (The Euler potentials are continuous, their normal derivatives may jump in accordance with the jump of $\boldsymbol{B}_t$.)

So far we have discussed several possibilities of boundary conditions without reference to time scales. It is quite possible that the slow loading phase and the rapid release phase require different boundary conditions. It seems that the final decision needs a case by case consideration.

Fortunately, as seen below, in some cases there are comparison properties, which render a given boundary condition useful even outside its physical range of validity.

14.3 Arcade models of solar flares

Many solar flare models are based on magnetic arcade configurations (Fig. 14.1). The prototype is the two-ribbon flare (Section 2.2).

14.3.1 Footpoint motion

Consider an ideal MHD arcade with line-tying, so that B_n and v_t is to be prescribed. Energy loading requires that the field line footpoints move with different velocities. Let us illustrate this for force-free fields with translational invariance in the z-direction (Fig. 14.3 and Section 5.2.1) and a rigid-wall boundary.

Energy supply

If the magnetic field strength decays sufficiently fast for large x and y the rate of change of the magnetic energy per unit length in the z-direction, W_m is given by

$$\frac{\mathrm{d}W_m}{\mathrm{d}t} = \int \boldsymbol{S} \cdot \boldsymbol{n}\mathrm{d}x\mathrm{d}z$$

$$= \frac{1}{\mu_0} \int_{B_n>0} \Delta v_z B_z \mathrm{d}A \ . \qquad (14.1)$$

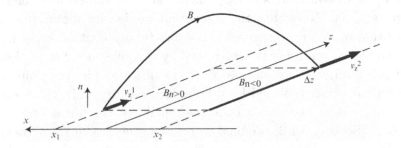

Fig. 14.3 Footpoint motion of a line-tied magnetic field. The relative velocity $v_z^2 - v_z^1$ causes the footpoint displacement Δz; the z-axis is the polarity reversal line, where the normal field component B_n changes its sign.

Here, it is used that $B_n dx dy = dA d\beta$ (Sections 5.1.2, 5.2.1), the length in the z-direction is set to 1 and $\boldsymbol{S}$ is the Poynting vector and $\boldsymbol{n}$ the surface normal pointing into the corona. A and $\beta = z + \tilde\beta$ are Euler potentials. The integrand in the second line of (14.1) has different contributions from the different sides of the reversal line (Fig. 14.3), which leads to the velocity difference $\Delta v_z = v_z^2 - v_z^1$ in the integral. Equation (14.1) confirms that energy input requires relative footpoint motion.

With the help of Section 5.2.1 one immediately finds a well-known relation between the footpoint displacement Δz and B_z. As the Euler potential β is constant on field lines one finds

$$\Delta z = -\Delta \tilde\beta = B_z(A) V(A) , \tag{14.2}$$

where $V(A)$ is the (differential) flux tube volume (Section 8.3.4).

For simplicity let Δv_z be time independent and $\Delta z = 0$ at $t = 0$, so that $\Delta z = \Delta v_z t$. With this simplification and using (14.2) one finds

$$\frac{dW_m}{dt} = \frac{t}{\mu_0} \int_{B_n > 0} \frac{[\Delta v_z(A)]^2}{V(A,t)} dA . \tag{14.3}$$

Note that flux tubes with large volumes give only small contributions to the input. The reason is that by (14.2), for a given Δz, large V corresponds to small $|B_z|$ so that the Poynting flux $\boldsymbol{n} \cdot \boldsymbol{S} = -B_n B_z v_z$ is small.

Loss of equilibrium and stability

In the simple examples with one degree of freedom of Section 12.1 there is a close relationship between bifurcation points and stability properties. A corresponding property was found for the Grad–Shafranov equation under the conditions studied in Section 12.1.1. For example, without further stability analysis we could conclude that the lowest branch of models sketched in Fig. 12.5 is F-stable. The occurrence of the stability functional F_2 given by (10.58) can be understood in the following way. In simplified terms, the equilibrium is determined by the vanishing of the first variation of the functional F, given by (10.62). The second variation comes in, when one considers an infinitesimal step along the solution curve. The existence of a unique neighbouring state requires that the minimum of the second variation has a definite sign (Zwingmann, 1987). So, stability, as defined by the sign of second variation of F cannot change unless the evolution has reached a point at which the uniqueness is lost. This happens at a bifurcation point. So, if one knows that a branch starts stable, this property implies that stability can be inferred for the entire branch before the next bifurcation point is reached. Apart from F-stability, this argument was shown to be correct also under more general conditions (see the model by Zwingmann (1987), discussed below).

In both cases, the stability functional has a simple interpretation in terms of the MHD energy principle (Section 10.2.1). The stability functional F_2 is obtained if one imposes the model constraints regarding symmetry, boundary conditions and global constraints on $\mathcal{V}_2$, given by (10.29). In particular, F_2 is obtained from $\mathcal{V}_2$ by confining the perturbations to two dimensions and by allowing for normal flow at the boundary. The open boundary implies that the compressibility term $\gamma p (\nabla \cdot \boldsymbol{\xi})^2$ minimizes to zero. In other words, one can use $\nabla \cdot \boldsymbol{\xi} = 0$ as a boundary condition for $\boldsymbol{\xi}$ (Schindler *et al.*, 1983). This means that positive definite F_2 is sufficient for 2D ideal MHD stability in the sense of (10.57). Note that, if pressure exchange is still possible on the time scale of the instability, the F_2 criterion becomes necessary and sufficient for perturbations subject to the translational invariance restriction.

The numerical iteration scheme of Zwingmann (1987) automatically provides the minimum of the stability functional so that stability information is obtained without additional effort.

Zwingmann's model

Earlier arcade models were based on the formal equivalence between force-free and pressure supported systems (Low, 1977; Birn *et al.*, 1978), see Section 5.2.1. One finds a catastrophe indicating loss of equilibrium (see Fig. 12.6). However, as already mentioned this approach prescribes $\boldsymbol{B}_\mathrm{t}$, which is inappropriate. An acceptable magnetic boundary condition for equilibrium sequences is line-tying, realized by prescribing the footpoint displacement Δz, see Fig. 14.3 (e.g., Jockers, 1978; Zwingmann, 1987; Klimchuk and Sturrock, 1989; Finn and Chen, 1990). Let us look at Zwingmann's model as an illustrative example.

Zwingmann (1987) assumed line-tying and prescribed footpoint locations and plasma pressure. He also included gravity. Thus, the photosphere was assumed to be open to parallel flow, keeping the pressure constant (see the discussion on boundary conditions above). The boundary condition for the poloidal flux function α at the coronal base ($y = 0$), fixing B_n, corresponds to a line dipole located below the photosphere at $x = 0$, $y = y_0$, so that on the boundary $\alpha \propto 1/(x^2 + y_0^2)$. The function $\tilde{\beta}$ was chosen proportional to $\lambda_s \tanh(x/L)$ so that $\Delta z = 2\lambda_s L \tanh(x_1/L)$ (Fig. 14.3). The pressure was described by $p_0(\alpha) \exp(-y/H)$, where H is the gravity scale height (see (7.27)) and $p_0(\alpha)$ is chosen from

$$\frac{dp_0}{d\alpha} = \begin{cases} \frac{\lambda_p}{\mu_0 \alpha_c^3 L^2} \left(1 - \frac{\alpha}{\alpha_c}\right) & \text{for } 0 < \alpha < \alpha_c, \\ 0 & \text{otherwise.} \end{cases} \qquad (14.4)$$

Here L and α_c are fixed constants, the control parameters are λ_s and λ_p, controlling shear and pressure. On the lateral boundaries the normal magnetic field component is prescribed and on the upper boundary Neumann conditions are applied, so that one can consider the field lines that cross the upper boundary as open, they remain unsheared.

Fig. 14.4 shows a set of solution curves. Importantly, the force-free case $\lambda_p = 0$ gives a unique total energy W profile, there is no turning point so that there is no loss of equilibrium. The shear considerably enlarges the flux tubes. Fig. 14.5 compares a strongly sheared field with the corresponding unsheared case.

There is a finite region in the $\lambda_s - \lambda_p$-plane for which the solution curves become S-shaped so that three solutions exist. The turning points indicate possible loss of equilibrium. However, the corresponding pressures are too high to be realistic for the lower corona (Zwingmann, 1987). All branches,

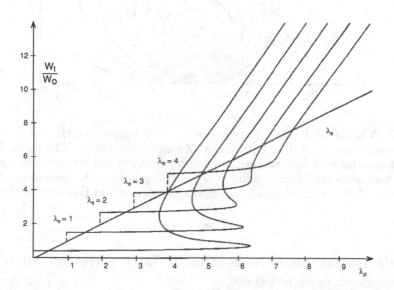

Fig. 14.4 Solution curves of the model by Zwingmann (1987); shown is (normalized) total energy vs. the pressure parameter λ_p for several values of the shear parameter λ_s; the scale height H is $5L$. (From Zwingmann (1987) with permission of Springer Science and Business Media.)

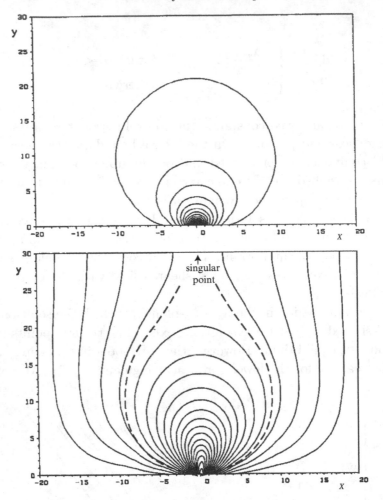

Fig. 14.5 Poloidal field lines of the cases $\lambda_s = 0$ (upper panel) and $\lambda_s = 8.5$ (lower panel) and $\lambda_p = 1$ in both cases (Zwingmann, 1987). The dashed lines (inserted qualitatively into the original figure) indicate the separatrices between sheared and unsheared flux, joining at a neutral point on the boundary. (From Zwingmann (1987) with permission of Springer Science and Business Media.)

except the intermediate branches of the S-shaped curves, are MHD stable by the argument explained above.

Although we do not wish to enter details of the numerical procedure, it seems appropriate to point out that special measures were taken to ensure that unstable solution branches are computed without difficulties. Straightforward iteration schemes often do not have that capability.

Platt and Neukirch (1994) have pointed out that the upper branch of the S-shaped curves is sensitive to the position of the lateral boundaries, and that there are thin current layers along the separatrices, which can be interpreted as tangential discontinuities (see Section 15.1 below).

It is to be expected that the existence of multiple solutions in Fig. 14.4 will disappear if, instead of prescribing the pressure as a boundary condition, rigid-wall conditions are used so that the pressure is derived from entropy conservation (Finn and Chen, 1990).

We conclude from Zwingmann's arcade study that under realistic coronal parameters this model does not give any indication concerning an ideal eruptive process, neither by loss of equilibrium nor by (ideal) instability. This result is consistent with conclusions reached by Klimchuk and Sturrock (1989), Finn and Chen (1990) and others. However, as seen in the following section, this does not mean that shearing of a single 2D magnetic arcade in the coronal plasma must be excluded entirely as a process leading to an eruption.

Analytical arcade models are largely two-dimensional. An interesting class of 3D models was presented by Neukirch (1997), who solved the nonlinear MHS equations analytically.

14.3.2 Dynamic arcade models

An alternative to quasi-static modelling is solving the dynamic MHD equations with a boundary condition prescribing v_t. The 2D models to be discussed here have either translational or rotational invariance including a magnetic field component in the invariant direction (often referred to as *2 1/2*-dimensional fields (Section 5.2)).

A sometimes wanted, but sometimes unwanted typical feature of MHD simulations is that even without a nominal resistivity, the effect of numerical diffusion may lead to magnetic reconnection of thin current layers that develop during the evolution; this often makes it difficult to decide whether a current layer would develop into a state where physical reconnection is to be expected (see Section 15.1 below). However, several authors (e.g., Mikić and Linker, 1994; Hu, 2004) have found ways to avoid numerical diffusion in a current layer.

An example is the simulation by Mikić and Linker (1994), which has taken up earlier work by Barnes and Sturrock (1972) with new methods. Fig. 14.6 shows a typical result. The field has rotational invariance, i.e., it is independent of ϕ in spherical coordinates r, θ, ϕ. The computational domain has an outer boundary at $r = 200 R_0$ (R_0 being the solar radius) with an

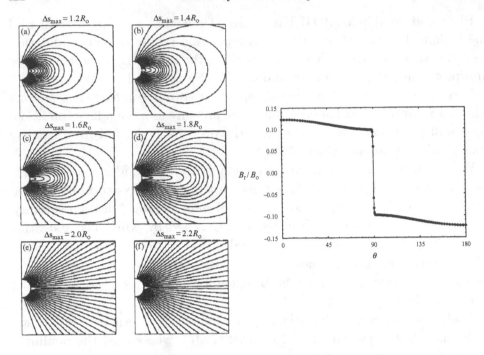

Fig. 14.6 Sheared arcade model by Mikić and Linker (1994) under ideal force-free conditions. The magnetic field has rotational invariance. The left panel shows the poloidal field lines for increasing toroidal footpoint displacement with maximum Δs_{max}. Near $\Delta s_{max} \approx 1.8R_0$ the field rapidly opens and a tangential discontinuity develops in a time-dependent process. The right panel shows the corresponding jump in the radial magnetic field at $r = 3R_0$ for $\Delta s_{max} = 2.2R_0$. (Reproduced by permission of the AAS.)

upper boundary condition that allows for outward flow. The initial field is generated by a dipole singularity located at the centre of the Sun. On the surface the footpoints move in the ϕ direction, with velocity depending on θ, the maximum velocity being 0.94 km/s. The parameter Δs_{max} represents half the maximum displacement applied in each case. When that value is reached, the velocity is turned off. At a critical point near $\Delta s_{max} = 1.8R_0$ there is a rapid transition from solutions with essentially all field lines closed to states with large open fluxes. The supercritical development is a rapid (non-quasi-static) evolution, during which an equatorial thin current layer develops (on the left of Fig. 14.6), associated with a large gradient in B_r (on the right). The authors point out that the current layer will rapidly develop into a true tangential discontinuity. Emphasizing the formation of current sheets by nonequilibrium (Parker, 1972, 1994), the transition at the critical displacement is termed *magnetic nonequilibrium*.

It is not clear whether this phenomenon corresponds to loss of equilibrium in the sense of a turning point of the quasi-static solution curve (Section 12.1.1). A clear distinction between loss of equilibrium in the sense of a catastrophe and instability is difficult to assess in dynamical computations. The main distinguishing element between the cases b and c of Fig. 12.1 is the presence or absence of an unstable branch in the supercritical regime, and it takes special measures to include unstable branches (Zwingmann, 1987; Isenberg *et al.*, 1993; Schröer *et al.*, 1994). Although this distinction is important from a basic theoretical point of view, what counts observationally is *loss of stable equilibrium*. We will use that term when a further distinction is not available.

The addition of a small resistivity (physical or due to numerical diffusion) gives rise to a sudden onset of a fast reconnection process and plasmoid formation, qualitatively similar to the process of Section 11.2.10. Violent eruption, however, occurs only after the critical point was reached.

Although the system opens, the energy constraint (Aly, 1984; Sturrock, 1991) is satisfied, for large shear the energy of the open configuration is approached from below but not reached. The reason is that a small amount of the arcade flux remains closed. Note that energy release occurs through the final fast reconnection process.

In another version of their model Mikić and Linker (1994) included a small pressure. These studies show a behaviour quite similar to that of the pressureless case described above.

Traditionally, arcade structures have been considered appropriate models for two-ribbon flares. Nevertheless, there is ample evidence for eruptions arising from magnetic fields with a more complicated structure. Accordingly, double and triple arcades as well as periodic arcade sequences have been studied (e.g., Mikić *et al.*, 1988; Biskamp and Welter, 1989; Finn *et al.*, 1992). The work by Finn *et al.* (1992) provides a detailed picture of the response of double arcades with translational invariance to shearing footpoint motion. In typical cases the field complexity leads to an interplay of instability, current layer formation and fast reconnection with plasmoid formation. Their study includes cases where the strength of the shearing was shifted from one arcade to the other in the course of the evolution. Some eruptions were found to be due to loss of equilibrium in the sense of a multidimensional form of the cusp catastrophe (Chapter 12). Plasmoids that formed inside the structure find their way through an overlaying separatrix by a process, the essential part of which is topologically equivalent to the case illustrated by Fig. 11.11.

The tendency of an arcade to erupt can be strengthened by adding to the shearing motion a converging flow component, i.e., on both sides directed toward the reversal line. This effect was studied by Inhester *et al.* (1992) in a configuration with translational invariance. They considered a case that was found stable for pure shear flow. The addition of a suitable amount of convergence led to current layer enhancement, resistive instability and plasmoid formation.

Birn *et al.* (2000) generalized this work by dropping the constraint of translational invariance. They found that one of the parameters that had a major influence on stability was the fanning of the field lines as viewed horizontally and perpendicular to the reversal line. Less fanning gave more pronounced eruptions. Their 3D configuration allowed them to determine the parallel electric potential U associated with the reconnection process (Section 11.5.1). They found values of the order of several 100 MV.

In most of the cases discussed so far, and in others with similar objectives, the eruption is initiated by a reconnection process and, typically, by the formation and ejection of a plasmoid, which in the presence of a magnetic field component in the invariant direction, is a flux rope. The reconnection also energizes several phenomena below the reconnection site. This includes *ablation* (also called *evaporation*) of heated chromospheric material (Fig. 2.6); for details see articles in Strong *et al.* (1999) and for corresponding modelling, e.g., Forbes *et al.* (1989) and Yokoyama and Shibata (2001).

14.4 Coronal mass ejections

The flare scenario discussed in the previous section needs a modification for large flares that are associated with CMEs (Gosling, 1993). At first sight, the flare process as described so far can explain not only the emissions from below the reconnection site, but also the CMEs, if one identifies them with the expelled plasmoid. However, this picture would imply that the flare-associated heating of coronal loops starts before the beginning of the CME, which is in conflict with the observational constraint stated in Section 14.2.1, demanding the opposite sequence.

This dilemma is removed by models, in which the CME is initiated by some largely ideal process or a process which involves only moderate reconnection (which could be interpreted as *preflare brightening* (Sterling and Moore, 2005)). Such processes should generate an upward motion. The rising structure may be a flux rope or the top of an arcade (Fig. 14.8). Typically, there is magnetic flux on top of this structure, which will be stretched by the upward motion, so that a current layer forms below it. If reconnection tears that layer, a flare process is initiated similar to that described above (see also Magara *et al.* (1996)).

If this picture applies, CME models must explain rapid upward motion with no or only moderate reconnection. Several models have been constructed that have this property. The simulation method of Mikić and Linker (1994), already discussed for arcade shearing (Fig. 14.6), also has been applied to a helmet streamer configuration. Again, magnetic nonequilibrium leads to a fast upward motion (Fig. 14.8). The structure contains a current layer, which via reconnection could initiate the flare. Linker *et al.* (2003) concentrated on the effect of *flux cancellation* caused by footpoint motion toward the reversal line in a streamer configuration, including a solar wind flow. They obtained a rapid transition from a stable flux rope configuration to an eruption, when a critical threshold of flux cancellation is surpassed.

A flux rope that has risen by some MHD process may assume an equilibrium state by the magnetic stress exerted by overlying flux, still rooted in the photosphere. That flux assumes the role of a *tether* that keeps the flux rope from further rising (Kopp and Pneuman, 1976). Reconnection below the rope then has the effect of tether cutting (Sturrock, 1989; Moore and Roumeliotis, 1992), which eventually initiates the upward motion and eruption of the flux rope (Lin *et al.*, 2001; Sterling and Moore, 2005).

A CME model based on a more complex field is the *magnetic breakout model* by Antiochos *et al.* (1999). In its simplest version the initial field is generated by the superposition of a dipole and an octopole at the centre of the Sun, generating a quadrupolar field on the surface (Fig. 14.7). Magnetic shear is applied to the central region of the inner arcade, which rises and pushes against the overlying flux. This generates currents as the angle between the separatrices decreases (Section 11.2.10). In this picture the eruption is started by reconnection at the stressed neutral point configuration, which reduces the overlying unsheared flux, so that the central arcade can burst open. The eruption has been seen in MHD computations (MacNeice *et al.*, 2004). The results comprise the initiation, the plasmoid formation and ejection, and the transition of the coronal field to a more relaxed state. This process satisfies the Aly–Sturrock energy constraint due to partial opening.

In this context a question has been raised about the appropriate minimum energy state reached by relaxation via reconnection. As mentioned above, Taylor relaxation would result in the linear force-free field (see also Section 11.5.3) consistent with the boundary condition. However, this requires that all flux tubes participate in the reconnection. It was pointed out

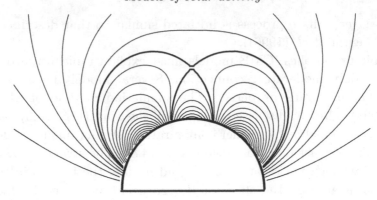

Fig. 14.7 Field lines of a quadrupolar field in the corona, generated by the superposition of a dipole and an octopole at the centre of the Sun, suggested by Antiochos *et al.* (1999).

by Antiochos *et al.* (2002) that in cases where the reconnection can occur only near a neutral point, the minimum energy state is a nonlinear force-free field and the minimum energy is larger than the Taylor value. In simulations even that minimum was not reached, instead Antiochos *et al.* (2002) found highly complex end states, which led them to introduce the notion of *reconnection-driven current filamentation*.

Considerable attention has been devoted to the question whether there is a suitable quasi-static evolution that leads to loss of equilibrium in the presence of a pre-existing flux rope. A largely analytical approach is based on the combination of potential fields with current sheets in translationally invariant geometry (Van Tend and Kuperus, 1978; Démoulin and Priest, 1988; Forbes and Isenberg, 1991; Forbes and Priest, 1995; and others). In simple terms, ignoring gravity, the quasi-static evolution starts under conditions where the flux rope, often carrying a filament, is in a stable equilibrium. One assumes that the flux rope has formed on a time scale that does not allow its field to penetrate the photosphere, treated as a perfectly conducting boundary. This requires shielding currents in the boundary, which generate a magnetic field that causes an upward Lorentz force at the flux rope. For an axial flux rope current the shielding effects are the same as in the case of an oppositely directed mirror current, symmetrically located below the boundary. Here the upward Lorentz force is particularly obvious as the opposite currents repel each other. The downward force stems from the tension of overlying fluxes anchored in the photosphere.

It has been shown for several different configurations of that type that a suitable quasi-static evolution can reach a catastrophe point. A number of earlier models suffered from the restriction that loss of equilibrium was available only for unrealistically small flux rope radii (Anzer and Ballester, 1990). The reason for this property was analysed by Lin (2001), who concluded that such radius limitation is not necessarily a problem, if there is no current sheet in the pre-eruption field. In one of the explicit earlier approaches Isenberg *et al.* (1993) used boundary conditions that correspond to a magnetic quadrupole with strength σ located at a distance d below the photosphere. The quasi-static sequence is generated by gradually reducing σ, until a point of loss of equilibrium is reached. An example of the bifurcation diagram is shown in Fig. 14.9. The diagram indicates that at the critical point the flux rope suddenly rises. The upper equilibrium (which would probably not be reached in an overshooting dynamic process) has a current layer beneath the flux rope as shown in the right panel of Fig. 14.8. Note that σ controls the boundary condition and therefore can be regarded

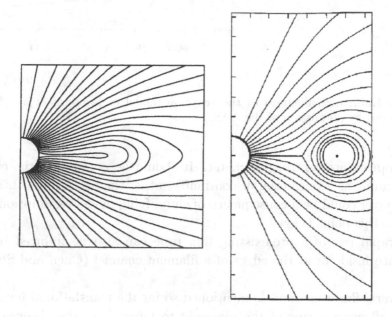

Fig. 14.8 Ideal phase of CME eruptions. On the left the upward motion is the consequence of footpoint shearing of a helmet streamer configuration (Pneuman and Kopp, 1971) leading to magnetic nonequilibrium in a system with rotational invariance (Linker and Mikić, 1995). On the right a quasi-static evolution of an ideal MHD configuration with translational invariance including a pre-existing low-lying flux rope experiences loss of equilibrium, so that the rope rises (Forbes, 2000a). In both cases a current layer develops beneath the rising structure. (Reproduced by permission of the American Astronomical Society (left part) and the American Geophysical Union (right part).)

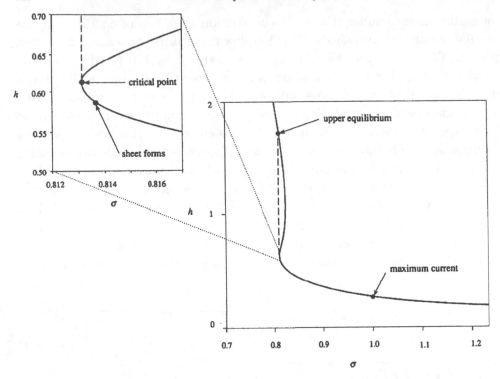

Fig. 14.9 Bifurcation diagram of the eruption model of Isenberg *et al.* (1993). (Reproduced by permission of the AAS.)

as an appropriate control parameter. It should be noted that the current sheet forming as a tangential discontinuity rather than as a thin continuous current layer could be a consequence of simplifying model assumptions. We return to this point below.

The rapid rise of a pre-existing flux rope can also be induced by flux emergence inside or at the edges of a filament channel (Chen and Shibata, 2000).

In several flux rope models mentioned so far the translational invariance did not allow an escape of the plasmoid to infinity. In the absence of re-connection in the current layer forming below the rope an escape to infinity would require infinite energy. This is different for systems with rotational symmetry, where the energy remains finite. Also, there is an additional radial force on the flux rope, which in rotational symmetry has the shape of a torus around the Sun. In the absence of any external forces the torus cannot be in equilibrium (Section 5.1.7), so that it expands. The Aly–Sturrock constraint is not applicable, because not all field lines are anchored at the base.

It takes 3D modelling to anchor the flux rope at its ends. A flux system may also emerge from below the photosphere as a magnetically contained structure (Low, 2001).

Hu *et al.* (2003) used a relaxation method to find force-free equilibria numerically. A low-lying flux rope is introduced into a dipolar or partly open potential field. The configuration either relaxes into a stable equilibrium without significant rising of the flux rope or the flux rope erupts catastrophically to infinity forming a corresponding current layer. What type of behaviour takes place depends on the energy; the eruption occurs if the process is energetically possible.

Zhang *et al.* (2005) considered a low-lying flux tube suspended in the quadrupole field suggested by Antiochos *et al.* (1999), see Fig. 14.7, using the same numerical method as Hu *et al.* (2003). Other than Hu *et al.* (2003), they found a parameter regime where increasing toroidal flux of the rope led to a critical stage where the flux rope rises catastrophically but relaxes into an equilibrium at a finite height, with a current layer below it (see Fig. 14.10). Increasing the toroidal flux in the rope further causes a second catastrophe resulting in an equilibrium at a still larger height. In the ideal dynamics, the neutral point of the potential field without flux rope turns into an overlying current layer, which seems to prevent an escape to infinity. It is expected that a sufficient amount of reconnection at the current layers will cause the flux rope to escape.

There are also 3D computations concerning twisted flux ropes. As already mentioned, this offers the possibility of anchoring the rope ends in the photosphere. The interesting question is, what happens if the twist is gradually increased by footpoint motion. Several authors (e.g., Amari *et al.*, 2000) have studied this case by ideal MHD simulations. It was found that after a quasi-static phase a dynamical phase develops, characterized by a fast expansion. Maximum energies can amount to a considerable fraction of the energy of the corresponding fully open configuration. The transition from a quasi-static to a dynamic evolution may be due to loss of equilibrium or instability (Chapter 12).

Several eruption models are based on the kink instability (Section 10.2.4) of a twisted flux tube (Roussev *et al.*, 2003; Kliem *et al.*, 2004). In the simulation by Kliem *et al.* (2004), a flux rope that is helically distorted by the kink instability deforms the ambient coronal field. Current layers form at the interface with the surrounding medium and below the rising part of the rope. This approach profits from the 3D flux rope model of Titov and Démoulin (1999), which includes a background field. As in the 2D case, reconnection is expected to drive a flare. Without reconnection the eruption

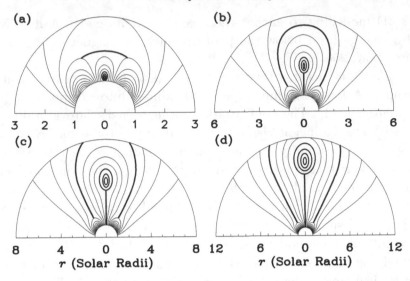

Fig. 14.10 Two step catastrophe model from Zhang *et al.* (2005). A flux rope embedded in the quadrupole field shown in Fig. 14.7 relaxes into different equilibria for increasing toroidal flux. Shown are the equilibria: (a) right before, (b) right after the first catastrophe point, (c) right after and (d) at a distance away from the second catastrophe point. (Reproduced by permission of the AAS.)

saturates at a finite amplitude. In the study by Roussev *et al.* (2003), also based on the equilibria of Titov and Démoulin (1999), the regime of loss of stable equilibrium was investigated. For the flux rope to escape, the arcade field had to be eliminated.

Fan and Gibson (2004) simulated the emergence of a flux rope. The rope becomes kink-unstable when a sufficient amount of twist is transported into the corona. Also in this case a thin current layer forms. The results indicate that this layer is consistent with being a tangential discontinuity smoothed by finite numerical resolution.

Some of the flux rope CME simulations, which were already mentioned (e.g., Linker and Mikić, 1995; Linker *et al.*, 2003; Roussev *et al.*, 2004), include the presence of the solar wind. A further, largely analytical approach uses a 3D selfsimilar MHD expansion model (Gibson and Low, 1998). This model concentrates on the CME propagation into the solar wind, however, leaving open the process that leads from a quiescent state to the dynamic evolution. Envisaged possibilities are the loss of prominence material or increasing twist of the flux rope (Low, 2001). As in other flux rope models a two-ribbon flare is caused by reconnection of a current layer that forms beneath the rising structure.

14.5 Further aspects

14.5.1 Particle acceleration

In this book the discussion focusses on large-scale loading/release processes. Corresponding fluid models, as widely used for describing solar eruptions, do not directly address the acceleration of nonthermal particles. This is an unfortunate limitation, because observations show that in solar eruptions particles are efficiently accelerated. It has been stated that a considerable fraction of the magnetic energy, released in an (impulsive) flare, goes into energetic particles (Miller *et al.*, 1997) with energies far above the thermal level. The presence of high energy particles is manifested by continuum emission extending into the GeV regime, as well as by γ-ray line emission. These observations set strong constraints to flare models (Miller *et al.*, 1997), but tests would require the inclusion of single particle aspects. The existing modelling suffers from a severe gap between the MHD eruption models on the one hand and single particle models or corresponding kinetic theories on the other (Neukirch *et al.*, 2006).

In an approximate approach one studies single particle orbits in electromagnetic fields obtained from MHD simulation, a method that has been successfully applied to magnetospheric particle acceleration (Chapter 13). For solar applications Turkmani *et al.* (2005) have used the fields from MHD braiding simulations for test particle calculations.

Other test particle studies are based on prescribed field geometries expected to represent reconnection sites (e.g., Martens and Young, 1990; Wood and Neukirch, 2005). For a 2D reconnection geometry with a guide field Wood and Neukirch (2005) included the parallel electric field associated with finite-B reconnection (Section 11.5.1). As qualitatively expected from Figs. 11.36 and 11.38, the particle energies were strongly peaked for particles staying near separatrices. This behaviour was demonstrated by Wood and Neukirch (2005) for electrons reaching energies above 5 keV.

It is a general property of this and similar models that the number of accelerated particles remains too small to explain observed fluxes. Shock waves or MHD turbulence can accelerate more particles than reconnection fields, but they often suffer from other deficiencies (Miller *et al.*, 1997; Neukirch *et al.*, 2006). Interestingly, in simulations by Roussev *et al.* (2004) the eruption caused a fast MHD shock wave (Section 3.9), for which diffusive shock acceleration theory (see, e.g., Schlickeiser, 2003) predicts a distribution of solar energetic protons with a cutoff energy near 10 GeV.

Despite such promising results, a satisfactory picture of particle acceleration in solar eruptions is not yet available.

14.5.2 Sun–Earth connection

As a typical CME expands into the interplanetary medium it develops a magnetic substructure called *magnetic cloud* (Burlaga *et al.*, 1981). It is characterized by an enhanced magnetic field strength and low ion temperature and by a rotation of the magnetic field direction as seen by a stationary observer. The association between CMEs and magnetic clouds is well established. There is evidence even for the presence of prominence material in magnetic clouds (Burlaga *et al.*, 1998).

Magnetic clouds also are known to play an important role in geomagnetic activity (Chapter 13). Due to the magnetic field rotation the cloud provides long periods of southward-pointing IMF, which leads to enhanced magnetospheric activity.

So here we encounter a fascinating connection between solar and magnetospheric activity, which has been referred to as *Sun–Earth connection*. This connection aroused considerable research interest. A strong CME eruption in January 1997 and its consequences were tracked *from cradle to grave* (Fox *et al.*, 1998, and references therein). Consistent with the expected travel time, the CME was observed to leave the lower corona about 4 days before the magnetic cloud was seen near the Earth. The magnetosphere was strongly disturbed resulting in a magnetic storm. During a period of about 12 hours with southward-pointing IMF there was intensified substorm and auroral activity. The fact that the solar surface activity remained low and no pronounced flare signatures could be associated with the CME underlines the fact that large flares are not necessarily associated with CMEs. In terms of the models discussed in this chapter, we might speculate that the absence of a flare might be a consequence of the overlying flux being small, so that a significant current layer did not form or a significant current layer may have formed but it did not reconnect. Alternatively, a flare did occur but it was difficult to observe (Reeves and Forbes, 2005).

The investigations of Sun–Earth connections are motivated not only from a basic science point of view but also by its relevance for the safety of human life and of technical systems. This has led to the notion of *space weather* defined as *conditions on the Sun and in the solar wind, magnetosphere, ionosphere and thermosphere that can influence the performance and reliability of space-borne and ground-based technological systems and can endanger human life or health* (US National Space Weather Programme, 1995).

15

Discussion

Having considered two major applications of space plasma activity, it seems appropriate to close this part with a discussion of their differences and similarities. We first address the role of magnetic reconnection and then suggest a general eruption scheme that covers both magnetospheric and solar activity.

15.1 The reconnection problem

We begin by considering reconnection in solar activity and then bring in magnetospheric reconnection for comparison.

As we have seen, models of solar activity involve magnetic reconnection in one way or the other. In some models reconnection is involved in the eruption process itself. In addition, reconnection is considered generally as a potential release process for a field configuration with a thin current layer below a fast rising object. A CME-associated flare would be the consequence (e.g., Amari *et al.*, 2000).

Here, and more generally in the context of solar activity, the difficult question arises of what are the quantitative criteria for reconnection to occur. Unfortunately, we can only narrow down the problem, a clear-cut answer is not yet available. We limit the discussion to fast reconnection (Chapter 11).

As magnetic reconnection cannot take place under ideal MHD conditions, the first point to address is whether the necessary nonideal process is collisional or collisionless. Using the values of the middle column of Table 9.1 as an example for coronal plasma conditions, we see that the plasma parameter is of the order of 10^8, which is a first indication that collisions are extremely rare. This is confirmed by the large value of the Lundquist number S, which is larger than 10^{14}. In fact, it takes more than 10 million years for resistive

433

diffusion (based on (11.1)) to become significant on the assumed overall scale length of 3×10^7 m.

What about smaller scales? Let us assume a reconnection rate of $0.1 \, v_A B_0$ (Section 11.3.3). With Ohm's law (11.3) and (11.32) this means that the current layer width is of the order of $\delta = 10L/S$, which is smaller than 10^{-6} m. However, here we have left the regime of validity of resistive MHD, because that scale is much smaller than the electron mean free path, and the corresponding time scale (using velocity v_A) is much smaller than the collision time, so that the standard transport theory, based on small mean free paths, is no longer applicable. It has to be realized that below, say, 600 km, collisional reconnection can safely be excluded for the present parameters. (The temporal condition gives the same order of magnitude, which is a consequence of the thermal electron velocity being of the same order of magnitude as the Alfvén velocity.) So, in the collisional regime resistive effects are much too small to allow for fast magnetic reconnection; in other words, the reconnection must be collisionless. (There is a complicated transition region between the two regimes, but under coronal conditions that region can be ignored.)

For laminar processes this immediately puts the length scale in the regime below the ion inertial length $c/\omega_{pi} \approx 7$ m or the proton gyroradius $r_{gi} \approx 0.2$ m. To avoid stabilization of an undisturbed current sheet with a normal component B_n, the ratio B_n/B must not exceed r_{ge}/L (Section 10.5.5), which is of the order of 10^{-10}. If the reconnection is based on microturbulence, the scale condition can be less restrictive. For considering the regime of the LHD instability (Section 9.3.3), let us assume that the drift velocity equals the ion thermal velocity. This gives a length scale of the order of 100 m. So, under the assumed conditions, current layer scales of the order of or smaller than 100 m are necessary for reconnection.

As we have seen, in the MHD picture an attractive possibility for the generation of small current scales would be the formation of tangential discontinuities, where formally the current layer becomes infinitely thin (Section 3.9). It has been suggested that the shuffling of footpoints of loops or arcades by subphotospheric motions would lead to nonequilibrium and relaxation into states that contain tangential discontinuities (Parker, 1972, 1994; Low, 1987). Although the general nonequilibrium argument has been challenged (Van Ballegooijen, 1985; Zweibel and Li, 1987; Arendt and Schindler, 1988), there is ample evidence for thin current layers to form under a variety of circumstances.

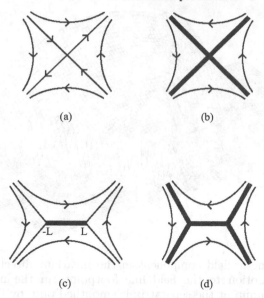

Fig. 15.1 Current sheets arising from footpoint motion, (a) initial potential field, (b) magnetic shear causes separatrix sheets, (c) motion toward the separator causes a separator sheet, (d) more general motions can cause separatrix and separator sheets simultaneously. (From Priest *et al.* (2005) by permission of the AAS.)

Tangential discontinuities can be expected to be associated with separatrices (Section 8.6.2), resulting from a change in magnetic field connectivity. In particular, connectivity changes typically arise in fields that contain a neutral line or a null. If one assumes a process where the end state has relaxed into a potential field, one can see that, typically, the end states involve singular current sheets (Longcope, 2001; Priest *et al.*, 2005). For a system with translational invariance this is illustrated in Fig. 15.1. If one starts with a simple x-line configuration (with a guide field admitted), suitable footpoint motion leads to separatrix or separator sheets or both.

A separator sheet was discussed in Section 8.6.1, see Fig. 8.10. The flux function (8.130), written as $A(x, y, L)$, can be understood as describing current layer formation in response to flux transfer. The flux $\psi = A(0, 1, L) - A(0, 0, L)$ increases monotonically with L, so that L increases monotonically with the added flux ψ (Priest *et al.*, 2005). Qualitatively the same process occurs at separators connecting two nulls (Priest and Forbes, 2000).

Under suitable conditions separatrix sheets can be obtained also in the cases that include pressure and distributed currents. An example by Zwingmann (1987) was discussed above, see Fig. 14.5. The separatrix current

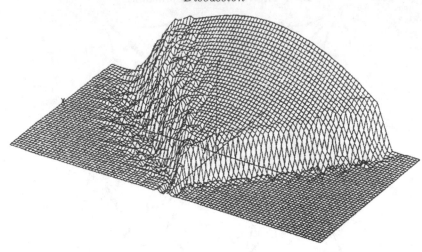

Fig. 15.2 The magnetic field component in the invariant direction arising from applying a shear motion to the field line footpoints in the model by Zwingmann (1987). The jump at the separatrix is smoothed only by finite grid effects. (From Platt and Neukirch (1994) with permission of Springer Science and Business Media.)

layer of that model was analysed in detail by Platt and Neukirch (1994). It results from the fact that the 'open' field lines (intersecting the upper boundary) do not bend in the direction of the footpoint motion. Fig. 15.2 gives an example showing the profile of the magnetic field component in the invariant direction (B_z) resulting from footpoint shear obtained by using Zwingmann's numerical method. The width of the current layer (rapid variation of B_z) was found to be determined by the scale of the grid spacing. So, the authors suggest, that the observed rapid transition approximates a tangential discontinuity. A corresponding result was found by Fan and Gibson (2004), as discussed in Section 14.4.

Even the presence of an MHD discontinuity is not sufficient for reconnection to occur. We have to look inside. If one does so in a particle picture, one sees a continuous structure, which, however, may be extremely complicated, possibly involving a significant level of fluctuations. The reason is that (possibly with a few unphysical exceptions) the motion of particles with a random velocity spread will not be consistent with a sharp field discontinuity. Note that even a discontinuous distribution function does not necessarily imply a discontinuous current density, for example see (6.13). So we have to expect that the build-up of the current density will stop before a real discontinuity is reached.

It has been argued that pressure effects will not stop the current build-up because of motions of plasma inside the current layer arising from a gradient parallel to the sheet. This picture is intuitively attractive, but as we have seen in Section 8.6.3, it is not necessarily applicable in all cases. The example discussed there shows how a counteracting pressure gradient can be sustained in an arbitrarily thin current layer, independent of the presence of gradients along the sheet. On the other hand, pressure gradient reduction by parallel flow may well be possible in other situations. Adiabatic processes with losses or longitudinal expansion can also lead to extremely thin current sheets (Section 8.2). A general conclusion does not seem to be available without specifying details of the configuration.

This can be illustrated for the Earth's magnetosphere. The magnetopause, which is a tangential discontinuity in the MHD descriptions (Chapter 13), has the possibility to adjust the pressure, at least on the open field lines, in the outer parts of the sheet. As a consequence, the thickness reduces down to the intrinsic particle scales, see Table 2.2, which is necessary for magnetopause reconnection to occur. The near-Earth tail current layer is different. It is located on closed field lines and, due to pressure forces, it settles down to thicknesses that, under quiescent conditions, can be much larger than the particle scales. Further flux transfer eventually leads to thin current layers, forming inside the plasma sheet, at which eventually the reconnection starts.

In resistive MHD computations the choice of the Lundquist number S and the grid spacing is a compromise between physical and numerical requirements. If the physical resistivity is set to zero, the resistive dissipation results from the finite grid (unless special measures are taken (Mikić and Linker, 1994; Hu, 2004)). Thus, for the collisionless plasmas discussed here, resistive reconnection simulations must be understood as representing collisionless reconnection. That this is possible, at least in some crude sense, has been demonstrated by comparative studies (Section 11.3.3). Although the details depend on the circumstances (e.g., whether or not the resistivity is localized), a rough guide line for an appropriate regime of S is obtained by choosing the value of S, such that reconnection would take place at a sheet width of the critical length scale, i.e., 100 m for the corona. This would set S to 3×10^5 and would require that the grid spacing must resolve this length at least in the region where reconnection takes place. For numerical dissipation the effective Lundquist number becomes L/L_{grid}, as the scale of a thinning structure reaches the grid scale L_{grid} (Antiochos et al., 1999).

In cases where a current layer settles down at a scale that is too large to allow for significant reconnection, external perturbations might still initiate the reconnection locally. Note that under suitable conditions an external pulse of finite duration can initiate an eruptive reconnection process, which continues to evolve well after the end of the pulse (Section 11.3.3). Also, at bifurcation points small perturbations can have large effects (see Section 12.3).

In the Earth's magnetosphere the understanding of reconnection, although still incomplete, enormously profited from *in situ* observations, resolving particle scales (Chapter 13), while it has proven much more difficult to assess the large-scale dynamics from spacecraft observations. Regarding reconnection in the solar atmosphere, one is faced with the opposite situation: there are the excellent global pictures (Section 2.2) on the one hand, and severely limited knowledge about the extremely small scales on which reconnection is expected to be active on the other.

15.2 A general eruption scheme

The two application areas that we have discussed in some detail, magnetospheric and solar activity, are quite different in many ways. Not only do the plasma properties show substantial differences (Table 9.1), but also the overall geometry, as well as the role of gravity. As we have seen, these differences manifest themselves in different properties of reconnection. Nevertheless, in both cases the models use the same basic building blocks, namely quasi-static MHD evolution, formation of thin current layers, MHD instability or loss of equilibrium, and magnetic reconnection. So, let us now concentrate on the common aspects.

As we have seen, a common property of magnetospheric and coronal plasmas is that, to a very good approximation, they are ideal on the overall length and time scales (Table 9.1). We have also discussed that this poses significant constraints on the energy release from closed flux regions on those scales (see Sections 3.8.2 and 11.4.2). Efficient release requires breakdown of magnetic line conservation, which most efficiently is provided by reconnection (Chapter 11). But reconnection, requiring nonideal effects, works only on small scales, typically, at thin current layers (Chapter 8). Viewed in this way the common relevance of thin layers and reconnection for magnetospheric and solar activity is less surprising.

Nevertheless, there remain important consequences of the fact that the coronal magnetic field has a much more involved large scale structure than the magnetosphere. One of the consequences is that ideal instabilities and ideal loss of equilibrium is more commonly available on the Sun than in the magnetosphere (Section 10.2.5).

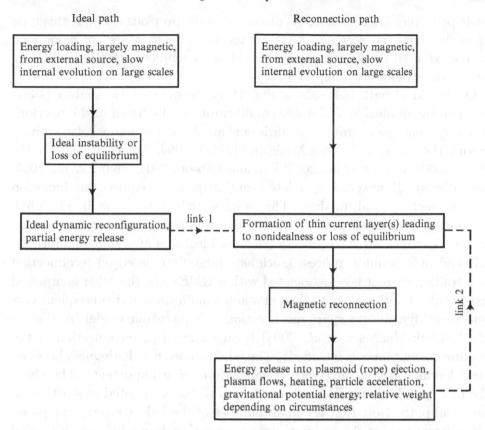

Fig. 15.3 Scheme illustrating possible paths of plasma eruptions.

Fig. 15.3 is an attempt to put the magnetospheric and solar eruption models, discussed in Chapters 13 and 14, into a unifying scheme. The path shown on the right illustrates the prototype of eruptions based on reconnection. It applies to the magnetospheric substorm model discussed in Section 13.3. The loading phase is the substorm growth phase (which is ideal by considering magnetopause reconnection as an external source process); the second step addresses thin current sheet formation in the plasma sheet, which leads to the onset of reconnection, initiating energy release in the substorm expansion phase. As we concentrate here on loading and eruption, the recovery phase is not included in the scheme.

The reconnection path also applies to several models of solar flares (e.g., Inhester *et al.*, 1992), including models based on tether cutting (e.g., Moore and Roumeliotis, 1992). The essential difference with respect to the magnetospheric case is that the loading in the magnetosphere is due to flux transfer while it is largely footpoint motion for coronal eruptions, but the basic principles are the same. As discussed in the previous chapter, this type of model, however, seems to be restricted to flares that are not associated with CMEs. (If it was, the process would be in conflict with the CME/flare timing constraint, see Chapter 14.)

On the ideal path (left side of Fig. 15.3) the quasi-static loading is terminated by an ideal loss of stable equilibrium. As discussed in the previous chapter, ideal loss of stable equilibrium is an element of many solar activity models (Isenberg *et al.*, 1993; Mikić and Linker, 1994; Amari *et al.*, 2000; Hu *et al.*, 2003; Roussev *et al.*, 2003; Fan and Gibson, 2004; Zhang *et al.*, 2005; and others). However, a complete eruption process requires the inclusion of reconnection in all models. This is indicated by the links in Fig. 15.3, generating compound paths. One of the compound paths, generated by link 1, describes an ideal process (such as ideal loss of stable equilibrium) followed by a nonideal process (such as a large flare based on reconnection of a trailing current layer associated with a CME). On the other compound path (link 2) both processes (loss of stable equilibrium and subsequent current sheet dynamics) require reconnection. The breakout model (Antiochos *et al.*, 1999; MacNeice *et al.*, 2004) is an example (if reconnection of the trailing current layer is included). The scheme does not distinguish between ideal loading and loading with reconnection being important. Therefore, the flux cancellation model of Linker *et al.* (2003) is counted as starting on the ideal path. Note that the main purpose of the links concerns the possible reconnection at the trailing current layer that forms behind an ejected structure in a solar eruption (Fig. 14.8).

Observations indicate that for CMEs the links are not always of equal importance. For instance, the well studied Sun–Earth connection event (Fox *et al.*, 1998) did not exhibit a large flare (Section 14.5.2). This may be due to the fact that, as we have seen in Section 15.1, the occurrence of reconnection requires rather restrictive conditions. Alternatively, the magnetic field strength in the erupting region may play a decisive role (Reeves and Forbes, 2005).

Hopefully, the reader found that the theoretical tools of Parts II and III were helpful for providing a deeper understanding of major aspects of space plasma activity.

However, important questions remained unanswered. This applies, for instance, to the competition between quasi-steady convection and substorms in the magnetosphere and to the precise conditions under which current sheets erupt by reconnection in both magnetospheric and solar contexts. So, in this fascinating research area there remain great challenges for future observations, analytical theory and simulations.

Appendix 1
Unified theory: details and derivations

Here we consider the theory of steady states which includes pressure anisotropy, parallel flow and a gravity force. We first treat the momentum equation in a general form covering these cases and derive the associated equations for the magnetic field. Then we show that the field equations possess a variational principle. Finally, we incorporate mass and entropy conservation. In all cases the magnetic field is represented by Euler potentials

$$\boldsymbol{B} = \nabla\alpha \times \nabla\beta \qquad (A1.1)$$

(see Section 5.1.2).

The momentum equation

We consider a momentum equation that has the form

$$\nabla \cdot \mathcal{M} - \rho\nabla\psi = 0 \qquad (A1.2)$$

with the tensor $\mathcal{M}$ given by

$$\mathcal{M} = R\mathcal{I} + S\boldsymbol{B}\boldsymbol{B}, \qquad (A1.3)$$

where R and S are scalar functions, and $\mathcal{I}$ is the unit tensor, ρ density and ψ the external gravity potential.

This form of momentum conservation is sufficiently general to allow for incorporating pressure anisotropy and plasma flow parallel to the magnetic field, in addition to the magnetic stresses.

In the first step the external gravity force is ignored, such that we are dealing with a momentum equation of the form

$$\nabla \cdot \mathcal{M} = 0 . \qquad (A1.4)$$

441

In the second step the gravity force is taken into account.

The steady state potential and the field equations

Using the form (A1.3) of $\mathcal{M}$ in (A1.4) we find

$$\nabla R + \boldsymbol{B} \cdot \nabla(S\boldsymbol{B}) = 0 \,. \tag{A1.5}$$

With the vector identity

$$\boldsymbol{B} \times (\nabla \times S\boldsymbol{B}) = B^2\nabla S + SB\nabla B - \boldsymbol{B} \cdot \nabla(S\boldsymbol{B}) \tag{A1.6}$$

we obtain from (A1.5)

$$\nabla R - (\nabla\alpha \times \nabla\beta) \times (\nabla \times S\boldsymbol{B}) + B^2\nabla S + SB\nabla B = 0, \tag{A1.7}$$

where in one instant (A1.1) was used.

We now reinterpret the functions R and S locally as functions of α, β, B such that, for instance, ∇R may be expressed as

$$\nabla R = R_\alpha\nabla\alpha + R_\beta\nabla\beta + R_B\nabla B, \tag{A1.8}$$

where the subscripts indicate partial differentiation. Then, expanding the triple vector product, we find from (A1.7)

$$R_\alpha\nabla\alpha + R_\beta\nabla\beta + R_B\nabla B - \nabla\beta\,\nabla\alpha \cdot (\nabla \times S\boldsymbol{B}) + \nabla\alpha\nabla\beta \cdot (\nabla \times S\boldsymbol{B}) +$$
$$B^2S_\alpha\nabla\alpha + B^2S_\beta\nabla\beta + B^2S_B\nabla B + SB\nabla B = 0. \tag{A1.9}$$

Excluding the trivial case where $\boldsymbol{B} \cdot \nabla B$ vanishes identically (this case may be treated in a more direct way), the gradients of α, β, B can be regarded as linearly independent, which implies

$$R_\alpha + \nabla\beta \cdot (\nabla \times S\boldsymbol{B}) + B^2S_\alpha \;=\; 0 \tag{A1.10}$$

$$R_\beta - \nabla\alpha \cdot (\nabla \times S\boldsymbol{B}) + B^2S_\beta \;=\; 0 \tag{A1.11}$$

$$R_B + B^2S_B + SB \;=\; 0 \,. \tag{A1.12}$$

These are two equations for α and β and a condition on the choice of R and S. The latter implies that we can replace R and S by a single independent function

$$T = R + B^2S \,, \tag{A1.13}$$

which is related to $\mathcal{M}$ by $T = \boldsymbol{B} \cdot \mathcal{M} \cdot \boldsymbol{B}/B^2$. Then S and R can be expressed by T as

$$S = \frac{T_B}{B} \tag{A1.14}$$

$$R = T - BT_B . \tag{A1.15}$$

In fact, (A1.12) and (A1.13) are equivalent to (A1.15) and (A1.14). The function T plays a central role in the present formulation and we refer to T as the *steady state potential*. In the following we will express the steady state problem in terms of T alone. Equation (A1.14) will then pose a constraint on the form of the function $T(\alpha, \beta, B)$. The relation (A1.15) is then satisfied automatically. There may be further constraints, e.g., resulting from additional conservation laws.

Using the final form of $T(\alpha, \beta, B)$, where all constraints are incorporated, we can rewrite the field equations (A1.10) and (A1.11) to obtain the final field equations for the Euler potentials as functions of r:

$$\nabla\beta \cdot \left[\nabla \times (\frac{T_B}{B}\nabla\alpha \times \nabla\beta) \right] + T_\alpha = 0$$

$$\tag{A1.16}$$

$$-\nabla\alpha \cdot \left[\nabla \times (\frac{T_B}{B}\nabla\alpha \times \nabla\beta) \right] + T_\beta = 0 ,$$

where B is understood as $|\nabla\alpha \times \nabla\beta|$.

We now incorporate the gravity force $-\rho\nabla\psi$. For this purpose we replace R, S formally by $\hat{R}, \hat{S}$, which are functions of the four variables α, β, B, ψ. We proceed in essentially the same way as we did above without gravity. Instead of (A1.9) we obtain

$$\hat{R}_\alpha\nabla\alpha + \hat{R}_\beta\nabla\beta + \hat{R}_B\nabla B - \nabla\beta\,\nabla\alpha \cdot (\nabla \times \hat{S}B) + \nabla\alpha\nabla\beta \cdot (\nabla \times \hat{S}B) +$$

$$B^2\hat{S}_\alpha\nabla\alpha + B^2\hat{S}_\beta\nabla\beta + B^2\hat{S}_B\nabla B + \hat{S}B\nabla B +$$
$$(\hat{R}_\psi + B^2\hat{S}_\psi - \rho)\nabla\psi = 0 . \tag{A1.17}$$

Defining $\hat{T} = \hat{R} + B^2\hat{S}$ and specifying the ψ-dependence of $\hat{T}$ by

$$\frac{\partial\hat{T}}{\partial\psi} = \rho , \tag{A1.18}$$

we find analogous to (A1.14) and (A1.15)

$$\hat{S} = \frac{\hat{T}_B}{B} \tag{A1.19}$$

$$\hat{R} = \hat{T} - B\hat{T}_B \tag{A1.20}$$

and the field equations in the form

$$\nabla\beta \cdot \left[\nabla \times \left(\frac{\hat{T}_B}{B} \nabla\alpha \times \nabla\beta \right) \right] + \hat{T}_\alpha = 0$$

$$-\nabla\alpha \cdot \left[\nabla \times \left(\frac{\hat{T}_B}{B} \nabla\alpha \times \nabla\beta \right) \right] + \hat{T}_\beta = 0 \,.$$

(A1.21)

The constraints on $\hat{T}$ are (A1.18), (A1.19) together with any additional constraints (conservation laws etc.). Using $\hat{T}$ with (A1.19), the relation (A1.20) is satisfied automatically.

The field equations (A1.21) are the same as (A1.16) except that T is replaced by $\hat{T}$. As before, they are partial differential equations for $\alpha(\boldsymbol{r}), \beta(\boldsymbol{r})$, where B is understood as $|\nabla\alpha \times \nabla\beta|$ and ψ as the known gravity potential $\psi(\boldsymbol{r})$.

The derivation of (A1.21) from (A1.17) requires a remark. In (A1.17) only three of the four gradients $\nabla\alpha, \nabla\beta, \nabla B, \nabla\psi$ can be regarded as independent. Accordingly, we set the coefficients of $\nabla\alpha, \nabla\beta$ and ∇B to zero on the account of independency. The coefficient of $\nabla\psi$ is set to zero for a different reason: the vanishing of that coefficient specifies the ψ-dependence of $\hat{T}$, which was open up to that point. The idea behind that procedure was borrowed from the technique of using Lagrange multipliers.

Variational principle

The steady state potential T, or $\hat{T}$ in the presence of gravity, not only governs the differential representation of the fields, but also a variational formulation. For a negligible gravity force, one finds that the steady states satisfying (A1.16) can be derived from the variational principle

$$\delta \int T \mathrm{d}^3 r = 0,$$

(A1.22)

where the independent functions to be varied are the Euler potentials α, β, which assume fixed values on the boundary of the domain of integration. Thus, the variations $\delta\alpha$ and $\delta\beta$ vanish at the boundary. The magnetic field magnitude B is understood as $|\nabla\alpha \times \nabla\beta|$. All constraints are assumed to be incorporated.

We briefly verify that the Euler–Lagrange equations associated with (A1.22) are the field equations (A1.16). Since T is given as a function of α, β, B we obtain from (A1.22)

$$\int \left(\frac{\partial T}{\partial \alpha} \delta\alpha + \frac{\partial T}{\partial \beta} \delta\beta + \frac{\partial T}{\partial B} \delta B \right) d^3 r = 0 \,. \tag{A1.23}$$

Here, δB has to be expressed by the independent variations $\delta\alpha$ and $\delta\beta$. To do this we write

$$\begin{aligned}
\frac{\partial T}{\partial B} \delta B &= \frac{T_B}{B} \delta \left(\frac{B^2}{2} \right) \\
&= \frac{T_B}{B} \boldsymbol{B} \cdot \delta\boldsymbol{B} \\
&= \frac{T_B}{B} \boldsymbol{B} \cdot \nabla\delta\alpha \times \nabla\beta + \frac{T_B}{B} \boldsymbol{B} \cdot \nabla\alpha \times \nabla\delta\beta \,, \tag{A1.24}
\end{aligned}$$

where we have used (A1.1).

With the help of the identity

$$\nabla \cdot (\delta\alpha \frac{T_B}{B} \boldsymbol{B} \times \nabla\beta) = \delta\alpha\nabla\beta \cdot \nabla \times \frac{T_B}{B} \boldsymbol{B} - \frac{T_B}{B} \boldsymbol{B} \cdot \nabla\delta\alpha \times \nabla\beta \tag{A1.25}$$

and a corresponding identity for terms involving $\delta\beta$, we find from (A1.24)

$$\begin{aligned}
\frac{\partial T}{\partial B} \delta B &= \delta\alpha\nabla\beta \cdot \nabla \times \frac{T_B}{B} \boldsymbol{B} - \nabla \cdot (\delta\alpha \frac{T_B}{B} \boldsymbol{B} \times \nabla\beta) \\
&\quad - \delta\beta\nabla\alpha \cdot \nabla \times \frac{T_B}{B} \boldsymbol{B} + \nabla \cdot (\delta\beta \frac{T_B}{B} \boldsymbol{B} \times \nabla\alpha). \tag{A1.26}
\end{aligned}$$

Inserting (A1.26) into (A1.23) and using Gauss's theorem, we get

$$\int_D \left[\left(T_\alpha + \nabla\beta \cdot \nabla \times \frac{T_B}{B} \boldsymbol{B} \right) \delta\alpha + \left(T_\beta - \nabla\alpha \cdot \nabla \times \frac{T_B}{B} \boldsymbol{B} \right) \delta\beta \right] d^3 r$$

$$+ \oint_S \left(\frac{T_B}{B} \boldsymbol{n} \cdot \boldsymbol{B} \times \nabla\alpha \, \delta\beta - \frac{T_B}{B} \boldsymbol{n} \cdot \boldsymbol{B} \times \nabla\beta \, \delta\alpha \right) d^2 S = 0 \,, \tag{A1.27}$$

where $\boldsymbol{n}$ denotes the outward-pointing unit normal vector of the surface S bounding the domain D. The surface integral vanishes because of the boundary conditions. Since α and β are varied independently, the brackets multiplying $\delta\alpha$ and $\delta\beta$ in the integrand of the volume integral have to vanish, which gives the field equations (A1.16).

In the presence of an external gravity potential ψ one proceeds in the same way, however using the variation functional $\hat{T}(\alpha, \beta, B, \psi)$ instead of $T(\alpha, \beta, B)$. Since the potential ψ is externally prescribed, it is not included in the variation procedure. As before, the condition (A1.18) is simply added to determine the ψ-dependence of $\hat{T}$. Thus, one finds the field equations in the form (A1.21).

Examples

For several simple plasma models the steady state potential T is readily obtained from the momentum equation. For instance, in magnetohydrostatics without gravity one finds from the momentum equation (A1.2) that

$$R = -p - \frac{B^2}{2\mu_0}, \qquad S = \frac{1}{\mu_0}, \qquad T = \frac{B^2}{2\mu_0} - p. \qquad \text{(A1.28)}$$

The relation (A1.14) gives $p_B = 0$, such that $p = p(\alpha, \beta)$. The field equations (A1.16) reduce to (5.23) and (5.24).

For a static case with a (one-fluid) CGL pressure tensor (3.69) one obtains

$$R = -p_\perp - \frac{B^2}{2\mu_0}, \qquad S = \frac{1}{\mu_0} - \frac{p_\parallel - p_\perp}{B^2}, \qquad T = \frac{B^2}{2\mu_0} - p_\parallel. \qquad \text{(A1.29)}$$

Here, (A1.14) gives

$$p_\parallel - p_\perp = B \frac{\partial p_\parallel}{\partial B}. \qquad \text{(A1.30)}$$

The field equations are obtained from (A1.16).

In the presence of parallel plasma flow or gravity, the problem of finding T or $\hat{T}$, respectively, is more complicated due to further constraints. Here we illustrate the procedure of determining $\hat{T}$ for the case of scalar pressure, parallel plasma flow, entropy conservation and gravity. This is the model (4.3)–(4.9) specialized for flow parallel to the magnetic field.

The presence of parallel plasma flow requires supplementing the steady state momentum equation by conservation of mass

$$\nabla \cdot (\rho \boldsymbol{v}) = 0, \qquad \text{(A1.31)}$$

where $\boldsymbol{v} = v_\parallel \boldsymbol{B}/B$ is the flow velocity, and by an energy or entropy equation, which for an ideal plasma may have the form of the adiabatic law

$$\boldsymbol{v} \cdot \nabla \left(\frac{p}{\rho^\gamma} \right) = 0 . \qquad \text{(A1.32)}$$

These additional constraints must be incorporated into the formalism described above. To do that, we write the momentum equation (A1.2) as

$$\nabla \cdot \left(-\left(p + \frac{B^2}{2\mu_0} \right) \mathcal{I} + \left(1 - \frac{\mu_0 \rho v_\parallel^2}{B^2} \right) \frac{\boldsymbol{B}\boldsymbol{B}}{\mu_0} \right) - \rho \nabla \psi = 0 , \qquad \text{(A1.33)}$$

such that

$$\hat{R} = -p - \frac{B^2}{2\mu_0}, \qquad \hat{S} = \frac{1}{\mu_0} - \frac{\rho v_\parallel^2}{B^2} \qquad \text{(A1.34)}$$

and thus

$$\hat{T} = \frac{B^2}{2\mu_0} - p - \rho v_{\|}^2. \tag{A1.35}$$

All observables are understood as functions of α, β, B, ψ with

$$\rho = \hat{T}_\psi, \tag{A1.36}$$

which determines the function $\rho(\alpha, \beta, B, \psi)$. We integrate (A1.31) to obtain

$$\frac{\rho v_{\|}}{B} = m(\alpha, \beta), \tag{A1.37}$$

where $m(\alpha, \beta)$ is an arbitrary function. Similarly, (A1.32) implies that p/ρ^γ is an arbitrary function of α and β. For convenience, let us use a more general pressure law of the form

$$p = P(\rho, \alpha, \beta), \tag{A1.38}$$

where for the present purpose $P(\rho, \alpha, \beta)$ does not have to be specified. Inserting (A1.37) and (A1.38) into (A1.35) we obtain

$$\hat{T} = \frac{B^2}{2\mu_0} - P(\rho, \alpha, \beta) - \frac{m^2 B^2}{\rho}. \tag{A1.39}$$

Now (A1.36) is used to find

$$-P_\rho \rho_\psi + \frac{m^2 B^2}{\rho^2} \rho_\psi = \rho. \tag{A1.40}$$

This equation is integrated to give

$$\psi + \frac{B^2 m^2}{2\rho^2} + \int^\rho \frac{P_\rho}{\rho} d\rho = C(\alpha, \beta, B). \tag{A1.41}$$

The condition (A1.19), evaluated by using (A1.39) and (A1.41) gives $\partial C/\partial B = 0$ such that (A1.41) finally reads

$$\psi + \frac{B^2 m^2}{2\rho^2} + \int^\rho \frac{P_\rho}{\rho} d\rho = C(\alpha, \beta), \tag{A1.42}$$

from which one determines $\rho(\alpha, \beta, B, \psi)$. Equation (A1.42) is Bernoulli's equation, which holds in each flux tube. Inserting $\rho(\alpha, \beta, B, \psi)$ into (A1.35) gives $T(\alpha, \hat{\beta}, \beta, \psi)$ in its final form. By this procedure the problem is reduced to solving the field equations (A1.21), just as in the simple MHS case.

Symmetric states

Here we explore how the theory of symmetric magnetohydrostatic states with an isotropic pressure tensor generalizes in the present unified picture. In particular, we are interested in the unified form of the Grad–Shafranov equation for translational and rotational invariance and for the more general case of helical invariance. For simplicity, we do not take into account an external gravity force. However, at the end it is pointed out, how such a force is readily included.

We begin with translational invariance with respect to the Cartesian z-coordinate. In Section 5.2.1 it is shown that in this case the magnetic field may be written as

$$\boldsymbol{B} = \nabla A(x, y) \times \boldsymbol{e}_z + B_z(x, y)\boldsymbol{e}_z \,, \tag{A1.43}$$

which can be expressed by the Euler potentials

$$\alpha \;=\; A(x, y) \tag{A1.44}$$

$$\beta \;=\; z + \tilde{\beta}(x, y) \,, \tag{A1.45}$$

where

$$B_z = \boldsymbol{e}_z \cdot \nabla A \times \nabla \tilde{\beta} \,. \tag{A1.46}$$

The invariance requires that the steady state potential T is independent of z which implies $T = T(A, B)$.

To discuss the field equations, we consider the identity

$$\nabla \times \frac{T_B}{B} \boldsymbol{B} = -\nabla \cdot \left(\frac{T_B}{B} \nabla A \right) \boldsymbol{e}_z + \nabla \left(\frac{T_B}{B} B_z \right) \times \boldsymbol{e}_z \,. \tag{A1.47}$$

Using (A1.47) in the field equations (A1.16) gives

$$-\nabla \cdot \left(\frac{T_B}{B} \nabla A \right) + \nabla \tilde{\beta} \cdot \nabla \left(\frac{T_B}{B} B_z \right) \times \boldsymbol{e}_z + T_A \;=\; 0 \tag{A1.48}$$

$$\nabla A \cdot \nabla \left(\frac{T_B}{B} B_z \right) \times \boldsymbol{e}_z \;=\; 0 \,. \tag{A1.49}$$

Here, (A1.49) implies that $B_z T_B / B$ is a function only of A,

$$B_z = \frac{BG(A)}{T_B} \,, \tag{A1.50}$$

where $G(A)$ is arbitrary. Using (A1.50) in (A1.48) gives the unified Grad–Shafranov-type equation for translational invariance in unified form,

$$-\nabla \cdot \left(\frac{T_B}{B}\nabla A\right) - \frac{B}{T_B}G(A)G'(A) + T_A = 0 . \tag{A1.51}$$

For magnetohydrostatics this equation reduces to (5.78).

In an analogous way one finds the unified Grad–Shafranov equation for rotational invariance

$$-\frac{1}{r}\frac{\partial}{\partial r}\left(\frac{T_B}{rB}\frac{\partial A}{\partial r}\right) - \frac{1}{r^2}\frac{\partial}{\partial z}\left(\frac{T_B}{B}\frac{\partial A}{\partial z}\right) - \frac{BK(A)K'(A)}{r^2 T_B} + T_A = 0 \tag{A1.52}$$

and

$$B_\phi = \frac{BK(A)}{rT_B} \tag{A1.53}$$

where K(A) is arbitrary.

The corresponding results for helical invariance (Section 5.2.4) are

$$\frac{1}{r}\frac{\partial}{\partial r}\left(\frac{r}{\nu^2 + \omega^2 r^2}\frac{T_B}{B}\frac{\partial A}{\partial r}\right) + \frac{1}{r^2}\frac{\partial}{\partial \eta}\left(\frac{T_B}{B}\frac{\partial A}{\partial \eta}\right)$$

$$-\frac{2\nu\omega L(A)}{(\nu^2 + \omega^2 r^2)^2} + \frac{BL(A)}{T_B(\nu^2 + \omega^2 r^2)}\frac{dL(A)}{dA} - T_A = 0 \tag{A1.54}$$

and

$$B_3 = \frac{BL(A)}{T_B} , \tag{A1.55}$$

L(A) arbitrary.

The inclusion of an external gravity force is easily possible in the same way as in the general 3D case. The potential $\hat{T}$ receives an additional dependence on the gravity potential ψ, indicated by the hat label. The generalized Grad–Shafranov equation (A1.54) holds with T replaced by $\hat{T}$.

In all cases treated so far the plasma flow is aligned with the magnetic field. In a symmetric case a generalization is possible, allowing for a perpendicular component also. In Cartesian coordinates one replaces the stress tensor (A1.3) by

$$\mathcal{M} = R\mathcal{I} + S\mathbf{B}_\mathrm{p}\mathbf{B}_\mathrm{p} + V(\mathbf{B}_\mathrm{p}\mathbf{e}_z + \mathbf{e}_z\mathbf{B}_\mathrm{p}) + U\mathbf{e}_z\mathbf{e}_z , \tag{A1.56}$$

where $\mathbf{B}_\mathrm{p}$ is the poloidal component (projection in the x, y-plane). The functions R, S, V, U are understood as functions of A, B_p or A, B_p, ψ, when an external gravity potential ψ is taken into account. One defines $T = R + B_\mathrm{p}^2 S$ and obtains

$$-\nabla \cdot \left(\frac{T_{B_\mathrm{p}}}{B_\mathrm{p}}\nabla A\right) + T_A = 0 \tag{A1.57}$$

together with the condition that V depends on A only. As before, explicit constraints can be incorporated into the final form of T. A necessary constraint is $T_{B_p}/B_p = S$ with $T_\psi = \rho$ added in the presence of gravity, the $\wedge$ symbol being ignored here for simplicity. The stress tensor (A1.56), for instance, allows choosing a velocity field with $\nabla \times (\boldsymbol{v} \times \boldsymbol{B}) = 0$, such that $\boldsymbol{v}$ is not necessarily parallel to $\boldsymbol{B}$.

Appendix 2

Variational principle for collisionless plasmas

Here we use the energy method described in Section 10.2.1 to obtain sufficient stability criteria for two-dimensional collisionless plasmas.

All observables are y-independent and the equilibrium distribution function of particle species s is of the form of

$$f_{s0} = F_s(H_{s0}, P) , \tag{A2.1}$$

where $H_{s0} = (m_s w_x^2 + m_s w_z^2)/2 + \psi_{s0}$ with

$$\psi_{s0} = \frac{1}{2m_s}(P - q_s A_0)^2 + q_s \phi \tag{A2.2}$$

is the one-particle Hamiltonian and P the y-component of the canonical
momentum, which is treated as an independent variable. Number density is
$n_s = \int f_s d\tau_s$ with $d\tau_s = dw_x dw_z d(P/m_s)$. We assume $\partial F_s / \partial H_{s0} < 0$ and
consider a quasi-neutral plasma.

The equilibrium magnetic field has the form $\boldsymbol{B}_0 = \nabla A_0(x, z) \times \boldsymbol{e}_y$, where
A_0 satisfies the Grad–Shafranov equation (6.24). Unlike the equilibrium,
the perturbations may involve a non-vanishing B_y-component. There is no
energy flux across the boundary.

Consider energy conservation (3.40), specialized for quasi-neutrality and
integrated over the selected domain

$$W = \sum_s \int \left(\frac{m_s}{2}(w_x^2 + w_z^2) + \frac{1}{2m_s}(P - q_s A)^2 \right) f_s d\Omega_s$$

$$+ \frac{1}{2\mu_0} \int B^2 \, dx dz , \tag{A2.3}$$

where $d\Omega_s = dx dz d\tau_s$ and W is a constant.

Expanding the time-dependent quantities to second order in a power series
with respect to a parameter that measures the linear perturbation A_1 of the

451

flux function, we find

$$
\sum_s \int \left[\left(\frac{m_s}{2}(w_x^2 + w_z^2) + \frac{1}{2m_s}(P - q_s A_0)^2 \right)(F_s + f_{s1} + f_{s2}) \right.
$$
$$
\left. - \frac{q_s}{m_s}(P - q_s A_0) A_1 (F_s + f_{s1}) + \frac{q_s^2}{2m_s} A_1^2 F_s \right] d\Omega_s
$$
$$
+ \frac{1}{2\mu_0} \int \left(B_0^2 + 2 B_0 \cdot B_1 + B_1^2 \right) dx dz = W_0 + W_1 + W_2 . \quad \text{(A2.4)}
$$

The zeroth order of (A2.4) simply gives the equilibrium energy W_0. The first order contribution vanishes implying $W_1 = 0$. This is shown in two steps, with the second step being postponed to after the equation (A2.9). In the first step we demonstrate that two of the linear terms cancel each other:

$$
\sum_s \int \left(-\frac{q_s}{m_s}(P - q_s A_0) A_1 F_s \right) d\Omega_s + \frac{1}{\mu_0} \int B_0 \cdot B_1 dx dz
$$
$$
= \int \left(-j_0 A_1 + \frac{1}{\mu_0} \nabla A_0 \cdot \nabla A_1 \right) dx dz \quad \text{(A2.5)}
$$
$$
= \int \left(-j_0 A_1 - \frac{1}{\mu_0} \Delta A_0 \, A_1 \right) dx dz = 0 .
$$

Here, the expression for the equilibrium current density was used in the first term, integration by parts (with $A_1 = 0$ on the boundary) was applied to the second term and the Grad–Shafranov equation was used to obtain the final result. Thus, we find from (A2.4)

$$
\sum_s \int \left(H_{s0}(f_{s1} + f_{s2}) - \frac{q_s}{m_s}(P - q_s A_0) A_1 f_{s1} + \frac{q_s^2}{2m_s} A_1^2 F_s \right) d\Omega_s
$$
$$
+ \frac{1}{2\mu_0} \int \left((\nabla A_1)^2 + B_{y1}^2 \right) dx dz = W_2 . \quad \text{(A2.6)}
$$

The equilibrium kinetic energy of a particle of species s was replaced by H_{s0}, which is possible because the assumed quasi-neutrality implies

$$
\sum_s q_s \int f_s d\tau_s = 0 \quad \text{(A2.7)}
$$

to all orders, such that the term involving ϕ_0 drops out of (A2.6).

The term involving H_{s0} in (A2.6) can be put in a more appropriate form by using the fact that f_s is a solution of the Vlasov equation. Since y is an ignorable coordinate, the Vlasov equation and thus Liouville's theorem holds in four-dimensional phase space (x, z, w_x, w_z) separately. This implies that during the time-evolution the values of the distribution function $f_s(x, z, w_x, w_z, P, t)$ are incompressibly redistributed in that four-dimensional phase space for any value of P. This property implies that

$$\int G(f_s, P) \mathrm{d}\Omega_s = \int G(F_s, P) \mathrm{d}\Omega_s \qquad (A2.8)$$

for arbitrary functions $G(f_s, P)$.

In first order (A2.8) gives

$$\int X_s(H_{s0}, P) f_{s1} \mathrm{d}\Omega_s = 0, \quad X_s(H_{s0}, P) \text{ arbitrary}, \qquad (A2.9)$$

implying that the remaining linear term of (A2.6) vanishes. In second order (A2.8) gives

$$\int \left(\frac{\partial G(F_s, P)}{\partial F_s} f_{s2} + \frac{1}{2} \frac{\partial^2 G(F_s, P)}{\partial F_s^2} f_{s1}^2 \right) \mathrm{d}\Omega_s = 0 . \qquad (A2.10)$$

The particular choice $\partial G(F_s, P)/\partial F_s = H_{s0}(F_s, P)$, where $H_{s0}(F_s, P)$ is obtained by solving $F_s(H_{s0}, P)$ for H_{s0} (uniqueness being ensured by $\partial F_s/\partial H_{s0} < 0$), leads to

$$\int H_{s0} f_{s2} \mathrm{d}\Omega_s = - \int \frac{f_{s1}^2}{2 \partial F_s/\partial H_{s0}} \mathrm{d}\Omega_s , \qquad (A2.11)$$

such that from (A2.6) one finds second-order energy conservation as

$$W_2 = \frac{1}{2} \sum_s \int \left(-\frac{f_{s1}^2}{F_s'} - 2\frac{q_s}{m_s}(P - q_s A_0) A_1 f_{s1} + \frac{q_s^2}{m_s} A_1^2 F_s \right) \mathrm{d}\Omega_s$$

$$+ \frac{1}{2\mu_0} \int \left((\nabla A_1)^2 + B_{y1}^2 \right) \mathrm{d}x \mathrm{d}z , \qquad (A2.12)$$

where we have used the abbreviation

$$F_s' = \frac{\partial F_s}{\partial H_{s0}} . \qquad (A2.13)$$

The final expression for W_2 is obtained by rewriting (A2.12) as

$$W_2 = -\frac{1}{2} \sum_s \int \frac{1}{F_s'} \left(f_{s1} + \frac{q_s}{m_s}(P - q_s A_0) A_1 F_s' \right)^2 \mathrm{d}\Omega_s$$

$$+ \frac{1}{2} \int \left(\frac{(\nabla A_1)^2}{\mu_0} - \frac{\partial j_0}{\partial A_0} A_1^2 + \frac{B_{y1}^2}{\mu_0} \right) \mathrm{d}x \mathrm{d}z , \qquad (A2.14)$$

where in view of (6.13) we have used

$$\frac{\partial j_0}{\partial A_0} = -\sum_s \int \left(\frac{q_s^2}{m_s^2} (P - q_s A_0)^2 F_s' + \frac{q_s^2}{m_s} F_s \right) d\tau_s \,. \tag{A2.15}$$

The form of W_2 as given by (A2.14) allows us to derive stability criteria as described in Section 10.2.1 where the energy perturbation is decomposed into two terms (see (10.26))

$$W_2 = T_2 + V_2 \,. \tag{A2.16}$$

Here, it is not appropriate to identify T with the kinetic energy of bulk flow because the bulk flow velocity is not a variable appearing in the functional W_2. Fortunately, the only formal property that is needed for applying the method described in Section 10.2.1 is that T is non-negative. For the stability definition (10.6) to make sense, T must be a sizable part of the energy involved in the dynamic evolution. In the present case a suitable decomposition of the form (A2.16) is obtained by minimizing W_2 with respect to f_{s1} and B_{y1} for a fixed perturbation A_1, which generates the functional $V_2(A_1)$,

$$V_2(A_1) = \min_{(f_1, B_{y1})} W_2 \tag{A2.17}$$

and T_2 then follows from (A2.16). By construction, T_2 is non-negative.

The further procedure depends on the constraints that one imposes on the minimization.

No constraints

In the absence of any constraints the minimization of W_2 simply sets the terms involving f_{s1} and B_{y1} in (A2.14) to zero, which gives

$$V_2 = \frac{1}{2} \int \left(\frac{(\nabla A_1)^2}{\mu_0} - \frac{\partial j_0}{\partial A_0} A_1^2 \right) dx dz \tag{A2.18}$$

$$T_2 = -\frac{1}{2} \sum_s \int \frac{1}{F_s'} \left(f_{s1} + \frac{q_s}{m_s} (P - q_s A_0) A_1 F_s' \right)^2 d\Omega_s$$

$$+ \frac{1}{2\mu_0} \int B_{y1}^2 \, dx dz \,. \tag{A2.19}$$

By construction T_2 is positive (note that $F_s' < 0$). Positive definite V_2 implies stability.

Phase space condition and quasi-neutrality

Here the constraints are the phase space condition (A2.9) together with the quasi-neutrality condition (A2.7) evaluated to first order. The former will be rewritten by expressing the integration over velocity space by an integration over H_{s0} and P,

$$
\int X_s(H_{s0}, P) f_{s1} \mathrm{d}\Omega_s
$$

$$
= \frac{2\pi}{m_s^2} \int X_s(H_{s0}, P) \mathrm{d}H_{s0} \mathrm{d}P \int_{\psi_{s0} \leq H_{s0}} f_{s1} \mathrm{d}x \mathrm{d}z = 0 . \quad \text{(A2.20)}
$$

The H_{s0}, P-integration can be stripped because $X(H_{s0}, P)$ is arbitrary. The resulting condition is written as $\langle f_{s1} \rangle_s = 0$, where $\langle ... \rangle_s$ denotes an average defined as

$$
\langle ... \rangle_s = \frac{\int_{\psi_{s0} \leq H_{s0}} ... \mathrm{d}x \mathrm{d}z}{\int_{\psi_{s0} \leq H_{s0}} \mathrm{d}x \mathrm{d}z} . \quad \text{(A2.21)}
$$

The average $\langle ... \rangle_s$ depends on H_{s0}, P and s. Thus, the functions f_{s1} are subject to the constraints

$$
\langle f_{s1} \rangle_s = 0 \quad \text{(A2.22)}
$$

$$
\sum_s \int q_s f_{s1} \mathrm{d}\tau_s = 0 . \quad \text{(A2.23)}
$$

The minimization of (A2.14) with respect to f_{s1} and B_{y1} subject to constraints (A2.22) and (A2.23) is expressed as

$$
\delta \Bigg(-\frac{1}{2} \sum_s \int \frac{1}{F_s'} \Big(f_{s1} + \frac{q_s}{m_s}(P - q_s A_0) F_s' A_1\Big)^2 \mathrm{d}\Omega_s
$$

$$
+ \sum_s \int \big(Y_s(H_{s0}, P) f_{s1} + \phi_1(x, z) q_s f_{s1}\big) \mathrm{d}\Omega_s
$$

$$
+ \frac{1}{2} \int \Big(\frac{(\nabla A_1)^2}{\mu_0} - \frac{\partial j_0}{\partial A_0} A_1^2 + \frac{B_{y1}^2}{2\mu_0}\Big) \mathrm{d}x \mathrm{d}z \Bigg) = 0 . \quad \text{(A2.24)}
$$

Here, $Y_s(H_{s0}, P)$ and $\phi_1(x, z)$ play the role of Lagrangian multipliers associated with the constraints. That these parameters are functions rather than constants reflects the fact that we are dealing with infinite sets of constraints.

Carrying out the variation procedure with A_1 kept fixed we find from (A2.24) for the minimizing functions $f_{s1}^{(m)}$ and $B_{y1}^{(m)}$

$$f_{s1}^{(m)} = F_s'(\psi_{s1} + Y_s) \tag{A2.25}$$

$$B_{y1}^{(m)} = 0 , \tag{A2.26}$$

with

$$\psi_{s1} = -\frac{q_s}{m_s}(P - q_s A_0)A_1 + q_s\phi_1 . \tag{A2.27}$$

The Lagrangian multipliers Y and ϕ_1 are determined by inserting (A2.25) into the constraints (A2.22) and (A2.23). From (A2.22) one obtains

$$Y_s = -\langle\psi_{s1}\rangle_s , \tag{A2.28}$$

such that

$$f_{s1}^{(m)} = F_s'(\psi_{s1} - \langle\psi_{s1}\rangle_s) . \tag{A2.29}$$

Using (A2.29) in (A2.23) gives an equation for ϕ_1,

$$\sum_s \int q_s F_s'(\psi_{s1} - \langle\psi_{s1}\rangle_s)\mathrm{d}\tau_s = 0 . \tag{A2.30}$$

Inserting (A2.26) and (A2.29) into (A2.14) we find

$$\mathcal{V}_2 = \frac{1}{2}\sum_s \int \left(-F_s'q_s^2\phi_1^2 + 2F_s'q_s\phi_1\langle\psi_{s1}\rangle_s - F_s'\langle\psi_{s1}\rangle_s^2\right)\mathrm{d}\Omega_s$$

$$+ \frac{1}{2}\int\left(\frac{(\nabla A_1)^2}{\mu_0} - \frac{\partial j_0}{\partial A_0}A_1^2\right)\mathrm{d}x\mathrm{d}z , \tag{A2.31}$$

where ϕ_1 is determined by (A2.30).

By considering marginal states it can be shown that ϕ_1 can be interpreted as the perturbation of the electric potential (Schindler *et al.*, 1973), which explains the notation.

Using (A2.30) we rewrite (A2.31) in the form

$$\mathcal{V}_2 = \frac{1}{2}\int\left(\frac{(\nabla A_1)^2}{\mu_0} - dA_1^2 + 2bA_1\phi_1 + a\phi_1^2\right)\mathrm{d}x\mathrm{d}z$$

$$- \frac{1}{2}\sum_s \int F_s'\langle\psi_{s1}\rangle_s^2\mathrm{d}\Omega_s , \tag{A2.32}$$

where again for each A_1 the potential ϕ_1 is determined by (A2.30). The quantities a, b, d are defined as

$$a = \frac{\partial \sigma_0}{\partial \phi_0} = \sum_s a_s, \quad a_s = \int q_s^2 F_s' \mathrm{d}\tau \tag{A2.33}$$

$$b = \frac{\partial \sigma_0}{\partial A_0} = -\frac{\partial j_0}{\partial \phi_0} = \sum_s b_s, \quad b_s = -\int \frac{q_s^2}{m_s}(P - q_s A_0) F_s' \mathrm{d}\tau \tag{A2.34}$$

$$d = \frac{\partial j_0}{\partial A_0}, \tag{A2.35}$$

with $\partial j_0 / \partial A_0$ given by (A2.15).

So far only first-order variation calculus was applied to W_2. It remains to confirm that $\mathcal{V}_2$ corresponds to a minimum of W_2 rather than to a maximum or a saddle point. In fact, the difference $W_2(f_{s1}^{(m)} + \delta f_{s1}, B_{y1}^{(m)} + \delta B_{y1}) - \mathcal{V}_2$ is easily shown to be positive for $(\delta f_{s1})^2 + (\delta B_{y1})^2 \neq 0$. Remembering that the linear term in (A2.24) vanishes, one finds for that difference $-\frac{1}{2}\sum_s \int (\delta f_{s1})^2 / F_s' \mathrm{d}\Omega_s + \int (\delta B_{y1})^2 / 2\mu_0 \mathrm{d}x\mathrm{d}z$ which is positive because of $F_s' < 0$. This confirms that W_2 assumes a minimum at $(f_{s1}^{(m)}, B_{y1}^{(m)})$ and that $\mathcal{T}_2 = W_2 - \mathcal{V}_2$ is positive.

The expression (A2.32) can be simplified further by substituting

$$\phi_1 = \frac{\mathrm{d}\phi_0}{\mathrm{d}A_0} A_1 + \varphi_1, \tag{A2.36}$$

where the fact has been used that ϕ_0 is a function of A_0 alone, which is implied by quasi-neutrality. The derivative $\mathrm{d}\phi_0/\mathrm{d}A_0$ is found from the unperturbed quasi-neutrality condition $\sigma_0(A_0, \phi_0) = 0$. Differentiating this equation with respect to A_0 gives

$$\frac{\mathrm{d}\phi_0}{\mathrm{d}A_0} = -\frac{b}{a}. \tag{A2.37}$$

Using (A2.36) with (A2.37) in (A2.32) gives

$$\mathcal{V}_2 = \frac{1}{2}\int \left(\frac{(\nabla A_1)^2}{\mu_0} - \frac{\mathrm{d}j_0}{\mathrm{d}A_0} A_1^2 + a\varphi_1^2\right) \mathrm{d}x\mathrm{d}z$$

$$-\frac{1}{2}\sum_s \int F_s' \langle \psi_{s1}\rangle_s^{\,2} \mathrm{d}\Omega_s, \tag{A2.38}$$

where $\mathrm{d}j_0/\mathrm{d}A_0 = d + b^2/a$ was used. The perturbed quasi-neutrality condition (A2.7) becomes

$$a\varphi_1 - \sum_s q_s \int F_s' \langle \psi_{s1}\rangle_s \mathrm{d}\tau = 0. \tag{A2.39}$$

At this point it is convenient to introduce the following definitions:

$$\Psi_{s1} = \frac{\psi_{s1}}{q_s} \, , \tag{A2.40}$$

$$[\![\ldots]\!] = \frac{\sum_s \int \ldots q_s^2 F_s' \mathrm{d}\tau}{\sum_s \int q_s^2 F_s' \mathrm{d}\tau} \, , \tag{A2.41}$$

$$[\![\ldots]\!]_s = \frac{\int \ldots q_s^2 F_s' \mathrm{d}\tau}{\int q_s^2 F_s' \mathrm{d}\tau} \, . \tag{A2.42}$$

With these definitions (A2.38) and (A2.39) assume the form

$$\mathcal{V}_2 = \frac{1}{2} \int \left(\frac{(\nabla A_1)^2}{\mu_0} - \frac{\mathrm{d}j_0}{\mathrm{d}A_0} A_1^2 + a\varphi_1^2 - a[\![\langle \Psi_{s1} \rangle_s^2]\!] \right) \mathrm{d}x \mathrm{d}z \tag{A2.43}$$

and

$$\varphi_1 - [\![\langle \Psi_{s1} \rangle_s]\!] = 0 \, , \tag{A2.44}$$

which implies that φ_1 is a function of A_0 alone.

An alternative form of (A2.43) is obtained by eliminating the explicit occurrence of φ_1 from (A2.43) and (A2.44):

$$\mathcal{V}_2 = \frac{1}{2} \int \left(\frac{(\nabla A_1)^2}{\mu_0} - \frac{\mathrm{d}j_0}{\mathrm{d}A_0} A_1^2 - a\left[\!\!\left[\left(\langle \Psi_{s1} \rangle_s - [\![\langle \Psi_{s1} \rangle_s]\!] \right)^2 \right]\!\!\right] \right) \mathrm{d}x \mathrm{d}z \, . \tag{A2.45}$$

Kiessling and Krallmann (1998) solved (A2.30) for ϕ_1 using a Neumann series, expressing ϕ_1 by A_1. Inserting that result into $\mathcal{V}_2$, they found a representation of the form

$$\mathcal{V}_2 = \frac{1}{2} \int \left(\frac{(\nabla A_1)^2}{\mu_0} - \frac{\partial j_0}{\partial A_0} A_1^2 \right) \mathrm{d}x \mathrm{d}z$$

$$+ \frac{1}{2} \int \int A_1(\boldsymbol{r}) K(\boldsymbol{r}, \boldsymbol{r}') A_1(\boldsymbol{r}') \mathrm{d}^2 r \mathrm{d}^2 r' \, , \tag{A2.46}$$

where the kernel K was expressed explicitly in terms of the solution of (A2.30).

The limit $m_e \to 0$ for two-species plasmas

Here we specialize for a plasma consisting of electrons and a single ion species with $q_i = e$. The equilibrium is chosen such that in the domain considered the width of the region accessible to an electron is much smaller than the smallest scale of the equilibrium (Fig. A2.1). In that case we can describe the electrons in the formal limit $m_e \to 0$, however keeping its kinetic energy finite.

The first step is to write (A2.43) and (A2.44) explicitly for ions and electrons $(s = \mathrm{i}, \mathrm{e})$, using that

$$a[\![...]\!] = a_\mathrm{i}[\![...]\!]_\mathrm{i} + a_\mathrm{e}[\![...]\!]_\mathrm{e} \tag{A2.47}$$

and

$$[\![\langle\Psi_{\mathrm{i}1}\rangle_\mathrm{i}^2]\!]_\mathrm{i} = [\![(\langle\Psi_{\mathrm{i}1}\rangle_\mathrm{i} - [\![\langle\Psi_{\mathrm{i}1}\rangle_\mathrm{i}]\!]_\mathrm{i})^2]\!]_\mathrm{i} + [\![\langle\Psi_{\mathrm{i}1}\rangle_\mathrm{i}]\!]_\mathrm{i}^2 . \tag{A2.48}$$

Eliminating $[\![\langle\Psi_{\mathrm{i}1}\rangle]\!]_\mathrm{i}^2$ with the help of the quasi-neutrality condition gives

$$\mathcal{V}_2 = \frac{1}{2}\int\left(\frac{(\nabla A_1)^2}{\mu_0} - \frac{\mathrm{d}j_0}{\mathrm{d}A_0}A_1^2 - a_\mathrm{i}\left[\!\!\left[(\langle\Psi_{\mathrm{i}1}\rangle_\mathrm{i} - [\![\langle\Psi_{\mathrm{i}1}\rangle_\mathrm{i}]\!]_\mathrm{i})^2\right]\!\!\right]_\mathrm{i}\right.$$
$$\left. - a_\mathrm{e}\left[\!\!\left[(\langle\Psi_{\mathrm{e}1}\rangle_\mathrm{e} - \varphi_1)^2\right]\!\!\right]_\mathrm{e} - \frac{a_\mathrm{e}^2}{a_\mathrm{i}}([\![\langle\Psi_{\mathrm{e}1}\rangle_\mathrm{e}]\!]_\mathrm{e} - \varphi_1)^2\right)\mathrm{d}x\mathrm{d}z . \tag{A2.49}$$

Consider an electron with constants of motion $(P, H_{\mathrm{e}0})$. For sufficiently small m_e the domain D_e accessible to the electron $(\psi_{\mathrm{e}0} \leq H_{\mathrm{e}0})$ becomes a narrow flux tube centred at the field line $A_0 = A_\mathrm{c}$, where $\psi_{\mathrm{e}0}$ assumes a minimum. The domain is bounded by the field lines $A_\mathrm{c} \pm \delta A$ corresponding to $\psi_{\mathrm{e}0} = H_{\mathrm{e}0}$; Fig. A2.1 gives an example. Here, A_c is a function of P and satisfies the equation

$$\frac{1}{m_\mathrm{e}}(P + eA_\mathrm{c}) - \phi'(A_\mathrm{c}) = 0 , \tag{A2.50}$$

which states that at the centre field line w_y equals the y-component of the $\boldsymbol{E} \times \boldsymbol{B}$ drift velocity. (Note that $-\nabla\phi_0 \times \boldsymbol{B}_0 \cdot \boldsymbol{e}_y/B_0^2 = \nabla\phi_0 \cdot \nabla A_0/B_0^2 = \phi'(A_0)$.) For δA one finds to lowest order in m_e

$$\delta A = \frac{1}{e}\sqrt{2m_\mathrm{e}(H_{\mathrm{e}0} + e\phi_0(A_\mathrm{c}))} . \tag{A2.51}$$

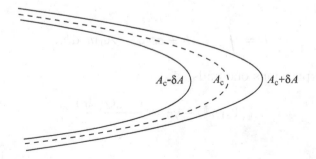

Fig. A2.1 Magnetic flux tube accessible to an electron.

We now turn to the average $\langle \Psi_{e1} \rangle_e$, which, with the integrals over D_e written explicitly, takes the form

$$\langle \Psi_{e1} \rangle_e = \int_{A_c - \delta A}^{A_c + \delta A} \mathrm{d}A_0 \int_{A_0} \frac{\mathrm{d}s_0}{B_0} \left(-\frac{P + eA_0}{m_e} A_1 + \phi_1 \right) \Bigg/ \int_{A_c - \delta A}^{A_c + \delta A} \mathrm{d}A_0 \int_{A_0} \frac{\mathrm{d}s_0}{B_0} \ . \tag{A2.52}$$

To lowest order in m_e this expression becomes

$$\langle \Psi_{e1} \rangle_e = -\frac{2}{3e} Q(A_c)(H_{e0} + e\phi_0(A_c)) + \varphi_1(A_c) \ , \tag{A2.53}$$

where the integrand was expanded in powers of $A_0 - A_c$ and (A2.50) and (A2.51) were used; $Q(A)$ is defined as

$$Q(A) = \frac{1}{V(A)} \frac{\mathrm{d}}{\mathrm{d}A} \int_A A_1 \frac{\mathrm{d}s_0}{B_0} \ , \tag{A2.54}$$

where $V(A_0) = \int_{A_0} \mathrm{d}s_0/B_0$ is the flux tube volume associated with the field line A_0.

The $[\![\,]\!]$-averages involve integrals of the form

$$\int G(H, P)\mathrm{d}\tau = \frac{2\pi}{m_e^2} \int_{-e\phi_0}^{\infty} \mathrm{d}H \int_{-eA_0 - \delta P}^{-eA_0 + \delta P} \mathrm{d}P G(H, P) \tag{A2.55}$$

where $\delta P = (2m_2(H + e\phi_0)^{1/2})$. To lowest order in the electron mass (A2.55) becomes

$$\int G(H, P)\mathrm{d}\tau = \frac{4\pi}{m_e^2} \int_0^{\infty} G(h - e\phi_0, -eA_0)\sqrt{2m_e h}\,\mathrm{d}h \ . \tag{A2.56}$$

For the equilibrium electron density n_{e0} one finds (after integration by parts)

$$n_{e0} = -\frac{8\pi\sqrt{2}}{3m_e^{3/2}} J_{3/2} \tag{A2.57}$$

with

$$J_r = \int_0^{\infty} F_e'(h - e\phi_0, -eA_0)h^r\,\mathrm{d}h \ . \tag{A2.58}$$

Using these expressions one finds

$$[\![\langle \Psi_{e1} \rangle_e]\!]_e - \varphi_1 = \frac{en_{e0}Q(A_0)}{a_e} \tag{A2.59}$$

and

$$a_e[\![(\langle \Psi_{e1} \rangle_e - \varphi_1)^2]\!]_e = -\frac{2Q(A_0)^2 n_{e0} J_{5/2}}{3J_{3/2}} \ . \tag{A2.60}$$

With (A2.59) and (A2.60) one then finds from (A2.49)

$$V_2 = \frac{1}{2} \int \Big(\frac{(\nabla A_1)^2}{\mu_0} - \frac{\mathrm{d}j_0}{\mathrm{d}A_0} A_1^2 + |a_i| [\![(\langle \Psi_{i1} \rangle_i - [\![\langle \Psi_{s1} \rangle_i]\!]_i)^2]\!]_i$$

$$+ n_0 Q^2 \big(\tfrac{5}{3} k_B T_e + \frac{e^2 n_0}{|a_i|} \big) \Big) \, \mathrm{d}x\mathrm{d}z \, , \quad \text{(A2.61)}$$

where $n_0 = n_{e0} = n_{i0}$ in view of quasi-neutrality and T_e is the kinetic electron temperature

$$k_B T_e = \frac{2}{3n_0} \int \frac{m_e}{2} (\boldsymbol{w} - \boldsymbol{v})^2 F_e \mathrm{d}\tau_e \, , \quad \text{(A2.62)}$$

which to lowest order in m_e equals $2J_{5/2}/(5J_{3/2})$. In the case of exponential distribution functions (6.27) $e^2 n_0/|a_i|$ becomes $k_B T_i$.

The validity of the formal limit $m_e \to 0$ can be judged by carrying the expansion one step farther. One finds that the first corrections are of the order of m_e. Typical corrections have the form $m_e (\mathrm{d}\phi_0/\mathrm{d}A_0)^2/k_B T_e$ or $m_e (H_{e0} + e\phi_0 I(A_0)'''/(e^2 I(A_0)')$, where $I(A_0) = \int_{A_0} A_1 \mathrm{d}s_0/B_0$. Assuming that ϕ_0 is of order $k_B T_e/e$ and replacing derivatives with respect to A_0 by $1/A^*$, where A^* is a typical scale of the corresponding A_0-dependence, both expressions assume the form $m_e k_B T_e/(e^2 A^{*2})$.

A lower bound on W_2

A lower estimate of W_2 can be found in the following way (Pellat *et al.*, 1991). Writing (A2.14) as

$$W_2 = \frac{1}{2} \sum_s \int \frac{1}{|F_s'|} \tilde{f}_{s1}^2 \mathrm{d}\Omega_s + \frac{1}{2} \int \Big(\frac{(\nabla A_1)^2}{\mu_0} - \frac{\partial j_0}{\partial A_0} A_1^2 + \frac{B_{y1}^2}{\mu_0} \Big) \, \mathrm{d}x\mathrm{d}z \, , \quad \text{(A2.63)}$$

where

$$\tilde{f}_{s1} = f_{s1} + \frac{q_s}{m_s} (P - q_s A_0) A_1 F_s' \quad \text{(A2.64)}$$

and applying Schwarz's inequality to the first term of (A2.63), one finds

$$W_2 \geq \frac{1}{2} \int \Big(\frac{(\nabla A_1)^2}{\mu_0} - \frac{\partial j_0}{\partial A_0} A_1^2 + \sum_s \frac{q_s^2 \tilde{n}_{s1}^2}{|a_s|} \Big) \, \mathrm{d}x\mathrm{d}z \quad \text{(A2.65)}$$

with

$$\tilde{n}_{s1} = \int \tilde{f}_{s1} \mathrm{d}\tau \, , \quad \text{(A2.66)}$$

such that

$$n_{s1} = \tilde{n}_{s1} + \frac{b_s}{q_s} A_1 . \qquad (A2.67)$$

In view of the inequality sign in (A2.65) it was legitimate to drop the B_{y1} term.

Let us now consider the small electron mass regime, which allows expressing $\overline{\tilde{n}_{e1}}$ by A_1. In that limit the number of electrons confined to a flux interval dA' at $A = A'$ remains invariant. To lowest order in m_e this means

$$dA' \int_{A=A'} n_e(A, s) \frac{ds}{B(A, s)} = dA' \int_{A_0=A'} n_{e0}(A_0) \frac{ds_0}{B_0(A_0, s_0)} . \qquad (A2.68)$$

From this equation one obtains

$$\int dA \, g(A) \int_A n_e(A, s) \frac{ds}{B(A, s)} = \int dA_0 \, g(A_0) \int_{A_0} n_{e0}(A_0) \frac{ds_0}{B_0(A_0, s_0)} , \qquad (A2.69)$$

where an arbitrary function $g(A)$ was introduced and the integration variables were renamed. On the left side of (A2.69) we now transform the integration variables (A, s) to (A_0, s_0), noting that $dA ds / B = dA_0 ds_0 / B_0$,

$$\int \frac{dA_0 ds_0}{B_0} g(A_0 + A_1) n_e(A_0 + A_1, s_0 + s_1) = \int \frac{dA_0 ds_0}{B_0} g(A_0) n_{e0}(A_0) . \qquad (A2.70)$$

After linearizing in the perturbations and integration by parts (which eliminates dg/dA), one finds

$$\int dA_0 g(A_0) \left(-n_{e0} \frac{d}{dA_0} \int_{A_0} A_1 \frac{ds_0}{B_0} + \int_{A_0} \tilde{n}_{e1} \frac{ds_0}{B_0} \right) = 0 . \qquad (A2.71)$$

As g is arbitrary, the A_0 integration can be stripped, which gives

$$\overline{\tilde{n}_{e1}} = \frac{n_{e0}(A_0)}{V(A_0)} \int_{A_0} A_1 \frac{ds_0}{B_0} , \qquad (A2.72)$$

and with the help of (A2.54)

$$\overline{\tilde{n}_{e1}} = n_{e0}(A_0) Q(A_0) . \qquad (A2.73)$$

With this result we rewrite (A2.65) as

$$W_2 \geq \frac{1}{2} \int \left(\frac{(\nabla A_1)^2}{\mu_0} - \frac{\partial j_0}{\partial A_0} A_1^2 + \sum_{\text{ions}} \frac{q_s^2 \overline{\tilde{n}_{s1}}^2}{|a_s|} + \frac{e^2}{|a_e|} n_{e0}^2 Q^2 \right) dx dz . \qquad (A2.74)$$

Specializing for one ion species with $q_i = e$, choosing exponential distribution functions and the frame of exact neutrality ($n_{i0} = n_{e0} = n_0$), one finds $\overline{\tilde{n}_{i1}} = \overline{\tilde{n}_{e1}}$ and $q_s^2/|a_s| = k_B T_s/n_0$ such that (A2.74) gives

$$W_2 \geq \frac{1}{2} \int \left(\frac{(\nabla A_1)^2}{\mu_0} - \frac{\partial j_0}{\partial A_0} A_1^2 + n_0 k_B T_0 Q^2 \right) \mathrm{d}x \mathrm{d}z , \qquad (A2.75)$$

where $T_0 = T_e + T_i$.

Note that (A2.75) is consistent with (A2.61).

Appendix 3
Symbols and fundamental constants

Symbols

General symbols (additional and deviating definitions are declared locally):

α, β	Euler potentials
β	ratio of kinetic and magnetic pressures
γ	polytropic index, ratio of specific heats, growth rate
δ, Δ	length scales of current sheets
Δ	Laplace operator
ϵ	parameter measuring smallness
η	resistivity
λ, Λ	eigenvalue
λ	control parameter, wavelength
λ_{D}	Debye length
$\boldsymbol{\xi}$	displacement of fluid element
$\xi(A)$	vertex position, see also $v(A)$
ρ	mass density
σ	charge density
ϕ	electric potential
Ψ	gravity potential, term in the single-particle Hamiltonian
ω	angular frequency
$\omega_{\mathrm{g}}, \Omega$	gyrofrequency
ω_{p}	plasma frequency
ω_s	plasma frequency of species s
∇	nabla-operator (gradient)
A	magnetic flux function
$\boldsymbol{B}$	magnetic field
D	stream function
D	domain
D_{R}	nonideal region (or diffusion region) of a reconnection site
$\boldsymbol{e}$	unit vector
$\boldsymbol{E}$	electric field
f, F	distribution function

F	stability functional
F_2	second variation of F
$\mathcal{F}$	free energy, electromagnetic field tensor
$\boldsymbol{g}$	gravity acceleration
H	Hamiltonian
$\mathcal{I}$	unit tensor
$\boldsymbol{j}$	electric current density
$\boldsymbol{k}$	wave number
K	helicity
$\boldsymbol{K}$	surface current density
L	scale length
m	particle mass
n	number density
$\boldsymbol{n}$	unit normal vector
N	number of particles
$\boldsymbol{P}$	canonical momentum
$\mathcal{P}$	pressure tensor
p	pressure
q	particle charge
r_{g}	gyroradius
s	arclength (e.g., on magnetic field lines)
S	Lundquist number, entropy, unit step function
$\boldsymbol{S}$	Poynting vector
T	temperature, potential of unified steady state theory
$\mathcal{T}$	generalized kinetic energy
t	time
U	potential (in several different contexts)
$U_{\mathrm{c}}, V_{\mathrm{c}}$	curvature potential
V	(differential) flux tube volume
$\boldsymbol{v}$	bulk plasma velocity
$v(A)$	vertex position, see also $\xi(A)$
v_{A}	Alfvén velocity
$v_s, v_{\mathrm{t}s}$	thermal velocity of species s
$\mathcal{V}$	generalized potential energy
W	total energy
$\boldsymbol{w}$	particle velocity, transport velocity of magnetic field lines
$\boldsymbol{w}^*$	transport velocity as $\boldsymbol{w}$, but coinciding with $\boldsymbol{v}$ in the ideal region

General subscripts:

    e   electron
    g   gyration
    i   ion
    n   normal
    p   proton, poloidal
    t   toroidal, thermal
    s   species

Brackets:

    $[\,]$ difference
$\langle\,\rangle, [\![\,]\!], \langle\,\rangle$ averages

Fundamental constants

Boltzmann's constant	$k_B = 1.3806 \cdot 10^{-23}$ VAs/deg
elementary charge	$e = 1.6022 \cdot 10^{-19}$ As
gravitational constant	$G = 6.6726 \cdot 10^{-11}$ m^3/(kg s^2)
mass of electron	$m_e = 9.1094 \cdot 10^{-31}$ kg
mass of proton	$m_p = 1.6726 \cdot 10^{-27}$ kg
Planck's constant	$h = 6.6261 \cdot 10^{-34}$ VAs^2
vacuum dielectric constant	$\epsilon_0 = 8.8542 \cdot 10^{-12}$ As/(Vm)
vacuum permeability	$\mu_0 = 1.2566 \cdot 10^{-6}$ Vs/(Am)
velocity of light	$c = 2.9979 \cdot 10^8$ m/s

References

Abramowitz M., Stegun I. A. (1965). *Handbook of Mathematical Functions.* New York: Dover Publications.

Akasofu S.-I. (1968). *Polar and Magnetospheric Substorms.* Dordrecht: D. Reidel.

Akasofu S., Deforest S., McIlwain C. E. (1974). Auroral displays near the "foot" of the field line of the ATS-5 satellite. *Planet. Space Sci.* **22**, 25.

Alfvén H. (1972). Relations between cosmic and laboratory plasma physics. In *Cosmic Plasma Physics*, K. Schindler, ed. New York: Plenum Press, pp. 1–14.

Ali F., Sneyd A. D. (2002). Twisting together magnetic flux tubes under helical symmetry. *Solar Phys.* **205**, 279–301.

Aly J. J. (1984). On some properties of force-free magnetic fields in infinite regions of space. *Astrophys. J.* **283**, 349–362.

Aly J. J. (1991). How much energy can be stored in a three-dimensional force-free magnetic field? *Astrophys. J.* **375**, L61–L64.

Aly J. J. (1995). Nonequilibrium in sheared axisymmetric force-free magnetic fields. *Astrophys. J.* **439**, L63–L66.

Amann H. (1976). Fixed point equations and nonlinear eigenvalue problems in ordered Banach spaces. *SIAM Review* **18**, 620–709.

Amari T., Luciani J. F., Mikić Z., Linker J. (2000). A twisted flux rope model for coronal mass ejections and two-ribbon flares. *Astrophys. J.* **529**, L49–L52.

An C.-H. (1984). Comments on the MHD stability of coronal plasmas with line-tying. *Astrophys. J.* **281**, 419–425.

Andrews M. D. (2003). A search for CMEs associated with big flares. *Sol. Phys.* **218**, 261–279.

Angelopoulos V., Baumjohann W., Kennel C. F. *et al.* (1992). Bursty bulk flows in the inner central plasma sheet. *J. Geophys. Res.* **97**, 4027–4039.

Angelopoulos V., Kennel C. F., Coroniti F. V. *et al.* (1993). Characteristics of ion flow in the quiet state of the inner plasma sheet. *Geophys. Res. Lett.* **20**, 1711–1714.

Angelopoulos V., Mitchell D. G., McEntire R. W. *et al.* (1996). Tailward progression of magnetotail acceleration centers: Relationship to substorm current wedge. *J. Geophys. Res.* **101**, 24599–24620.

Angelopoulos V., Mozer F. S., Lin R. P. *et al.* (1999). Comment on "Geotail survey of ion flow in the plasma sheet: Observations between 10 and 50 R_E" by W. R. Paterson *et al.* *J. Geophys. Res.* **104**, 17521–17525.

Antiochos S. K., DeVore C. R., Klimchuk J. A. (1999). A model of solar coronal mass ejections. *Astrophys. J.* **510**, 485–493.

Antiochos S. K., Karpen J. T., DeVore C. R. (2002). Coronal magnetic field relaxation by null-point reconnection. *Astrophys. J.* **575**, 578–584.

Anzer U., Ballester J. L. (1990). Prominence models with line currents: Stabilisation by flux conservation. *Astron. Astrophys.* **238**, 365–368.

Arendt U., Schindler K. (1988). On the existence of three-dimensional magnetohydrostatic equilibria. *Astron. Astrophys.* **204**, 229–234.

Arnol'd V. I. (1979). *Mathematical Methods of Classical Mechanics.* New York: Springer.

Arnoldy R. L. (1971). Signatures in the interplanetary medium for substorms. *J. Geophys. Res.* **76**, 5189–5201.

Asai A., Yokoyama T., Shimojo M., Masuda S., Kurokawa H., Shibata K. (2004). Flare ribbon expansion and energy release rate. *Astrophys. J.* **611**, 557–567.

Asano Y., Mukai T., Hoshino M., Saito Y., Hayakawa H., Nagai T. (2004). Current sheet structure around the near-Earth neutral line observed by Geotail. *J. Geophys. Res.* **109**. doi:10.1029/2003JA010114.

Ashour-Abdalla M., El-Alaoui M., Coroniti F. V., Walker R. J., Peroomian V. (2002). A new convection state at substorm onset: Results from an MHD study. *Geophys. Res. Lett.* **29**. doi:10.1029/2002GL015787.

Axford W. I. (1984). Magnetic field reconnection. In *Magnetic Reconnection in Space and Laboratory Plasmas*, E. W. Hones, ed. Geophys. Monogr. Ser., vol. 30. Washington, DC: AGU, pp. 1–8.

Axford W. I., Hines C. O. (1961). A unifying theory of high-latitude geophysical phenomena and geomagnetic storms. *Canad. J. Phys.* **39**, 1433–1464.

Axford W. I., McKenzie J. F. (2002). Coronal heating. *Adv. Space Res.* **30**, 505.

Bak P., Tang C., Wiesenfeld K. (1988). Self-organized criticality. *Phys. Rev. A* **38**, 364–374.

Baker D. N., Fritz T. A., McPherron R. L., Fairfield D. H., Kamide Y., Baumjohann W. (1985). Magnetotail energy storage and release during the CDAW 6 substorm analysis intervals. *J. Geophys. Res.* **90**, 1205–1216.

Baker D. N., Klimas A. J., McPherron R. L., Büchner J. (1990). The evolution from weak to strong geomagnetic activity: An interpretation in terms of deterministic chaos. *Geophys. Res. Lett.* **17**, 41–44.

Balescu R. (1975). *Equilibrium and Nonequilibrium Statistical Mechanics.* New York: John Wiley & Sons.

Balescu R. (1988). *Transport Processes in Plasmas.* New York: North-Holland.

Bargatze L. F., Baker D. N., Hones E. W., McPherron R. L. (1985). Magnetospheric impulse response for many levels of geomagnetic activity. *J. Geophys. Res.* **90**, 6387–6394.

Bargatze L. F., Ogino T., McPherron R. L., Walker R. J. (1999). Solar wind magnetic field control of magnetospheric response delay and expansion phase onset timing. *J. Geophys. Res.* **104**, 14583–14599.

Barnes C. W., Sturrock P. A. (1972). Force-free magnetic field structures and their role in solar activity. *Astrophys. J.* **174**, 659–670.

Baumjohann W. (1982). Ionospheric and field-aligned current systems in the auroral zone: A concise review. *Adv. Space Res.* **2**, 55–62.

Baumjohann W. (2002). Special topics section: Modes of convection in the magnetotail. *Phys. Plasmas* **9**. doi:10.1063/1.1499116.

Baumjohann W., Paschmann G., Lühr H. (1990). Characteristics of high-speed ion flows in the plasma sheet. *J. Geophys. Res.* **95**, 3801–3809.

Bennett W. H. (1934). Magnetically self-focussing streams. *Phys. Rev.* **45**, 890–897.

Berge P., Pomeau Y., Vidal C. (1986). *Order Within Chaos - Towards a Deterministic Approach to Turbulence.* New York: John Wiley & Sons.

Berger M. A. (1984). Rigorous new limits on magnetic helicity dissipation in the solar corona. *Geophys. Astrophys. Fluid Dynamics* **30**, 79–104.

Berger M. A. (1991). Generation of coronal magnetic fields by random surface motions I. Mean square twist and current density. *Astron. Astrophys.* **252**, 369–376.

Berger M. A., Field G. B. (1984). The topological properties of magnetic helicity. *J. Fluid Mech.* **147**, 133–148.

Bernstein I. B., Freeman E. A., Kruskal M. D., Kulsrud R. M. (1958). An energy principle for hydromagnetic stability problems. *Proc. Roy. Soc. Lond., Ser. A* **244**, 17.

Bhattacharjee A., Ma Z. W., Wang X. (1998). Ballooning instability of a thin current sheet in the high-Lundquist-number magnetotail. *Geophys. Res. Lett.* **25**, 861–864.

Bhattacharjee A., Germaschewski K., Ng C. S. (2005). Current singularities: Drivers of impulsive reconnection. *Phys. Plasmas* **12**, 42305–42315.

Birk G. T., Otto A. (1991). The resistive tearing instability for generalized resistivity models: Applications. *Physics of Fluids B* **3**, 1746–1754.

Birn J. (1980). Computer studies of the dynamic evolution of the geomagnetic tail. *J. Geophys. Res.* **85**, 1214–1222.

Birn J. (1991). The boundary value problem of magnetotail equilibrium. *J. Geophys. Res.* **96**, 19441–19450.

Birn J., Hesse M. (1996). Details of current disruption and diversion in simulations of magnetotail dynamics. *J. Geophys. Res.* **101**, 15345–15358.

Birn J., Hones E. W. (1981). Three-dimensional computer modelling of dynamic reconnection in the geomagnetic tail. *J. Geophys. Res.* **86**, 6802–6808.

Birn J., Schindler K. (1981). Two-ribbon flares: Magnetostatic equilibria. In *Solar Flare Magnetohydrodynamics*, E. R. Priest, ed. New York: Gordon and Breach, pp. 337–378.

Birn J., Schindler K. (1983). Self-consistent theory of three-dimensional convection in the geomagnetic tail. *J. Geophys. Res.* **88**, 6969–6980.

Birn J., Schindler K. (2002). Thin current sheets in the magnetotail and the loss of equilibrium. *J. Geophys. Res.* **107**. doi:10.1029/2001JA000291.

Birn J., Sommer R. R., Schindler K. (1975). Open and closed magnetospheric tail configurations and their stability. *Astrophys. Space Sci.* **35**, 389–402.

Birn J., Goldstein H., Schindler K. (1978). A theory of the onset of solar eruptive processes. *Solar Phys.* **57**, 81–101.

Birn J., Browning P. K., Sakurai T., Schindler K., Zwingmann W. (1984). Energy conversion in active astrophysical plasmas: Evolution of equilibrium fields. *Rapport d'activité scientifique du CECAM*, 52.

Birn J., Gary S. P., Hesse M. (1995). Microscale anisotropy reduction and macroscale dynamics of the magnetotail. *J. Geophys. Res.* **100**, 19211–19220.

Birn J., Hesse M., Schindler K. (1996). MHD simulations of magnetotail dynamics. *J. Geophys. Res.* **101**, 12939–12954.

Birn J., Thomsen M. F., Borovsky J. E. *et al.* (1997). Substorm ion injections: Geosynchronous observations and test particle orbits in three-dimensional dynamic MHD fields. *J. Geophys. Res.* **102**, 2325–2342.

Birn J., Hesse M., Schindler K. (1998a). Formation of thin current sheets in space plasmas. *J. Geophys. Res.* **103**, 6843–6852.

Birn J., Thomsen M. F., Borovsky J. E. *et al.* (1998b). Substorm electron injections: Geosynchronous observations and test particle simulations. *J. Geophys. Res.* **103**, 9235–9248.

Birn J., Gosling J. T., Hesse M., Forbes T. G., Priest E. R. (2000). Simulations of three-dimensional reconnection in the solar corona. *Astrophys. J.* **541**, 1078–1095.

Birn J., Drake J. F., Shay M. A. *et al.* (2001). Geospace Environmental Modeling (GEM) magnetic reconnection challenge. *J. Geophys. Res.* **106**, 3715–3720.

Birn J., Schindler K., Hesse M. (2003). Formation of thin current sheets in the magnetotail: Effects of propagating boundary deformations. *J. Geophys. Res.* **108**. doi:10.1029/2002JA009641.

Birn J., Dorelli J.C., Hesse M., Schindler K. (2004). Thin current sheets and loss of equilibrium: Three-dimensional theory and simulations. *J. Geophys. Res.* **109**. doi:10.1029/2003JA010275.

Birn J., Galsgaard K., Hesse M. *et al.* (2005). Forced magnetic reconnection. *Geophys. Res. Lett.* **32**. doi:10.1029/2004GL022058.

Biskamp D. (1982). Dynamics of a resistive sheet pinch. *Z. Naturforsch. A* **37A**, 840–847.

Biskamp D. (1986). Magnetic reconnection via current sheets. *Phys. Fluids* **29**, 1520–1531.

Biskamp D. (1993). *Nonlinear Magnetohydrodynamics*. Cambridge: Cambridge University Press.

Biskamp D. (2000). *Magnetic Reconnection in Plasmas*. Cambridge: Cambridge University Press.

Biskamp D., Welter H. (1977). Numerical studies of resistive instabilities. In *Proceedings of the Sixth Conference on Plasma Physics and Controlled Nuclear Fusion Research*. Vol. 2. Vienna: Intern. Atomic Energy Agency, pp. 579–589.

Biskamp D., Welter H. (1980). Coalescence of magnetic islands. *Phys. Rev. Lett.* **44**, 1069–1072.

Biskamp D., Welter H. (1989). Magnetic arcade evolution and instability. *Solar Phys.* **120**, 49–77.

Bopp B. W., Moffett T. J. (1973). High time resolution studies of UV Cet. *Astrophys. J.* **185**, 239–240.

Borovsky J. E., Funsten H. O. (2003). MHD turbulence in the Earth's plasma sheet: Dynamics, dissipation, and driving. *J. Geophys. Res.* **108**. doi:10.1029/2002JA009625.

Borovsky J. E., Thomsen M. F., Elphic R. C., Cayton T. E., McComas D. J. (1998). The transport of plasma sheet material from the distant tail to geosynchronous orbit. *J. Geophys. Res.* **103**, 20297–20331.

Boyd T. J. M., Sanderson J. J. (2003). *The Physics of Plasmas*. Cambridge: Cambridge University Press.

Braginskii S. I. (1965). Transport processes in a plasma. In *Reviews of Plasma Physics*, M. A. Leontovich, ed. Vol. 1. New York: Consultants Bureau, pp. 205–311.

Brinkmann R. P. (1987). Thermodynamic stability analysis of current-carrying plasmas. *Phys. Fluids* **30**, 3713–3723.

Brio M., Wu C. C. (1988). An upwind differencing scheme for the equations of ideal magnetohydrodynamics. *J. Comp. Phys.* **75**, 400–422.

Brittnacher M., Quest K. B., Karimabadi H. (1994). On the energy principle and ion tearing in the magnetotail. *Geophys. Res. Lett.* **21**, 1591–1594.

Brittnacher M., Quest K. B., Karimabadi H. (1995). A new approach to the linear theory of single-species tearing in two-dimensional quasi-neutral sheets. *J. Geophys. Res.* **100**, 3551–3562.

Browning P. K., Jain R. (2004). Coronal heating by forced magnetic reconnection with multi-pulse driving. In *SOHO 15 Workshop: Coronal heating*, R. W. Walsh, J. Ireland, D. Danesy, and B. Fleck, eds. Paris: ESA, pp. 474–478.

Brueckner G. E., Delaboudiniere J., Howard R. A. *et al.* (1998). Geomagnetic storms caused by coronal mass ejections (CMEs): March 1996 through June 1997. *Geophys. Res. Lett.* **25**, 3019–3022.

Büchner J., Daughton W. S. (2006). Role of current-aligned instabilities. In *Reconnection of Magnetic Fields*, J. Birn and E. R. Priest, eds. Cambridge: Cambridge University Press. In press.

Büchner J., Zelenyi L. M. (1989). Regular and chaotic charged particle motion in magnetotaillike field reversals: I - Basic theory of trapped motion. *J. Geophys. Res.* **94**, 11821–11842.

Bungey T. N., Priest E. R. (1995). Current sheet configurations in potential and force-free fields. *Astron. Astrophys.* **293**, 215–224.

Burke W. J., Weimer D. R., Maynard N. C. (1999). Geoeffective interplanetary scale sizes derived from regression analysis of polar cap potentials. *J. Geophys. Res.* **104**, 9989–9994.

Burlaga L. F., Sittler E., Mariani F., Schwenn R. (1981). Magnetic loop behind an interplanetary shock: Voyager, Helios and IMP-8 observations. *J. Geophys. Res.* **86**, 6673–6684.

Burlaga L. F., Fitzenreiter R., Lepping R. *et al.* (1998). A magnetic cloud containing prominence material: January 1997. *J. Geophys. Res.* **103**, 277–286.

Burton R. K., McPherron R. L., Russell C. T. (1975). An empirical relationship between interplanetary conditions and *Dst*. *J. Geophys. Res.* **80**, 4204–4214.

Chandrasekhar S. (1961). *Hydrodynamic and Hydromagnetic Stability*. Mineola, N. Y.: Dover Publications.

Chang T. (1999). Self-organized criticality, multi-fractal spectra, sporadic localized reconnections and intermittent turbulence in the magnetotail. *Phys. Plasmas* **6**, 4137–4145.

Chapman S., Bartels J. (1940). *Geomagnetism*. Oxford: Clarendon.

Chapman S. C., Watkins N. W. (1996). Scaling parameters and parametric coordinates in static and time dependent magnetic reversals. *Advances in Space Research* **18**, 285–289.

Chapman S. C., Watkins N. W., Dendy R. O., Helander P., Rowlands G. (1998). A simple avalanche model as an analogue for magnetospheric activity. *Geophys. Res. Lett.* **25**, 2397–2400.

Chen C. X., Wolf R. A. (1993). Interpretation of high-speed flows in the plasma sheet. *J. Geophys. Res.* **98**, 21409–21420.

Chen C. X., Wolf R. A. (1999). Theory of thin-filament motion in Earth's magnetotail and its application to bursty bulk flows. *J. Geophys. Res.* **104**, 14613–14626.

Chen J., Palmadesso P. J. (1986). Chaos and nonlinear dynamics of single-particle orbits in a magnetotail-like magnetic field. *J. Geophys. Res.* **91**, 1499–1508.

Chen P. F., Shibata K. (2000). An emerging flux trigger mechanism for coronal mass ejections. *Astrophys. J.* **545**, 524–531.

Cheng C. Z., Zaharia S. (2004). MHD ballooning instability in the plasma sheet. *Geophys. Res. Lett.* **31**. doi:10.1029/2003GL018823.

Chew G. F., Goldberger M. L., Low F. E. (1956). The Boltzmann equation and the one-fluid hydromagnetic equations in the absence of particle collisions. *Proc. Roy. Soc. Lond., Ser. A* **236**, 112–118.

Choudhary D. P., Moore R. L. (2003). Filament eruption without coronal mass ejection. *Geophys. Res. Lett.* **30**. doi:10.1029/2003GL018332.

Cowley S. W. H. (1973). A qualitative study of the reconnection between the Earth's magnetic field and an interplanetary field of arbitrary orientation. *Radio Science* **8**, 903–913.

Cowley S. W. H. (1978). The effect of pressure anisotropy on the equilibrium structure of magnetic current sheets. *Planet. Space Sci.* **26**, 1037–1061.

Cowley S. W. H. (1984). Evidence for the occurrence and importance of reconnection between the Earth's magnetic field and the interplanetary magnetic field. In *Magnetic Reconnection in Space and Laboratory Plasmas*, E. W. Hones, ed. Geophys. Monogr. Ser., vol. 30. Washington, DC: AGU, pp. 375–378.

Craig I. J. D., Fabling R. B. (1996). Exact solutions for steady state, spine, and fan magnetic reconnection. *Astrophys. J.* **462**, 969–976.

Craig I. J. D., Henton S. M. (1995). Exact solutions for steady state incompressible magnetic reconnection. *Astrophys. J.* **450**, 280–288.

Cravens T. E. (2004). *Physics of Solar System Plasmas*. Cambridge: Cambridge University Press.

Crooker N. U. (1979). Dayside merging and cusp geometry. *J. Geophys. Res.* **84**, 951–959.

Daughton W. (1999). The unstable eigenmodes of a neutral sheet. *Phys. Plasmas* **6**, 1329–1343.

Daughton W., Lapenta G., Ricci P. (2004). Nonlinear evolution of the lower-hybrid drift instability in a current sheet. *Phys. Rev. Lett.* **93**. doi:10.1103/PhysRevLett.93.105004.

Davidson R. C. (1972). *Methods in Nonlinear Plasma Theory*. New York: Academic Press.

Davidson R. C., Gladd N. T. (1975). Anomalous transport properties associated with the lower-hybrid-drift instability. *Phys. Fluids* **18**, 1327–1335.

Davis T. N., Sugiura M. (1966). Auroral electrojet activity index AE and its universal time variations. *J. Geophys. Res.* **71**, 785.

de Bruyne P., Hood A. W. (1989). Bounds on the ideal MHD stability of line-tied 2-D coronal magnetic fields. *Solar Physics* **123**, 241–269.

de Hoffmann F., Teller E. (1950). Magneto-hydrodynamic shocks. *Phys. Rev.* **80**, 692–703.

Démoulin P., Priest E. R. (1988). Instability of a prominence supported in a linear force-free field. *Astron. Astrophys.* **206**, 336–347.

Dendy R. O., Helander P. (1998). Appearance and non-appearance of self-organized criticality in sandpiles. *Phys. Rev. E* **57**, 3641–3644.

Drake J. F., Guzdar P. N., Huba J. D. (1983). Saturation of the lower-hybrid-drift instability by mode coupling. *Phys. Fluids* **26**, 601–604.

Drake J. F., Guzdar P. N., Hassam A. B., Huba J. D. (1984). Nonlinear mode coupling theory of the lower-hybrid-drift instability. *Phys. Fluids* **27**, 1148–1159.

Dreicer H. (1960). Electron and ion runaway in a fully ionized gas. II. *Phys. Rev.* **117**, 329–342.

Duncan R., Thompson C. (1992). Formation of very strongly magnetized neutron stars: Implications for gamma-ray bursts. *Astrophys. J.* **392**, L9–L13.

Dungey J. W. (1953). Conditions for the occurrence of electrical discharges in astrophysical systems. *Phil. Mag.* **44**, 725–738.

Dungey J. W. (1961). Interplanetary magnetic field and the auroral zones. *Phys. Rev. Lett.* **6**, 47–48.

Eckhaus W. (1973). *Matched Asymptotic Expansions and Singular Perturbations.* Amsterdam: North-Holland Pub. Co.

Edenstrasser J. W. (1980a). The only three classes of symmetric MHD equilibria. *J. Plasma Phys.* **24**, 515–518.

Edenstrasser J. W. (1980b). Unified treatment of symmetric MHD equilibria. *J. Plasma Phys.* **24**, 299–313.

Einaudi G., van Hoven G. (1983). The stability of coronal loops: Finite-length and pressure-profile limits. *Solar Phys.* **88**, 163–177.

Elsasser W. M. (1956). Hydromagnetic dynamo theory. *Rev. Mod. Phys.* **28**, 135.

Erickson G. M., Wolf R. A. (1980). Is steady convection possible in the Earth's magnetotail? *Geophys. Res. Lett.* **7**, 897–900.

Erickson G. M., Spiro R. W., Wolf R. A. (1991). The physics of the Harang discontinuity. *J. Geophys. Res.* **96**, 1633–1645.

Fairfield D. H., Otto A., Mukai T. *et al.* (2000). Geotail observations of the Kelvin-Helmholtz instability at the equatorial magnetotail boundary for parallel northward fields. *J. Geophys. Res.* **105**, 21159–21173.

Falle S. A. E. G., Komissarov S. S. (2004). On the inadmissibility of non-evolutionary shocks. *J. Plasma Phys.* **65**, 29–58.

Fan Y., Gibson S. E. (2004). Numerical simulations of three-dimensional coronal magnetic fields resulting from the emergence of twisted magnetic flux tubes. *Astrophys. J.* **609**, 1123–1133.

Feldman U., Laming J. M., Doschek G. A. (1995). The correlation of solar flare temperature and emission measure extrapolated to the case of stellar flares. *Astrophys. J. Lett.* **451**, L79–L82.

Finn J. M., Antonsen T. M. (1985). Magnetic helicity: what is it and what is it good for? *Comments Plasma Phys. Control. Fusion* **9**, 111–120.

Finn J. M., Chen J. (1990). Equilibrium of solar coronal arcades. *Astrophys. J.* **349**, 345–361.

Finn J. M., Guzdar P. N., Chen J. (1992). Fast plasmoid formation in double arcades. *Astrophys. J.* **393**, 800–814.

Forbes T. G. (2000a). A review on the genesis of coronal mass ejections. *J. Geophys. Res.* **105**, 23153–23165.

Forbes T. G. (2000b). Solar and stellar flares. *Phil. Trans. Roy. Soc. Lond.* **358**, 711–727.

Forbes T. G., Isenberg P. A. (1991). A catastrophe mechanism for coronal mass ejections. *Astrophys. J.* **373**, 294–307.

Forbes T. G., Priest E. R. (1983). A numerical experiment relevant to line-tied reconnection in two-ribbon flares. *Solar Phys.* **84**, 169–188.

Forbes T. G., Priest E. R. (1995). Photospheric magnetic field evolution and eruptive flares. *Astrophys. J.* **446**, 377–389.

Forbes T. G., Malherbe J. M., Priest E. R. (1989). The formation of flare loops by magnetic reconnection and chromospheric ablation. *Solar Phys.* **120**, 285–307.

Fox N. J., Peredo M., Thompson B. J. (1998). Cradle to grave tracking of the January 6–11, 1997 Sun-Earth connection event. *Geophys. Res. Lett.* **25**, 2461–2464.

Freeman M. P., Farrugia C. J. (1999). Solar wind input between substorm onsets during and after the October 18-20, 1995, magnetic cloud. *J. Geophys. Res.* **104**, 22729–22744.

Freidberg J. P. (1987). *Ideal Magnetohydrodynamics*. New York: Plenum Press.

Fujita H. (1969). On the nonlinear equations $\delta u + e^u = 0$ and $\partial v / \partial t = \delta v + e^v$. *Bull. Am. Math. Soc.* **75**, 132–135.

Furth H. P., Killeen J., Rosenbluth M. N. (1963). Finite resistivity instabilities of a sheet pinch. *Phys. Fluids* **6**, 459–484.

Galeev A. A. (1984). Spontaneous reconnection of magnetic field lines in a collisionless plasma. In *Basic Plasma Physics*, A. A. Galeev and R. N. Sudan, eds. Vol. 2. Amsterdam: North-Holland Pub. Co., pp. 305–334.

Galeev A. A., Rosner R., Vaiana G. S. (1979). Structured coronae of accretion disks. *Astrophys. J.* **229**, 318–326.

Galsgaard K., Nordlund Å. (1996). Heating and activity of the solar corona 1. Boundary shearing of an initially homogeneous magnetic field. *J. Geophys. Res.* **101**, 13445–13460.

Galsgaard K., Moreno-Insertis F., Archontis V., Hood A. (2005). A three-dimensional study of reconnection, current sheets, and jets resulting from magnetic flux emergence in the sun. *Astrophys. J.* **618**, L153–L156.

Garner T. W., Wolf R. A., Spiro R. W. (2003). Pressure balance inconsistency exhibited in a statistical model of magnetospheric plasma. *J. Geophys. Res.* **108**. doi:10.1029/2003JA009877.

Gary S. P. (1993). *Theory of Space Plasma Microinstabilities.* New York: Cambridge University Press.

Gary S. P., Moldwin M. B., Thomsen M. F., Winske D., McComas D. J. (1994). Hot proton anisotropies and cool proton temperatures in the outer magnetosphere. *J. Geophys. Res.* **99**, 23603–23615.

Gibson S. E., Low B. C. (1998). A time-dependent, three-dimensional, magnetohydrodynamic model of the coronal mass ejection. *Astrophys. J.* **493**, 460–473.

Giovanelli R. G. (1946). A theory of chromospheric flares. *Nature* **158**, 81–82.

Goedbloed J. P. (2004). Variational principles for stationary one- and two-fluid equilibria of axisymmetric laboratory and astrophysical plasmas. *Phys. Plasmas* **11**, L81–L84.

Gold T. (1964). Magnetic energy shedding in the solar atmosphere. In *The Physics of Solar Flares*, W. N. Hess, ed. Washington, DC: NASA, pp. 389–408.

Gold T., Hoyle F. (1960). On the origin of solar flares. *Mon. Not. Roy. Astron. Soc.* **120**, 89.

Goldstein H. (2002). *Classical Mechanics.* Reading, Massachusetts: Addison-Wesley.

Goldstein H., Schindler K. (1982). Large-scale collision-free instability of two-dimensional plasma sheets. *Phys. Rev. Lett.* **48**, 1468–1471.

Gosling J. T. (1990). Coronal mass ejections and magnetic flux ropes in interplanetary space. In *Physics of Magnetic Flux Ropes*, C. T. Russell, E. R. Priest, and L. C. Lee, eds. Washington, DC: AGU, pp. 343–364.

Gosling J. T. (1993). The solar flare myth. *J. Geophys. Res.* **98**, 18937–18950.

Gosling J. T., Birn J., Hesse M. (1995). Three-dimensional magnetic reconnection and the magnetic topology of coronal mass ejection events. *Geophys. Res. Lett.* **22**, 869–872.

Grad H. (1964). Some new variational properties of hydromagnetic equilibria. *Phys. Fluids* **7**, 1283–1292.

Grad H., Rubin H. (1958). Hydromagnetic equilibria and force-free fields. In *Proceedings of the Second United Nations International Conference on the Peaceful Uses of Atomic Energy*. Vol. 31. Geneva: United Nations, pp. 190–197.

Greene J. M. (1988). Geometrical properties of three-dimensional reconnecting magnetic fields with nulls. *J. Geophys. Res.* **93**, 8583–8590.

Guckenheimer J., Holmes P. (1983). *Nonlinear Oscillations, Dynamical Systems, and Bifurcations of Vector Fields*. New York: Springer.

Guenther D. B., Demarque P., Kim Y., Pinsonneault M. H. (1992). Standard solar model. *Astrophys. J.* **387**, 372–393.

Gurnett D. A., Frank L. A. (1977). A region of intense plasma wave turbulence on auroral field lines. *J. Geophys. Res.* **82**, 1031–1050.

Hahm T. S., Kulsrud R. M. (1985). Forced magnetic reconnection. *Phys. Fluids* **28**, 2412–2418.

Hain K., Lüst R., Schlüter A. (1957). Zur Stabilität eines Plasmas. *Z. Naturforschg. a* **12**, 833–841.

Harris E. G. (1962). On a plasma sheath separating regions of oppositely directed magnetic field. *Nuovo Cimento* **23**, 115.

Hasegawa A. (1975). *Plasma Instabilities and Nonlinear Effects*. Berlin: Springer-Verlag.

Hayashi M. R., Shibata K., Matsumoto R. (1996). X-ray flares and mass outflows driven by magnetic interaction between a protostar and its surrounding disk. *Astrophys. J.* **468**, L37–L40.

Hesse M. (2005). On the components of the electron pressure tensor in magnetic reconnection. Manuscript.

Hesse M., Birn J. (1991a). On dipolarization and its relation to the substorm current wedge. *J. Geophys. Res.* **96**, 19417–19426.

Hesse M., Birn J. (1991b). Plasmoid evolution in an extended magnetotail. *J. Geophys. Res.* **96**, 5683–5696.

Hesse M., Birn J. (1993). Parallel electric fields as acceleration mechanisms in three-dimensional magnetic reconnection. *Adv. Space Res.* **13**, 249–252.

Hesse M., Schindler K. (1986). Bifurcation of current sheets in plasmas. *Phys. Fluids* **29**, 2484–2492.

Hesse M., Schindler K. (1988). A theoretical foundation of general magnetic reconnection. *J. Geophys. Res.* **93**, 5559–5567.

Hesse M., Schindler K., Birn J., Kuznetsova M. (1999). The diffusion region in collisionless magnetic reconnection. *Phys. Plasmas* **6**, 1781–1795.

Hesse M., Birn J., Kuznetsova M. (2001). Collisionless magnetic reconnection: Electron processes and transport modeling. *J. Geophys. Res.* **106**, 3721–3735.

Hesse M., Kuznetsova M., Birn J. (2004). The role of electron heat flux in guide-field magnetic reconnection. *Phys. Plasmas* **11**, 5387.

Hesse M., Forbes T. G., Birn J. (2005a). On the relation between reconnected magnetic flux and parallel electric fields in the solar corona. *Astrophys. J.* **631**, 1227–1238.

Hesse M., Kuznetsova M., Schindler K., Birn J. (2005b). Three-dimensional modeling of electron quasiviscous dissipation in guide-field magnetic reconnection. *Phys. Plasmas* **12**. DOI:10.1063/1.2114350.

Heyn M. F., Biernat H. K., Rijnbeek R. P., Semenov V. S. (1988). The structure of reconnection layers. *J. Plasma Phys.* **40**, 235–252.

Heyvaerts J., Priest E. R. (1984). Coronal heating by reconnection in DC current systems: A theory based on Taylor's hypothesis. *Astron. Astrophys.* **137**, 63–78.

Heyvaerts J., Priest E. R., Rust D. M. (1977). An emerging flux model for the solar flare phenomenon. *Astrophys. J.* **216**, 123–137.

Heyvaerts J., Lasry J. M., Schatzmann M., Witomsky P. (1982). Blowing up of two-dimensional magnetohydrostatic equilibria by an increase of electric current or pressure. *Astron. Astrophys.* **111**, 104–112.

Holm D. D., Marsden J. E., Ratiu T., Weinstein A. (1985). Nonlinear stability of fluid and plasma equilibria. *Phys. Reports* **123**, 1–116.

Hones E. W. (1973). Plasma flow in the plasma sheet and its relation to substorms. *Radio Sci.* **8**, 979–990.

Hones E. W. (1977). Substorm processes in the magnetotail. *J. Geophys. Res.* **82**, 5633–5640.

Hones E. W. (1979). Transient phenomena in the magnetotail and their relation to substorms. *Space Sci. Rev.* **23**, 393–410.

Hones E. W., Pytte T., West H. I. (1984). Associations of geomagnetic activity with plasma sheet thinning and expansion: A statistical study. *J. Geophys. Res.* **89**, 5471–5478.

Hood A. W., Priest E. R. (1979). Kink instability of solar coronal loops as the cause of solar flares. *Solar Phys.* **64**, 303–321.

Hood A. W., Priest E. R. (1981). Critical conditions for magnetic instabilities in force-free coronal loops. *Geophys. Astrophys. Fluid Dynamics* **17**, 297–318.

Horiuchi R., Sato T. (1994). Particle simulation study of driven magnetic reconnection in a collisionless plasma. *Phys. Plasmas* **1**, 3587–3597.

Hornig G. (1997a). The covariant transport of electromagnetic fields and its relation to magnetohydrodynamics. *Phys. Plasmas* **4**, 646–654.

Hornig G. (1997b). Zur kovarianten Formulierung der magnetischen Rekonnexion. Ph.D. thesis, Ruhr–Universität Bochum.

Hornig G. (2006). Fundamental concepts. In *Reconnection of Magnetic Fields*, J. Birn and E. R. Priest, eds. Cambridge: Cambridge University Press. In press.

Hornig G., Priest E. R. (2003). Evolution of magnetic flux in an isolated reconnection process. *Phys. Plasmas* **10**, 2712–2721.

Hornig G., Schindler K. (1996a). Magnetic topology and the problem of its invariant definition. *Phys. Plasmas* **3**, 781–791.

Hornig G., Schindler K. (1996b). The problem of magnetic topology and reconnection in relativistic systems. *Astro. Lett. Comm.* **34**, 231.

Horton W., Doxas I. (1996). A low-dimensional energy-conserving state space model for substorm dynamics. *J. Geophys. Res.* **101**, 27223–27237.

Horton W., Doxas I. (1998). A low-dimensional dynamical model for the solar wind driven geotail-ionosphere system. *J. Geophys. Res.* **103**, 4561–4572.

Hoshino M. (1987). The electrostatic effect for the collisionless tearing mode. *J. Geophys. Res.* **92**, 7368–7380.

Howard R. A., Michels D. J., Sheeley N. R., Koomen M. J. (1982). The observation of a coronal transient directed at Earth. *Astrophys. J.* **263**, L101–L104.

Hu Y. Q. (2004). Energy buildup of multipolar magnetic fields by photospheric shear motion. *Astrophys. J.* **607**, 1032–1038.

Hu Y. Q., Li G. Q., Xing X. Y. (2003). Equilibrium and catastrophe of coronal flux ropes in axisymmetrical magnetic field. *J. Geophys. Res.* **108**. doi:10.1029/2002JA009419.

Huba J. D., Gladd N. T., Papadopoulos K. (1977). The lower-hybrid-drift instability as a source of anomalous resistivity for magnetic field line reconnection. *Geophys. Res. Lett.* **4**, 125–128.

Huba J. D., Gladd N. T., Papadopoulos K. (1978). Lower-hybrid-drift wave turbulence in the distant magnetotail. *J. Geophys. Res.* **83**, 5217–5226.

Hudson P. D. (1970). Discontinuities in an anisotropic plasma and their identification in the solar wind. *Planet. Space Sci.* **18**, 1611–1622.

Hughes W. J., Sibeck D. G. (1987). On the 3-dimensional structure of plasmoids. *Geophys. Res. Lett.* **14**, 636–639.

Hundhausen A. J. (1999). Coronal mass ejections. In *The Many Faces of the Sun*, K. T. Strong, J. L. R. Saba, B. M. Haisch and J. T. Schmelz, eds. New York: Springer, pp. 143–200.

Hundhausen A. J., Burkepile J. T., St. Cyr O. C. (1994). Speeds of coronal mass ejections: SMM observations from 1980 and 1984–1989. *J. Geophys. Res.* **99**, 6543–6552.

Hurley K., Boggs S. E., Smith D. M. *et al.* (2005). An exceptionally bright flare from SGR 1806–20 and the origins of short-duration γ-ray bursts. *Nature* **434**, 1098–1103.

Hurricane O. A. (1997). MHD ballooning stability of a sheared plasma sheet. *J. Geophys. Res.* **102**, 19903–19911.

Ichimaru S. (1973). *Basic Principles of Plasma Physics.* Reading, Massachusetts: Benjamin.

Ieda A., Machida S., Mukai T. *et al.* (1998). Statistical analysis of the plasmoid evolution with Geotail observations. *J. Geophys. Res.* **103**, 4453–4465.

Ieda A., Fairfield D. H., Mukai T. *et al.* (2001). Plasmoid ejection and auroral brightenings. *J. Geophys. Res.* **106**, 3845–3858.

Inhester B., Birn J., Hesse M. (1992). The evolution of line-tied coronal arcades including a converging footpoint motion. *Solar Phys.* **138**, 257–281.

Isenberg P. A., Forbes T. G., Démoulin P. (1993). Catastrophic evolution of a force-free flux rope: A model for eruptive flares. *Astrophys. J.* **417**, 368–386.

Jackson J. D. (1998). *Classical Electrodynamics.* New York: John Wiley & Sons.

Jamitzky F., Scholer M. (1995). Steady state magnetic reconnection at high magnetic Reynolds number. *J. Geophys. Res.* **100**, 19277–19286.

Janicke L. (1980). Resistive tearing mode in weakly two-dimensional neutral sheets. *Phys. Fluids* **23**, 1843–1849.

Jeffrey A., Taniuti T. (1964). *Non-linear Wave Propagation.* New York: Academic Press.

Jockers K. (1978). Bifurcation of force-free solar magnetic fields: A numerical approach. *Solar Phys.* **56**, 37–53.

Kamide Y., Baumjohann W., Daglis I. A. *et al.* (1998). Current understanding of magnetic storms: Storm-substorm relationships. *J. Geophys. Res.* **103**, 17705–17728.

Kan J. R. (1973). On the structure of the magnetotail current sheet. *J. Geophys. Res.* **78**, 3773–3781.

Karimabadi H., Pritchett P. L., Daughton W., Krauss-Varban D. (2003). Ion-ion kink instability in the magnetotail: 2. Three-dimensional full particle and hybrid simulations and comparison with observations. *J. Geophys. Res.* **108**. doi:10.1029/2003JA010109.

Karimabadi H., Krauss-Varban D., Huba J. D., Vu H. X. (2004). On magnetic reconnection regimes and associated three-dimensional asymmetries: Hybrid, Hall-less hybrid, and Hall-MHD simulations. *J. Geophys. Res.* **109**. doi:10.1029/2004JA010478.

Karimabadi H., Daughton W., Quest K. B. (2005). Antiparallel versus component merging at the magnetopause: Current bifurcation and intermittent reconnection. *J. Geophys. Res.* **110**. doi:10.1029/2004JA010750.

Karpen J. T., Antiochos S. K., DeVore C. R. (1990). On the formation of current sheets in the solar corona. *Astrophys. J.* **356**, L67–L70.

Kaufmann R. L. (1987). Substorm currents: Growth phase and onset. *J. Geophys. Res.* **92**, 7471–7486.

Kiessling M., Schindler K. (1987). Analytical stability analysis for a two-dimensional self-consistent magnetotail model by use of statistical mechanics. *J. Geophys. Res.* **92**, 5795–5806.

Kiessling M. K.-H. (1989). On the equilibrium statistical mechanics of isothermal classical self-gravitating matter. *J. Stat. Phys.* **55**, 203–257.

Kiessling M. K.-H. (1993). Statistical mechanics of classical particles with logarithmic interactions. *Comm. Pure Appl. Math.* **47**, 27–56.

Kiessling M. K.-H. (1995). Statistical mechanics of weakly dissipative current-carrying plasmas. Habilitationsschrift, Faculty of Physics and Astronomy, Ruhr University, Bochum, Germany.

Kiessling M. K.-H., Krallmann T. (1998). Quasi-neutral Vlasov stability. *Physica Scripta* **T74**, 20–25.

Kiessling M., Brinkmann R. P., Schindler K. (1986). Statistical-mechanics approach to stability of current-carrying plasmas. *Phys. Rev. Lett.* **56**, 143–146.

Kippenhahn R., Schlüter A. (1957). Eine Theorie der solaren Filamente. *Zs. Ap.* **43**, 36–62.

Kivelson M. G., Spence H. E. (1988). On the possibility of quasi-static convection in the quiet magnetotail. *Geophys. Res. Lett.* **15**, 1541–1544.

Kliem B., Titov V. S., Török T. (2004). Formation of current sheets and sigmoidal structure by the kink instability of a magnetic loop. *Astron. Astrophys.* **413**, L23–L26.

Klimas A. J., Baker D. N., Roberts D. A., Fairfield D. H., Büchner J. (1992). A nonlinear dynamical analogue model of geomagnetic activity. *J. Geophys. Res.* **97**, 12253–12266.

Klimas A. J., Vassiliadis D., Baker D. N., Roberts D. A. (1996). The organized nonlinear dynamics of the magnetosphere. *J. Geophys. Res.* **101**, 13089–13113.

Klimas A. J., Valdivia J. A., Vassiliadis D., Baker D. N., Hesse M., Takalo J. (2000). Self-organized criticality in the substorm phenomenon and its relation to localized reconnection in the magnetospheric plasma sheet. *J. Geophys. Res.* **105**, 18765–18780.

Klimchuk J. A., Sturrock P. A. (1989). Force-free magnetic fields: Is there a "loss of equilibrium"? *Astrophys. J.* **345**, 1034–1041.

Kopp A., Schindler K. (1991). Singular perturbation theory applied to magnetohydrostatic equilibria: proof of convergence. *J. Math. Phys.* **32**, 1437–1439.

Kopp R. A., Pneuman G. W. (1976). Magnetic reconnection in the corona and the loop prominence phenomenon. *Solar Phys.* **50**, 85–98.

Koskinen H. E. J., Lopez R. E., Pellinen R. J., Pulkkinen T. I., Baker D. N., Bsinger T. (1993). Pseudobreakup and substorm growth phase in the ionosphere and magnetosphere. *J. Geophys. Res.* **98**, 5801–5813.

Kouveliotou C., Dieters S., Strohmayer T. *et al.* (1998). An X-ray pulsar with a superstrong magnetic field in the soft gamma-ray repeater SGR 1806–20. *Nature* **393**, 235–237.

Krall N. A., Trivelpiece A. W. (1973). *Principles of Plasma Physics*. New York: McGraw-Hill.

Kronberg E. A., Woch J., Krupp N., Lagg A., Khurana K. K., Glassmeier K. (2005). Mass release at Jupiter: Substorm-like processes in the Jovian magnetotail. *J. Geophys. Res.* **110**. doi:10.1029/2004JA010777.

Kropotkin A. P., Trubachev O. O., Schindler K. (2002a). Nonlinear mechanisms for the substorm explosion in the geomagnetic tail. *Geomagnetism and Aeronomy, Russia* **42**, 277–285.

Kropotkin A. P., Trubachev O. O., Schindler K. (2002b). Substorm onset: Fast reconfiguration of the magnetotail caused by explosive growth of the turbulence level. *Geomagnetism and Aeronomy, Russia* **42**, 286–294.

Kulsrud R. M. (2001). Magnetic reconnection: Sweet-Parker versus Petschek. *Earth Planets Space* **53**, 417–422.

Kuznetsova M., Hesse M., Winske D. (1998). Kinetic quasi-viscous and bulk flow inertia effects in collisionless magnetotail reconnection. *J. Geophys. Res.* **103**, 199–214.

Landau L. D., Lifshitz E. M. (1963). *Course of Theoretical Physics*. New York: Pergamon Press.

Lang K. R. (1974). *Astrophysical Formulae*. Berlin: Springer-Verlag.

Lapenta G., Brackbill J. U., Daughton W. S. (2003). The unexpected role of the lower-hybrid drift instability in magnetic reconnection in three dimensions. *Phys. Plasmas* **10**, 1577–1587.

Lau Y. T., Finn J. M. (1990). Three-dimensional kinematic reconnection in the presence of field nulls and closed field lines. *Astrophys. J.* **350**, 672–691.

Laval G., Mercier C., Pellat R. M. (1965). Necessity of the energy principle for magnetostatic stability. *Nuclear Fusion* **5**, 156–158.

Laval G., Pellat R., Vuillemin M. (1966). Instabilités électromagnétiques des plasmas sans collisions. In *Proceedings of the Second Conference on Plasma Physics and Controlled Nuclear Fusion Research*. Vol. 2. Vienna: Intern. Atomic Energy Agency, 259.

Lee D. Y., Wolf R. A. (1992). Is the Earth's magnetotail balloon unstable? *J. Geophys. Res.* **97**, 19251–19257.

Lee L. C., Fu Z. F. (1986). Multiple x-line reconnection. 1. A criterion for the transition from a single x-line to a multiple x-line reconnection. *J. Geophys. Res.* **91**, 6807–6815.

Lembège B., Pellat R. (1982). Stability of a thick twodimensional quasineutral sheet. *Phys. Fluids* **25**, 1995–2004.

Lemon C., Wolf R. A., Hill T. W. *et al.* (2004). Magnetic storm ring current injection modeled with the Rice Convection Model and a self-consistent magnetic field. *Geophys. Res. Lett.* **31**. doi:10.1029/2004GL020914.

Lichtenberg A. J., Liebermann M. A. (1983). *Regular and Stochastic Motion*. New York: Springer.

Lin H. (2003). Measuring coronal magnetic fields with coronal emission line polarimetry. *AGU Fall Meeting Abstracts*. Abstract number SH42D-02.

Lin J. (2001). Theoretical mechanisms for solar eruptions. Ph.D. thesis, University of New Hampshire.

Lin J., Forbes T. G., Isenberg P. A. (2001). Prominence eruptions and coronal mass ejections triggered by newly emerging flux. *J. Geophys. Res.* **106**, 25053–25073.

Linker J. A., Mikić Z. (1995). Disruption of a helmet streamer by photospheric shear. *Astrophys. J.* **438**, L45–L48.

Linker J. A., Mikić Z., Lionello R., Riley P. (2003). Flux cancellation and coronal mass ejections. *Plasma Phys.* **10**, 1971–1978.

Liouville J. (1853). Sur l'équation aux différences partielles $\partial^2 log\lambda/\partial u\partial v \pm \lambda/2/a^2 = 0$. *J. de Math. Pures Appl.* **18**, 71–72.

Litvinenko Y. E., Forbes T. G., Priest E. R. (1996). A strong limitation on the rapidity of flux-pile-up reconnection. *Solar Phys.* **167**, 445–448.

Longcope D. W. (2001). Separator current sheets: Generic features in minimum-energy magnetic fields subject to flux constraints. *Phys. Plasmas* **8**, 5277–5290.

Lopez R. E., Lui A. T. Y. (1990). A multisatellite case study of the expansion of a substorm current wedge in the near-Earth magnetotail. *J. Geophys. Res.* **95**, 8009–8017.

Lopez R. E., Koskinen H. E. J., Pulkkinen T. I., Bösinger T., McEntire R. W., Potemra T. A. (1993). Simultaneous observations of the poleward expansion of substorm electrojet activity and the tailward expansion of current sheet disruption in the near-earth magnetotail. *J. Geophys. Res.* **98**, 9285–9295.

Lottermoser R.-F., Scholer M. (1997). Undriven magnetic reconnection in magnetohydrodynamics and Hall magnetohydrodynamics. *J. Geophys. Res.* **102**, 4875–4892.

Louarn P., Fedorov A., Budnik E. *et al.* (2004). Cluster observations of complex 3D magnetic structures at the magnetopause. *Geophys. Res. Lett.* **31**. doi:10.1029/2004GL020625.

Low B. C. (1977). Evolving force-free magnetic fields. I - The development of the preflare stage. *Astrophys. J.* **212**, 234–242.

Low B. C. (1982). Nonlinear force-free magnetic fields. *Rev. Geophys. Space Phys.* **20**, 145–159.

Low B. C. (1987). Electric current sheet formation in a magnetic field induced by continuous magnetic footpoint displacements. *Astrophys. J.* **323**, 358–367.

Low B. C. (1993). Force-free magnetic fields with singular current-density surfaces. *Astrophys. J.* **409**, 798–808.

Low B. C. (1996). Solar activity and the corona. *Solar Phys.* **167**, 217–265.

Low B. C. (1999). Coronal mass ejections, flares and prominences. In *Solar Wind Nine*, S. R. Habbal, R. Esser, J. V. Hollweg and P. A. Isenberg, eds. Woodbury, N. Y.: AIP, pp. 109–114.

Low B. C. (2001). Coronal mass ejections, magnetic flux ropes, and solar magnetism. *J. Geophys. Res.* **106**, 25141–25164.

Lu E. T. (1995). Avalanches in continuum driven dissipative systems. *Phys. Rev. Lett.* **74**, 2511–2514.

Lui A. T. Y. (1996). Current disruption in the Earth's magnetosphere: Observations and models. *J. Geophys. Res.* **101**, 13067–13088.

Lui A. T. Y. (2004). Potential plasma instabilities for substorm expansion onsets. *Space Sci. Rev.* **113**, 127–206.

Lüst R., Schlüter A. (1957). Axialsymmetrische magnetohydrodynamische Gleichgewichtskonfigurationen. *Z. Naturforschung* **12a**, 850–854.

MacNeice P., Antiochos S. K., Phillips A., Spicer D. S., DeVore C. R., Olson K. (2004). A numerical study of the breakout model for coronal mass ejection initiation. *Astrophys. J.* **614**, 1028–1041.

Magara T., Mineshige S., Yokoyama T., Shibata K. (1996). Numerical simulation of magnetic reconnection in eruptive flares. *Astrophys. J.* **466**, 1054–1066.

Mandrini C. H., Démoulin P., Klimchuk J. A. (2000). Magnetic field and plasma scaling laws: Their implications for coronal heating models. *Astrophys. J.* **530**, 999–1015.

Martens P. C. H., Young A. (1990). Neutral beams in two-ribbon flares and in the geomagnetic tail. *Astrophys. J. Supp.* **73**, 333–342.

McPherron R. L. (1970). Growth phase of magnetospheric substorms. *J. Geophys. Res.* **75**, 5592–5599.

McPherron R. L., Russell C. T., Aubry M. (1973). Satellite studies of magnetospheric substorms on August 15, 1968, IX. Phenomenological model for substorms. *J. Geophys. Res.* **78**, 3131–3149.

McPherron R. L., Nishida A., Russell C. T. (1987). Is near-Earth current sheet thinning the cause of auroral substorm onset? In *Quantitative Modeling of Magnetosphere-Ionosphere Coupling Processes*, Y. Kamide and R. A. Wolf, eds. Kyoto: Kyoto Sangyo Univ., p. 252.

Melrose D. B. (1986). *Instabilities in Space and Laboratory Plasmas*. Cambridge: Cambridge University Press.

Mende S. B., Frey H. U., Carlson C. W. *et al.* (2002). IMAGE and FAST observations of substorm recovery phase aurora. *Geophys. Res. Lett.* **29**. doi:10.1029/2001GL013027.

Mikhailovskii A. B. (1974). *Theory of Plasma Instabilities*. New York: Consultants Bureau.

Mikić Z., Linker J. A. (1994). Disruption of coronal magnetic field arcades. *Astrophys. J.* **430**, 898–912.

Mikić Z., Barnes D. C., Schnack D. D. (1988). Dynamical evolution of a solar coronal magnetic field arcade. *Astrophys. J.* **328**, 830–847.

Miller J. A., Cargill P. J., Emslie A. G. *et al.* (1997). Critical issues for understanding particle acceleration in impulsive solar flares. *J. Geophys. Res.* **102**, 14631–14659.

Mitchell D. G., Williams D. J., Huang C. Y., Frank L. A., Russell C. T. (1990). Current carriers in the near-Earth cross-tail current sheet during substorm growth phase. *Geophys. Res. Lett.* **17**, 583–586.

Miura A. (2000). Conditions for the validity of the incompressible assumption for the ballooning instability in the long-thin magnetospheric equilibrium. *J. Geophys. Res.* **105**, 18793–18806.

Miura A., Ohtani S., Tamao T. (1989). Ballooning instability and structure of diamagnetic hydromagnetic waves in a model magnetosphere. *J. Geophys. Res.* **94**, 15231–15242.

Moffatt H. K. (1978). *Magnetic Field Generation in Electrically Conducting Fluids*. Cambridge: Cambridge University Press.

Mok Y., van Hoven G. (1995). The solar-surface boundary conditions of coronal magnetic loops. *Solar Phys.* **161**, 67–81.

Moldwin M. B., Hughes W. J. (1992). On the formation and evolution of plasmoids: A survey of ISEE 3 Geotail data. *J. Geophys. Res.* **97**, 19259–19282.

Moon G., Ahn B., Kamide Y., Reeves G. D. (2004). Correlation between particle injections observed at geosynchronous orbit and the *Dst* index during geomagnetic storms. *J. Geophys. Res.* **109**. doi:10.1029/2004JA010390.

Moore R. L., Roumeliotis G. (1992). Triggering of eruptive flares–destabilization of the preflare magnetic field configuration. In *Eruptive Solar Flares*, Z. Švestka, B. V. Jackson and M. E. Machado, eds. Vol. 399. New York: Springer-Verlag, pp. 69–78.

Morse P. M., Feshbach H. (1953). *Methods of Theoretical Physics*. New York: McGraw-Hill.

Nakai H., Kamide Y. (2000). Substorm currents associated with magnetotail magnetic dipolarization: GEOTAIL observations. *J. Geophys. Res.* **105**, 18781–18792.

Nakamura M., Scholer M. (2000). Structure of the magnetopause reconnection layer and of flux transfer events: Ion kinetic effects. *J. Geophys. Res.* **105**, 23179–23191.

Ness N. F. (1969). Direct measurements of interplanetary magnetic fields and plasmas. *Annals of the IQSY* **4**, 88–109.

Neukirch T. (1997). Nonlinear self-consistent three-dimensional arcade-like solutions of the magnetohydrostatic equations. *Astron. Astrophys.* **325**, 847–856.

Neukirch T., Giuliani P., Wood P. (2006). Particle acceleration in flares, theory. In *Reconnection of Magnetic Fields*, J. Birn and E. R. Priest, eds. Cambridge: Cambridge University Press. In press.

Newcomb W. A. (1958). Motion of magnetic lines of force. *Ann. Phys.* **3**, 347–385.

Nishida A. (1983). Reconnection in the Jovian magnetosphere. *Geophys. Res. Lett.* **10**, 451–454.

Nishida A., Hones E. W. (1974). Association of plasma sheet thinning with neutral line formation in the magnetotail. *J. Geophys. Res.* **79**, 535.

Nishida A., Nagayama N. (1973). Synoptic survey of the neutral line in the magnetotail during the substorm expansion phase. *J. Geophys. Res.* **78**, 3782–3789.

Northrop T. G. (1963). *The Adiabatic Motion of Charged Particles.* New York: John Wiley & Sons.

Nötzel A., Schindler K., Birn J. (1985). On the cause of approximate pressure isotropy in the quiet near-Earth plasma sheet. *J. Geophys. Res.* **90**, 8293–8300.

Ogino T., Walker R. J., Ashour-Abdalla M. (1990). Magnetic flux ropes in 3-dimensional MHD simulations. In *Physics of Magnetic Flux Ropes*, C. T. Russell, E. R. Priest and L. C. Lee, eds. Washington, DC: AGU, pp. 669–678.

Ohtani S.-I., Raeder J. (2004). Tail current surge: New insights from a global MHD simulation and comparison with satellite observations. *J. Geophys. Res.* **109**. doi:10.1029/2002JA009750.

Opgenoorth H. J., Persson M. A. L., Pulkkinen T. I., Pellinen R. J. (1994). Recovery phase of magnetospheric substorms and its association with morning-sector aurora. *J. Geophys. Res.* **99**, 4115–4129.

Otto A. (1987). Zur linearen und nichtlinearen Analyse resistiver Instabilitätsprozesse in schwach zwei-dimensionalen Gleichgewichten. Ph.D. thesis, Ruhr-Universität Bochum, Germany.

Otto A. (1991). The resistive tearing instability for generalized resistivity models - Applications. *Phys. Fluids B* **3**, 1746–1754.

Otto A. (1995). Forced three-dimensional magnetic reconnection due to linkage of magnetic flux tubes. *J. Geophys. Res.* **100**, 11863–11874.

Otto A. (2001). Geospace environment modeling (GEM) magnetic reconnection challenge: MHD and Hall MHD - constant and current dependent resistivity models. *J. Geophys. Res.* **106**, 3751–3757.

Otto A., Fairfield D. H. (2000). Kelvin-Helmholtz instability at the magnetotail boundary: MHD simulation and comparison with Geotail observations. *J. Geophys. Res.* **105**, 21175–21190.

Paris R. B. (1987). The resistive tearing mode in a weakly two-dimensional sheet pinch. *Phys. Fluids* **30**, 102–107.

Parker E. N. (1963a). *Interplanetary Dynamical Processes.* New York: Interscience.

Parker E. N. (1963b). The solar flare phenomenon and the theory of reconnection and annihilation of magnetic fields. *Astrophys. J. Suppl.* **8**, 177–211.

Parker E. N. (1972). Topological dissipation and the small-scale fields in turbulent gases. *Astrophys. J.* **174**, 499–510.

Parker E. N. (1979). *Cosmical Magnetic Fields.* Oxford: Clarendon Press.

Parker E. N. (1983). Magnetic neutral sheets in evolving fields. II. Formation of the solar corona. *Astrophys. J.* **264**, 642–647.

Parker E. N. (1988). Nanoflares and the solar X-ray corona. *Astrophys. J.* **330**, 474–479.

Parker E. N. (1994). *Spontaneous Current Sheets in Magnetic Fields: With applications to stellar X-rays.* New York: Oxford University Press.

Parnell C. E., Smith J., Neukirch T., Priest E. R. (1996). The structure of three-dimensional magnetic neutral points. *Phys. Plasmas* **3**, 759–770.

Partamies N., Amm O., Kauristie K., Pulkkinen T. I., Tanskanen E. (2003). A pseudo-breakup observation: Localized current wedge across the post-midnight auroral oval. *J. Geophys. Res.* **108**. doi:10.1029/2002JA009276.

Paschmann G., Sonnerup B. U. Ö., Papamastorakis I. *et al.* (1979). Plasma acceleration at the Earth's magnetopause: Evidence for reconnection. *Nature* **282**, 243–246.

Paterson W. R., Frank L. A., Kokubun S., Yamamoto T. (1998). Geotail survey of ion flow in the plasma sheet: Observations between 10 and 50 R_E. *J. Geophys. Res.* **103**, 11811–11826.

Paterson W. R., Frank L. A., Kokubun S., Yamamoto T. (1999). Reply. *J. Geophys. Res.* **104**, 17527–17529.

Pellat R., Coroniti F. V., Pritchett P. L. (1991). Does ion tearing exist? *Geophys. Res. Lett.* **18**, 143–146.

Perreault P. D., Akasofu S. (1978). A study of geomagnetic storms. *Geophys. J. Roy. Astron. Soc.* **54**, 547–573.

Petschek H. E. (1964). Magnetic field annihilation. In *The Physics of Solar Flares*, W. N. Hess, ed. Washington, DC: NASA, pp. 425–440.

Phillips K. J. H. (2004). The solar flare 3.8-10 keV X-ray spectrum. *Astrophys. J.* **605**, 921–930.

Platt U., Neukirch T. (1994). Theoretical study of onset conditions for solar eruptive processes: Influence of the boundaries. *Solar Phys.* **153**, 287–306.

Pneuman G. W., Kopp R. A. (1971). Gas-magnetic field: Interactions in the solar corona. *Solar Phys.* **18**, 258–270.

Pontin D. I., Hornig G., Priest E. R. (2004). Kinematic reconnection at a magnetic null point: Spine reconnection. *Geophys. Astrophys. Fluid Dynamics* **98**, 407–428.

Pontin D. I., Hornig G., Priest E. R. (2005). Kinematic reconnection at a magnetic null point: Fan aligned current. *Geophys. Astrophys. Fluid Dynamics* **99**, 77–93.

Pontius, Jr. D. H., Wolf R. A. (1990). Transient flux tubes in the terrestrial magnetosphere. *Geophys. Res. Lett.* **17**, 49–52.

Priest E. R. (1981). Current sheets. In *Solar Flare Magnetohydrodynamics*, E. R. Priest, ed. New York: Gordon and Breach, pp. 139–215.

Priest E. R. (1982). *Solar Magnetohydrodynamics*. Dordrecht: D. Reidel.

Priest E. R., Démoulin P. (1995). Three-dimensional reconnection without null points. *J. Geophys. Res.* **100**, 23443–23463.

Priest E. R., Forbes T. G. (1986). New models for fast steady-state magnetic reconnection. *J. Geophys. Res.* **91**, 5579–5588.

Priest E. R., Forbes T. G. (1992a). Does fast magnetic reconnection exist? *J. Geophys. Res.* **97**, 16757–16772.

Priest E. R., Forbes T. G. (1992b). Magnetic flipping - reconnection in three dimensions without null points. *J. Geophys. Res.* **97**, 1521–1531.

Priest E. R., Forbes T. G. (2000). *Magnetic Reconnection*. Cambridge: Cambridge University Press.

Priest E. R., Milne A. M. (1980). Force-free magnetic arcades relevant to two-ribbon solar flares. *Solar Phys.* **65**, 315–346.

Priest E. R., Titov V. S. (1996). Magnetic reconnection at three-dimensional null points. *Phil. Trans. Roy. Soc. Lond.* A **354**, 2951–2992.

Priest E. R., Titov V. S., Grundy R. E., Hood A. W. (2000). Exact solutions for reconnective magnetic annihilation. *Proc. Roy. Soc. Lond.* A **456**, 1821–1849.

Priest E. R., Longcope D. W., Heyvaerts J. (2005). Coronal heating at separators and separatrices. *Astrophys. J.* **624**, 1057–1071.

Pritchett P. L. (1994). Effect of electron dynamics on collisionless reconnection in two-dimensional magnetotail equilibria. *J. Geophys. Res.* **99**, 5935–5942.

Pritchett P. L., Coroniti F. V., Pellat R., Karimabadi H. (1991). Collisionless reconnection in two-dimensional magnetotail equilibria. *J. Geophys. Res.* **96**, 11523–11538.

Pulkkinen T. I., Baker D. N., Mitchell D. G., McPherron R. L., Huang C. Y., Frank L. A. (1994). Thin current sheets in the magnetotail during substorms: CDAW 6 revisited. *J. Geophys. Res.* **99**, 5793–5803.

Pulkkinen T. I., Baker D. N., Pellinen R. J., Murphree J. S., Frank L. A. (1995). Mapping of the auroral oval and individual arcs during substorms. *J. Geophys. Res.* **100**, 21987–21994.

Quest K. B., Karimabadi H., Brittnacher M. (1996). Consequences of particle conservation along a flux surface for magnetotail tearing. *J. Geophys. Res.* **101**, 179–183.

Raeder J., McPherron R. L., Frank L. A. *et al.* (2001). Global simulation of the Geospace Environment Modeling substorm challenge event. *J. Geophys. Res.* **106**, 381–395.

Rastätter L., Voge A., Schindler K. (1994). On current sheets in two-dimensional ideal magnetohydrodynamics caused by pressure perturbations. *Phys. Plasmas* **1**, 3414–3424.

Reeves K. K., Forbes T. G. (2005). Predicted light curves for a model of solar eruptions. *Astrophys. J.* **610**, 1133–1147.

Reeves G. D., Henderson M. G. (2001). The storm-substorm relationship: Ion injections in geosynchronous measurements and composite energetic neutral atom images. *J. Geophys. Res.* **106**, 5833–5844.

Reeves G. D., Fritz T. A., Cayton T. E., Belian R. D. (1990). Multi-satellite measurements of the substorm injection region. *Geophys. Res. Lett.* **17**, 2015–2018.

Retinò A., Bavassano Cattaneo M. B., Marcucci M. F. *et al.* (2005). Cluster multispacecraft observations at the high-latitude duskside magnetopause: Implications for continuous and component magnetic reconnection. *Ann. Geophys.* **23**, 461–473.

Ricci P., Lapenta G., Brackbill J. U. (2002). GEM reconnection challenge: Implicit kinetic simulations with the physical mass ratio. *Geophys. Res. Lett.* **29**. doi:10.1029/2002GL015314.

Rosner R., Knobloch E. (1982). On perturbations of magnetic field configurations. *Astrophys. J.* **262**, 349–357.

Rostoker G. (1983). Triggering of expansive phase intensifications of magnetospheric substorms by northward turnings of the interplanetary magnetic field. *J. Geophys. Res.* **88**, 6981–6993.

Roussev I. I., Forbes T. G., Gombosi T. I., Sokolov I. V., DeZeeuw D. L., Birn J. (2003). A three-dimensional flux-rope model for coronal mass ejections based on a loss of equilibrium. *Astrophys. J.* **588**, L45–L48.

Roussev I. I., Sokolov I. V., Forbes T. G. *et al.* (2004). A numerical model of a coronal mass ejection: Shock development with implications for the acceleration of GeV protons. *Astrophys. J.* **605**, L73–L77.

Roux A., Perraut S., Robert P. *et al.* (1991). Plasma sheet instability related to the westward traveling surge. *J. Geophys. Res.* **96**, 17697–17714.

Runov A., Nakamura R., Baumjohann W. *et al.* (2003). Current sheet structure near magnetic X-line observed by Cluster. *Geophys. Res. Lett.* **30**. doi:10.1029/2002GL016730.

Russell C. T., Elphic R. C. (1978). Initial ISEE magnetometer results: Magnetopause observations. *Space Sci. Rev.* **22**, 681–715.

Russell C. T., Elphic R. C. (1979). ISEE observations of flux transfer events at the dayside magnetopause. *Geophys. Res. Lett.* **6**, 33–36.

Russell C. T., McPherron R. L. (1973). The magnetotail and substorms. *Space Sci. Rev.* **15**, 205–266.

Russell C. T., Khurana K. K., Kivelson M. G., Huddleston D. E. (2000). Substorms at Jupiter: Galileo observations of transient reconnection in the near tail. *Adv. Space Sci.* **26**, 1499–1504.

Sagdeev R. Z., Galeev A. A. (1969). *Nonlinear Plasma Theory.* New York: Benjamin.

Schindler K. (1966). A variational principle for one-dimensional plasmas. In *Proceedings of the Seventh International Conference on Phenomena in Ionized Gases, Vol. II.* Belgrade: Gradevinska Knjiga Publishing House, p. 736.

Schindler K. (1972). A self-consistent theory of the tail of the magnetosphere. In *Earth's Magnetospheric Processes*, B. M. McCormac, ed. Dordrecht: D. Reidel, pp. 200–209.

Schindler K. (1974). A theory of the substorm mechanism. *J. Geophys. Res.* **79**, 2803–2810.

Schindler K., Birn J. (1982). Self-consistent theory of time-dependent convection in the Earth's magnetotail. *J. Geophys. Res.* **87**, 2263–2275.

Schindler K., Birn J. (1987). On the generation of field-aligned plasma flow at the boundary of the plasma sheet. *J. Geophys. Res.* **92**, 95–107.

Schindler K., Birn J. (1993). On the cause of thin current sheets in the near-Earth magnetotail and their possible significance for magnetospheric substorms. *J. Geophys. Res.* **98**, 15477–15485.

Schindler K., Birn J. (1999). Thin current sheets and magnetotail dynamics. *J. Geophys. Res.* **104**, 25001–25010.

Schindler K., Birn J. (2002). Models of two-dimensional embedded thin current sheets from Vlasov theory. *J. Geophys. Res.* **107**. doi:10.1029/2001JA000304.

Schindler K., Birn J. (2004). MHD stability of magnetotail equilibria including a background pressure. *J. Geophys. Res.* **109**. doi:10.1029/2004JA010537.

Schindler K., Goldstein H. (1983). A nonlinear kinetic energy principle for two-dimensional collision-free plasmas. *Phys. Fluids* **26**, 2222–2226.

Schindler K., Pfirsch D., Wobig H. (1973). Stability of two-dimensional collison-free plasmas. *Plasma Phys.* **15**, 1165–1184.

Schindler K., Birn J., Janicke L. (1983). Stability of two-dimensional pre-flare structures. *Solar Phys.* **87**, 103–133.

Schindler K., Hesse M., Birn J. (1988). General magnetic reconnection, parallel electric fields, and helicity. *J. Geophys. Res.* **93**, 5547–5557.

Schindler K., Birn J., Hesse M. (1991). Magnetic field-aligned electric potentials in nonideal plasma flows. *Astrophys. J.* **380**, 293–301.

Schlickeiser R. (2003). *Cosmic Ray Astrophysics*. Berlin: Springer-Verlag.

Schödel R., Baumjohann W., Nakamura R., Sergeev V. A., Nakamura R., Mukai V. A. T. (2001). Rapid flux transport in the central plasma sheet. *J. Geophys. Res.* **106**, 301–313.

Schröer A., Neukirch T., Kiessling M., Hesse M., Schindler K. (1994). Numerical bifurcation study of a nonlinear current sheet model. *Phys. Plasmas* **1**, 213–215.

Scudder J. D., Mozer F. S. (2005). Electron demagnetization and collisionless magnetic reconnection in $\beta_e \ll 1$ plasmas. *Phys. Plasmas* **12**. DOI:10.1063/1.2046887.

Scudder J. D., Mozer F. S., Maynard N. C., Russell C. T. (2002). Fingerprints of collisionless reconnection at the separator, I. Ambipolar-Hall signatures. *J. Geophys. Res.* **107**. doi:10.1029/2001JA000126.

Sergeev V. A., Tanskanen P., Mursula K., Korth A., Elphic R. C. (1990). Current sheet thickness in the near-Earth plasma sheet during substorm growth phase. *J. Geophys. Res.* **95**, 3819–3828.

Sergeev V. A., Pellinen R., Pulkkinen T. (1996). Steady magnetospheric convection: A review of recent results. *Space Sci. Rev.* **75**, 551–604.

Sergeev V. A., Kubyshkina M. V., Liou K. *et al.* (2001). Substorm and convection bay compared: auroral and magnetotail dynamics during convection bay. *J. Geophys. Res.* **106**, 18843–18855.

Sergeev V. A., Runov A., Baumjohann W. *et al.* (2004). Orientation and propagation of current sheet oscillations. *Geophys. Res. Lett.* **31**. doi:10.1029/2003GL019346.

Shafranov V. D. (1958). Magnetohydrodynamical equilibrium configurations. *Sov. Phys. JETP* **6**, 545–554.

Shay M. A., Drake J. F., Denton R. E., Biskamp D. (1998). Structure of the dissipation region during collisionless magnetic reconnection. *J. Geophys. Res.* **103**, 9165–9176.

Shay M. A., Drake J. F., Rogers B. N. (1999). The scaling of collisionless magnetic reconnection for large systems. *Geophys. Res. Lett.* **26**, 2163–2166.

Shibata K., Yokoyama T. (2002). A Hertzsprung-Russell-like diagram for solar/stellar flares and corona: Emission measure versus temperature diagram. *Astrophys. J.* **577**, 422–432.

Shuo J., Wolf R. A. (2003). Double-adiabatic MHD theory for motion of a thin magnetic filament and possible implications for bursty bulk flows. *J. Geophys. Res.* **108**. doi:10.1029/2002JA009655.

Silin I., Büchner J. (2005). Small-scale reconnection due to lower-hybrid drift instability in current sheets with sheared fields. *Phys. Plasmas* **12**, 35–50.

Siscoe G. L. (1972). Consequences of an isotropic static plasma sheet in models of the geomagnetic tail. *Planet. Space Sci.* **20**, 937–953.

Siscoe G. L., Sanchez E. (1987). An MHD model for the complete open magnetotail boundary. *J. Geophys. Res.* **92**, 7405–7412.

Siscoe G. L., Erickson G. M., Sonnerup B. U. Ö. *et al.* (2001). Global role of $E_\parallel$ in magnetopause reconnection: An explicit example. *J. Geophys. Res.* **106**, 13015–13022.

Sitnov M. I., Malova H. V., Sharma A. S. (1998). Role of the temperature ratio in the linear stability of the quasi-neutral sheet tearing mode. *Geophys. Res. Lett.* **25**, 269–272.

Slavin J. A., Baker D. N., Fairfield D. H. *et al.* (1989). CDAW 8 observations of plasmoids in the geotail: An assessment. *J. Geophys. Res.* **94**, 15153–15175.

Slavin J. A., Fairfield D. H., Lepping R. P. *et al.* (1997). WIND, GEOTAIL, and GOES 9 observations of magnetic field dipolarization and bursty bulk flows in the near-tail. *Geophys. Res. Lett.* **24**, 971–974.

Slinker S. P., Fedder J. A., Lyon J. G. (1995). Plasmoid formation and evolution in a numerical simulation of a substorm. *Geophys. Res. Lett.* **22**, 859–862.

Solanki S. K., Unruh Y. C. (2004). Spot sizes on sun-like stars. *Mon. Not. Roy. Astron. Soc.* **348**, 307–315.

Solov'ev L. S. (1975). Hydromagnetic stability of closed plasma configurations. In *Reviews of Plasma Physics*, M. A. Leontovich, ed. Vol. 6. New York: Consultants Bureau, p. 239.

Song P., DeZeeuw D. L., Gombosi T. I., Groth C. P. T., Powell K. G. (1999).

A numerical study of solar wind–magnetosphere interaction for northward interplanetary magnetic field. *J. Geophys. Res.* **104**, 28361–28378.

Sonnerup B. U. Ö. (1970). Magnetic field reconnection in a highly conducting incompressible fluid. *J. Plasma Phys.* **4**, 161–174.

Sonnerup B. U. Ö. (1971). Adiabatic particle orbits in a magnetic null sheet. *J. Geophys. Res.* **76**, 8211–8222.

Sonnerup B. U. Ö. (1974). Magnetopause reconnection rate. *J. Geophys. Res.* **79**, 1546–1549.

Sonnerup B. U. Ö. (1988). On the theory of steady state reconnection. *Computer Phys. Communications* **49**, 143–159.

Sonnerup B. U. Ö., Priest E. R. (1975). Resistive MHD stagnation-point flows at a current sheet. *J. Plasma Phys.* **14**, 283–294.

Sonnerup B. U. Ö., Paschmann G., Papamastorakis I. *et al.* (1981). Evidence for magnetic field reconnection at the Earth's magnetopause. *J. Geophys. Res.* **86**, 10049–10067.

Sonnerup B. U. Ö., Hasegawa H., Paschmann G. (2004). Anatomy of a flux transfer event seen by Cluster. *Geophys. Res. Lett.* **31**. doi:10.1029/2004GL020134.

Speiser T. W. (1965). Particle trajectories in model current sheets. 1. Analytical solutions. *J. Geophys. Res.* **70**, 4219–4226.

Spitzer, Jr. L. (1962). *Physics of Fully Ionized Gases*. New York: Interscience.

Spreiter J. R., Summers A. L., Alskne A. Y. (1966). Hydromagnetic flow around the magnetosphere. *Planet. Space Sci.* **14**, 223–250.

Sterling A. C., Moore R. L. (2005). Slow-rise and fast-rise phases of an erupting solar filament, and flare emission onset. *Astrophys. J.* **630**, 1148–1159.

Stern D. P. (1966). The motion of magnetic field lines. *Space Sci. Rev.* **6**, 147–173.

Stern D. P. (1970). Euler potentials. *Am. J. Phys.* **38**, 494–501.

Stiles G. S., Hones E. W., Bame S. J., Asbridge J. R. (1978). Plasma sheet pressure anisotropies. *J. Geophys. Res.* **83**, 3166–3172.

Strong K. T., Saba J. L. R., Haisch B. M., Schmelz J. T. (1999). *The Many Faces of the Sun: A summary of the results from NASA's Solar Maximum Mission*. New York: Springer.

Sturrock P. A. (1989). The role of eruption in solar flares. *Solar Phys.* **121**, 387–397.

Sturrock P. A. (1991). Maximum energy of semi-infinite magnetic field configurations. *Astrophys. J.* **380**, 655–659.

Sturrock P. A. (1994). *Plasma Physics*. Cambridge: Cambridge University Press.

Švestka Z. (2001). Varieties of coronal mass ejections and their relation to flares. *Space Sci. Rev.* **95**, 135–146.

Sweet P. A. (1958). The production of high-energy particles in solar flares. *Nuovo Cimento Suppl.* **8**, 188–196.

Syrovatskii S. I. (1971). Formation of current sheets in a plasma with a frozen-in strong magnetic field. *Sov. Phys. JETP* **33**, 933–940.

Takalo J., Timonen J., Koskinen H. (1994). Properties of *AE* data and bicolored noise. *J. Geophys. Res.* **99**, 13239–13249.

Tandberg-Hanssen E. (1994). *The Nature of Solar Prominences*. Dordrecht: Kluwer.

Tandberg-Hanssen E., Emslie A. G. (1988). *The Physics of Solar Flares*. Cambridge: Cambridge University Press.

Tange T., Ichimaru S. (1974). Theory of anomalous resistivity and turbulent heating in plasmas. *J. Phys. Soc. Japan* **36**, 1437–1445.

Tanskanen E. I., Slavin J. A., Fairfield D. H. *et al.* (2005). Magnetotail response to prolonged southward IMF b_z intervals: Loading, unloading, and continuous magnetospheric dissipation. *J. Geophys. Res.* **110**. doi:10.1029/2004JA010561.

Tassi E., Titov V. S., Hornig G. (2002). Exact solutions for magnetic annihilation in curvilinear geometry. *Phys. Lett. A* **302**, 313–317.

Tassi E., Titov V. S., Hornig G. (2003). New classes of exact solutions for magnetic reconnective annihilation. *Phys. Lett. A* **315**, 382–388.

Taylor J. B. (1974). Relaxation of toroidal plasma and generation of reverse magnetic fields. *Phys. Rev. Lett.* **33**, 1139–1141.

Thompson C., Lyutikov M., Kulkarni S. (2002). Electrodynamics of magnetars: Implications for the persistent X-ray emission and spin-down of the soft gamma repeaters and anomalous X-ray pulsars. *Astrophys. J.* **574**, 332–355.

Titov V. S., Démoulin P. (1999). Basic topology of twisted magnetic configurations in solar flares. *Astron. Astrophys.* **351**, 707–720.

Treumann R. A., Baumjohann W. (1997). *Advanced Space Plasma Physics*. London: Imperial College Press.

Trubnikov B. A. (1965). Particle interactions in a fully ionized plasma. In *Reviews of Plasma Physics, Vol. 1*, M. A. Leontovich, ed. New York: Consultants Bureau, p. 105.

Tsinganos K. C. (1981). Magnetohydrodynamic equilibrium. I - Exact solutions of the equations. *Astrophys. J.* **245**, 764–782.

Tsinganos K. C. (1982). Magnetohydrodynamic equilibrium. IV - Nonequilibrium of nonsymmetric hydrodynamic topologies. *Astrophys. J.* **259**, 832–843.

Tsurutani B. T., Lakhina G. S., Ho C. M. *et al.* (1998). Broadband plasma waves observed in the polar cap boundary layer: Polar. *J. Geophys. Res.* **103**, 17351–17366.

Turkmani R., Vlahos L., Galsgaard K., Cargill P. J., Isliker H. (2005). Particle acceleration in stressed coronal magnetic fields. *Astrophys. J.* **620**, L59–L62.

Ugai M. (1982). Spontaneously developing magnetic reconnections in a current sheet system under different sets of boundary conditions. *Phys. Fluids* **25**, 1027–1036.

Ugai M. (1992). Computer studies on development of the fast reconnection mechanism for different resistivity models. *Phys. Fluids B* **4**, 2953–2963.

Van Ballegooijen A. A. (1985). Electric currents in the solar corona and the existence of magnetostatic equilibrium. *Astrophys. J.* **298**, 421–430.

Van Ballegooijen A. A. (1986). Cascade of magnetic energy as a mechanism of coronal heating. *Astrophys. J.* **311**, 1001–1014.

Van Tend W., Kuperus M. (1978). The development of coronal electric current systems in active regions and their relation to filaments and flares. *Solar Phys.* **59**, 115–127.

Vasyliunas V. M. (1970). Mathematical models of magnetospheric convection and its coupling to the ionosphere. In *Particles and Field in the Magnetosphere*, B. M. McCormac and A. Renzini, eds. Dordrecht: D. Reidel, pp. 60–74.

Vasyliunas V. M. (1972). Nonuniqueness of magnetic field line motion. *J. Geophys. Res.* **77**, 6271–6274.

Vasyliunas V. M. (1975). Theoretical models of magnetic field line merging. *Rev. Geophys. Space Phys.* **13**, 303–336.

Vasyliunas V. M. (1976). An overview of magnetospheric dynamics. In *Magnetospheric Particles and Fields*, B. M. McCormac, ed. Dordrecht: D. Reidel, pp. 99–110.

Vekstein G., Katsukawa Y. (2000). Scaling laws for a nanoflare-heated solar corona. *Astrophys. J.* **541**, 1096–1103.

Voge A., Otto A., Schindler K. (1994). Nonlinear current-sheet formation in ideal plasmas. *J. Geophys. Res.* **99**, 21241–21248.

Voigt G. (1986). Magnetospheric equilibrium configurations and slow adiabatic convection. In *Solar Wind and Magnetosphere Coupling*, Y. Kamide and J. A. Slavin, eds. Tokyo: Terra Scientific, pp. 233–273.

Walthour D. W., Gosling J. T., Sonnerup B. U. Ö., Russell C. T. (1994). Observation of anomalous slow-mode shock reconnection in the dayside magnetosphere. *J. Geophys. Res.* **99**, 23705–23722.

Wang C.-P., Lyons L. R., Chen M. W., Wolf R. A. (2001). Modeling the quiet time inner plasma sheet protons. *J. Geophys. Res.* **106**, 6161–6178.

Wang C.-P., Lyons L. R., Chen M. W., Toffoletto F. R. (2004). Modeling the transition of the inner plasma sheet from weak to enhanced convection. *J. Geophys. Res.* **109**. doi:10.1029/2004JA010591.

Weitzner H. (1983). Linear wave propagation in ideal magnetohydrodynamics. In *Basic Plasma Physics I*, A. A. Galeev and R. N. Sudan, eds. Amsterdam: North-Holland Pub. Co., pp. 201–242.

Weygand J. M., Kivelson M. G., Khurana K. K. *et al.* (2005). Plasma sheet turbulence observed by Cluster II. *J. Geophys. Res.* **110**. doi:10.1029/2004JA010581.

Whang Y. C. (2004). Theory and observation of double discontinuities. *Nonlinear Processes in Geophysics* **11**, 259–266.

White R. B., Monticello D. A., Rosenbluth M. N., Wadell B. V. (1977). Saturation of the tearing mode. *Phys. Fluids* **20**, 800–805.

Whittaker E. T., Watson G. N. (1973). *A Course of Modern Analysis*. New York: Cambridge University Press.

Wiechen H., Büchner J., Otto A. (1997). Driven reconnection in the near-earth plasma sheet. *Adv. Space Res.* **19**, 1939–1942.

Wiegelmann T., Schindler K. (1995). Formation of thin current sheets in a quasistatic magnetotail model. *Geophys. Res. Lett.* **22**, 2057–2060.

Wild J. A., Milan S. E., Cowley S. W. H. *et al.* (2005). Simultaneous in-situ observations of the signatures of dayside reconnection at the high- and low-latitude magnetopause. *Annales Geophysicae* **23**, 445–460.

Wolf R. A. (1983). The quasi-static (slow-flow) region of the magnetosphere. In *Solar-Terrestrial Physics: Principles and Theoretical Foundations*, R. L. Carovillano and J. M. Forbes, eds. Dordrecht: D. Reidel, pp. 303–368.

Wolf R. A., Freeman J. W., Hausman B. A., Spiro R. W., Hilmer R. V., Lambour R. L. (1997). Modeling convection effects in magnetic storms. In *Magnetic Storms*, B. T. Tsurutani, W. D. Gonzales, Y. Kamide and J. K. Arballo, eds. Geophys. Monogr. Ser., Vol. 98. Washington, DC: AGU, pp. 161–172.

Woltjer L. (1958). A theorem on force-free magnetic fields. *Proc. Nat. Acad. Sci. USA* **44**, 489–491.

Wood P. D., Neukirch T. (2005). Electron acceleration in reconnecting current sheets. *Solar Phys.* **226**, 73–95.

Wu C. C. (1987). On MHD intermediate shocks. *Geophys. Res. Lett.* **14**, 668–671.

Yan M., Lee L. C., Priest E. R. (1992). Fast magnetic reconnection with small shock angles. *J. Geophys. Res.* **97**, 8277–8293.

Yeh T., Axford W. I. (1970). On the reconnection of magnetic field lines in conducting fluids. *J. Plasma Phys.* **4**, 207–229.

Yokoyama T., Shibata K. (2001). Magnetohydrodynamic simulation of a solar flare with chromospheric evaporation effect based on the magnetic reconnection model. *Astrophys. J.* **549**, 1160–1174.

Zhang M., Low B. C. (2002). Magnetic flux emergence into the solar corona. II. Global magnetic fields with current sheets. *Astrophys. J.* **576**, 1005–1017.

Zhang Y. Z., Hu Y. Q., Wang J. X. (2005). Double catastrophe of coronal flux rope in quadrupolar magnetic field. *Astrophys. J.* **626**, 1096–1101.

Zhu P., Bhattacharjee A., Ma Z. W. (2003). Hall magnetohydrodynamic ballooning instability in the magnetotail. *Phys. Plasmas* **10**, 249–258.

Zubov V. I. (1964). *Methods of A. M. Lyapunov and Their Application.* Groningen: P. Noordhoff.

Zweibel E. G. (1985). Application of the MHD energy principle to magnetostatic atmospheres. *Geophys. Astrophys. Fluid Dynamics* **32**, 317–331.

Zweibel E. G., Boozer A. H. (1985). Evolution of twisted magnetic fields. *Astrophys. J.* **295**, 642–647.

Zweibel E. G., Li H.-S. (1987). The formation of current sheets in the solar atmosphere. *Astrophys. J.* **312**, 423–430.

Zwingmann W. (1983). Self-consistent magnetotail theory: Equilibrium structure including arbitrary variations along the tail axis. *J. Geophys. Res.* **88**, 9101–9108.

Zwingmann W. (1987). Theoretical study of onset conditions for solar eruptive processes. *Solar Phys.* **111**, 309–331.

Index

503